Machine Learning for
Radar Signal Processing

IEEE Press

445 Hoes Lane

Piscataway, NJ 08854

Machine Learning for Radar Signal Processing

Edited by

Yujie Gu
Aptiv, USA

Yimin D. Zhang
Temple University, USA

IEEE Press
WILEY

Published by John Wiley & Sons, Inc., Hoboken, New Jersey.

Library of Congress Cataloging-in-Publication Data has been applied for:

Hardback ISBN: 9781394339952
ePDF ISBN: 9781394339976
ePub ISBN: 9781394339969
oBook ISBN: 9781394339983

Cover Design: Wiley
Cover Image: sanee/stock.adobe.com

To my devoted family for their endless love, unwavering support, and constant encouragement—Yujie Gu

To my loving family whose unwavering support and boundless love have illuminated my entire academic career—Yimin D. Zhang

Contents

List of Contributors

Aya Mostafa Ahmed
Corporate Research
Robert Bosch GmbH
Renningen
Germany

Batu K. Chalise
Department of Electrical and
Computer Engineering
New York Institute of Technology
Old Westbury, NY
USA

Davide Dardari
Department of Electrical, Electronic,
and Information Engineering
"Guglielmo Marconi"
University of Bologna
Cesena
Italy

Petar M. Djurić
Department of Computer and
Electrical Engineering
Stony Brook University
Stony Brook, NY
USA

Jahi Douglas
Department of Electrical and
Computer Engineering
New York Institute of Technology
Old Westbury, NY
USA

Arian Eamaz
Department of Electrical and
Computer Engineering
University of Illinois Chicago
Chicago, IL
USA

Yonina C. Eldar
Faculty of Mathematics and Computer
Science
Weizmann Institute of Science
Rehovot
Israel

Tara Esmaeilbeig
AI Research Lab
Nokia Bell Labs
Murray Hill, NJ
USA

Gianluca Fontanesi
Radio Systems Research
Nokia Bell Labs
Stuttgart
Germany

Fulvio Gini
Department of Information Engineering
University of Pisa
Pisa
Italy

Maria Sabrina Greco
Department of Information Engineering
University of Pisa
Pisa
Italy

Anna Guerra
Department of Electrical, Electronic, and Information Engineering "Guglielmo Marconi"
University of Bologna
Cesena
Italy

Francesco Guidi
Institute of Electronics, Information Engineering and Telecommunications
National Research Council of Italy
Bologna
Italy

Qian Guo
School of Information Science and Technology
Fudan University
Shanghai
China

Yujie Gu
Advanced Safety & User Experience
Aptiv
Agoura Hills, CA
USA

Sevgi Zubeyde Gurbuz
Department of Electrical and Computer Engineering
North Carolina State University
Raleigh, NC
USA

Yunqiao Hu
Department of Electrical and Computer Engineering
The University of Alabama
Tuscaloosa, AL
USA

Benjamin H. Kirk
DEVCOM Army Research Laboratory
U.S. Army
Adelphi, MD
USA

Visa Koivunen
Department of Information and Communications Engineering
Aalto University
Espoo
Finland

Junho Kweon
Department of Electrical Engineering
Pennsylvania State University
University Park, PA
USA

David McNamara
DEVCOM Army Research Laboratory
U.S. Army
Adelphi, MD
USA

Vishal Monga
Department of Electrical Engineering
Pennsylvania State University
University Park, PA
USA

Xiaoyong Ni
School of Engineering
EPFL Campus Biotech
Geneva
Switzerland

Petteri Pulkkinen
Department of Information and
Communications Engineering
Aalto University
Espoo
Finland

Saab Finland
Helsinki
Finland

Tirza Routtenberg
School of Electronic and Computer
Engineering
Ben-Gurion University of the Negev
Be'er Sheva
Israel

Aydin Sezgin
Faculty of Electrical Engineering and
Information Technology
Ruhr University Bochum
Bochum
Germany

Zhiguo Shi
College of Information Science and
Electronic Engineering
Zhejiang University
Hangzhou, Zhejiang
China

Nir Shlezinger
School of Electronic and Computer
Engineering
Ben-Gurion University of the Negev
Be'er Sheva
Israel

Ruud J. G. van Sloun
Department of Electronic Engineering
Eindhoven University of Technology
Eindhoven
The Netherlands

Mojtaba Soltanalian
Department of Electrical and
Computer Engineering
University of Illinois Chicago
Chicago, IL
USA

Shunqiao Sun
Department of Electrical and
Computer Engineering
The University of Alabama
Tuscaloosa, AL
USA

Adam Umra
Faculty of Electrical Engineering and
Information Technology
Ruhr University Bochum
Bochum
Germany

Sergiy A. Vorobyov
Department of Information and Communications Engineering
Aalto University
Espoo
Finland

Haipeng Wang
School of Information Science and Technology
Fudan University
Shanghai
China

Feng Xu
School of Information Science and Technology
Fudan University
Shanghai
China

Farhang Yeganegi
Department of Electrical and Computer Engineering
University of Illinois Chicago
Chicago, IL
USA

Siwei Zhang
Institute of Communications and Navigation
German Aerospace Center
Wessling
Germany

Yimin D. Zhang
Department of Electrical and Computer Engineering
Temple University
Philadelphia, PA
USA

Hang Zheng
College of Information Science and Electronic Engineering
Zhejiang University
Hangzhou, Zhejiang
China

Ruxin Zheng
Department of Electrical and Computer Engineering
The University of Alabama
Tuscaloosa, AL
USA

About the Editors

Yujie Gu received his PhD degree from Zhejiang University, Hangzhou, China, in 2008. He currently works at Aptiv as a Senior Radar Scientist focusing on research and development of advanced automotive radar systems for next-generation intelligent vehicles. His research experience lies in statistical and information-theoretic signal processing, with significant contributions to robust adaptive beamforming, information-theoretic compressive sensing, spatial spectrum estimation, and performance bound analysis. Dr. Gu is a Senior member of the Institute of Electrical and Electronics Engineers (IEEE). He is an elected member of the IEEE Sensor Array and Multichannel (SAM) Technical Committee and an elected member of the IEEE Signal Processing Theory and Methods (SPTM) Technical Committee. Dr. Gu is an Associate Editor for *IEEE Transactions on Signal Processing*. He was a Special Sessions Co-chair of the 2020 IEEE Sensor Array and Multichannel Signal Processing Workshop. Dr. Gu was the recipient of the 2019 IET Communications Premium Award and the 2023 Excellent Scientific Paper Award from China Association for Science and Technology. He is the corresponding author of a paper that received the 2021 IEEE Signal Processing Society Young Author Best Paper Award.

Yimin D. Zhang received his PhD degree from the University of Tsukuba, Tsukuba, Japan, in 1988. He is currently a Professor with the Department of Electrical and Computer Engineering, Temple University, Philadelphia, PA, USA. His research interests include array signal processing, compressive sensing, machine learning, information theory, convex optimization, and time-frequency analysis, with applications to radar, wireless communications, and satellite navigation. Dr. Zhang currently serves as a Senior Area Editor for *IEEE Transactions on Signal Processing* and an Editor for *Signal Processing*. He was a Technical Co-chair of the 2018 IEEE Sensor Array and Multichannel Signal Processing Workshop. He also serves on the IEEE Sensor Array and Multichannel (SAM) Technical Committee and the EURASIP Signal Processing for Multisensor Systems (SPMuS) Technical

Area Committee. He was the recipient of the 2016 IET Radar, Sonar & Navigation Premium Award, the 2017 IEEE Aerospace and Electronic Systems Society Harry Rowe Mimno Award, the 2019 IET Communications Premium Award, and the 2021 EURASIP Best Paper Award for Signal Processing. He coauthored two papers that respectively received the 2018 and 2021 IEEE Signal Processing Society Young Author Best Paper Awards. Dr. Zhang was a Distinguished Lecturer of the IEEE Signal Processing Society. He is a Fellow of IEEE, a Fellow of EURASIP, and a Fellow of SPIE.

Preface

Radar signal processing has undergone a profound evolution since its inception, driven by the need to accurately extract target information from reflected electromagnetic waves. Since World War II, radar has served as a foundational pillar of modern sensing technologies, enabling critical capabilities in defense, surveillance, weather monitoring, automotive safety, and remote sensing. Traditionally, radar systems have relied on well-established electromagnetic propagation models and rigorous statistical signal processing frameworks to perform essential tasks such as target detection, localization, tracking, and classification in complex, dynamic, and cluttered environments.

However, with the advent of data-rich sensing platforms and increasingly dynamic, heterogeneous, and unpredictable electromagnetic environments, the limitations of purely model-based approaches have become more evident. In many practical scenarios, conventional models may be inaccurate, overly idealized, difficult to calibrate, or computationally prohibitive, particularly when confronted with strong nonidealities, environmental uncertainties, and stringent real-time operational constraints.

Over the past decade, the radar community has increasingly embraced machine learning techniques, recognizing their powerful capabilities for data-driven learning, adaptation, and inference. Initially developed for domains such as computer vision and natural language processing, these methods have rapidly expanded into radar applications—evolving from early work in target recognition and classification to more recent advances in waveform design, detection, parameter estimation, filtering, tracking, and adaptive resource management. Approaches spanning classical supervised and unsupervised learning, deep neural networks, reinforcement learning, and generative models now provide flexible and scalable tools for extracting structure from high-dimensional measurements. By learning complex nonlinear relationships directly from data and operating effectively under incomplete or uncertain prior knowledge, machine learning enables radar systems to meet the demands of increasingly sophisticated and data-intensive sensing environments.

In parallel, growing interest has emerged in *model-based machine learning*, a hybrid paradigm that seeks to synergistically integrate physical principles and domain expertise with the representational power and adaptability of data-driven methods. By embedding structure, constraints, and prior knowledge into learning architectures, this approach aims to combine interpretability and theoretical rigor with the flexibility of modern machine learning. A prominent example is algorithm unrolling, in which iterative optimization or signal processing procedures are unfolded into structured deep network architectures with learnable parameters. Such frameworks retain clear connections to their underlying mathematical foundations while achieving improved learning efficiency, robustness, and empirical performance, particularly in low signal-to-noise ratio (SNR) conditions and limited-snapshot regimes that frequently arise in practical radar scenarios.

Taken together, these developments reflect a historic transition in radar signal processing—from a discipline long dominated by handcrafted analytical models to one increasingly enriched by data-driven methodologies, learning-based adaptation, and hybrid intelligence. This book aims to provide a comprehensive overview of recent advances at the intersection of machine learning and radar signal processing, with particular emphasis on both purely data-driven techniques and hybrid, model-informed approaches.

From a methodological perspective, this book spans a broad spectrum of modern machine learning paradigms and network architectures tailored to radar applications. These include model-based and physics-informed learning frameworks, such as algorithm unrolling and hybrid model-driven/data-driven framework designs, which explicitly embed knowledge and physical constraints into learnable architectures. Reinforcement learning methodologies, encompassing model-based and deep reinforcement learning, are explored for adaptive sensing, waveform optimization, and dynamic resource allocation in complex and uncertain environments.

The book further presents a variety of deep neural network architectures, including fully connected and convolutional neural networks, tensorized networks, transformer-based models, and attention mechanisms, with an emphasis on exploiting signal structures, array geometry, and frequency-domain representations. Practical considerations, such as low-resolution quantization, limited-snapshot regimes, and low-SNR conditions, are systematically addressed. Collectively, these methodological developments illustrate the diverse ways in which machine learning can be carefully designed, structured, and integrated to meet the unique theoretical and operational challenges of radar signal processing.

This book is intended not only for researchers in radar signal processing but also for postgraduate and Ph.D. students, as well as scientists, engineers, and technical managers working across a broad range of signal processing disciplines and applications. These include radar, sonar, wireless communications, medical imaging, acoustic systems, and other sensing and information systems. Readers are expected to possess a foundational background in linear algebra, probability and statistics, detection and estimation theory, control and optimization, information theory, and the basic principles of radar systems.

The book comprises eleven self-contained chapters, each devoted to a specific topic and offering an in-depth exploration of recent advances, emerging trends, and open challenges in the application of machine learning to radar signal processing. Authored by leading experts from academia, industry, and government laboratories, the chapters cover a broad range of machine learning-enabled radar applications, including waveform design, low-bit sampling, detection and environmental mapping, array signal processing, direction-of-arrival (DOA) estimation and beamforming, synthetic aperture radar (SAR) image interpretation, robust filtering and tracking, human activity recognition, integrated sensing and communications (ISAC), and adaptive resource allocation.

In Chapter 1, the authors (Junho Kweon and Vishal Monga) introduce deep algorithm unrolling as a principled framework for radar waveform design that bridges classical optimization and deep learning. By unrolling iterative algorithms into neural network architectures, domain knowledge is embedded into learnable models, thus improving interpretability, convergence speed, and generalization robustness under reduced training data. The chapter demonstrates the effectiveness of unrolled networks for beampattern shaping, single-target signal-to-interference-plus-noise ratio (SINR) maximization, and multi-target worst-case SINR optimization.

In Chapter 2, the authors (Petteri Pulkkinen and Visa Koivunen) detail a model-based reinforcement learning framework for ISAC waveform design and resource allocation with explicit awareness of radio environment dynamics to improve learning efficiency and interpretability over conventional model-free approaches. This framework enables flexible and efficient waveform adaptation and resource allocation in dynamic ISAC environments with time-, frequency-, and space-varying channels and interference. Applications include multicarrier waveform optimization and allocating subcarriers, resource blocks, beams, and their powers to support coordinated communication and sensing objectives.

In Chapter 3, the authors (Farhang Yeganegi, Arian Eamaz, Tara Esmaeilbeig, and Mojtaba Soltanalian) address DOA estimation using dithered one-bit measurements, where dithering enables recovery of covariance information from heavily quantized data. The chapter reviews covariance estimation techniques supporting sparse DOA recovery within a compressed sensing framework and highlights recent theoretical advances. It also introduces model-based deep learning approaches, including algorithm unrolling methods, such as the learned iterative shrinkage-thresholding algorithm, which provides efficient and interpretable solutions to inverse sensing problems.

In Chapter 4, the authors (Adam Umra, Aya Mostafa Ahmed, Aydin Sezgin, Maria Sabrina Greco, and Fulvio Gini) develop a reinforcement learning–based framework for intelligent target detection in massive multiple-input multiple-output (MMIMO) radar. Integrating colocated MMIMO architectures, Wald-type statistical detection, and reinforcement learning-driven adaptive waveform selection and beamforming, the approach enables real-time adaptation without prior environmental knowledge. Analytical modeling and simulation results across diverse target scenarios demonstrate improved robustness, adaptability, beamforming, and detection performance compared with traditional detectors relying on prior noise and clutter models.

In Chapter 5, the authors (Anna Guerra, Siwei Zhang, Gianluca Fontanesi, Francesco Guidi, Davide Dardari, and Petar M. Djurić) review recent advances in learning-based methods for target detection and environment mapping in networks of autonomous agents. The chapter examines cooperative sensing, communication, inference, and decision-making in unknown environments. Two complementary paradigms are presented: model-based learning leveraging analytical environment models for control optimization and data-driven learning based on reinforcement learning. Case studies illustrate the effectiveness in time-critical applications involving integrated sensing, tracking, and mapping.

In Chapter 6, the authors (Yunqiao Hu, Ruxin Zheng, Shunqiao Sun, and Yimin D. Zhang) present advanced techniques for sparse antenna array interpolation and DOA estimation using model-based and deep learning techniques. Linear and two-dimensional sparse array designs enable high angular resolution with reduced hardware complexity. An unrolled iterative hard-thresholding network improves sparse array interpolation accuracy, while a frequency-domain attention framework enhances adaptability and robustness. Numerical results demonstrate superior reconstruction accuracy and DOA performance under practical constraints.

In Chapter 7, the authors (Hang Zheng, Zhiguo Shi, Sergiy A. Vorobyov, and Yujie Gu) introduce two tensorized neural network architectures for efficient DOA estimation. A tensorized deep neural network designed for uniform rectangular arrays employs inverse Tucker decomposition to compress trainable parameters, while a tensorized convolutional neural network for sparse planar arrays uses canonical polyadic decomposition to reduce convolutional complexity. Both approaches significantly lower computational and storage requirements while maintaining accurate and robust DOA estimation compared with conventional network implementations.

In Chapter 8, the authors (Qian Guo, Feng Xu, and Haipeng Wang) explore SAR image interpretation using deep neural networks that incorporate electromagnetic scattering knowledge. The microwave vision framework integrates radar imaging principles with learning-based models. A scattering information-enhanced attention pyramid network supports improved target detection, while a causal adversarial auto-encoder enables data generation and classification. An evaluation criterion based on recognition versus substitution rate is established for objective assessment of SAR datasets.

In Chapter 9, the authors (Xiaoyong Ni, Nir Shlezinger, Ruud J. G. van Sloun, Tirza Routtenberg, and Yonina C. Eldar) examine data-driven Kalman filters through the KalmanNet framework, which embeds trainable components within the recursive structure of traditional Kalman filters. After reviewing classical and model-based estimation methods, the chapter presents the KalmanNet architectures performing effective filtering, smoothing, and adaptive tracking. The hybrid design offers interpretability, uncertainty quantification, unsupervised learning, structure awareness, and end-to-end optimization, with qualitative and empirical demonstrations for practical target-tracking scenarios.

In Chapter 10, the author (Sevgi Zubeyde Gurbuz) designs accurate and robust deep neural networks for radar-based human activity recognition under dynamic operating conditions. Because radar data are time-varying complex signals, the chapter examines trade-offs among signal processing, data dimensionality, and network architecture to balance performance, complexity, and efficiency. Physics-aware artificial intelligence/machine learning approaches are emphasized, integrating physics models with data-driven deep learning to improve reliability, interpretability, and robustness across diverse sensing environments.

In Chapter 11, the authors (Batu K. Chalise, Jahi Douglas, Benjamin H. Kirk, and David McNamara) study detection and tracking in distributed radar networks with emphasis on optimal bandwidth allocation. Model-based optimization is first introduced for joint radar and communication bandwidth, carrier frequency, and waveform design, followed by a long short-term memory network-based supervised learning approach to reduce runtime complexity. For equal subband allocation under external and inter-node interference, a deep Q-network is applied, demonstrating improved performance-complexity trade-offs compared with exhaustive search methods.

The organization of the book is designed to help readers develop both a comprehensive and in-depth understanding of core radar signal processing problems through the lens of machine learning, while clearly illustrating how learning-based methodologies can be leveraged to design efficient, adaptive, and high-performance radar systems. By integrating theoretical foundations with practical applications, the book provides a coherent perspective on emerging paradigms in intelligent sensing.

As such, this book serves as an advanced reference for academic researchers and industry professionals alike. At the same time, the self-contained structure of each chapter offers flexibility for instructional use, enabling instructors to select individual chapters or curated subsets for graduate-level courses, specialized tutorials, or short-term seminars.

Acknowledgments

Yimin D. Zhang would like to acknowledge funding from the National Science Foundation (NSF) under Grant ECCS-2236023 and the Air Force Office of Scientific Research (AFOSR) under Grant FA9550-23-1-0255, which supported his editorial work on this book.

1

Deep Algorithm Unrolling for Radar Waveform Design

Junho Kweon and Vishal Monga*

Department of Electrical Engineering, Pennsylvania State University, University Park, PA, USA

*Corresponding author: junho.kweon@gmail.com

1.1 Introduction

Radar waveform design is a fundamental problem in radar signal processing, as the choice of waveform significantly affects system performance, including target detection, clutter suppression, and signal-to-interference-plus-noise ratio (SINR) [1]. There are various practical constraints to be considered in radar waveform design, such as spectral compatibility and hardware-related constraints, which introduce difficulties from optimization perspective. One such critical constraint is the constant modulus constraint (CMC) on waveforms, which is essential due to nonlinear power amplifiers working in saturation mode to maximize energy efficiency. The CMC ensures that transmitted signals preserve signal quality by reducing nonlinear distortion [2].

Over time, radar systems have evolved from traditional phased-array architectures to more flexible and capable systems such as multiple-input multiple-output (MIMO) radar. MIMO radar allows the transmission of independent waveforms from different antennas, offering enhanced spatial degrees of freedom. Unlike phased-array systems that transmit phase-shifted replicas of the same signal, MIMO radar can leverage waveform diversity to achieve improved target resolution, better interference mitigation, and more adaptable beampattern control [3–5].

1.1.1 From Iterative Optimization to Deep Learning in Radar Waveform Design

Over the past few decades, researchers have developed numerous optimization-based methods to design radar waveforms that satisfy practical constraints while achieving desired performance. Specifically, due to the nonconvex CMC on the waveform, waveform design under the CMC is a hard nonconvex problem, where its global minimum is elusive. The early efforts in waveform design focused on iterative numerical algorithms to solve the problem. These existing algorithms seek to iteratively refine the waveform to optimize metrics such as beampattern shaping, SINR maximization, and sidelobe suppression. However, despite their effectiveness, these methods are often computationally expensive, making real-time deployment infeasible, especially as the problem scales with the number of antennas and time samples.

1.1.1.1 Deep Learning as a Computationally Efficient Alternative

To address the computational burden of iterative algorithms, deep learning (DL) has emerged as a promising alternative for radar waveform design [6]. DL-based methods for waveform design leverage neural networks (NNs) to approximate complex optimization mappings, enabling rapid inference of optimal waveforms once trained. Therefore, the computational burden of DL is mainly moved to offline training, yielding fast inference of waveforms as a solution. Figure 1.1

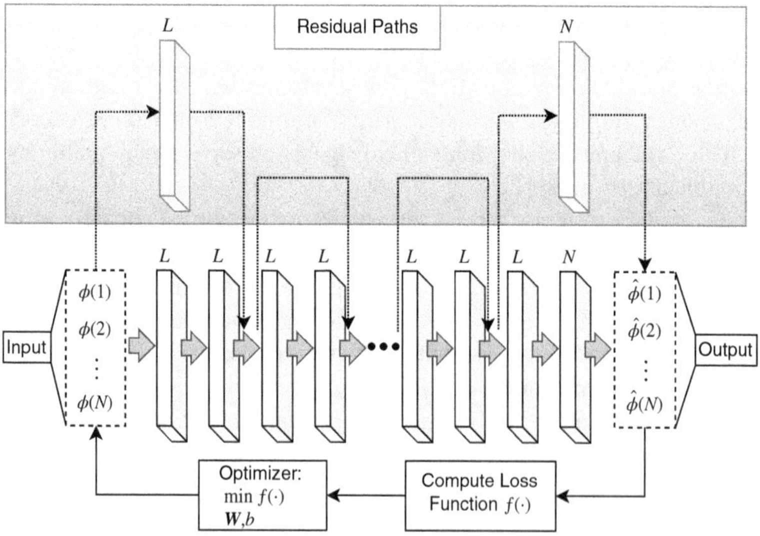

Figure 1.1 Diagram that generalizes the network architectures used in [7–10]. The network produces phase codes $\hat{\phi}(n)$ for $n = 1, \ldots, N$ to guarantee the waveforms satisfy the CMC. The residual paths do not exist in [9, 10].

shows the architecture of the NN proposed for radar waveform design [7–10]. These networks receive input specifications such as desired beampatterns or target information and produce a phase-coded waveform to generate constant modulus (CM) waveforms. In general, DL methods can produce quality solutions when significant training data is available.

While DL offers significant speed advantages, conventional DL models suffer from a major limitation: they often function as black-box systems. Traditional DL approaches do not provide interpretability or insight into the underlying optimization process, making it difficult to ensure robustness and generalizability across diverse radar scenarios. Additionally, DL typically requires large-scale training data, which can be impractical to obtain in radar applications where acquiring labeled data is costly and time-consuming.

1.1.2 Algorithm Unrolling: Bridging Iterative Optimization and Deep Learning

To overcome the black-box nature of DL while retaining computational efficiency, *algorithm unrolling* (also known as *algorithm unfolding* or *model-based learning*) has emerged as an innovative framework for radar waveform design. Rather than building an arbitrary (generic) network structure for DL, an unrolled network (also called a model-based network) mimics an existing iterative algorithm such that each iteration corresponds to a custom-designed network step. The trained unrolled network is then naturally interpreted as a parameter-optimized iterative algorithm [11].

Figure 1.2 shows a high-level visualization of unrolling an iterative algorithm. In a traditional iterative algorithm, each update step is denoted by $\tilde{\mathcal{J}}(\cdot)$ and operates with fixed key parameters $\tilde{\mathcal{X}}$. In contrast, the unrolled network emulates

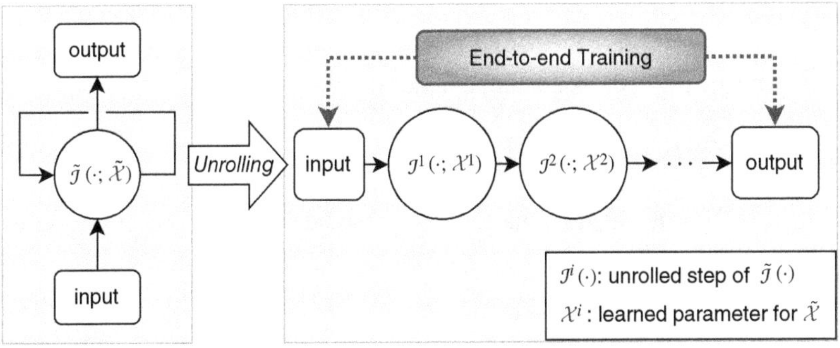

(a) **(b)**

Figure 1.2 High-level visualization of algorithm unrolling. (a) An iterative algorithm $\tilde{\mathcal{J}}(\cdot)$ with fixed parameters $\tilde{\mathcal{X}}$. (b) An unrolled network using $\mathcal{J}^i(\cdot)$ emulating its base-model update $\tilde{\mathcal{J}}(\cdot)$ with *learnable* key parameters $\mathcal{X}^i$ for $i = 1, 2, \ldots$.

these update steps as $\mathcal{J}^i(\cdot)$ for $i = 1, 2, \dots$, but allows the parameters $\tilde{\mathcal{X}}^i$ to vary and be learned from data. Specifically, the unrolled network is trained end-to-end using backpropagation to minimize a given loss function. Once trained, the inference phase of the network can be viewed as a *parameter-optimized algorithm* derived from its iterative counterpart.

A great example that showcases the power of unrolling is learned iterative shrinkage-thresholding algorithm (LISTA), which is a transformation of the iterative shrinkage-thresholding algorithm (ISTA) into a trainable NN. LISTA laid the foundation for unrolling-based approaches, showing how learning can be integrated into iterative structures.

1.1.3 From Iterations to Learning: How LISTA Sparked Algorithm Unrolling

One of the earliest and most influential demonstrations of algorithm unrolling is the LISTA model, introduced by Gregor and LeCun in [12]. LISTA reinterprets the ISTA—commonly used for solving sparse recovery problems—as a trainable deep network. By learning key parameters such as step sizes and thresholds, LISTA accelerates convergence while preserving the interpretability of its iterative roots.

The fundamental objective is to estimate a sparse signal $\boldsymbol{x} \in \mathbb{R}^{N\times 1}$ from a set of noisy linear measurements $\boldsymbol{y} \in \mathbb{R}^{M\times 1}$, formulated as the following least absolute shrinkage and selection operator (LASSO) problem

$$\min_{\boldsymbol{x}} \frac{1}{2}\|\boldsymbol{y} - \boldsymbol{A}\boldsymbol{x}\|_2^2 + \lambda\|\boldsymbol{x}\|_1, \tag{1.1}$$

where $\boldsymbol{A} \in \mathbb{R}^{M\times N}$ is a given sensing matrix. Here, M is the number of measurements and N is the signal dimension. The role of $\boldsymbol{A}$ is to map the high-dimensional sparse signal $\boldsymbol{x}$ into a lower dimensional measurement space, often in the context of compressed sensing.

This optimization problem promotes sparsity in $\boldsymbol{x}$ through the ℓ_1-norm regularization term, controlled by the regularization parameter λ. ISTA solves this problem iteratively by using the following update equation:

$$\boldsymbol{x}^{(i+1)} = \mathcal{S}_\lambda\left\{\left(\boldsymbol{I}_N - \frac{1}{\mu}\boldsymbol{A}^{\mathrm{T}}\boldsymbol{A}\right)\boldsymbol{x}^{(i)} + \frac{1}{\mu}\boldsymbol{A}^{\mathrm{T}}\boldsymbol{y}\right\}, \quad i = 0,\ 1, \dots, \tag{1.2}$$

where $\boldsymbol{I}_N \in \mathbb{R}^{N\times N}$ is the identity matrix, μ is a positive parameter that controls the iteration step size, and

$$\mathcal{S}_\lambda(x) = \mathrm{sign}(x)\max\{|x| - \lambda,\ 0\} \tag{1.3}$$

is the soft-thresholding function, where x is a scalar. $\mathcal{S}_\lambda(\cdot)$ operates elementwise on vectors and matrices.

Although ISTA provides a systematic approach to sparse signal recovery, its convergence rate is often slow due to the fixed step size and thresholding parameter.

To address this limitation, LISTA *unrolls* the ISTA iterations into a fixed-depth NN. First, ISTA update can be rewritten as

$$\boldsymbol{x}^{(i+1)} = \tilde{\mathcal{J}}(\cdot,\ \tilde{\mathcal{X}}) = \mathcal{S}_{\lambda}\{\tilde{\boldsymbol{W}}_t\boldsymbol{x}^{(i)} + \tilde{\boldsymbol{W}}_e\boldsymbol{y}\}, \tag{1.4}$$

where $\tilde{\boldsymbol{W}}_t = \boldsymbol{I}_N - (1/\mu)\boldsymbol{A}^{\mathrm{T}}\boldsymbol{A}$ and $\tilde{\boldsymbol{W}}_e = (1/\mu)\boldsymbol{A}^{\mathrm{T}}$. Instead of using fixed $\tilde{\boldsymbol{W}}_t$ and $\tilde{\boldsymbol{W}}_e$, LISTA learns these matrices and thresholding parameter λ through data-driven training. Substituting these fixed parameters $\tilde{\mathcal{X}} = \{\tilde{\boldsymbol{W}}_t,\ \tilde{\boldsymbol{W}}_e,\ \lambda\}$ with the corresponding learnable parameters $\mathcal{X} = \{\boldsymbol{W}_t,\ \boldsymbol{W}_e,\ \theta\}$, the unrolled update can be expressed as:

$$\boldsymbol{x}^{(i+1)} = \mathcal{J}^{(i)}(\cdot, \mathcal{X}) = \mathcal{S}_{\theta}(\boldsymbol{W}_t\boldsymbol{x}^{(i)} + \boldsymbol{W}_e\boldsymbol{y}), \quad i = 0,\ 1, \ldots, \kappa - 1, \tag{1.5}$$

where κ is a predefined number of steps.

The forward propagation of LISTA can be visualized in Figure 1.3. Instead of relying on fixed parameters, LISTA optimizes (learns) these parameters ($\mathcal{X} = \{\boldsymbol{W}_t, \boldsymbol{W}_e, \theta\}$) using supervised learning, minimizing the reconstruction error across training data pairs $(\boldsymbol{x}, \boldsymbol{y})$. After offline training, LISTA can approximate the solution of the sparse coding problem in significantly fewer iterations compared to ISTA.

LISTA exemplifies the core principles of *algorithm unrolling*, demonstrating how iterative optimization techniques can be reformulated as structured deep networks. By preserving the algorithmic interpretability of ISTA while leveraging the representational power of DL, LISTA established a blueprint for integrating model-based methods into trainable architectures. This work laid the foundation for unrolling techniques in broader applications like radar waveform design.

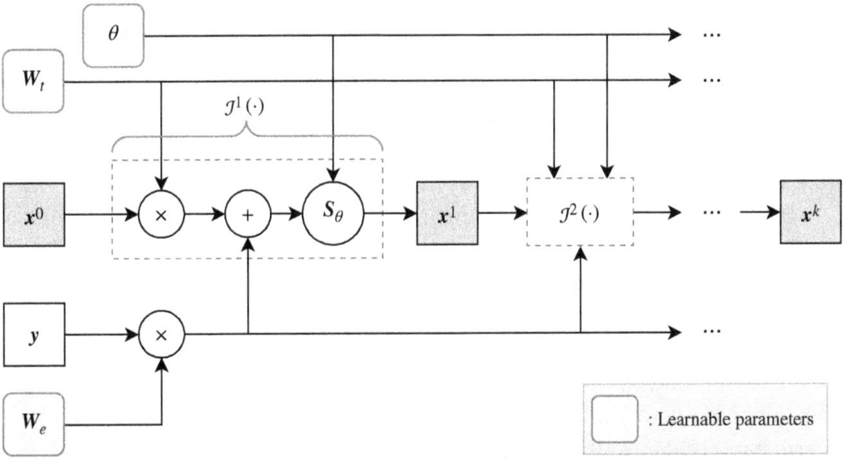

Figure 1.3 Visualized forward propagation of LISTA. Across LISTA steps, learnable parameters $\mathcal{X} = \{\boldsymbol{W}_t,\ \boldsymbol{W}_e,\ \theta\}$ are shared as the same values.

1.1.4 Unrolling in Radar Waveform Design

The primary advantage of algorithm unrolling lies in its ability to preserve the interpretability and domain-inspired structure of iterative solvers while enhancing performance through data-driven learning. This synergy between domain knowledge and learning has shown great promise in tackling key radar waveform design challenges.

1.1.4.1 Overview of the Chapter

This chapter explores deep algorithm unrolling for radar waveform design, with a focus on three central applications: beampattern shaping, single-target SINR maximization, and multi-target SINR maximization. In the sections that follow, the traditional iterative formulations for each problem are first introduced, and their computational and modeling limitations are identified. The subsequent discussions demonstrate how deep unrolling offers favorable solutions at reduced complexity by combining model-based structure with learning flexibility.

1.2 Beampattern Shaping Under Spatial and Spectral Constraints

In MIMO radar, a fundamental objective is to synthesize transmit waveforms that produce a desired beampattern—concentrating radiated power in specific spatial directions. Beyond spatial focusing, a carefully designed beampattern also contributes to enhanced SINR at target locations, thereby improving detection performance [5, 13].

The ability of MIMO radar to assign independent waveforms to each transmit antenna offers substantial design flexibility [5]. In wideband settings, this flexibility becomes especially useful for jointly managing spatial and spectral objectives—an approach referred to as spatio-spectral design [14, 15]. The goal involves spatially suppressing reflections from unwanted directions while spectrally reducing interference with coexisting communication systems or other spectrum users [16, 17].

In the context of MIMO beampattern design, several studies have formulated the problem as minimizing the deviation from an idealized beampattern, constructed based on known environmental parameters [14, 18–20]. To enforce the CMC, different strategies have been explored, such as using more relaxed constraints [19] or approximating the problem [15]. However, these methods often do not guarantee the strict modulus condition or require substantial computational resources.

To address these limitations while maintaining algorithmic structure and interpretability, Metwaly et al. proposed a deep unrolled network tailored for beampattern design under the CMC [21, 22]. This approach specifically unrolls iterative projection-descent-retraction (PDR) introduced in [20], transforming it into a trainable model. PDR ensures convergence and feasibility with respect to the CMC at every iteration. This property makes it well-suited for unrolling, where intermediate outputs may be used during training [15, 20].

1.2.1 Problem Formulation

Assume a MIMO radar system with a uniform linear array (ULA) that has N_t transmit antennas with inter-element spacing d. Let $x_m(t) \in \mathbb{C}$ be the waveform of the mth transmit antenna at time t, and its discrete-time sample is expressed as $x_m(n) = x_m(t = nT_s)$ for $n = 0, \ldots, N-1$, where $T_s(= 1/f_s)$ is sampling period (f_s is sampling frequency). Let $\boldsymbol{x}_m = [x_m(0), x_m(1), \ldots, x_m(N-1)]^{\mathrm{T}} \in \mathbb{C}^{N\times 1}$ denote the waveform vector for the mth antenna. $\boldsymbol{x} \in \mathbb{C}^{L\times 1}$ is the concatenated waveforms (L = MN) such that

$$\boldsymbol{x} = [\boldsymbol{x}_1^{\mathrm{T}}, \boldsymbol{x}_2^{\mathrm{T}}, \ldots, \boldsymbol{x}_M^{\mathrm{T}}]^{\mathrm{T}} = [x_1, \ldots, x_L]^{\mathrm{T}}. \tag{1.6}$$

Let $\boldsymbol{a}(\theta, p) \in \mathbb{C}^{M\times 1}$ be the steering vector at frequency $\dfrac{p}{NT_s} + f_c$ defined as

$$\boldsymbol{a}(\theta, p) = \left[1, e^{j2\pi\left(\frac{p}{NT_s}+f_c\right)\frac{d\cos\theta}{c}}, \ldots, e^{j2\pi\left(\frac{p}{NT_s}+f_c\right)\frac{(M-1)d\cos\theta}{c}}\right]^{\mathrm{T}}, \tag{1.7}$$

where f_c is center frequency. Dividing angle $[0°, 180°]$ into S sub-intervals, the beampattern at the sth angle θ_s can be expressed on the discrete angle-frequency grid as

$$P_{sp} = \left|\boldsymbol{g}_{s,p}^{\mathrm{H}}\boldsymbol{x}\right|^2 = \left|\boldsymbol{a}_{sp}^{\mathrm{H}}(\boldsymbol{e}_p^{\mathrm{T}} \otimes \boldsymbol{I}_M)\boldsymbol{x}\right|^2, \tag{1.8}$$

where $\boldsymbol{a}_{sp} = \boldsymbol{a}(\theta_s, p)$ and $\boldsymbol{e}_p = \left[1, \quad e^{-j2\pi\frac{p}{N}}, \quad \ldots, \quad e^{-j2\pi\frac{(N-1)p}{N}}\right]^{\mathrm{T}} \in \mathbb{C}^{N\times 1}$, which is for discrete Fourier transform (DFT). Herein, $[\,\cdot\,]^{\mathrm{H}}$ denotes Hermitian (complex conjugate transpose), and $\otimes$ stands for Kronecker product.

The beampattern shaping aims to minimize the deviation between the desired beampattern and the actual beampattern resulting from the given waveforms. The problem can be expressed as

$$\begin{aligned} \min_{\boldsymbol{x}} \quad & f(\boldsymbol{x}) = \sum_{s=1}^{S} \sum_{p=-\frac{N}{2}}^{\frac{N}{2}-1} \left[d_{sp} - \left|\boldsymbol{g}_{s,p}^{\mathrm{H}}\boldsymbol{x}\right|\right]^2 = \|\boldsymbol{d} - |\boldsymbol{H}\boldsymbol{x}|\|_2^2 \\ \text{s.t.} \quad & |x_l| = 1, \ \forall l, \end{aligned} \tag{1.9}$$

where the constraint denotes the CMC. $d_{sp} \in \mathbb{R}_{\geq 0}$ is the given desired beampattern, which can be determined in cognitive paradigm [16, 17]. $\boldsymbol{d}$ is vector form of

$\{d_{s,p}\}$ such that $\boldsymbol{d} = \left[d_{1,-\frac{N}{2}}, \ldots, d_{1,\frac{N}{2}-1}, \ldots, d_{S,\frac{N}{2}-1}\right]^{\mathrm{T}} \in \mathbb{R}^{NS\times 1}, \boldsymbol{g}_{s,p} \in \mathbb{C}^{L\times 1}\ \forall s,p,$ and

$$\boldsymbol{H} = \left[\boldsymbol{g}_{1,-\frac{N}{2}}, \ldots, \boldsymbol{g}_{1,\frac{N}{2}-1}, \ldots, \boldsymbol{g}_{S,\frac{N}{2}-1}\right]^{\mathrm{T}} \in \mathbb{C}^{NS\times L}. \tag{1.10}$$

1.2.2 Iterative Algorithm: Projection-descent-retraction

The CM waveform set can be regarded as the direct sum of L unit complex circle manifolds (CCMs) (each denoted as $\mathcal{S}$), which can be represented as $\mathcal{S}^L$. In [20], Alhujaili et al. proposed an iterative **p**rojection-**d**escent-**r**etraction (**PDR**) algorithm to minimize $f(\boldsymbol{x})$ under the CMC following these steps (update $\boldsymbol{x}_{(k)}$ to $\boldsymbol{x}_{(k+1)}$):

1. **P**roject search direction $\tilde{\boldsymbol{\eta}}_{(k)} = -\nabla_{\boldsymbol{x}^*} f(\boldsymbol{x}_{(k)})$ onto the tangent space of $\mathcal{S}^L$, $\mathcal{T}_{\boldsymbol{x}_{(k)}}\mathcal{S}^L$, at $\boldsymbol{x}_{(k)}$ as follows[1] :

$$\boldsymbol{P}_{\mathcal{T}_{\boldsymbol{x}_{(k)}}\mathcal{S}^L}\left(\tilde{\boldsymbol{\eta}}_{(k)}\right) = \tilde{\boldsymbol{\eta}}_{(k)} - \Re\{\tilde{\boldsymbol{\eta}}_{(k)}^{*} \odot \boldsymbol{x}_{(k)}\} \odot \boldsymbol{x}_{(k)}, \tag{1.11}$$

   where $\odot$ denotes Hadamard product.
2. A **D**escent update of the current value of $\boldsymbol{x}_{(k)}$ on the tangent space $\mathcal{T}_{\boldsymbol{x}_{(k)}}\mathcal{S}^L$ as

$$\bar{\boldsymbol{x}}_{(k)} = \boldsymbol{x}_{(k)} + \beta \boldsymbol{P}_{\mathcal{T}_{\boldsymbol{x}_{(k)}}\mathcal{S}^L}\left(\tilde{\boldsymbol{\eta}}_{(k)}\right). \tag{1.12}$$

3. **R**etraction of $\bar{\boldsymbol{x}}_{(k)}$ to the CCM:

$$\boldsymbol{x}_{(k+1)} = \boldsymbol{R}\left(\bar{\boldsymbol{x}}_{(k)}\right). \tag{1.13}$$

Despite the advancements made by iterative algorithms in addressing radar waveform design under the CMC, several critical challenges remain. A major limitation is that most iterative algorithms rely on fixed or heuristic-based parameter selection. For example, the PDR algorithm uses a fixed descent direction derived from the Wirtinger gradient theorem [24] and a fixed step size to ensure monotonic decrease of the cost function. However, many gradient-based radar waveform design algorithms still leave open questions regarding the optimal choice of descent direction and step size. Notably, Zhang et al. [25] demonstrate that replacing the *fixed* step size in PDR with an *adaptive* one can lead to more efficient cost minimization, thereby significantly accelerating convergence [26] and accomplishing better minima than in [20, 27].

1 The tangent space $\mathcal{T}_{\boldsymbol{x}_{(k)}}\mathcal{S}^L$ at $\boldsymbol{x}_{(k)}$ is the direct sum of L tangent spaces of L unit complex circles, i.e., $\mathcal{T}_{\boldsymbol{x}_{(k)}}\mathcal{S}^L = \mathcal{T}_{x_{1(k)}}\mathcal{S} \times \mathcal{T}_{x_{2(k)}}\mathcal{S} \ldots \times \mathcal{T}_{x_{L(k)}}\mathcal{S}$ where the tangent space of $\mathcal{S}$ at the point $x_{l(k)} \in \mathcal{S}$ is defined as [23] $\mathcal{T}_{x_{l(k)}}\mathcal{S} = \{y \in \mathbb{C} : \Re\{y^* x_{l(k)}\} = 0\}$.

1.2.3 Unrolling PDR into FLED

Metwaly et al. unrolled iterative PDR algorithm, and the unrolled network is named fast, learned, and explainable deep (FLED) [21]. One of the key features of FLED is the *learnable* descent direction and step size, which are fixed in its base model, PDR. During forward propagation, FLED performs waveform updates that resemble PDR iterations. One update step that emulates a single PDR iteration is called FLED-step, denoted as $\mathcal{J}_i(\cdot)$, and κ is the predetermined number of FLED-steps.

In the beginning of the forward propagation of FLED, **initialization** $\mathcal{I}(\cdot)$ generates initial waveform candidates $\{\boldsymbol{x}_0^j\}_{j=1}^{N_e}$ for given desired beampattern $\boldsymbol{d}$. These multiple candidates go through FLED steps, each FLED step $\mathcal{J}_i(\cdot)$ performs the following:

1. Multiple solution candidates $\{\boldsymbol{x}_i^j\}_{j=1}^{N_e}$ are fed to $\mathcal{J}_i(\cdot)$ from the previous FLED-step (or initialization block $\mathcal{I}(\cdot)$ for the first FLED step).
2. Learned descent direction $\boldsymbol{\eta}_i$ and step size β_i are computed by neural network blocks (NNBs).
3. **Projection** $\mathcal{P}(\cdot)$ projects the learned descent direction $\boldsymbol{\eta}_i$ to the tangent space of the CMC:
$$\mathcal{P}(\boldsymbol{\eta}_i^{(j)}) = \boldsymbol{\eta}_i^{(j)} - \Re\{\boldsymbol{\eta}_i^{(j)} \odot \boldsymbol{x}_{i-1}^{(j)}\} \odot \boldsymbol{x}_{i-1}^{(j)}. \tag{1.14}$$
4. The waveforms are updated to a **descent** direction with learned step size:
$$\bar{\boldsymbol{x}}_i^{(j)} = \boldsymbol{x}_{i-1}^{(j)} + \beta_i \mathcal{P}(\boldsymbol{\eta}_i^{(j)}). \tag{1.15}$$
5. **Retract** $\mathcal{R}(\cdot)$ ensures that the output vectors $\tilde{\boldsymbol{x}}_i^{(j)}$ satisfy the CMC:
$$\tilde{\boldsymbol{x}}_i^{(j)} = \bar{\boldsymbol{x}}_i^{(j)} \odot \frac{1}{|\bar{\boldsymbol{x}}_i^{(j)}|} \quad \text{for } j = 1, \ldots, N_e, \tag{1.16}$$
and $\tilde{\boldsymbol{x}}_i^{(j)}$ is fed to Prune $\mathcal{N}(\cdot)$ as input.
6. **Prune** $\mathcal{N}(\cdot)$, composed of a fully connected layer (FCL), is learned to remove $N_e - N_p$ waveform candidates ($N_p < N_e$) that have less potential to yield a favorable solution, defining the remaining N_p candidates as $\{\widehat{\boldsymbol{x}}_i^{(j)}\}_{j=1}^{N_p}$.
7. Taking $\{\widehat{\boldsymbol{x}}_i^{(j)}\}_{j=1}^{N_p}$ as input, **Expand** $\mathcal{E}(\cdot)$ is learned to generate $N_e - N_p$ waveform candidates, denoting the output as $\{\boldsymbol{x}_i^{(j)}\}_{j=1}^{N_e}$, which is also the output of the FLED step $\mathcal{J}_i(\cdot)$ for $i = 1, \ldots, \kappa$.

The forward propagation of FLED is visualized as in Figure 1.4.

Initialization $\mathcal{I}(\cdot)$ and Prune/Expand $\mathcal{E}(\mathcal{N}(\cdot))$ are newly introduced, which do not exist in PDR. The *initialization block* learns to provide good waveform

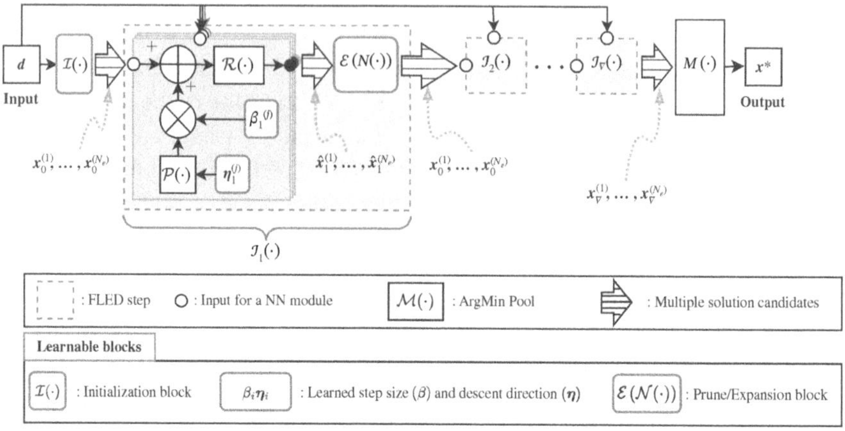

Figure 1.4 Forward propagation of FLED. In the architecture, there are κ FLED-steps, each emulates a single PDR iteration. In contrast to iterative PDR, FLED has learnable blocks: Initialization $\mathcal{I}(\cdot)$, step size β_i, descent direction η_i, and Prune/Expand $\mathcal{E}(\mathcal{N}(\cdot))$. In forward propagation, multiple waveform candidates are updated through FLED steps.

candidates for a given desired beampattern[2], which is denoted as $\boldsymbol{d}$. Prune/Expand block refines the solution space by pruning unfavorable candidates via $\mathcal{N}$ and expanding promising waveform candidates that may achieve better minima via $\mathcal{E}$. This Prune/Expand block is visualized in Figure 1.5.

Ideally, at the end of FLED steps, the solution waveform can be selected by taking one with the smallest cost function value among the candidates, i.e.,

$$\boldsymbol{x}^{\star} = \underset{\boldsymbol{x}_{\kappa}^{(j)},\, j\in\{1,\dots,N_e\}}{\operatorname{argmin}} \; f(\boldsymbol{x}_{\kappa}^{(j)}). \tag{1.17}$$

However, to enable backpropagation from the output waveform during FLED training, ArgMin Pool $\mathcal{M}(\cdot)$ is used as

$$\boldsymbol{x}^{\star} = \frac{1}{N_e}\sum_{j=1}^{N_e} \boldsymbol{x}_{\kappa}^{(j)}, \tag{1.18}$$

allowing the gradient of the training loss function to be spread to the network blocks.

2 The authors trained the initialization block can produce waveforms that have small cost function values in (1.9) while the waveform candidates have substantial disparity, i.e., distinct from each other for better generalizability.

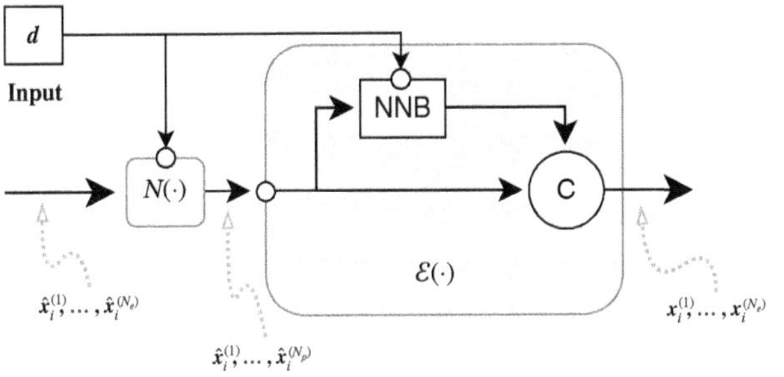

Figure 1.5 Architecture of Prune/Expand block. Prune $\mathcal{N}(\cdot)$ removes unfavorable waveform candidates, and Expand $\mathcal{E}(\cdot)$ explores potential waveform solutions for the problem.

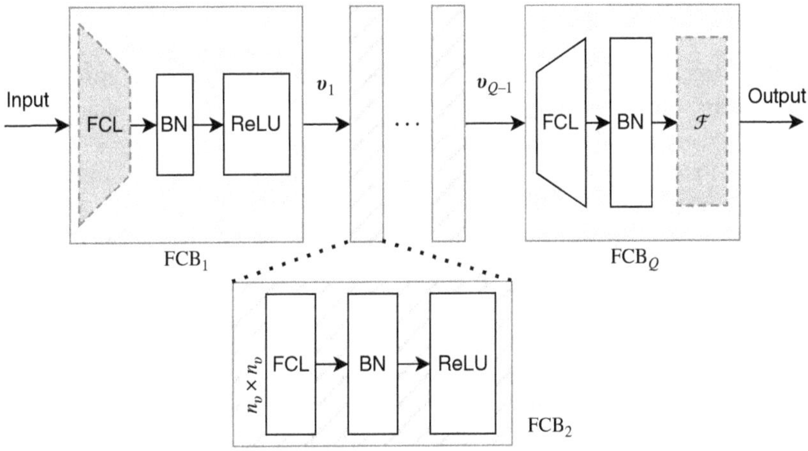

Figure 1.6 Detailed network architecture for FLED's building blocks: Initialization $\mathcal{I}(\cdot)$, Direction Evaluation $\eta(\cdot)$, Prune $\mathcal{N}(\cdot)$, and Expand $\mathcal{E}(\cdot)$. The building blocks consist of FCBs, each has FCL, BN, and an activation function.

1.2.4 Detailed Network Architecture in FLED

FLED employs a custom-designed deep architecture to implement the regression functions associated with its four primary building blocks: Initialization $\mathcal{I}(\cdot)$, Direction Evaluation $\eta(\cdot)$, Prune $\mathcal{N}(\cdot)$, and Expand $\mathscr{E}(\cdot)$. These modules share a common architectural structure illustrated in Figure 1.6. Each module is composed of multiple fully connected blocks (FCBs), each consists of an FCL, followed by batch normalization (BN) and a nonlinear activation function, such as ReLU.

Let Q denote the number of FCBs in a module. The intermediate output of the rth block is denoted by $\boldsymbol{v}_r \in \mathbb{C}^{n_v}$ for $r = 1, \ldots, Q$, and the final activation function is represented by $\mathcal{F}$. Detailed input/output specifications for each module, along

Table 1.1 Architectural specifications of the FLED network.

Type	Input	$\mathcal{F}$	Output
$\mathcal{I}(\cdot)$	$\boldsymbol{d}\,[1 \times K]$	$\mathcal{R}(\cdot)$	$\boldsymbol{x}_0^{(j)}\,[N_e \times L]$
$\eta_i(\cdot)$	$\{\boldsymbol{x}_{i-1}^{(j)}\}_j\,[N_e \times L]$	None	$\{\boldsymbol{\eta}_i^{(j)}\}_j\,[N_e \times L]$
$\mathcal{N}_i(\cdot)$	$\{\widetilde{\boldsymbol{x}}_i^{(j)}\}_j\,[N_e \times L]$	ReLU	$\{\widehat{\boldsymbol{x}}_i^{(j)}\}_j\,[N_p \times L]$
$\mathcal{E}_i(\cdot)$	$\{\widehat{\boldsymbol{x}}_i^{(j)}\}_j\,[N_p \times L]$	ReLU	$\{\boldsymbol{x}_i^{(j)}\}_j\,[N_e \times L]$

with the corresponding activation function $\mathcal{F}$, are summarized in Table 1.1. The bracketed expressions $[\cdot]$ in the table indicate the dimensionality of the inputs and outputs.

1.2.5 Numerical Results

This subsection summarizes the numerical results from the study that introduced FLED, focusing on the training methodology, hyperparameter configuration, and experimental setup used to evaluate the model.

Training Procedure: The FLED network was trained using a self-supervised approach on a dataset consisting of 20,000 synthetically generated desired beampatterns. Since ground-truth waveform solutions are generally unavailable for supervised training, the network was trained to minimize a loss function that reflects the deviation from the desired beampattern. The diversity of the generated beampatterns enabled the network to generalize well to unseen test cases.

The hyperparameters are decided via cross-validation, which is a common practice in DL [28]. The learning rate was initialized at 10^{-2} and decayed by a factor of 0.1 every 10 epochs. The batch size was set to 40. The values for N_e and N_p were selected as 5 and 2, respectively. Regularization hyperparameters were determined as follows: $\lambda_{f\star} = 1$, $\lambda_{f_0} = 1$, $\lambda_{f_i} = 0.5$, $\lambda_{\eta_i} = 10^{-4}$, and $\lambda_{d_0} = 10^{-5}$. The number of neurons between FCBs was set to $n_v = 1000$, and the numbers of FCBs used in the submodules $\mathcal{I}(\cdot)$, $\eta(\cdot)$, $\mathcal{N}(\cdot)$, and $\mathcal{E}(\cdot)$ were set to 7, 4, 4, and 4, respectively.

Given a fixed system configuration, such as the number of antennas and time/frequency samples, FLED is trained offline. Once trained, it can generate optimized waveforms for any new desired beampattern during inference.

Experimental Setup: The experiments assume a ULA MIMO radar system configuration. The number of transmit antennas was set to $M = 40$ and the number of time samples to $N = 32$. The system operated at a carrier frequency of $f_c = 1$ GHz with a bandwidth of $B = 200$ MHz. The sampling period was set to $T_s = 1/B$, and the inter-element spacing was defined as $d = c/2(f_c + B/2)$, where c is the speed of light. The spatial angle range was discretized into $S = 90$ angular bins.

The authors of the study [21] conducted an ablation study in the paper, demonstrating that the main components of FLED—learned descent direction, regularization, and prune/expansion—contribute to the final performance and robustness of the model.

1.2.5.1 Evaluation: Fidelity to Desired Beampattern

The performance of FLED was also evaluated by comparing it with several existing state-of-the-art (SOTA) approaches in terms of how closely the produced waveforms matched the desired beampatterns under the CMC. As a baseline, the study included an unconstrained least-squares (LS) solution that ignores the modulus constraint, providing a practical lower bound on beampattern error.

FLED was compared against several well-established iterative algorithms for constrained waveform design, including WBFIT [18], SDR [29, 30], IA-CPC [31], ADMM [32], PDR [20], and VSGP [25]. These methods were all applied to solve the same constrained beampattern design problem.

Additionally, FLED was compared to DL-based approaches, including comprehensive optimization network (CON) [8], which employs a residual NN for waveform design. To evaluate generalization performance on unseen data, the authors of the FLED study [21] ensured that the beampatterns used during testing were excluded from the training set for both unrolled and DL-based models.

To ensure a fair comparison, all the counterpart methods used the same randomly initialized CM waveform as input. Performance was assessed based on the achieved cost function value and computational time, either until convergence (for iterative algorithms) or during inference (for DL models), following standard evaluation practices in the field [8, 33].

Scenario 1) a typical beampattern in previous literature

The study of [21] shows numerical results for a widely used desired beampattern defined as:

$$d(\theta, f) = \begin{cases} 0, & \theta \in [10^\circ, 80^\circ], \quad -\frac{B}{2} + f_c \leq f \leq f_c, \\ 0, & \theta \in [95^\circ, 145^\circ], \quad f_c \leq f \leq \frac{B}{2} + f_c, \\ 1, & \text{Otherwise.} \end{cases} \tag{1.19}$$

Herein, nulls are inserted to attenuate the radiated energy in the region of interference or disinterest and facilitate coexistence of communication systems or other users within a frequency band [14, 32, 34].

As summarized in Table 1.2, FLED outperforms both iterative and DL-based methods in terms of beampattern accuracy, while maintaining a computational

Table 1.2 Comparison of FLED with iterative, DL, and unrolled baselines for the desired beampattern in (1.19) (scenario 1).

Method	Type	Error (dB)	Time (sec)	# Iter
Unconstrained		14.36	—	—
WBFIT [18]	Iterative methods	32.10	120.05	211
SDR [29, 30]		31.89	1275	33
IA-CPC [31]		27.54	11.03	686
ADMM [32]		28.67	20.36	208
PDR [20]		26.69	7.38	115
VSGP [25]		26.43	13.15	493
CON [8]	Deep learning	27.15	**0.13**	—
FLED (unrolled) [21]		**24.82**	**0.15**	5

Note: Bold-faced values indicate the best performing results (e.g., minimum error and minimum iterations). For computation time, both deep learning methods are bolded to highlight their substantially faster performance compared to iterative methods.

footprint suitable for near-real-time inference. The reported error metric is defined as $10\log_{10}(f(\boldsymbol{x}^\star))$ in dB, where the unconstrained solution—computed without enforcing the CMC—serves as a theoretical lower bound. For a fair comparison, the conventional DL (CON) was trained using the same dataset as FLED, following the architecture proposed in [8].

Further analysis in Figure 1.7 compares the error trajectories of FLED against SOTA iterative methods such as PDR and VSGP. The solid line, the dotted line, and the dashed lines are FLED, PDR, and VSGP, respectively. FLED achieves smaller error, while the counterparts converge to local minima that are not preferred.

Scenario 2) Training generalizability and scalability

To assess the generalization capability of FLED against a conventional DL method (CON) across a diverse range of design targets, a set of 80 ideal beampatterns was synthesized, each characterized by distinct spatio-spectral profiles. Figure 1.8 compares the performance of FLED and the CON model under reduced training data conditions. Specifically, only 50% of the original training dataset used for Scenario 1 was retained for this evaluation. In the figure, the filled dots indicate the mean error values for each method, while the lines encompass the range of test sample errors. The bottom and top edges of the box indicate the 25th and 75th percentiles.

As observed, FLED consistently outperforms CON, both in average error and robustness to reduced training data. Its performance degrades only slightly—from 15.31 to 17.40 dB (2.09 dB difference)—while CON exhibits a more substantial

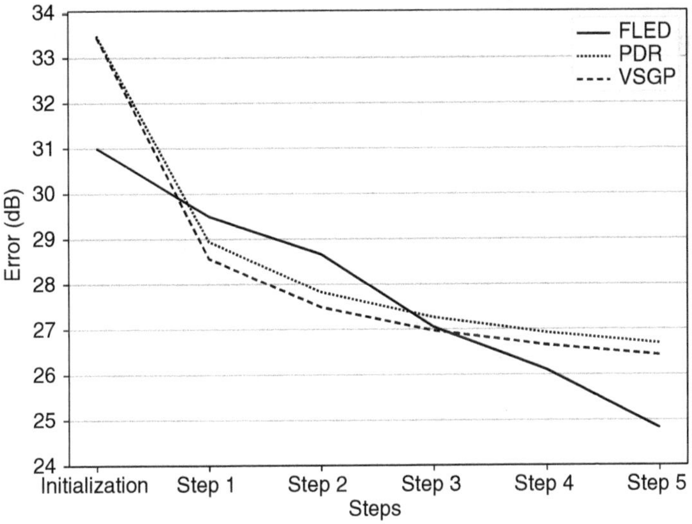

Figure 1.7 Empirical convergence analysis: the beampattern error in dB vs. update steps for scenario 1. Unrolled network FLED shows monotonically decreasing loss along steps.

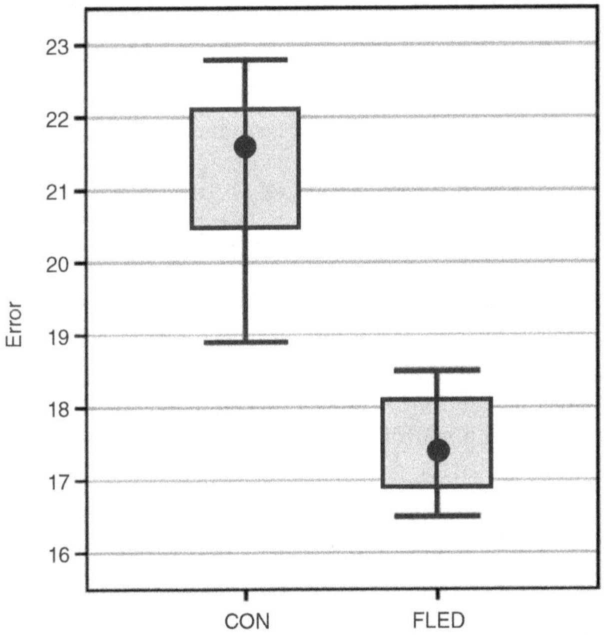

Figure 1.8 Box plots of beampattern error (deviation from desired beampattern) in dB for DL methods for small training (50%). Reducing the training set results in a notable 5.54 dB error increase for CON, whereas FLED shows a much smaller increase of 2.09 dB.

decline from 16.22 to 21.76 dB (5.54 dB difference). Moreover, the error variation across test samples is significantly smaller for FLED, indicating improved robustness to both the quantity and the specific composition of the training data. This reduction in variability reflects lower selection bias and stronger generalization capability.

1.2.6 Key Insights

FLED is derived by unrolling the iterative PDR algorithm, but enhances its performance through learning key parameters such as step size and descent direction. This results in a parameter-optimized version of the original iterative solver, achieving improved solution quality while enabling real-time inference. Unlike classical iterative algorithms with fixed hyperparameters, FLED adapts these elements through data-driven training, thereby improving convergence and robustness.

Furthermore, evaluation under reduced training conditions highlights a key advantage of algorithm unrolling over black-box DL. While conventional deep networks often suffer from high variance due to their unconstrained learning flexibility, FLED benefits from domain-informed architectural structure. This constrained learning framework helps guide the optimization process, leading to better generalization even with limited training data.

Since algorithm unrolling is an open-ended paradigm, there are no strict rules dictating which parameters should be learned. In practice, the best analytic descent direction may not be known or tractable, and learning from data offers a practical path toward improved descent behavior.

1.3 Single-target SINR Maximization Under the CMC

Designing waveforms to enhance the SINR is fundamental for reliable target detection in modern MIMO radar systems. Effective waveform design not only boosts detection accuracy but also suppresses unwanted interference and noise. In practical implementations, CM waveforms are preferred to maximize transmitter efficiency, adding a nonconvex constraint that complicates the optimization landscape.

Similar to the beampattern shaping, gradient projection methods like PDR algorithm, so-called Riemannian Conjugate Gradient (RCG) method, are utilized to design waveforms maximizing SINR [35, 36]. To address challenges of heavy computation while preserving interpretability, Zhong et al. [37] proposed a deep unfolding technique named as the learned complex circle manifold network (LCCM-Net). LCCM-Net directly addresses the CM-constrained SINR maximization problem by quadratic fractional programming defined over the CCM.

It then unrolls a gradient descent algorithm into a deep network architecture, where each layer performs a descent–retraction step on the manifold. Unlike traditional methods with fixed step sizes, LCCM-Net learns the step size adaptively at each iteration, enhancing convergence efficiency and generalization. Moreover, to accommodate practical hardware limitations, the framework is further extended to discrete phase design through a soft quantized version, termed QLCCM-Net, which employs a low-resolution quantizer with nonuniform intervals.

1.3.1 System Model

Let us consider a co-located MIMO radar system equipped with N_R receive antennas and N_T transmit antennas. Let $\boldsymbol{S} \in \mathbb{C}^{N_T \times M}$ denote the transmit waveform, where M is the number of time samples and $s_n(m)$ is the complex waveform value from the nth transmit antenna at time index m.

Define the waveform vector at time m as:

$$\boldsymbol{s}_m = [s_1(m),\ s_2(m), \dots,\ s_{N_T}(m)]^{\mathrm{T}}. \tag{1.20}$$

Then, the full waveform matrix can be written as $\boldsymbol{S} = [\boldsymbol{s}_1,\ \boldsymbol{s}_2, \dots,\ \boldsymbol{s}_M]$, and its vectorized form as $\boldsymbol{s} = \mathrm{vec}(\boldsymbol{S}) = [\boldsymbol{s}_1^{\mathrm{T}},\ \boldsymbol{s}_2^{\mathrm{T}}, \dots,\ \boldsymbol{s}_M^{\mathrm{T}}]^{\mathrm{T}}$.

The received signal at time m is modeled as:

$$\boldsymbol{x}_m = \gamma_0 \boldsymbol{A}(\theta_0)\boldsymbol{s}_m + \boldsymbol{d}_m + \boldsymbol{v}_m, \tag{1.21}$$

where γ_0 is the reflection coefficient of a target located at angle θ_0, with $\mathbb{E}[|\gamma_0|^2] = \beta_0$, where $\mathbb{E}[\cdot]$ denotes the expectation operator. $\boldsymbol{A}(\theta_0) = \boldsymbol{a}_r(\theta_0)\boldsymbol{a}_t^{\mathrm{H}}(\theta_0)$ is the spatial response matrix composed of the transmit and receive steering vectors. The transmit and receive steering vectors are defined, respectively, as:

$$\boldsymbol{a}_t(\theta) = \frac{1}{\sqrt{N_T}}\left[1, e^{j\pi \sin\theta}, \dots, e^{j\pi(N_T-1)\sin\theta}\right]^{\mathrm{T}}, \tag{1.22}$$

$$\boldsymbol{a}_r(\theta) = \frac{1}{\sqrt{N_R}}\left[1, e^{j\pi \sin\theta}, \dots, e^{j\pi(N_R-1)\sin\theta}\right]^{\mathrm{T}}. \tag{1.23}$$

The interference term $\boldsymbol{d}_m \in \mathbb{C}^{N_R}$ models the contributions from Q interfering scatterers. Let θ_q be the angle of the qth interferer and γ_q its reflection coefficient, with $\mathbb{E}[|\gamma_q|^2] = \beta_q$ for $q = 1, 2, \dots, Q$. Then, the interference component can be expressed as:

$$\boldsymbol{d}_m = \sum_{q=1}^{Q} \gamma_q \boldsymbol{A}(\theta_q)\boldsymbol{s}_m. \tag{1.24}$$

Finally, $\boldsymbol{v}_m \sim \mathcal{CN}(\boldsymbol{0}, \sigma_v^2 \boldsymbol{I}_{N_R})$ denotes the additive noise.

1.3.2 Problem Formulation

The objective is to maximize the SINR. To this end, the process begins by expressing the power of the received signal and the interference. The received signal power is given by:

$$\frac{1}{M}\sum_{m=1}^{M}\mathbb{E}\left[\|\gamma_0\boldsymbol{A}(\theta_0)\boldsymbol{s}_m\|^2\right]=\frac{\beta_0}{M}\sum_{m=1}^{M}\left\|\boldsymbol{a}_t^{\mathrm{H}}(\theta_0)\boldsymbol{s}_m\right\|^2=\frac{\beta_0}{M}\boldsymbol{s}^{\mathrm{H}}\boldsymbol{A}_0\boldsymbol{s}, \tag{1.25}$$

where $\boldsymbol{R}=\frac{1}{M}\sum_{m=1}^{M}\boldsymbol{s}_m\boldsymbol{s}_m^{\mathrm{H}}$ is the waveform covariance matrix and $\boldsymbol{A}_0=\boldsymbol{I}_M\otimes\left(\boldsymbol{a}_t(\theta_0)\boldsymbol{a}_t^{\mathrm{H}}(\theta_0)\right)$. Assuming that the interference sources are uncorrelated, the power of the received interference is expressed as:

$$\frac{1}{M}\sum_{m=1}^{M}\mathbb{E}\left[\left\|\sum_{q=1}^{Q}\gamma_q\boldsymbol{A}(\theta_q)\boldsymbol{s}_m\right\|^2\right]=\sum_{q=1}^{Q}\frac{\beta_q}{M}\boldsymbol{s}^{\mathrm{H}}\boldsymbol{A}_q\boldsymbol{s}, \tag{1.26}$$

where $\boldsymbol{A}_q=\boldsymbol{I}_M\otimes\left(\boldsymbol{a}_t(\theta_q)\boldsymbol{a}_t^{\mathrm{H}}(\theta_q)\right)$. Combining (1.25) and (1.26), and including additive noise with power $N_T\sigma_v^2$, the SINR is expressed as:

$$\mathrm{SINR}=\frac{\beta_0\boldsymbol{a}_t^{\mathrm{H}}(\theta_0)\boldsymbol{R}\boldsymbol{a}_t(\theta_0)}{\sum_{q=1}^{Q}\beta_q\boldsymbol{a}_t^{\mathrm{H}}(\theta_q)\boldsymbol{R}\boldsymbol{a}_t(\theta_q)+N_T\sigma_v^2}=\frac{\boldsymbol{s}^{\mathrm{H}}\boldsymbol{Y}_0\boldsymbol{s}}{\boldsymbol{s}^{\mathrm{H}}\boldsymbol{Z}_0\boldsymbol{s}}, \tag{1.27}$$

where

$$\boldsymbol{Y}_0=\frac{\beta_0}{M}\boldsymbol{A}_0, \tag{1.28}$$

$$\boldsymbol{Z}_0=\sum_{q=1}^{Q}\frac{\beta_q}{M}\boldsymbol{A}_q+\frac{1}{M}N_T\sigma_v^2\boldsymbol{I}_{MN_T}. \tag{1.29}$$

To maximize the SINR while ensuring CMC on the waveform entries, the problem becomes an unimodular quadratic program (UQP) as:

$$\begin{aligned}\max_{\boldsymbol{s}}\quad & f(\boldsymbol{s})=\frac{\boldsymbol{s}^{\mathrm{H}}\boldsymbol{Y}_0\boldsymbol{s}}{\boldsymbol{s}^{\mathrm{H}}\boldsymbol{Z}_0\boldsymbol{s}}\\ \text{s.t.}\quad & |\boldsymbol{s}(l)|=1,\quad l=1,\dots,L,\end{aligned} \tag{1.30}$$

where the constraint implies that $\boldsymbol{s}$ lies on the CCM, $\mathcal{S}^L$, as described in Section 1.2.2.

1.3.3 LCCM-net-based Waveform Design

One classical way to solve a UQFP in (1.30) is to use the following sequence:

1. Project the Euclidean gradient of the objective function onto the tangent space of the CCM.

2. Perform a descent step along the tangent direction using the step size learned via end-to-end training.
3. Retract the resulting point back onto the CCM.

The classical optimization algorithms taking the steps above often rely on fixed step sizes or use heuristic-based updates such as the Armijo line search [20, 25, 35, 36]. However, these techniques can be inefficient for nonconvex problems, as they are sensitive to initialization and require manual tuning, potentially increasing computational burden.

To overcome this issue, LCCM-Net replaces fixed (or heuristics-based) step sizes with learnable parameters, yielding one LCCM-Net step that corresponds to one iteration (one sequence above) of the classical projected-gradient descent. Since DL based on the TensorFlow platform cannot directly handle complex matrix operations, the authors of [37] convert the complex-valued mathematical forms into real-valued ones to describe the waveform update rule of LCCM-Net. The update rules are as follows:

1. Projecting the Euclidean Gradient

In the kth step, the gradient is projected onto the tangent space of the CCM, which can be given by

$$\nabla_{\mathcal{T}_{s_k}\mathcal{M}} f(\boldsymbol{s}_k^r) = \mathcal{P}_{\mathcal{T}_{s_k}\mathcal{M}}(\nabla f(\boldsymbol{s}_k^r)) = \nabla f(\boldsymbol{s}_k^r) - \begin{bmatrix} \Re(\boldsymbol{h}) \\ \Im(\boldsymbol{h}) \end{bmatrix}, \tag{1.31}$$

where $\boldsymbol{h} = \nabla f(\boldsymbol{s}_k^r) \odot \boldsymbol{s}_k^*$; and $\Re(\cdot)$, $\Im(\cdot)$ returns real, imaginary part, respectively. Here, the tangent space on to the CCM $\mathcal{T}_{s_k}\mathcal{M}$ is defined as:

$$\mathcal{T}_{s_k}\mathcal{M} = \{\boldsymbol{x} \in \mathbb{C}^L \mid \Re(\boldsymbol{x} \odot \boldsymbol{s}_k^*) = \boldsymbol{0}\}, \tag{1.32}$$

and the projection operator onto this space can be expressed as:

$$\mathcal{P}_{\mathcal{T}_{s_k}\mathcal{M}}(\boldsymbol{a}) = \boldsymbol{a} - \Re(\boldsymbol{a} \odot \boldsymbol{s}_k^*) \odot \boldsymbol{s}_k. \tag{1.33}$$

The real-valued representation of the gradient in (1.31) is given by:

$$\begin{aligned} \nabla f(\boldsymbol{s}_k^r) &= \frac{2(\boldsymbol{s}_k^{\mathrm{H}}\boldsymbol{Y}_0\boldsymbol{s}_k)\boldsymbol{Z}_0\boldsymbol{s}_k - 2(\boldsymbol{s}_k^{\mathrm{H}}\boldsymbol{Z}_0\boldsymbol{s}_k)\boldsymbol{Y}_0\boldsymbol{s}_k}{(\boldsymbol{s}_k^{\mathrm{H}}\boldsymbol{Y}_0\boldsymbol{s}_k)^2} \\ &= \frac{2(\boldsymbol{s}_k^{rT}\boldsymbol{Y}_0^r\boldsymbol{s}_k^r)\boldsymbol{Z}_0^r\boldsymbol{s}_k^r - 2(\boldsymbol{s}_k^{rT}\boldsymbol{Z}_0^r\boldsymbol{s}_k^r)\boldsymbol{Y}_0^r\boldsymbol{s}_k^r}{(\boldsymbol{s}_k^{rT}\boldsymbol{Y}_0^r\boldsymbol{s}_k^r)^2}, \end{aligned} \tag{1.34}$$

which can also be expressed in component form as:

$$\nabla f(\boldsymbol{s}_k^r) = \begin{bmatrix} \Re(\nabla f(\boldsymbol{s}_k)) \\ \Im(\nabla f(\boldsymbol{s}_k)) \end{bmatrix}. \tag{1.35}$$

The real-valued variables used above are defined as:

$$\boldsymbol{s}_k^r = \begin{bmatrix} \Re(\boldsymbol{s}_k) \\ \Im(\boldsymbol{s}_k) \end{bmatrix}, \quad \boldsymbol{Z}_0^r = \begin{bmatrix} \Re(\boldsymbol{Z}_0) & -\Im(\boldsymbol{Z}_0) \\ \Im(\boldsymbol{Z}_0) & \Re(\boldsymbol{Z}_0) \end{bmatrix}, \quad \boldsymbol{Y}_0^r = \begin{bmatrix} \Re(\boldsymbol{Y}_0) & -\Im(\boldsymbol{Y}_0) \\ \Im(\boldsymbol{Y}_0) & \Re(\boldsymbol{Y}_0) \end{bmatrix}. \tag{1.36}$$

2. Descent on the Tangent Space

The updated solution over the tangent space is computed by taking a gradient step along the projected direction:

$$\boldsymbol{s}_T^r = \boldsymbol{s}_k^r - \alpha_k \nabla_{\mathcal{T}_{\boldsymbol{s}_k}\mathcal{M}} f(\boldsymbol{s}_k^r), \tag{1.37}$$

where α_k denotes the step size for iteration k.

3. Retraction to the CCM

The final step involves retracting $\boldsymbol{s}_T$ back onto the CCM, which yields the next iterate:

$$\boldsymbol{s}_{k+1}^r = \boldsymbol{s}_T^r \odot \begin{bmatrix} |\boldsymbol{s}_T|^{-1} \\ |\boldsymbol{s}_T|^{-1} \end{bmatrix}. \tag{1.38}$$

The forward propagation of LCCM-Net is visualized in Figure 1.9, where $\mathcal{J}_k(\cdot)$ is the kth LCCM-Net step, $\mathcal{P}(\cdot)$ is projection described in (1.33), and $\mathcal{R}(\cdot)$ is retraction shown in (1.38). Distinct step sizes α_k are computed by the NN modules for given input $\boldsymbol{Y}_0$, $\boldsymbol{Z}_0$, and $\boldsymbol{s}_{k-1}$.

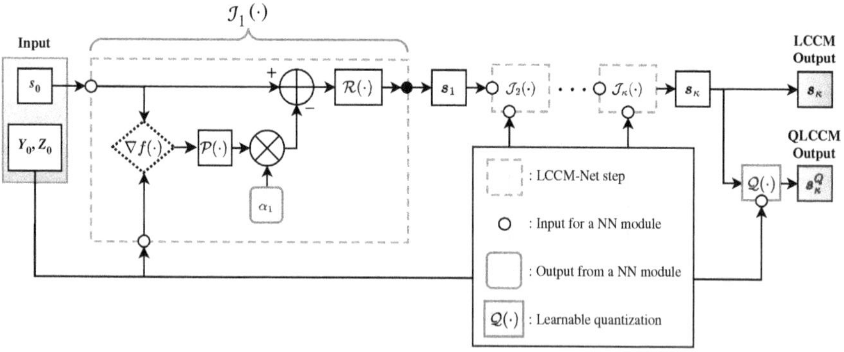

Figure 1.9 Generalized forward propagation of LCCM-Net and QLCCM-Net. While both networks share the same unrolled descent–retraction backbone, QLCCM-Net includes an additional quantization block at the output to enable discrete-phase radar waveform generation.

1.3.3.1 Learning the Step Sizes

During training, these step sizes are learned through backpropagation to minimize the loss of function, defined as

$$\mathcal{L} = f(\boldsymbol{s}_\kappa) = \frac{\boldsymbol{s}_\kappa^{\mathrm{H}} \boldsymbol{Z}_0 \boldsymbol{s}_\kappa}{\boldsymbol{s}_\kappa^{\mathrm{H}} \boldsymbol{Y}_0 \boldsymbol{s}_\kappa}. \tag{1.39}$$

As a result, the network adjusts the step size at each LCCM-Net step based on the training data, leading to faster convergence and better solution quality compared to the classical iterative algorithm.

This learning-based step size adaptation is particularly beneficial in challenging nonconvex scenarios, where the optimization surface may exhibit numerous local minima and sharp transitions. By tailoring the step size to each layer, LCCM-Net enhances the ability to escape suboptimal solutions and approach the global optimum more effectively.

1.3.4 QLCCM-net for Discrete Phase Waveform Design

In Section 1.3.3, waveform design was discussed under the assumption of continuous-phase control. However, practical radar systems often require low-resolution discrete-phase waveforms due to hardware limitations. Traditional approaches typically apply hard quantization using uniformly spaced thresholds to map continuous phases to discrete values. These heuristically chosen thresholds may lead to performance degradation, as they do not optimally account for waveform design objectives. Furthermore, hard quantization functions are non-differentiable, which obstructs gradient-based optimization methods in learning frameworks.

To address these challenges, the study in [37] introduced a soft quantization mechanism designed to be differentiable and compatible with backpropagation. This method replaces the non-differentiable hard quantization function with a smooth approximation based on a sum of shifted hyperbolic tangent functions. The resulting soft quantizer enables control over amplitude, smoothness, and nonuniformity of quantization intervals. By integrating this mechanism into the LCCM-Net architecture, the authors developed QLCCM-Net—a model that supports discrete-phase waveform optimization by learning both quantization thresholds and step sizes.

1.3.4.1 Soft Quantization Function

To enable end-to-end training with quantization, the QLCCM-Net architecture incorporates a differentiable quantization function defined as:

$$Q(\phi) = \sum_{i=1}^{B-1} a_i[\tanh(c_i(\phi - \rho_i)) + 1], \tag{1.40}$$

where a_i, c_i, and ρ_i denote the amplitude, smoothing coefficient, and quantization threshold associated with the ith level, respectively. The QLCCM-Net architecture, by combining LCCM-Net with this soft quantizer, offers a practical solution for generating discrete-phase radar waveforms in hardware-constrained systems, while still benefiting from model-based structure and learning-based adaptability.

The forward propagation of QLCCM-Net is visualized in Figure 1.9, where QLCCM-Net includes an additional quantization block with learnable quantization region thresholds $\rho_i\ \forall i$ at the output to enable discrete-phase in waveform. QLCCM-Net is trained based on the quantized output $\boldsymbol{s}_{\kappa}^{Q}$, computing the training loss as $\mathcal{L} = f(\boldsymbol{s}_{\kappa}^{Q})$, where $f(\cdot)$ is defined in (1.39).

1.3.5 Numerical Results

This subsection shows the performance of the proposed LCCM-Net and QLCCM-Net methods in terms of convergence behavior and SINR. To highlight the performance advantage, comparisons are made against several baseline methods, including Dinkel-ADMM [38, 39], Dinkel-PML [40], and Dinkel-MM acceleration method [41], PDR method (fixed step size) [20], RCG method [35, 36], and Res-Net method [8].

The simulation settings are as follows: $N_T = 16$ transmit antennas and $M = 16$ snapshots are used. The target is located at $\theta_0 = 15°$, while Q interfering sources are distributed over directions ranging from $-40°$ to $-30°$. The target signal-to-noise ratio (SNR) is set to $\beta_0 = 10$ dB and each interference source is assigned an interference-to-noise ratio (INR) of $\beta_q = 30$ dB. The additive noise is modeled as zero-mean Gaussian with variance $\sigma_v^2 = 0$ dB.

To ensure fairness in evaluation, all methods are initialized using an orthogonal linear frequency modulated (LFM) waveform defined as:

$$[\boldsymbol{S}_0]_{n_t,m} = \exp\{j2\pi n_t(m-1)/M\}\exp\left\{j\pi(m-1)^2/M\right\}, \tag{1.41}$$

for $n_t = 1, \ldots, N_T$ and $m = 1, \ldots, M$.

1.3.5.1 Convergence Behavior

The convergence characteristics of LCCM-Net are illustrated in Figure 1.10 by examining the cost function values across iterations. It is observed that for different network depths, the cost function converges to distinct levels, while deeper networks yield better SINR performance. These results suggest that increasing network depth may enhance solution quality while the computational burden for training increases.

1.3.5.2 Achieved SINR and Required Computational Cost

Table 1.3 shows a numerical comparison for SINR maximization. LCCM-Net (unrolled network) achieves the superior SINR in the shortest time and also

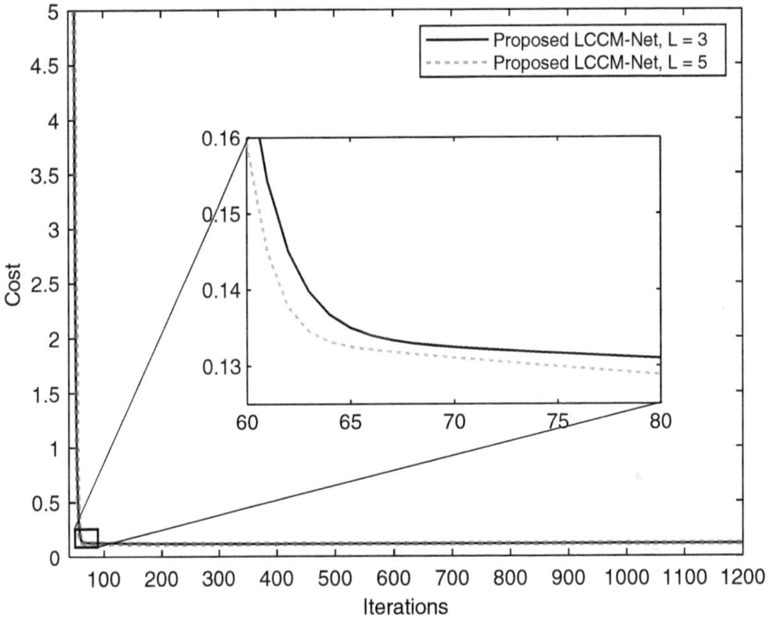

Figure 1.10 Convergence behavior: cost function values versus number of iterations with different network depth. For all evaluated depths, LCCM-Net consistently exhibits monotonic decay across iterations.

Table 1.3 Comparison of the computational time and achieved SINR.

Method	**Time (s)**	**SINR (dB)**	**Iterations**
LCCM-Net	9.8	9.394	1,200
Res-Net	79.3	8.804	40,000
RCG	18	8.741	2,800
PDR	121	9.186	30,000
PML	287.7	7.920	40,000
MM acceleration	134.6	8.968	40,000
ADMM	210	9.389	10,000

in the smallest number of iterations. Learning the step size from training data enables LCCM-Net works as a parameter-optimized version of PDR/RCG that uses fixed/heuristically determined step size, enhancing the optimization performance. LCCM-Net's learned step size adaptation enables more flexible navigation of the optimization surface, increasing the likelihood of reaching better solutions.

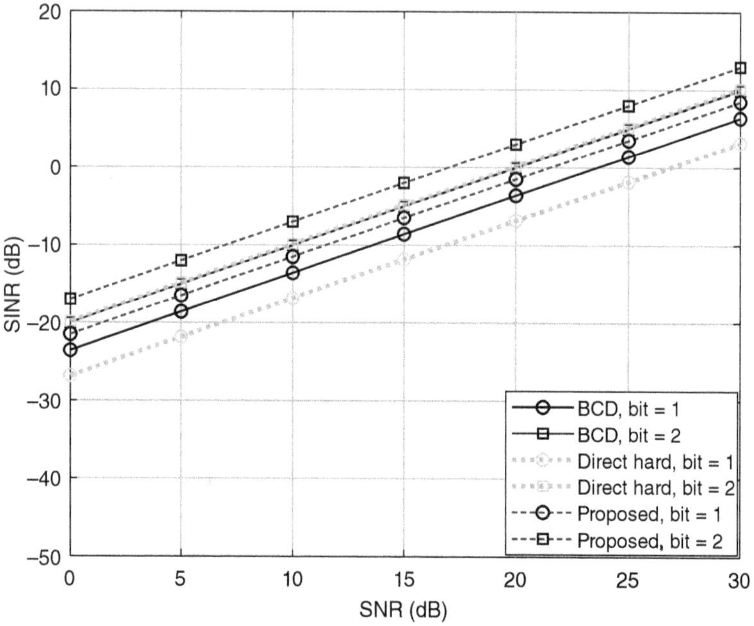

Figure 1.11 Comparisons of SINRs for each method under discrete phase constraints. The proposed methods (QLCCM-Net) demonstrate superior SINR compared to the counterparts under same bit constraint.

1.3.5.3 Performance of Waveform Quantization

QLCCM-Net is benchmarked against two baseline approaches: the direct hard quantization method and the block coordinate descent (BCD) algorithm. Figure 1.11 presents the SINR performance as a function of SNR under discrete phase constraints, specifically for 1- and 2-bit waveform cases. The results demonstrate that QLCCM-Net consistently outperforms both baselines. These results suggest that QLCCM-Net is capable of uncovering latent structure in the data, which allows it to learn effective step sizes and decision thresholds that better align with the underlying optimization objective.

1.3.6 Key Insights

LCCM-Net addresses the problem of single-target SINR maximization under the CMC by unrolling the RCG algorithm. A key innovation in LCCM-Net is the replacement of fixed step sizes with learnable parameters that are optimized end-to-end via backpropagation. This parameter learning improves convergence and enhances waveform performance relative to traditional iterative solvers.

An additional contribution is QLCCM-Net, which extends LCCM-Net to the waveforms having discrete phase. To support differentiable training under quantization, the authors introduced a soft quantization mechanism.

Together, LCCM-Net and QLCCM-Net illustrate how algorithm unrolling can be used to design interpretable, constraint-aware NNs while adapting to real-world hardware constraints. LCCM-Net preserves geometric structure through manifold-based learning, and QLCCM-Net bridges the gap to deployment by incorporating quantization-aware training.

1.4 Adaptive Single-target SINR Maximization in Dynamic Environments

While Section 1.3 addressed SINR maximization under the assumption of static environments with known interference statistics, real-world applications—particularly the integration of radar into autonomous vehicles—require robust perception in rapidly changing conditions. Traditional model-based approaches often rely on a priori environmental knowledge that is difficult to maintain in real-time, causing these methods to struggle as environmental complexity increases. Conversely, while data-driven DL methods can design waveforms directly from observed data, they often lack the interpretability and validation guarantees required for safety-critical automotive contexts.

To address the limitations of static optimization and enable practical, real-time deployment, Khobahi et al. proposed a hybrid architecture that combines model-driven and data-driven elements [42]. Their method adapts to time-varying environments and enables adaptive unimodular waveform design by leveraging observed data in conjunction with a structured, interpretable learning framework.

1.4.1 Problem Formulation

Consider a radar system that transmits unimodular codes to modulate a sequence of sub-pulses. Let $\boldsymbol{s} = [s_1\ s_2\ \cdots\ s_N]^{\mathrm{T}} \in \mathbb{C}^{N\times 1}$ denote the transmit waveform to be designed. The received discrete-time baseband signal after pulse compression and alignment with the target range cell can be expressed as follows [43]:

$$\boldsymbol{y} = \boldsymbol{A}^{\mathrm{H}}\boldsymbol{\alpha} + \boldsymbol{\epsilon}, \tag{1.42}$$

where

$$\boldsymbol{A}^{\mathrm{H}} = \begin{bmatrix} s_1 & 0 & \cdots & 0 & s_N & s_{N-1} & \cdots & s_2 \\ s_2 & s_1 & \cdots & 0 & 0 & s_N & \cdots & s_3 \\ \vdots & \vdots & \ddots & \vdots & \vdots & \vdots & \ddots & \vdots \\ s_N & s_{N-1} & \cdots & s_1 & 0 & 0 & \cdots & 0 \end{bmatrix}, \tag{1.43}$$

and

$$\boldsymbol{\alpha} = [\alpha_0, \alpha_1, \dots, \alpha_{N-1}, \alpha_{-N+1}, \dots, \alpha_{-1}]^{\mathrm{T}} \in \mathbb{C}^{2N-1}. \tag{1.44}$$

Here, α_0 denotes the scattering coefficient of the range cell of interest, while $\{\alpha_k\}_{k\neq 0}$ represents clutter contributions from neighboring range cells. The term ϵ includes measurement noise and other signal-independent interference such as jamming. The goal is to design the transmit waveform $\boldsymbol{s}$ to enable accurate estimation of the target scattering coefficient α_0 from $\boldsymbol{y}$.

In classical waveform design, the statistics of the interference and noise are typically assumed to be known, for example, through prior environmental measurements or estimation procedures. Under such conditions, waveform optimization reduces to a constrained quadratic or fractional quadratic program [40, 44]. A common design criterion is to maximize the SINR, which is defined as[3]:

$$f(\boldsymbol{s}) = \frac{|\boldsymbol{s}^{\mathrm{H}}\boldsymbol{y}|^2}{\sum_{k\neq 0} |\boldsymbol{s}^{\mathrm{H}}\boldsymbol{J}_k\boldsymbol{y}|^2} = \frac{\boldsymbol{s}^{\mathrm{H}}\boldsymbol{B}_1\boldsymbol{s}}{\boldsymbol{s}^{\mathrm{H}}\boldsymbol{B}_2\boldsymbol{s}} = \frac{n(\boldsymbol{s})}{d(\boldsymbol{s})}, \tag{1.45}$$

where $\boldsymbol{B}_1 = \boldsymbol{y}\boldsymbol{y}^{\mathrm{H}}$, $\boldsymbol{B}_2 = \sum_{k\neq 0} \boldsymbol{J}_k\boldsymbol{B}_1\boldsymbol{J}_k^{\mathrm{H}}$, and $\{\boldsymbol{J}_k\}$ are shift matrices satisfying $[\boldsymbol{J}_k]_{l,m} = \delta_{m-l-k}$. Herein, $d(\boldsymbol{s}) = \boldsymbol{s}^{\mathrm{H}}\boldsymbol{B}_2\boldsymbol{s} > 0$. This expression represents the ratio of the signal power to the total power of interference and noise after matched filtering.

The authors consider the CMC, that is, $s_k = e^{j\phi_k}$ with $\phi_k \in [0,\ 2\pi)$ for all $k \in \{1, \dots, N\}$ [2]. This leads to the following fractional optimization problem:

$$\max_{s} \frac{\boldsymbol{s}^{\mathrm{H}}\boldsymbol{B}_1\boldsymbol{s}}{\boldsymbol{s}^{\mathrm{H}}\boldsymbol{B}_2\boldsymbol{s}}, \quad \text{s.t. } |s_k| = 1, \quad k \in \{1, \dots, N\}. \tag{1.46}$$

1.4.1.1 Fractional Programming Reformulation

Because the numerator $n(\boldsymbol{s})$ and denominator $d(\boldsymbol{s})$ are both quadratic in $\boldsymbol{s}$, the optimization of $f(\boldsymbol{s})$ is well-suited for fractional programming techniques [45]. Following this approach, and given a current estimate $\boldsymbol{s}_\star$, the auxiliary function is defined as

$$\begin{aligned} e(\boldsymbol{s}) &\triangleq n(\boldsymbol{s}) - f(\boldsymbol{s}_\star)d(\boldsymbol{s}), \\ &= \boldsymbol{s}^{\mathrm{H}}\boldsymbol{B}_1\boldsymbol{s} - f(\boldsymbol{s}_\star)(\boldsymbol{s}^{\mathrm{H}}\boldsymbol{B}_2\boldsymbol{s}) = \boldsymbol{s}^{\mathrm{H}}\underbrace{(\boldsymbol{B}_1 - f(\boldsymbol{s}_\star)\boldsymbol{B}_2)}_{\triangleq \boldsymbol{\Phi}}\boldsymbol{s}. \end{aligned} \tag{1.47}$$

Let $\boldsymbol{s}_\dagger = \arg\max_{\boldsymbol{s}}\ e(\boldsymbol{s})$, then it is evident $e(\boldsymbol{s}_\dagger) \geq e(\boldsymbol{s}_\star) = 0$, which implies

$$f(\boldsymbol{s}_\dagger) \geq f(\boldsymbol{s}_\star). \tag{1.48}$$

Importantly, this inequality holds even if $\boldsymbol{s}_\dagger$ is not a global maximizer of $e(\boldsymbol{s})$, provided it satisfies $e(\boldsymbol{s}_\dagger) \geq e(\boldsymbol{s}_\star)$. This reformulated objective serves as the foundation for the unrolling-based learning method introduced next.

---

3 This is similar to the SINR derived in Section 1.3.2, which shows the final form in (1.27).

1.4.2 Iterative Algorithm: Power Method Like Iterations

To ensure that the matrix $\tilde{\boldsymbol{\Phi}}$ is positive definite, a diagonal loading procedure can be applied, resulting in a regularized matrix $\boldsymbol{\Phi} \triangleq \tilde{\boldsymbol{\Phi}} + \xi \boldsymbol{I}_N$, where $\xi \geq 0$ is decided to guarantee $\boldsymbol{\Phi} \succ 0$. With this modification, the fractional program in (4) can be recast as the following unimodular quadratic program (UQP) [40]:

$$\max_{\boldsymbol{s}} \ \boldsymbol{s}^{\mathrm{H}}\boldsymbol{\Phi}\boldsymbol{s}, \qquad \text{s.t. } |s_k| = 1, \qquad k \in \{1, \dots, N\}. \tag{1.49}$$

A class of iterative methods known as power method like iterations (PMLI) has been proposed to tackle such UQPs efficiently [40, 44]. These iterations update $\boldsymbol{s}$ by solving a nearest-vector problem:

$$\min_{\boldsymbol{s}^{n+1}} \|\boldsymbol{s}^{n+1} - \boldsymbol{\Phi}\boldsymbol{s}^n\|_2^2, \qquad \text{s.t. } |\boldsymbol{s}_k^{n+1}| = 1, \qquad k \in \{1, \dots, N\}. \tag{1.50}$$

The solution to this problem can be obtained in closed form:

$$\boldsymbol{s}^{n+1} = \tilde{\mathcal{J}}(\boldsymbol{s}^n, \boldsymbol{\Phi}) = e^{j \arg(\boldsymbol{\Phi}\boldsymbol{s}^n)}, \tag{1.51}$$

where n is the iteration index. Starting from an initial guess $\boldsymbol{s}^0$, this procedure is repeated either until convergence or for a fixed number of steps κ. These iterations are known to guarantee a monotonic increase in the quadratic objective, regardless of the signal constraints.

However, such methods often rely on full knowledge of the system model, including environmental parameters like clutter statistics or interference power. In practice, this information may not be available or may vary too rapidly to estimate accurately.

1.4.3 Unrolling PMLI into DECoR

To overcome the need for a priori environmental knowledge, Khobahi et al. proposed a hybrid framework that simultaneously designs adaptive transmit waveforms and infers the unknown parameters from observed data [42]. The key insight is that environmental information is implicitly encoded in the received signal $\boldsymbol{y}$, allowing both tasks to be performed jointly.

Khobahi et al. introduce a model-based deep NN named deep evolutionary cognitive radar (DECoR), which unrolls the iterative PMLI into an interpretable deep NN. The forward propagation of DECoR is drawn in Figure 1.12. Specifically, DECoR has a sequence of DECoR steps $\{\mathcal{J}_i(\cdot)\}_{i=1}^{\kappa}$, each mirrors one iteration of the PMLI update $\tilde{\mathcal{J}}(\cdot)$, maintaining the same nonlinear operations as the original optimization algorithm. As a result, the network preserves interpretability while gaining the flexibility of a learnable model. Moreover, DECoR allows data-driven tuning of network parameters through online learning, enabling real-time adaptation to changing environments.

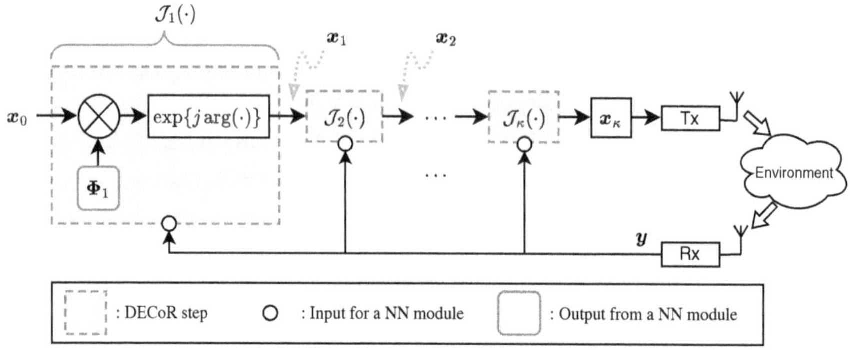

Figure 1.12 Forward propagation of DECoR. The parameters $\{\Phi_i\}_{i=1}^{\kappa}$ are learned online by collecting information pairs of transmitted waveforms s_κ and received signals y from dynamic environment.

In the original PMLI approach, the matrix $\mathbf{\Phi}$ remains fixed throughout all iterations. In contrast, DECoR introduces an over-parameterization by assigning a distinct matrix $\mathbf{\Phi}_i$ to each iteration index i. This modification leads to a more expressive deep architecture.

To enable real-time adaptation, DECoR employs an online learning strategy for updating the parameters $\{\mathbf{\Phi}_i\}_{i=1}^{\kappa}$ based on observed data. This is conceptually equivalent to learning an implicit model of the signal-dependent interference and noise characteristics embedded in the received signal $\boldsymbol{y}$.

1.4.3.1 Online Learning Strategy

To enable data-driven parameter adaptation, a random walk-based learning strategy is adopted for updating the trainable components of the DECoR architecture. DECoR performs the following sequential stages:

Stage 0 (Initialization): Begin with an arbitrary unimodular transmit sequence. Choose a search radius $\sigma > 0$, a reduction factor $\delta \in (0,\ 1]$. Initialize the set of weight matrices $\mathbf{\Omega}^{(0)} = \{\mathbf{\Phi}_i^{(0)}\}_{i=1}^{\kappa}$ such that $\mathbf{\Phi}_i^{(0)} \succ 0$. Set $t = 1$.

Stage 1 (Random walk): For each DECoR step $i \in \{1, \ldots, \kappa\}$, randomly perturb $\mathbf{\Omega}^{(t-1)}$ to make $\mathbf{\Omega}^{(t)}$. Compute the corresponding unimodular waveform

$$\boldsymbol{s}_{\kappa,i}^{(t)} = \mathcal{J}_\kappa(\cdot,\ \mathbf{\Phi}_\kappa^{(t)}) \circ \mathcal{J}_{\kappa-1}(\cdot,\ \mathbf{\Phi}_{\kappa-1}^{(t)}) \circ \cdots \circ \mathcal{J}_1(\boldsymbol{s}_0,\ \mathbf{\Phi}_1^{(t)}). \tag{1.52}$$

This online random walk-based learning procedure allows the DECoR architecture to continuously adapt to dynamic environments, improving performance under varying interference and noise conditions without requiring explicit prior knowledge of environmental parameters.

1.4.4 Numerical Results

The effectiveness of the DECoR and its associated online learning algorithm is evaluated. For all experiments, the number of DECoR steps is fixed at $\varkappa = 30$. The radar environment is modeled according to the formulation given in (1.42), with additional details provided in [43]. The simulation assumes a clutter power of $\beta = 1$ and a noise covariance matrix $\boldsymbol{\Gamma} = \boldsymbol{I}$. Importantly, this environmental information is not provided to the DECoR model during training and is used solely for data generation.

Figure 1.13 illustrates the objective function values $f(\boldsymbol{s}_\varkappa)$ from (1.45) across training iterations for a code length of $N = 10$. The results demonstrate that the DECoR leads to a monotonic increase in the SINR, confirming the stability of the learning process.

1.4.5 Key Insights

The DECoR framework introduces several practical and architectural innovations that distinguish it within the context of radar waveform design.

First, DECoR does not require access to ground-truth environmental parameters. This capability is crucial in automotive radar scenarios, where channel and clutter conditions change rapidly and unpredictably across time.

Second, the training procedure of DECoR employs a PMLI-inspired unrolled architecture that ensures CM waveforms and positive-definiteness of the learned

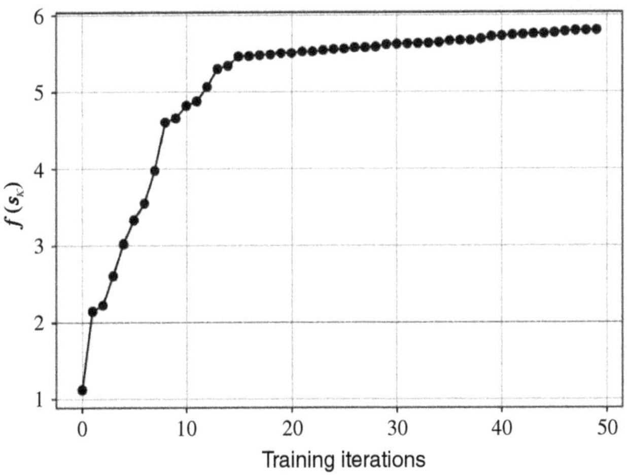

Figure 1.13 DECoR objective values $f(\boldsymbol{s}_\varkappa)$ versus training iterations for code length $N = 10$. DECoR demonstrates monotonic growth of the objective function across all iterations.

matrix. This addresses a common challenge in waveform design where physical constraints on the signal must be satisfied during inference. DECoR learns parameters χ_i rather than a fixing them (χ), allowing improved optimization performance.

1.5 Multi-target SINR Maximization

MIMO radar waveform design for maximizing SINR is frequently formulated as a constrained optimization problem, where the constraints include the CMC and the similarity constraint (SC). The CMC is required due to nonlinear amplifiers in radar systems [2]. Meanwhile, the SC is often enforced to preserve useful characteristics—such as range resolution and ambiguity function properties—of a reference waveform [46, 47]. However, enforcing the SC and the nonconvex CMC in radar waveform design introduces significant challenges.

Under these constraints, the joint design of transmit waveforms and receive filters in MIMO radar has been extensively studied for its benefits in enhancing detection performance. The SINR is often expressed as a quadratic fractional function, where the numerator quantifies the power of the desired target return and the denominator aggregates interference and noise. For single-target scenarios, this leads to a classical formulation based on the maximization of a quadratic rational function, as shown in Sections 1.3 and 1.4.

Compared to single-target problems, multi-target scenarios present a more practical yet complex challenge. Although critical in many applications, multi-target waveform design has received relatively limited attention in the literature [48–55]. A commonly adopted strategy in this context is the maximization of the worst-case SINR (a max-min optimization). Several techniques have been introduced to address this problem. For example, the nested iterative convex enhancement (NICE) and nested modified convex enhancement (NMICE) algorithms in [48] tackle convexified formulations. The SQCA method proposed in [49] leverages a semi-definite relaxation (SDR) framework with randomized post-processing. These approaches often reformulate the worst-case SINR problem by maximizing a lower bound, facilitating the conversion from a quadratic fractional to a quadratic program. An alternative approach in [50] applies first-order Taylor approximation and utilizes a minorization–maximization (MM) framework for iterative refinement.

It is important to note that they either relax the CMC or consider a constraint weaker than the CMC. They possibly lead to suboptimal practical performance due to the resulting non-CM waveforms [2].

To combine the computational efficiency of DL with the interpretability of iterative optimization, Kweon et al. introduced a model-based DL framework for multi-target detection in MIMO radar systems [56]. The study focuses on maximizing the worst-case SINR under practical constraints. As a first step, the authors propose a novel iterative optimization algorithm, termed sum-of-reciprocal exact descent (SRED). This method is subsequently extended into a learning-based variant, referred to as sum-of-reciprocal exact learning (SREL), which unrolls the iterative process and learns key parameters from training data. The resulting approach offers enhanced performance and generalization while maintaining a transparent connection to its optimization-theoretic foundation.

1.5.1 Problem Formulation

Consider a colocated MIMO radar system comprising N_t transmit antennas and N_r receive antennas, where both arrays follow a ULA with half-wavelength spacing between adjacent elements. Each transmit antenna emits a distinct waveform. Denote the waveform emitted from the n_tth transmit antenna as

$$\boldsymbol{s}_{n_t} = [s_{n_t}(1), s_{n_t}(2), \dots, s_{n_t}(N)]^{\mathrm{T}} \in \mathbb{C}^{N\times 1}, \tag{1.53}$$

where N is the number of discrete time samples per pulse. The complete transmit signal can be represented as a stacked vector $\boldsymbol{s} = [\boldsymbol{s}_1^{\mathrm{T}}, \boldsymbol{s}_2^{\mathrm{T}}, \dots, \boldsymbol{s}_{N_t}^{\mathrm{T}}]^{\mathrm{T}} \in \mathbb{C}^{N_t N\times 1}$.

Assume there are M targets located at azimuth angles θ_m and range bins l_m for $m = 1, \dots, M$, ordered such that $l_1 \le \cdots \le l_M$. Define $r_m := l_m - l_1$ to represent the relative range offset of the mth target. Additionally, K signal-dependent interference sources are positioned at azimuths θ'_k and range bins $l'_k = l_1 + r'_k$ for $k = 1, \dots, K$.

In practice, precise knowledge of the angles θ_m may be unavailable due to steering mismatches and environmental uncertainties [57]. Therefore, the angle of each target is modeled as a uniformly distributed random variable [48, 53]: $\theta_m \sim U(\bar{\theta}_m - \Delta\bar{\theta}_m, \bar{\theta}_m + \Delta\bar{\theta}_m)$, where $\bar{\theta}_m$ and $\Delta\bar{\theta}_m$ denote the nominal angle and its uncertainty bound, respectively.

The SINR for the mth target can be formulated in two equivalent forms, both expressed as quadratic fractional functions of the receive filter $\boldsymbol{w}_m$ and the transmit waveform $\boldsymbol{s}$ [48, 53, 58][4]:

$$\mathrm{SINR}_m(\boldsymbol{s}, \boldsymbol{w}_m) = \frac{\boldsymbol{w}_m^{\mathrm{H}} \boldsymbol{\Gamma}_m(\boldsymbol{s}) \boldsymbol{w}_m}{\boldsymbol{w}_m^{\mathrm{H}} \boldsymbol{\Psi}_m(\boldsymbol{s}) \boldsymbol{w}_m}, \tag{1.54a}$$

$$= \frac{\boldsymbol{s}^{\mathrm{H}} \boldsymbol{H}_m(\boldsymbol{w}_m) \boldsymbol{s}}{\boldsymbol{s}^{\mathrm{H}} \boldsymbol{G}_m(\boldsymbol{w}_m) \boldsymbol{s}}. \tag{1.54b}$$

4 This is similar to the SINR derived in Section 1.3.2, which shows the final form in (1.27).

The design objective is to maximize the minimum SINR across all M targets, i.e., improving the worst-case detection performance. $\boldsymbol{s}$ is constrained by the CMC, requiring each entry to have CM: $|s(i)| = \frac{1}{\sqrt{N_t N}}$ for all i. This leads to the following optimization problem:

$$\begin{aligned} \max_{\boldsymbol{s},\{\boldsymbol{w}_m\}_{m=1}^{M}} \quad & \min_{m} \ \mathrm{SINR}_m(\boldsymbol{s}, \boldsymbol{w}_m) \\ \text{s.t.} \quad & |s(i)| = \frac{1}{\sqrt{N_t N}}, \ \forall i, \\ & \|\boldsymbol{w}_m\|^2 = 1, \ \forall m. \end{aligned} \tag{1.55}$$

The norm constraint on each $\boldsymbol{w}_m$ eliminates any ambiguity related to scaling, as the SINR expression remains invariant to the norm of $\boldsymbol{w}_m$.

As shown in prior works [30, 49, 58, 59], for a fixed transmit waveform $\boldsymbol{s}$, the optimal receive filter $\boldsymbol{w}_{m,\mathrm{opt}}$ corresponds to the dominant eigenvector of the matrix $\boldsymbol{\Psi}_m^{-1}(\boldsymbol{s})\boldsymbol{\Gamma}_m(\boldsymbol{s})$:

$$\boldsymbol{\Psi}_m^{-1}(\boldsymbol{s})\boldsymbol{\Gamma}_m(\boldsymbol{s})\boldsymbol{w}_{m,\mathrm{opt}} = \lambda_{\max}\boldsymbol{w}_{m,\mathrm{opt}}, \tag{1.56}$$

where $\lambda_{\max}$ is the largest eigenvalue of the matrix product. Substituting this optimal receive filter back into the original formulation reduces the joint optimization to a transmit-only problem:

$$\max_{\boldsymbol{s}\in\mathcal{S}^{N_t N}} \ \min_{m} \ \mathrm{SINR}_m(\boldsymbol{s}, \boldsymbol{w}_{m,\mathrm{opt}}), \tag{1.57}$$

where $\mathcal{S}^{N_t N}$ denotes the feasible set of constant-modulus vectors in $\mathbb{C}^{N_t N\times 1}$.

1.5.2 Sum-of-reciprocal Exact Descent (SRED)

The optimization problem in (1.57) is a max–min fractional quadratic program, which is nonconvex [48]. The problem presents a dual challenge. First, the min–max structure is nonconvex. Second, the objective function itself is fractional and nonconvex. To address this, several studies have proposed reformulations or approximations. For instance, [48, 49, 52] simplify the max-min objective by maximizing a tight lower bound on the worst-case SINR. Another approach in [50] replaces the discrete minimum with a continuous formulation over a simplex, resulting in a nested optimization problem. The work in [58] employs Dinkelbach's method to eliminate the fractional structure by converting the objective into a difference of polynomials [60].

Despite these efforts, existing methods often encounter a trade-off between computational tractability and solution fidelity. To address these limitations, a recent approach reformulates the problem in (1.57) by directly enforcing the CMC through parameterization of the waveform in terms of its phase vector [61].

Let $\boldsymbol{\phi} = [\phi_1, \dots, \phi_{N_tN}]^{\mathrm{T}}$ denote the phase variables, and define the CM waveform as $s_i(\phi_i) = \frac{1}{\sqrt{N_tN}} e^{j\phi_i}$ for all $i = 1, \dots, N_tN$. Any update in $\boldsymbol{\phi}$ automatically preserves the CMC.

With this parameterization, the problem in (1.57) is equivalently rewritten as:

$$\min_{\boldsymbol{s}\in\mathcal{S}^{N_tN}} \max_{m} \frac{1}{\mathrm{SINR}_m(\boldsymbol{s}, \boldsymbol{w}_{m,\mathrm{opt}})}. \tag{1.58}$$

To obtain a smooth and differentiable surrogate for this minimax objective, the cost is reformulated as the sum of reciprocal SINRs, leading to the following unconstrained minimization:

$$\min_{\boldsymbol{\phi}} f(\boldsymbol{\phi}) = \sum_{m=1}^{M} f_m(\boldsymbol{\phi}) = \sum_{m=1}^{M} \frac{\boldsymbol{s}^{\mathrm{H}}(\boldsymbol{\phi})\boldsymbol{G}_m\boldsymbol{s}(\boldsymbol{\phi})}{\boldsymbol{s}^{\mathrm{H}}(\boldsymbol{\phi})\boldsymbol{H}_m\boldsymbol{s}(\boldsymbol{\phi})}, \tag{1.59}$$

where $f_m(\boldsymbol{\phi})$ represents the reciprocal of $\mathrm{SINR}_m\big(\boldsymbol{s}(\boldsymbol{\phi}), \boldsymbol{w}_{m,\mathrm{opt}}\big)$. Letting $\boldsymbol{s} = \boldsymbol{s}(\boldsymbol{\phi})$ for notational simplicity, the gradient of $f(\boldsymbol{\phi})$ can be written as:

$$\begin{aligned} -\tilde{\eta}(\boldsymbol{\phi}) &= -\nabla_{\boldsymbol{\phi}} f(\boldsymbol{\phi}) = -\sum_{m=1}^{M} \nabla_{\boldsymbol{\phi}} f_m(\boldsymbol{\phi}) \\ &= \sum_{m=1}^{M} \frac{1}{\{\mathrm{SINR}_m(\boldsymbol{s}, \boldsymbol{w}_m)\}^2} \nabla_{\boldsymbol{\phi}} \mathrm{SINR}_m(\boldsymbol{s}, \boldsymbol{w}_m). \end{aligned} \tag{1.60}$$

This expression reveals a weighted sum of individual gradients, where smaller SINR values contribute larger weights. As a result, the net descent direction is dominated by the component corresponding to the worst-case SINR, which aligns with the original design objective.

The gradient $\tilde{\eta}_m(\boldsymbol{\phi}) = \nabla_{\boldsymbol{\phi}} f_m(\boldsymbol{\phi})$ can be explicitly derived using the chain rule:

$$\tilde{\eta}_m(\cdot) = \frac{\boldsymbol{s}^{\mathrm{H}}\boldsymbol{H}_m\boldsymbol{s} \cdot \nabla_{\boldsymbol{\phi}}\boldsymbol{s}^{\mathrm{H}}\boldsymbol{G}_m\boldsymbol{s} - \boldsymbol{s}^{\mathrm{H}}\boldsymbol{G}_m\boldsymbol{s} \cdot \nabla_{\boldsymbol{\phi}}\boldsymbol{s}^{\mathrm{H}}\boldsymbol{H}_m\boldsymbol{s}}{(\boldsymbol{s}^{\mathrm{H}}\boldsymbol{H}_m\boldsymbol{s})^2}, \tag{1.61}$$

which leads to a closed-form expression for the exact gradient:

$$\tilde{\eta}_m(\boldsymbol{\phi}) = \nabla_{\boldsymbol{\phi}} f_m(\boldsymbol{\phi}) = \sigma_m \cdot \Im\big[\{(\boldsymbol{s}^{\mathrm{H}}\boldsymbol{H}_m\boldsymbol{s})\boldsymbol{G}_m\boldsymbol{s} - (\boldsymbol{s}^{\mathrm{H}}\boldsymbol{G}_m\boldsymbol{s})\boldsymbol{H}_m\boldsymbol{s}\} \odot \boldsymbol{s}^*\big], \tag{1.62}$$

where $\sigma_m = \frac{2}{(\boldsymbol{s}^{\mathrm{H}}\boldsymbol{H}_m\boldsymbol{s})^2} \in \mathbb{R}$. Using this exact descent direction $-\nabla_{\boldsymbol{\phi}} f(\boldsymbol{\phi})$, an iterative optimization scheme of SRED is defined as:

$$\boldsymbol{\phi}^{t+1} = \boldsymbol{\phi}^{t} - \rho^t \tilde{\eta}(\boldsymbol{\phi}^t) = \boldsymbol{\phi}^t - \rho^t \nabla_{\boldsymbol{\phi}} f(\boldsymbol{\phi}^t), \tag{1.63}$$

where ρ^t is the step size at the tth iteration. With appropriate step size selection, this method is guaranteed to converge to a stationary point that satisfies at least local optimality conditions [62].

1.5.3 Unrolling SRED into Sum-of-reciprocal Exact Learning

Most conventional gradient-based algorithms employ fixed or predefined choices for these parameters, which can hinder convergence speed or lead to suboptimal solutions. Further, identifying the most effective update direction remains an open research problem.

To resolve the issue that iterative SRED has, SREL was introduced. While SRED updates the waveform iteratively via (1.63), SREL unrolls this iterative process into a deep network whose parameters—including descent vectors and step sizes—are *learned* from data. In addition to learning these components, SREL incorporates an SC that encourages the designed waveform to resemble a known reference waveform with desirable characteristics such as favorable pulse compression and ambiguity function properties [46, 47].

1.5.4 Similarity Constraint

The SC ensures that the optimized waveform $\boldsymbol{s}$ maintains key features of a reference waveform $\boldsymbol{s}_r$. Mathematically, it is formulated as:

$$||\boldsymbol{s} - \boldsymbol{s}_r||_\infty \leq \zeta, \tag{1.64}$$

where $|| \cdot ||_\infty$ denotes the infinity norm, defined as the maximum absolute value of the vector elements. ζ controls the allowable deviation from the reference signal. Incorporating the CMC, this constraint can be re-expressed in terms of phase:

$$|| \exp(j\boldsymbol{\phi}) - \exp(j\boldsymbol{\phi}_r)||_\infty \leq \zeta_0, \tag{1.65}$$

with $\boldsymbol{\phi}_r = \angle \boldsymbol{s}_r$ and $\zeta_0 = \sqrt{N_t N}\zeta$. This inequality implies an elementwise phase constraint of the form:

$$-\varphi + \phi_{r,i} \leq \phi_i \leq \varphi + \phi_{r,i}, \qquad \text{for } i = 1, \dots, N_t N, \tag{1.66}$$

where $\varphi = \arccos\left(\frac{2 - \zeta_0^2}{2}\right)$ and ϕ_i, $\phi_{r,i}$ are the ith elements of $\boldsymbol{\phi}$ and $\boldsymbol{\phi}_r$, respectively. To enforce this constraint during waveform updates, a nonlinear projection function $\boldsymbol{p}(\cdot) : \mathbb{R}^{N_t N \times 1} \to \mathbb{R}^{N_t N \times 1}$ is defined. Its ith component is implemented using a sigmoid-based mapping:

$$p_i(\phi) = 2\varphi\left(\frac{1}{1 + e^{c_b - c_w(\phi - \phi_{r,i})}} - \frac{1}{2}\right) + \phi_{r,i}, \tag{1.67}$$

which ensures that $p_i(\phi)$ lies within the allowable range. When no SC is active (i.e., $\zeta_0 \geq 2$), $\boldsymbol{p}(\cdot)$ reduces to the identity function.

1.5.4.1 SREL Design and Forward Propagation

By incorporating learnable descent vectors and step sizes along with the SC-function, the update rule for SREL is expressed as:

$$\boldsymbol{\phi}^{t+1} = \boldsymbol{p}\left(\boldsymbol{\phi}^{t} - \boldsymbol{\phi}_{r} - \sum_{m=1}^{M} \mu_{m}^{t}\gamma_{m}^{t}\boldsymbol{\eta}_{m}^{t}\right) + \boldsymbol{\phi}_{r}, \tag{1.68}$$

where $\boldsymbol{\eta}_m^t \in \mathbb{C}^{N_t N}$ and γ_m^t are learnable descent vectors and step sizes, respectively. The weights μ_m^t are introduced to balance the contributions from different targets, offering finer control over the aggregated update direction. All three components—$\boldsymbol{\eta}_m^t$, γ_m^t, and μ_m^t—are learned from training data. Figure 1.14 illustrates the forward propagation of SREL, where the update in (1.68) is applied iteratively for κ steps. At each step, distinct NN modules generate the learned parameters based on shared and time-varying inputs.

To encode mutual-target interactions, a target-mapping matrix $\boldsymbol{Y}_m \in \mathbb{R}^{D_l \times D_\theta}$ is introduced, where $D_l = r_m + 1$ and $D_\theta = \theta_{\max} - \theta_{\min} + 1$. Each $\boldsymbol{Y}_m$ is initialized to zero and populated based on the relative strengths and locations of the mth and other targets:

$$\begin{aligned} Y_{r_m+1,i} &\leftarrow Y_{r_m+1,i} + \alpha_m, \\ Y_{l_{m'}-l_1+1,i'} &\leftarrow Y_{l_{m'}-l_1+1,i'} - \alpha_{m'}, \qquad \forall m' \neq m, \end{aligned}$$

where i and i' index the azimuth bins corresponding to each target's angular uncertainty range. Each module then receives an input vector $\boldsymbol{x}_m^t \in \mathbb{R}^{D_x}$ defined as:

$$\boldsymbol{x}_m^t = [\Re\{\hat{\boldsymbol{s}}^t\}^{\mathrm{T}},\ \Im\{\hat{\boldsymbol{s}}^t\}^{\mathrm{T}},\ \Re\{\boldsymbol{w}_m\}^{\mathrm{T}},\ \Im\{\boldsymbol{w}_m\}^{\mathrm{T}},\ \boldsymbol{y}_m^{\mathrm{T}}]^{\mathrm{T}}, \tag{1.69}$$

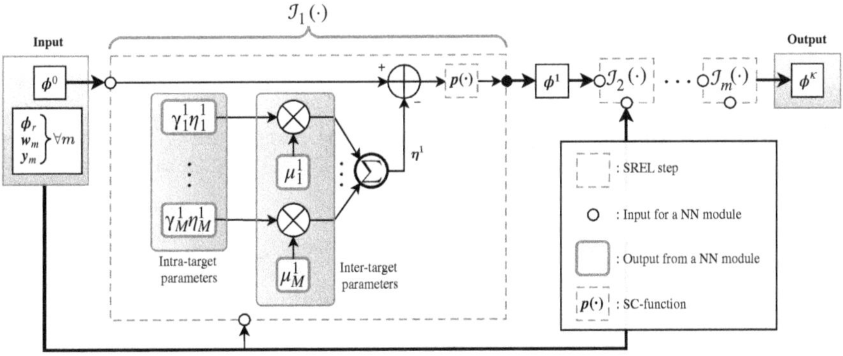

Figure 1.14 Forward propagation of SREL. During forward propagation, the SREL NNs modules receive $\boldsymbol{w}_m$ and $\boldsymbol{y}_m$ for $m = 1, \ldots, M$ as inputs, while $\boldsymbol{\phi}^t$ is updated at each step. Both intra- and inter-target features are leveraged to generate the descent direction $\boldsymbol{\eta}^t(\cdot)$.

where $\hat{\mathbf{s}}^t = \mathbf{s}^t - \mathbf{s}_r = \frac{1}{\sqrt{N_t N}} e^{j\phi^t} - \mathbf{s}_r$. Each NN module maps $\mathbf{x}_m^t$ to its corresponding outputs γ_m^t, $\boldsymbol{\eta}_m^t$, and μ_m^t.

1.5.4.2 Architectural Design of SREL

As shown in Figure 1.14, SREL is composed of three distinct types of NN modules, each taking the input $\mathbf{x}_m^t$ defined in (1.69). These include: (1) a module that generates the descent direction $\boldsymbol{\eta}_m^t$ corresponding to a single-target SINR reciprocal, (2) a module for producing the associated step size γ_m^t, and (3) a module that determines the inter-target balancing weight μ_m^t. The descent and step-size modules maintain separate trainable weights across the κ unrolled steps, while the balancing weight module uses a shared parameterization across steps.

Each of these NN modules is constructed using Q FCBs, each consisting of a FCL followed by a ReLU activation (Figure 1.15). A final layer, denoted $\mathcal{F}(\cdot)$, is then applied depending on the module type, as detailed in Table 1.4. The $\gamma^t(\cdot)$ and $\mu(\cdot)$ modules use the softplus function, $\log(1 + \exp(\cdot))$, to ensure nonnegative outputs [63].

Although increasing the number of unrolled steps κ can improve performance, it also increases memory and computational costs due to more intermediate variables and gradients during training.

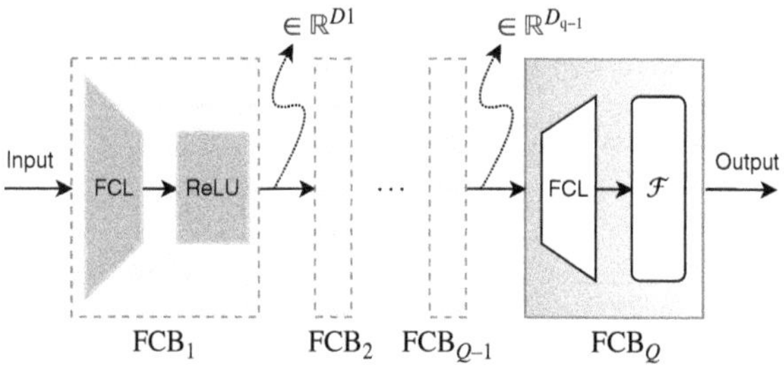

Figure 1.15 NN architecture for the SREL modules. The modules use Q FCBs, where each block contains an FCL and an activation function. D_q indicates the output dimension of the qth block for $q = 1, \ldots, Q$.

Table 1.4 Architectural specifications of the SREL network.

Type	Input	$\mathcal{F}$	Output
$\boldsymbol{\eta}^t(\cdot)$	$\mathbf{x}_m^t$	None	$\left[\Re\{\boldsymbol{\eta}_m^t\}^{\mathrm{T}}, \Im\{\boldsymbol{\eta}_m^t\}^{\mathrm{T}}\right]^{\mathrm{T}}$
$\gamma^t(\cdot)$		$\log(1 + \exp(\cdot))$	γ_m^t
$\mu(\cdot)$		$\log(1 + \exp(\cdot))$	μ_m^t

1.5.4.3 Training Strategy for SREL

To facilitate learning, a custom training strategy is introduced that decouples intra-target and inter-target parameter learning. Specifically, the modules for η_m^t and γ_m^t are trained first (intra-target phase), followed by learning the balancing weight μ_m^t (inter-target phase).

Intra-target Parameter Learning: The goal in this phase is to train the modules responsible for minimizing the reciprocal SINR cost for individual targets. The loss function used for this stage is:

$$\mathcal{L}_{\text{intra},m} = f_m^{\kappa} + \frac{\lambda_{f,\text{intra}}}{\kappa} \sum_{t=1}^{\kappa} e^{\lambda_{\text{MN}}(f_m^t - f_m^{t-1})} + \frac{\lambda_\eta}{\kappa} \sum_{t=1}^{\kappa} \mathcal{L}_{\eta,m}^t, \tag{1.70}$$

where f_m^t is the value of the reciprocal SINR cost at the tth step. The first term encourages convergence, the second enforces monotonic decrease, and the third aligns the learned descent direction with the exact gradient derived in (1.62), via:

$$\mathcal{L}_{\eta,m}^t = \|\eta_m^t(\boldsymbol{\phi}^t, \cdot) - \tilde{\eta}_m(\boldsymbol{\phi}^t, \cdot)\|_2. \tag{1.71}$$

Hyperparameters $\lambda_{f,\text{intra}}$, λ_{MN}, and λ_η control the relative weights of these terms.

Inter-target Parameter Learning: After intra-target modules are trained and frozen, the inter-target balancing weights μ_m^t are optimized. These weights govern how multiple descent directions $\gamma_m^t \eta_m^t$ are combined into a unified descent vector for optimizing the global cost. The corresponding loss function is:

$$\mathcal{L}_{\text{inter}} = f^{\kappa} + \frac{\lambda_{f,\text{inter}}}{\kappa} \sum_{t=1}^{\kappa} e^{\lambda_{\text{MN}}(f^t - f^{t-1})}, \tag{1.72}$$

where f^t denotes the global sum-of-reciprocals objective at iteration t, and $\lambda_{f,\text{inter}}$ governs the emphasis on monotonic cost reduction.

Self-supervised Learning in SREL: All NN modules in SREL—including those for $\gamma^t(\cdot)$, $\eta^t(\cdot)$, and $\mu(\cdot)$—are trained using backpropagation to minimize the loss functions in (1.70) and (1.72). This is achieved in a self-supervised manner, where no ground-truth waveform is required. Instead, training relies on minimizing cost functions derived directly from the problem formulation.

1.5.4.4 Step-by-step Algorithm

Algorithm 1.1 summarizes the complete procedure for jointly optimizing the transmit waveform and receive filter using either SRED or SREL. The process begins with a random initialization of the phase vector $\boldsymbol{\phi}$ from a uniform distribution over $[0, 2\pi]$. The algorithm then alternates between optimizing the waveform (inner loop, lines 8–11) and updating the receive filter based on the latest waveform (outer loop, lines 5–16). Convergence is declared once no further improvement in the worst-case SINR is observed. The final waveform and iteration count are denoted by $\boldsymbol{s}_c$ and i_c, respectively.

Algorithm 1.1: The worst-case SINR maximization using sum-of-reciprocal.

1: **Inputs**: ϵ, $\{l_m, \bar{\theta}_m, \Delta\bar{\theta}_m\}\ \forall m$, $\{l'_k, \bar{\theta}'_k, \Delta\bar{\theta}'_k\}\ \forall k$
2: **Outputs**: $\boldsymbol{s}_c$ and $\boldsymbol{w}_{m,\mathrm{opt}}$, $\forall m$
3: Initialize $\boldsymbol{s}^0 = \frac{1}{\sqrt{N_t N}} e^{j\boldsymbol{\phi}}$ with $\boldsymbol{\phi} \sim [0, 2\pi]$.
4: $i = 1$
5: Calculate $\boldsymbol{\Gamma}_m(\boldsymbol{s})$ and $\boldsymbol{\Psi}_m(\boldsymbol{s})\ \forall m$.
6: Derive receiver filter $\boldsymbol{w}^i_{m,\mathrm{opt}} \forall m$ as in (1.56).
7: Compute $\boldsymbol{G}_m(\boldsymbol{w}_{m,\mathrm{opt}})$, $\boldsymbol{H}_m(\boldsymbol{w}_{m,\mathrm{opt}})$.
8: **switch** *method* to find $\boldsymbol{s}^i$
9: **case** SRED: use (1.63) to find $\boldsymbol{s}^i$
10: **case** SREL: use (1.68) to find $\boldsymbol{s}^i$
11: **end switch**
12: **if** $|\min_m \mathrm{SINR}^i_m - \min_m \mathrm{SINR}^{i-1}_m| < \epsilon$
13: $\boldsymbol{s}_c = \boldsymbol{s}^i$, $i_c = i$, and STOP.
14: **else**
15: $i = i + 1$ and GO to step 5.
16: **end if**

1.5.5 Numerical Results

A colocated MIMO radar system is considered, comprising $N_t = 4$ transmit antennas and $N_r = 8$ receive antennas. Each antenna array is arranged as a ULA with inter-element spacing set to half the signal wavelength, i.e., $d_t = d_r = \frac{\lambda}{2}$. The system uses $N = 32$ time-domain samples per waveform.

Three targets are positioned at range bins $l_1 = 100$, $l_2 = 106$, and $l_3 = 109$, with corresponding nominal azimuth angles $\bar{\theta}_1 = 0°$, $\bar{\theta}_2 = -20°$, and $\bar{\theta}_3 = 30°$. Each target has an azimuth uncertainty of $\Delta\theta_m = 3°$ and a power level of $\delta^2_m = 20$ dB.

Signal-dependent interference is modeled using $K = 30$ sources grouped into four distinct categories as summarized in Table 1.5, with each interference source assumed to have an azimuth uncertainty of $\Delta\theta'_k = 3°$. Signal-independent noise is also included in the scenario, with a noise power of $\sigma^2_v = 0$ dB. Additionally, two spectral interference bands are introduced ($Q = 2$), each characterized by an interference power $\tilde{\delta}_q = 10$ dB. The occupied frequency ranges are $(f_{1,1}, f_{1,2}) = (0.2, 0.3)$ and $(f_{2,1}, f_{2,2}) = (0.7, 0.75)$, with respective incidence angles $\psi_1 = -15°$ and $\psi_2 = 20°$.

For both SRED and SREL, the optimization step size is set as $\rho^t = 1$ for all iterations, and the convergence tolerance for the cost function is set to $\epsilon_f = 10^{-7}$. The phase update tolerance ϵ is chosen via cross-validation [64], with $\epsilon = 0.03$ used in

Table 1.5 Parameters of signal-dependent interference sources.

Clutter #	Range bin (l'_k)	Azimuth (deg) ($\bar{\theta}'$)	Power (dB)
5	110	[20, 25, 30, 35, 40]	30
5	111	[20, 25, 30, 35, 40]	30
11	[95, 96, ⋯, 105]	−10	30
9	[100, 101 ⋯, 108]	−30	30

scenarios with the CMC constraint alone (Section 1.5.5.2), and $\epsilon = 0.1$ when both the CMC and the SC are applied (Section 1.5.5.3).

Under this shared scenario of worst-case SINR maximization, the performance of SRED is compared against several SOTA approaches applicable to this problem setting: dual ascent method (DAM) [53], BCD [58], MM [50], sequential quasi-convex-based algorithm (SQCA) [49], and the convex relaxation-based methods NICE and NMICE [48].

1.5.5.1 SREL's Setup and Training

To evaluate generalization capability, SREL is trained on datasets with varied target positions and radar cross-section (RCS) values, while all testing is conducted on previously unseen configurations. A total of 20,000 distinct training instances are generated through the following procedure: (1) specify physical configurations such as target and interference locations; (2) construct the target mapping vectors $\boldsymbol{y}_m \; \forall m$; (3) randomly sample initial phase codes $\boldsymbol{\phi}$ within the feasible domain; (4) compute $\boldsymbol{\Gamma}_m(\cdot)$ and $\boldsymbol{\Psi}_m(\cdot)$; (5) obtain the closed-form solution of the optimal receive filter $\boldsymbol{w}_{m,\,\mathrm{opt}}$; and (6) generate $\boldsymbol{G}_m(\cdot)$ and $\boldsymbol{H}_m(\cdot)$. This results in diverse sets of input vectors $\boldsymbol{x}^t_m$ as defined in (1.69) for training the neural modules of SREL, while $\boldsymbol{G}_m$ and $\boldsymbol{H}_m$ are used to compute loss functions in (1.70) and (1.72). The dataset is randomly partitioned into training and validation sets using an 80/20 split, following standard practice in machine learning [28].

1.5.5.1.1 Parameter Learning for SREL

Training of the SREL neural modules is carried out using the Adam optimizer [65]. During training, weights are updated to minimize the loss of functions defined in (1.70) for intra-target learning and (1.72) for inter-target learning.

1.5.5.1.2 Hyperparameter Tuning

Hyperparameters are selected through cross-validation [28]. The number of outer-loop iterations for waveform and filter updates is fixed to $i_c = 5$ (see Algorithm 1.1). The number of unrolling steps in SREL is set to $\kappa = 10$, which strikes a balance between model complexity and performance. Fewer unrolled steps result in degraded performance, while larger values ($\varkappa > 10$) increase the number of distinct modules ($\gamma^t(\cdot)$ and $\eta^t(\cdot)$) without significant gain in performance.

In the DL setup, the learning rate is set to 10^{-3} and the batch size to 5. The architectures of the NN modules used to estimate $\gamma^t(\cdot)$, $\eta^t(\cdot)$, and $\mu(\cdot)$ consist of 4, 5, and 4 FCL with ReLU, respectively. To regulate the influence of the regularization terms in the loss functions, the following hyperparameters are used: $\lambda_{f,\,\text{intra}} = 10^{-1}$, $\lambda_{\text{MN}} = 10^{-2}$, and $\lambda_\eta = 10^{-7}$. The effect of stronger regularization (larger λ_η) is explored in the next subsection, where it is shown to constrain the learned descent directions too closely to the analytical ones, which may impair performance.

1.5.5.2 Waveform Design Under the CMC

1.5.5.2.1 Comparison of SREL and SRED Performance

Figure 1.16 illustrates the evolution of the SINR values for each target as the phase vector $\boldsymbol{\phi}$ is iteratively updated using SRED, with fixed receive filters $\{\boldsymbol{w}_m\}$. In this example, the worst-case SINR corresponds to target 2, and its improvement over the iterations is more pronounced compared to the other targets. At early iterations, when SINR_1 is considerably lower than SINR_3, SINR_1 increases more rapidly. This behavior aligns with the design of SRED, which emphasizes updates in directions that most improve the lowest SINR values, as discussed in Section 1.5.2.

To further evaluate optimization efficiency, Figure 1.17 compares the performance of SRED and SREL in improving the worst-case SINR and reducing the cost function $f(\cdot)$ defined in (1.59). The comparison is based on 100 random

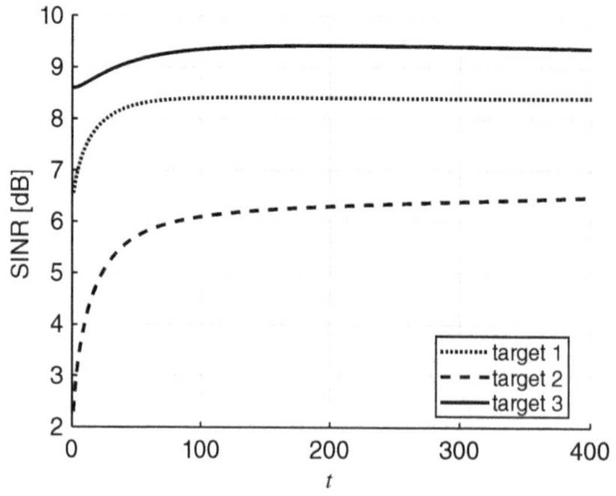

Figure 1.16 Evolution of SINR values for each target across iterations in SRED, showing the inner loop updates of $\boldsymbol{\phi}$ in Algorithm 1.1.

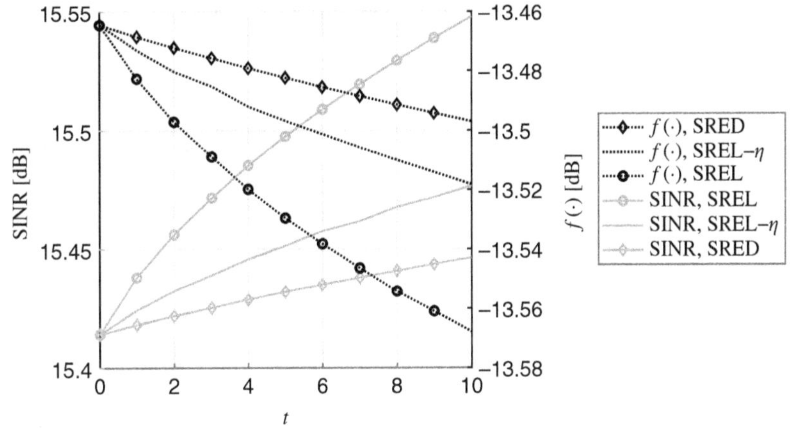

Figure 1.17 Comparison between SRED vs SREL-η vs SREL: worst-case SINR and $f(\cdot)$ values with respect to t. Solid lines are SINR, and dotted lines are $f(\cdot)$. The lines with diamonds are SRED, the lines with circles are SREL, and the others are SRED-η.

initializations of $\boldsymbol{\phi}$ drawn uniformly from $[0,\ 2\pi]$. Both methods use the same initialization set to ensure a fair comparison. Since SREL employs $\kappa = 10$ unrolled update steps, SRED is also evaluated after 10 steps for consistency.

In addition to the standard SREL model, a variant referred to as SREL-η is considered, where the regularization weight λ_η is increased to 2×10^{-6} (larger than the baseline setting). As shown in the figure, both SREL and SREL-η outperform SRED in terms of achieving higher worst-case SINR and lower values of the cost function $f(\cdot)$ within the same number of steps.

However, a comparison between SREL and SREL-η highlights the impact of over-regularization. When λ_η is too large, the learned descent direction $\eta^t(\cdot)$ becomes overly constrained to follow the exact gradient $\tilde{\eta}$ derived in (1.60), thereby limiting the flexibility of the network and slightly degrading performance. In contrast, the baseline SREL model selects λ_η via cross-validation, achieving a more favorable trade-off between gradient alignment and learning flexibility.

1.5.5.2.2 Comparison with State-of-the-art Methods

Table 1.6 summarizes the performance of SRED, SREL, and several SOTA methods in terms of the worst-case SINR, the number of outer-loop iterations (i_c), and total computation time. The best-performing values in each category are highlighted in bold.

The results demonstrate that both SRED and SREL achieve superior worst-case SINR compared to existing approaches, indicating their effectiveness in identifying favorable optima. Moreover, these methods exhibit faster convergence: SRED

Table 1.6 Worst-case SINR values (in decibel), iteration numbers, and global computational times [sec].

Algorithm	SINR [dB]	i_c	Time (sec])
DAM [53]	15.72	161	936.8
BCD [58]	14.10	295	641.1
MM [50]	15.33	100	154.6
SQCA [49]	15.55	56	6531.2
NICE [48]	15.69	210	674.8
NMICE [48]	15.71	240	871.5
SRED	**16.07**	**22**	**102.05**
SREL	**16.66**	**5**	**62.12**

Note: Bold-faced values indicate the top-two best-performing results for each metric (e.g., maximum SINR, minimum iterations, and shortest computation time).

converges with fewer iterations, and SREL further reduces computational cost by leveraging learned update steps through DL. The efficiency gain from unrolling is particularly evident in SREL, which achieves comparable or better performance at a fraction of the computational expense.

1.5.5.3 Waveform Design Under the CMC and SC

To promote desirable properties such as favorable pulse compression and ambiguity function behavior, the reference signal $\boldsymbol{s}_r$ used in the SC (1.64) is defined as an orthogonal LFM waveform, given by $\boldsymbol{s}_r = \text{vec}(\boldsymbol{S}_r)$ [30, 48]. The matrix $\boldsymbol{S}_r \in \mathbb{C}^{N_t \times N}$ is constructed as:

$$\boldsymbol{S}_r(n_t, n) = \frac{\exp\{j2\pi n_t(n-1)/N\}\exp\left\{j2\pi(n-1)^2/N\right\}}{\sqrt{N_t N}}.$$

By adopting the phase projection technique described in Sections 1.5.3 and 1.5.4, SREL ensures that the generated waveform lies within the feasible set defined by both the CMC and the SC. Table 1.7 presents a comparison between SREL and other SOTA methods under two different similarity thresholds: $\zeta_0 = 0.5$ and $\zeta_0 = 1.5$.

When the SC is tight ($\zeta_0 = 0.5$), SREL attains the highest worst-case SINR among all methods. In the more relaxed case ($\zeta_0 = 1.5$), SREL remains competitive, with performance closely matching that of NICE. Notably, even in this less restrictive setting, SREL demonstrates clear advantages in computational efficiency, highlighting the benefits of its unrolled, data-driven architecture.

Table 1.7 Worst-case SINR values (in decibel), iteration numbers, and global computational times (sec) under the CMC and the SC.

	$\zeta_0 = 0.5$			$\zeta_0 = 1.5$		
Algorithm	**SINR (dB)**	i_c	**Time (sec)**	**SINR (dB)**	i_c	**Time (sec)**
DAM [53]	12.2	142	8941.0	14.55	224	17677.9
BCD [58]	12.73	66	142.5	13.47	164	355.3
MM [50]	13.47	51	66.9	15.12	79	112
SQCA [49]	11.86	9	935.0	13.36	65	7858.1
NICE [48]	13.74	164	466.1	**15.51**	262	821.4
NMICE [48]	13.64	98	346.3	15.49	195	747.5
SREL	**14.05**	**5**	**61.76**	15.31	**6**	**78.62**

Note: Bold-faced values indicate the best-performing results for each metric (e.g., maximum SINR, minimum iterations, and shortest computation time).

1.5.6 Key Insights

The SREL framework offers several distinct contributions to the broader landscape of algorithm unrolling in radar waveform design. Unlike earlier unrolling methods, SREL is specifically tailored for multitarget scenarios, where the optimization must jointly balance the SINRs across multiple targets. This is achieved through the introduction of learnable inter-target balancing weights. These weights modulate the contributions of descent directions associated with each target. By incorporating these parameters, SREL enables the network to emphasize weaker SINR targets.

SREL unrolls the exact descent steps of a baseline iterative method (SRED), but with the flexibility of learning both descent vectors and step sizes. This learnable structure allows SREL to outperform its iterative counterpart in fewer steps, even when initialized identically.

Another unique feature of SREL is its treatment of the waveform SC. By introducing a differentiable phase-projection mechanism, SREL ensures that the designed waveform remains close to a desired reference signal. This capability preserves favorable properties like low sidelobes and desirable ambiguity characteristics—attributes often required in practical radar systems but rarely handled explicitly in learning-based waveform design.

Finally, the training strategy in SREL is designed with careful modularity. Intra-target parameters are trained separately from inter-target balancing weights, enabling stabilized learning and improved generalization to unseen scenarios. Together, these features make SREL a compelling example of how deep algorithm unrolling can be adapted for the unique challenges of multitarget radar waveform optimization, extending the frontiers of model-aware learning in radar signal processing.

1.6 Concluding Remarks and Future Research Directions

The referred studies demonstrate that an unrolled method offers a favorable balance between variance and bias in the achievable solution space. Figure 1.18 provides an intuitive illustration of the solution characteristics associated with an iterative algorithm, its deep unrolled network, and a conventional (black box) DL model. Iterative algorithms generally exhibit low variance but high bias due to their limited representational capacity, stemming from fixed update rules. In contrast, a generic NN achieves low bias, potentially approximating the target function accurately thanks to its learning capacity, but it often suffers from high variance [11], i.e., its performance is highly sensitive to the training data. An unrolled network, however, occupies an intermediate space. While preserving the algorithmic structure of its iterative counterpart, it extends representational power through learning, resulting in a more favorable bias–variance trade-off (low bias, low variance). These studies demonstrate that learning key algorithmic parameters can lead to superior performance compared to the base iterative algorithm.

Future work may aim to rigorously characterize the convergence behavior of deep unrolled algorithms under various practical constraints, offering deeper theoretical insights into their performance. Unlike traditional black-box DL methods, unrolled networks are built by unfolding iterative algorithms, which makes it possible to analyze their convergence properties more directly. For instance, the work by Zhao et al. [66] exemplifies this potential by providing a rigorous analytical

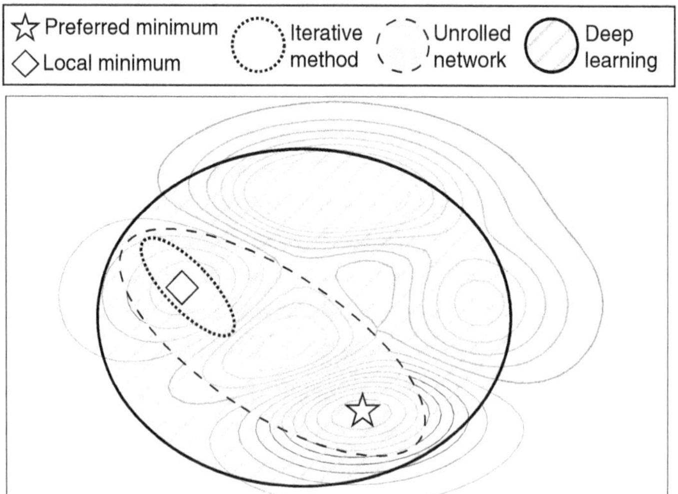

Figure 1.18 An intuitive illustration of achievable solutions for iterative method vs. unrolled network vs. black-box DL.

framework that establishes convergence guarantees and quantifies convergence rates for their proposed unrolled network.

The unrolling approach can also be extended to other SOTA iterative methods in radar waveform design and broader radar optimization problems. These extensions could incorporate practical constraints such as waveform similarity, spectral interference mitigation, or low peak-to-average power ratio, further enhancing the applicability of unrolled models in real-world systems.

Another important avenue for future research involves developing real-time implementations of these methods. Tackling computational bottlenecks, optimizing hardware resource utilization, and exploring parallelization strategies are key steps toward enabling faster and more efficient deployment. Moreover, in dynamic environments—such as scenarios involving moving targets—investigating strategies for when and how to retrain unrolled networks will be crucial for maintaining performance under changing physical conditions.

References

1 G. Cui, A. De Maio, A. Farina, and J. Li. *Radar Waveform Design Based on Optimization Theory*. SciTech Publishing, 2020.

2 L. K. Patton and B. D. Rigling. Modulus constraints in adaptive radar waveform design. In *2008 IEEE Radar Conference*, pages 1–6, 2008.

3 J. Li and P. Stoica. MIMO radar with colocated antennas. *IEEE Signal Processing Magazine*, 24(5):106–114, 2007.

4 P. Stoica, J. Li, and Y. Xie. On probing signal design for MIMO radar. *IEEE Transactions on Signal Processing*, 55(8):4151–4161, 2007.

5 J. Li and P. Stoica. *MIMO Radar Signal Processing*. John Wiley & Sons, 2008.

6 B. Kang, J. Kweon, M. Rangaswamy, and V. Monga. Deep learning for radar waveform design: Retrospectives and the road ahead. In *IEEE International Radar Conference*, pages 1–6. IEEE, 2023.

7 W. Zhang, J. Hu, Z. Wei, H. Ma, X. Yu, and H. Li. Constant modulus waveform design for MIMO radar transmit beampattern with residual network. *Signal Processing*, 177:107735, 2020.

8 J. Hu, Z. Wei, Y. Li, H. Li, and J. Wu. Designing unimodular waveform(s) for MIMO radar by deep learning method. *IEEE Transactions on Aerospace and Electronic Systems*, 57(2):1184–1196, 2020.

9 K. Zhong, W. Zhang, Q. Zhang, J. Hu, P. Wang, X. Yu, and Q. Zhou. MIMO radar waveform design via deep learning. In *IEEE Radar Conference*, pages 1–5. IEEE, 2021.

10 Y. Pei, J. Hu, K. Zhong, and J. Wu. MIMO radar waveform optimization by deep learning method. In *IEEE International Geoscience and Remote Sensing Symposium (IGARSS)*, pages 811–814. IEEE, 2022.

11 V. Monga, Y. Li, and Y. C. Eldar. Algorithm unrolling: Interpretable, efficient deep learning for signal and image processing. *IEEE Signal Processing Magazine*, 38(2):18–44, 2021.

12 K. Gregor and Y. LeCun. Learning fast approximations of sparse coding. In *Proceedings of the 27th International Conference on International Conference on Machine Learning (ICML)*, pages 399–406, 2010.

13 D. R. Fuhrmann and G. S. Antonio. Transmit beamforming for MIMO radar systems using signal cross-correlation. *IEEE Transactions on Aerospace and Electronic Systems*, 44(1):171–186, 2008.

14 B. Kang, O. Aldayel, V. Monga, and M. Rangaswamy. Spatio-spectral radar beampattern design for coexistence with wireless communication systems. *IEEE Transactions on Aerospace and Electronic Systems*, 55(2):644–657, 2018.

15 K. Alhujaili, X. Yu, G. Cui, and V. Monga. Spectrally compatible MIMO radar beampattern design under constant modulus constraints. *IEEE Transactions on Aerospace and Electronic Systems*, 56(6):4749–4766, 2020.

16 J. R. Guerci. Cognitive radar: A knowledge-aided fully adaptive approach. In *IEEE Radar Conference*, pages 1365–1370, 2010.

17 M. S. Greco, F. Gini, P. Stinco, and K. Bell. Cognitive radars: On the road to reality: Progress thus far and possibilities for the future. *IEEE Signal Processing Magazine*, 35(4):112–125, 2018.

18 H. He, P. Stoica, and J. Li. Wideband MIMO systems: Signal design for transmit beampattern synthesis. *IEEE Transactions on Signal Processing*, 59(2):618–628, 2010.

19 W. Fan, J. Liang, G. Yu, H. C. So, and G. Lu. MIMO radar waveform design for quasi-equiripple transmit beampattern synthesis via weighted ℓ_p-minimization. *IEEE Transactions on Signal Processing*, 67(13):3397–3411, 2019.

20 K. Alhujaili, V. Monga, and M. Rangaswamy. Transmit MIMO radar beampattern design via optimization on the complex circle manifold. *IEEE Transactions on Signal Processing*, 67(13):3561–3575, 2019.

21 K. Metwaly, J. Kweon, K. Alhujaili, F. Gini, M. S. Greco, M. Rangaswamy, and V. Monga. MIMO radar beampattern design via algorithm unrolling. *IEEE Transactions on Aerospace and Electronic Systems*, 60:9204–9220, 2024.

22 K. Metwaly, J. Kweon, K. Alhujaili, M. Greco, F. Gini, and V. Monga. Interpretable, unrolled deep radar beampattern design. In *Proceedings – ICASSP IEEE International Conference on Acoustics, Speech and Signal Processing*, pages 1–5. IEEE, 2023.

23 A. S. Bandeira, N. Boumal, and A. Singer. Tightness of the maximum likelihood semidefinite relaxation for angular synchronization. *Mathematical Programming*, 163:145–167, 2017.

24 D. H. Brandwood. A complex gradient operator and its application in adaptive array theory. In *IEE Proceedings F (Communications, Radar and Signal Processing)*, volume 130, pages 11–16. IET, 1983.

25 M. Zhang, J. Li, S. Zhu, and X. Chen. Fast and simple gradient projection algorithms for phase-only beamforming. *IEEE Transactions on Vehicular Technology*, 70(10):10620–10632, 2021.

26 S. S. Rao. *Engineering Optimization: Theory And Practice*. John Wiley & Sons, 2019.

27 J. Tranter, N. D. Sidiropoulos, X. Fu, and A. Swami. Fast unit-modulus least squares with applications in beamforming. *IEEE Transactions on Signal Processing*, 65(11):2875–2887, 2017.

28 I. Goodfellow, Y. Bengio, and A. Courville. *Deep Learning*. MIT Press, Cambridge, UK, 2016.

29 Z.-Q. Luo, W.-K. Ma, A. M.-C. So, Y. Ye, and S. Zhang. Semidefinite relaxation of quadratic optimization problems. *IEEE Signal Processing Magazine*, 27(3):20–34, 2010.

30 G. Cui, H. Li, and M. Rangaswamy. MIMO radar waveform design with constant modulus and similarity constraints. *IEEE Transactions on Signal Processing*, 62(2):343–353, 2013.

31 G. Cui, X. Yu, G. Foglia, Y. Huang, and J. Li. Quadratic optimization with similarity constraint for unimodular sequence synthesis. *IEEE Transactions on Signal Processing*, 65(18):4756–4769, 2017.

32 J. Liang, H. C. So, J. Li, and A. Farina. Unimodular sequence design based on alternating direction method of multipliers. *IEEE Transactions on Signal Processing*, 64(20):5367–5381, 2016.

33 Y. Li, M. Tofighi, J. Geng, V. M., and Y. C. Eldar. Efficient and interpretable deep blind image deblurring via algorithm unrolling. *IEEE Transactions on Computational Imaging*, 6:666–681, 2020.

34 C. Zhang, J. Pei, Y. Zhang, W. Huo, D. Mao, Y. Huang, and J. Yang. MIMO radar waveform design for simultaneous space–time–Doppler domain optimization: Framework and implementation. *IEEE Transactions on Geoscience and Remote Sensing*, 60:1–15, 2022.

35 J. Hu, W. Zhang, H. Zhu, K. Zhong, W. Xiong, Z. Wei, and Y. Li. Constant modulus waveform design for MIMO radar via manifold optimization. *Signal Processing*, 190:108322, 2022.

36 J. Li, G. Liao, Y. Huang, Z. Zhang, and A. Nehorai. Riemannian geometric optimization methods for joint design of transmit sequence and receive filter on MIMO radar. *IEEE Transactions on Signal Processing*, 68:5602–5616, 2020.

37 K. Zhong, J. Hu, Z. Zhao, X. Yu, G. Cui, B. Liao, and H. Hu. MIMO radar unimodular waveform design with learned complex circle manifold network. *IEEE Transactions on Aerospace and Electronic Systems*, 60(2):1798–1807, 2024.

38 Q. Li, C. Li, and J. Lin. Constant modulus secure beamforming for multicast massive MIMO wiretap channels. *IEEE Transactions on Information Forensics and Security*, 15:264–275, 2019.

39 Z. Cheng, Z. He, S. Zhang, and J. Li. Constant modulus waveform design for MIMO radar transmit beampattern. *IEEE Transactions on Signal Processing*, 65(18):4912–4923, 2017.

40 M. Soltanalian and P. Stoica. Designing unimodular codes via quadratic optimization. *IEEE Transactions on Signal Processing*, 62(5):1221–1234, 2014.

41 L. Wu, P. Babu, and D. P. Palomar. Transmit waveform/receive filter design for MIMO radar with multiple waveform constraints. *IEEE Transactions on Signal Processing*, 66(6):1526–1540, 2017.

42 S. Khobahi, A. Bose, and M. Soltanalian. Deep radar waveform design for efficient automotive radar sensing. In *IEEE Sensor Array and Multichannel Signal Processing Workshop*, pages 1–5. IEEE, 2020.

43 P. Stoica, J. Li, and M. Xue. Transmit codes and receive filters for radar. *IEEE Signal Processing Magazine*, 25(6):94–109, 2008.

44 M. Soltanalian, B. Tang, J. Li, and P. Stoica. Joint design of the receive filter and transmit sequence for active sensing. *IEEE Signal Processing Letters*, 20(5): 423–426, 2013.

45 W. Dinkelbach. On nonlinear fractional programming. *Management Science*, 13(7):492–498, 1967.

46 A. De Maio, S. De Nicola, Y. Huang, Z.-Q. Luo, and S. Zhang. Design of phase codes for radar performance optimization with a similarity constraint. *IEEE Transactions on Signal Processing*, 57(2):610–621, 2008.

47 M. A. Richards, J. A. Scheer, and W. A. Holm. *Principles of Modern Radar: Basic Principles*. IET, 2010.

48 X. Yu, K. Alhujaili, G. Cui, and V. Monga. MIMO radar waveform design in the presence of multiple targets and practical constraints. *IEEE Transactions on Signal Processing*, 68:1974–1989, 2020.

49 S. Imani and S. A. Ghorashi. Sequential quasi-convex-based algorithm for waveform design in colocated multiple-input multiple-output radars. *IET Signal Processing*, 10(3):309–317, 2016.

50 L. Zhao and D. P. Palomar. Maximin joint optimization of transmitting code and receiving filter in radar and communications. *IEEE Transactions on Signal Processing*, 65(4):850–863, 2016.

51 A. J. Duly, D. J. Love, and J. V. Krogmeier. Time-division beamforming for MIMO radar waveform design. *IEEE Transactions on Aerospace and Electronic Systems*, 49(2):1210–1223, 2013.

52 A. Aubry, A. De Maio, and M. M. Naghsh. Optimizing radar waveform and Doppler filter bank via generalized fractional programming. *IEEE Journal of Selected Topics in Signal Processing*, 9(8):1387–1399, 2015.

53 Z. Cheng, B. Liao, Z. He, Y. Li, and J. Li. Spectrally compatible waveform design for MIMO radar in the presence of multiple targets. *IEEE Transactions on Signal Processing*, 66(13):3543–3555, 2018.

54 A. M. Ahmed, A. A. Ahmad, S. Fortunati, A. Sezgin, M. S. Greco, and F. Gini. A reinforcement learning based approach for multitarget detection in massive MIMO radar. *IEEE Transactions on Aerospace and Electronic Systems*, 57(5):2622–2636, 2021.

55 Z. Qiu, K. Duan, Y. Wang, Z. Liao, and J. He. Designing constant modulus approximate binary phase waveforms for multi-target detection in MIMO radar using LSTM networks. *IEEE Transactions on Geoscience and Remote Sensing*, 62:162, 2024.

56 J. Kweon, F. Gini, M. S. Greco, M. Rangaswamy, and V. Monga. Model-based learning for MIMO radar waveform design in the presence of multiple targets. *IEEE Transactions on Aerospace and Electronic Systems*, 99:1–14, 2025.

57 A. Aubry, A. De Maio, and Y. Huang. MIMO radar beampattern design via PSL/ISL optimization. *IEEE Transactions on Signal Processing*, 64(15):3955–3967, 2016.

58 G. Cui, X. Yu, V. Carotenuto, and L. Kong. Space-time transmit code and receive filter design for colocated MIMO radar. *IEEE Transactions on Signal Processing*, 65(5):1116–1129, 2016.

59 A. Aubry, A. De Maio, A. Farina, and M. Wicks. Knowledge-aided (potentially cognitive) transmit signal and receive filter design in signal-dependent clutter. *IEEE Transactions on Aerospace and Electronic Systems*, 49(1):93–117, 2013.

60 A. I. Barros, JBG Frenk, S. Schaible, and S. Zhang. A new algorithm for generalized fractional programs. *Mathematical Programming*, 72:147–175, 1996.

61 J. Kweon, F. Gini, M. S. Greco, and V. Monga. Sum-of-reciprocal exact descent: waveform design for multi-target detection. In *IEEE Radar Conference*, pages 1–6. IEEE, 2024.

62 R. G. Strongin and Y. D. Sergeyev. *Global Optimization with Non-convex Constraints: Sequential and Parallel Algorithms*, volume 45. Springer Science & Business Media, 2013.

63 C. Dugas, Y. Bengio, F. Bélisle, C. Nadeau, and R. Garcia. Incorporating second-order functional knowledge for better option pricing. *Advances in Neural Information Processing Systems*, pages 451–457, 2000.

64 V. Monga. *Handbook of Convex Optimization Methods in Imaging Science.* Springer, 2017.

65 D. P. Kingma and J. Ba. Adam: A method for stochastic optimization. *arXiv preprint arXiv:1412.6980*, 2014.

66 Y. Zhao, Y. Li, H. Zhang, V. Monga, and Y. C. Eldar. Deep, convergent, unrolled half-quadratic splitting for image deconvolution. *IEEE Transactions on Computational Imaging*, 10:574–588, 2024.

2

Model-based Reinforcement Learning for Integrated Sensing and Communications

Petteri Pulkkinen[1,2] and Visa Koivunen[1,]*

[1] *Department of Information and Communications Engineering, Aalto University, Espoo, Finland*
[2] *Saab Finland, Helsinki, Finland*

*Corresponding author: visa.koivunen@aalto.fi

2.1 Introduction

Integrated sensing and communications (ISAC) technology is being developed for next-generation 6G wireless networks and multifunctional radio and radar systems [1–3]. ISAC systems are designed to operate in congested, shared, and even contested spectrum environments, where efficiently utilizing spectrum, power, transceiver hardware, and spatial resources is critical. The goal is to jointly design and optimize sensing and communication functionalities to achieve mutual benefits. This integration is supported by ongoing technological convergence in radio frequency (RF) systems, including multifunctional circuits, shared hardware components, and large-aperture antenna arrays. This chapter focuses on waveform co-design and resource allocation strategies for ISAC applications, primarily targeting radar sensing tasks such as target detection, parameter estimation, and tracking, while supporting high-throughput wireless communications.

Multicarrier waveforms are widely adopted in modern and emerging wireless standards, including 4G, 5G, and future 6G networks, as well as Wi-Fi, digital video broadcasting (DVB-T/T2), and digital audio broadcasting (DAB) [4]. Among these, the most widely used waveform is the orthogonal frequency division multiplexing (OFDM) and its multiuser variant, OFDMA. Multicarrier waveforms are also well-suited for ISAC due to their use as signals of opportunity in passive radar systems [5], as well as in dedicated OFDM-based radars [6] and multicarrier phase-coded (MCPC) radars [7]. Furthermore, multicarrier modulation is often combined with multi-antenna processing using fully adaptive multiple-input multiple-output (MIMO) architectures. The combination of multicarrier waveforms and multiantenna systems enables significant design flexibility via allocation of frequency resources (e.g., subcarriers [SCs] or resource blocks [RBs]) and spatial resources (e.g., beams, beamformers, or antennas) for different sensing and communication tasks [4, 8–15].

Waveform design and resource allocation problems can be formulated as structured optimization or machine learning problems. Structured approaches employ convex optimization, relaxations, and constraints to meet performance goals. Potentially, multi-objective optimization (MOO) frameworks are used to manage trade-offs among multiple sensing and communication key performance indicators (KPIs). However, these methods require accurate and timely information about the environment, system capabilities, and task requirements—an assumption that may not hold in distributed, large-scale, decentralized, and dynamic systems. Machine learning offers a data-driven alternative, particularly when precise modeling is difficult or when tractable solution methods are unavailable. RL is a sequential decision-making framework that learns optimal policies through interaction and reward feedback [16]. It enables agents to adapt online, making it well-suited for ISAC resource allocation in dynamic environments.

RL methods have been studied extensively in the context of wireless communications [17, 18]. More recently, their application in radar has also attracted increasing attention [8, 19–24]. However, RL in the ISAC context is a less explored area. Therefore, this chapter focuses on RL-based methods for waveform and resource co-design in multicarrier MIMO ISAC systems, with methods proposed in [9–15]. MFRL methods learn purely through trial and error, while MBRL approaches build predictive models of environment dynamics and rewards, enabling planning, improved interpretability, and better sample efficiency. The emphasis is on MBRL algorithms, with practical resource allocation examples, including frequency-domain SC or RB allocation and beamspace-domain beam allocation. The reward functions used in these RL approaches reflect typical sensing and communications KPIs.

Organization: Section 2.2 provides an overview of commonly used RL models and algorithms. Section 2.3 introduces a general framework for resource allocation in ISAC systems based on a constrained partially observable Markov decision process (C-POMDP) formulation and MBRL. Section 2.4 presents an application of MBRL to frequency-domain resource allocation in dynamic spectrum environments. Section 2.5 considers joint frequency and spatial-domain allocation using multicarrier signals and beamspace processing, focusing on dynamic target environments. Finally, Section 2.6 concludes the chapter and outlines directions for future research.

Notation: Vectors are denoted by bold lowercase letters (e.g., $\boldsymbol{x}$), and matrices by bold uppercase letters (e.g., $\boldsymbol{X}$). Time dependence is indicated with the subscript $\cdot_t$. The ith element of a generic vector $\boldsymbol{x}$ is written as x_i. Furthermore, the ith entry of a time-dependent vector $\boldsymbol{x}_t$ is denoted by $x_i(t)$. Random variables are typically represented by uppercase letters (e.g., X), while lowercase letters (e.g., x) denote their realizations. Occasionally, lowercase letters may also represent random variables when it is clear from the context. The notation $\mathbb{P}\{X \mid Y\}$ denotes the probability distribution of the random variable X conditioned on Y, while the corresponding probability density function (PDF) or probability mass function (PMF) is written as $p_{X|Y}\,(x \mid y)$, or simply $p(x \mid y)$ when unambiguous.

2.2 Overview of RL

Reinforcement learning (RL) is a family of methods for learning to act in stochastic sequential decision-making problems, formalized as Markov decision processes (MDPs) or partially observable Markov decision processes (POMDPs) [16]. The problem model consists of a decision-maker, called an agent, interacting with an environment. The agent takes actions in the environment, makes observations about the environment state, and receives scalar rewards that quantify the quality of the action and the environment state, as visualized in Figure 2.1. The goal of the agent is to maximize cumulative rewards. Generally, RL can be applied to problems where the environment model is unknown [16].

2.2.1 Preliminaries

2.2.1.1 Markov Decision Process

Markov decision process is comprised of a set of states $\mathcal{S}$ (state-space), a set of actions $\mathcal{A}$ (action-space), a model $\mathcal{M}$ defining the dynamics of the environment (transitions), rewards $u \in \mathcal{U}$ that assigns a scalar value to state $s \in \mathcal{S}$, next state

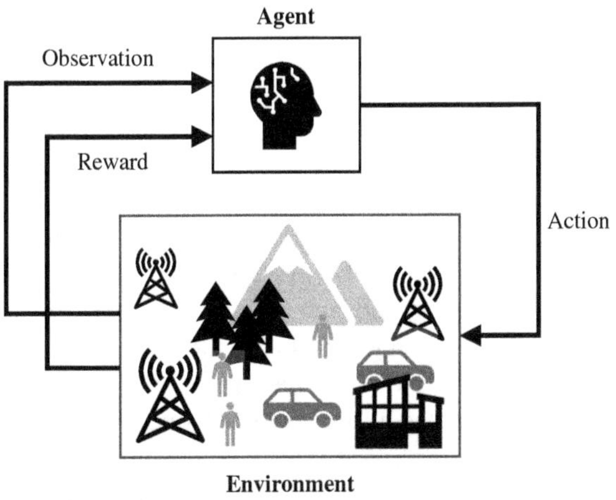

Figure 2.1 Reinforcement learning in ISAC systems. *Source*: [25] / with permission of Petteri Pulkkinen.

$s' \in \mathcal{S}$ and action $a \in \mathcal{A}$ pairs, and a discount factor $\gamma \in [0, 1)$ that models the importance of future rewards. Generally, the state and action spaces in MDPs can be finite or infinite sets. This chapter only considers discrete-time MDPs where time evolves in discrete steps, enumerated with indices $t \in \mathbb{N}$. In ISAC systems, discrete-time MDPs are well justified due to the time interval between the decisions and wide-spread use of quasi-stationary or piecewise stationary models.

A decision process to be an MDP, it needs to satisfy the Markov property:

$$\mathbb{P}\{S_{t+1}|S_t, \dots, S_0, A_t, \dots, A_0\} = \mathbb{P}\{S_{t+1}|S_t, A_t\}, \tag{2.1}$$

where $S_t \in \mathcal{S}$ and $A_t \in \mathcal{A}$ are the state and action at time instance t, respectively. If not else stated, time-homogeneous MDPs are considered where $p_{S_{t+1}|S_t,A_t}(s' \mid s, a) = p_{S_{t'+1}|S_{t'},A_{t'}}(s' \mid s, a)$ for all $t \in \mathbb{N}$ and $t' \in \mathbb{N}$. Consequently, the state transition dynamics for any $s' \in \mathcal{S}$, $s \in \mathcal{S}$, and $a \in \mathcal{A}$ can be written as

$$\mathcal{M}(s'|s, a) := p_{S_{t+1}|S_t,A_t}(s' \mid s, a), \tag{2.2}$$

which is independent of the time step t.

The reward $U_t \in \mathcal{U}$ in MDP is a scalar random variable in a subset of real numbers $\mathcal{U} \subseteq \mathbb{R}$. The complete model of the MDP dynamics can be written as

$$\mathcal{M}(s', u'|s, a) := p_{S_{t+1},U_{t+1}|S_t A_t}(s', u' \mid s, a). \tag{2.3}$$

Let us define the expected reward, also referred to as the reward function, as

$$u(s, a, s') := \mathbb{E}[U_{t+1}|S_t = s, A_t = a, S_{t+1} = s'], \tag{2.4}$$

where $\mathbb{E}[\cdot|\cdot]$ is the conditional expectation. The reward function can also be written as

$$u(s, a) := \mathbb{E}[U_{t+1}|S_t = s, A_t = a], \tag{2.5}$$

which marginalizes (2.4) with respect to the state dynamics.

In MDPs, the environment state $S_t \in \mathcal{S}$ is assumed to be fully observable at each time step t. Thus, the agent must determine a policy

$$\pi(a|s) := p_{A_t|S_t}(a \mid s), \tag{2.6}$$

which gives the probability of taking action $a \in \mathcal{A}$ given the state $s \in \mathcal{S}$. The time horizon of an MDP can be finite, with a limited number of steps, or infinite, continuing indefinitely. From the perspective of RL theory and resource allocation problems in ISAC, infinite horizon MDPs are more relevant than finite horizon MDPs. Hence, the focus is on infinite horizon MDPs, where the performance of the agent is evaluated by the expected sum of discounted future rewards defined as:

$$V_\pi(s) := \mathbb{E}_\pi\left[\sum_{i=0}^{\infty} \gamma^i U_{t+1+i} | S_t = s\right], \tag{2.7}$$

where $\gamma \in [0, 1)$ is the discount factor, and $\mathbb{E}_\pi[\cdot|\cdot]$ denotes the expectation subject to trajectories generated by the policy π. The equation (2.7) is called the *value function*. Similarly, a function that quantifies the quality of action $a \in \mathcal{A}$ given the state $s \in \mathcal{S}$ while following the policy π for the rest of the trajectory is defined as:

$$Q_\pi(s, a) := \underset{\pi}{\mathbb{E}}[U_{t+1} + \gamma V_\pi(S_{t+1})|S_t = s, A_t = a]. \tag{2.8}$$

It is called the *state-action* value or Q value. An optimal policy is defined as follows:

$$\pi^\star = \arg\max_\pi V_\pi(s) \quad \forall \quad s \in \mathcal{S}. \tag{2.9}$$

For $\gamma \ll 1$, future rewards are considered only a few steps ahead. Conversely, for γ values close to 1, the agent maximizes rewards over a longer term into the future. A policy maximizing only the immediate reward is known as a myopic policy. In general, myopic policies are suboptimal for $\gamma > 0$.

The widely used Bellman equation defines the optimal value function as:

$$V_{\pi^\star}(s) = \max_{a \in \mathcal{A}} \underbrace{\underset{\mathcal{M}}{\mathbb{E}}[U_{t+1} + \gamma V_{\pi^\star}(S_{t+1}) \mid S_t = s, A_t = a]}_{Q_{\pi^\star}(s,a)}, \tag{2.10}$$

indicating that the optimal policy is deterministic [26]. With *finite* state and action spaces and a known MDP model $\mathcal{M}$, MDPs can be solved optimally using *value iteration (VI)* [26] or *policy iteration (PI)* [27] methods. In VI, EQ. (2.10) gives the optimal values recursively, and the policy is derived from the optimal value function. The PI algorithm alternates between a policy evaluation step, where the value function for the current policy is computed using (2.7), and policy improvement, where the best actions based on the current value function are selected.

2.2.1.2 Partially Observable Markov Decision Process

A partially observable Markov decision process is an extension of the MDP framework in which the state is not fully observable [28]. In real-world environments, the state is usually observed by a sensor that introduces unobservable and random noise and gives only limited information about the entire state. Formally, POMDP framework adds to MDP a set of all possible observations $\mathcal{O}$ (finite or infinite set) called the observation space [28]. Observations $O_{t+1} \in \mathcal{O}$ are linked to states $S_{t+1} \in \mathcal{S}$ and actions $A_t \in \mathcal{A}$ via the observation model:

$$\mathcal{Z}(o'|s', a) := p_{O_{t+1}|S_{t+1},A_t}(o'|s', a), \tag{2.11}$$

which indicates that the likelihood of observation $O_{t+1} \in \mathcal{O}$ is conditional on the state $S_{t+1} \in \mathcal{S}$ and action $A_t \in \mathcal{A}$.

A policy for a POMDP depends on the complete history of past actions and observations [28], known as the information history. The information history up to time t is:

$$\mathcal{I}_t = \{A_0, O_1, A_1, O_2, \dots, A_{t-1}, O_t\}. \tag{2.12}$$

It is possible to utilize the information history $\mathcal{I}_t$ to calculate the posterior probability of the state defined as

$$b_t(s) := p_{S_t|\mathcal{I}_t}(s \mid I_t), \tag{2.13}$$

where $I_t \sim \mathbb{P}\{\mathcal{I}_t\}$ is a realization of the information history. The probability distribution b_t, also called the *belief* is a sufficient statistic for all previous observations and actions. We use $B_t \in \mathcal{B}$ to denote a random belief at time instance t, which is in the space of beliefs $\mathcal{B}$.

The belief can be updated assuming a known model $\mathcal{M}$ by using the Bayes' rule

$$b_{t+1}(s') \propto \mathcal{Z}(o_{t+1} \mid s', a_t)\,\mathbb{E}\left[\mathcal{M}(s' \mid S_t, a_t) \mid B_t = b_t\right], \tag{2.14}$$

where $O_{t+1} = o_{t+1}$ is the most recent observation [28]. Using (2.14) for finite state spaces is straightforward. However, Kalman filter (KF) algorithms and particle filters are practical realizations of (2.14) in continuous state spaces. We can write the belief update in (2.14) as

$$B_{t+1} = \mathcal{G}(O_{t+1}; B_t, A_t) \tag{2.15}$$

stating that the process $B_0, B_1, \dots, B_t$ satisfies the Markov property in (2.1). Therefore, POMDPs can be interpreted as a MDPs with state space $\mathcal{B}$, called belief state MDPs. The belief state transition probability is defined as

$$\mathbb{P}\{B_{t+1}|B_t, A_t\} = \mathop{\mathbb{E}}_{O_{t+1}}\Big[\underbrace{\mathbb{P}\{B_{t+1}|B_t, A_t, O_{t+1}\}}_{\text{Dirac delta PDF}} \mid B_t, A_t\Big] \tag{2.16}$$

$$= \mathbb{P}\{O_{t+1} \mid B_t, A_t\}, \tag{2.17}$$

where it is assumed that the mapping in (2.15) is injective. Equation (2.17) shows that the observation distribution $\mathbb{P}\{O_{t+1}|B_t, A_t\}$ directly defines the belief state transition probabilities.

For any belief state $b \in \mathcal{B}$ and action $a \in \mathcal{A}$, the expected reward is

$$u_{\mathcal{B}}(b_t, a_t) := \mathbb{E}[u(S_t, a_t) \mid B_t = b_t]. \tag{2.18}$$

The Bellman optimality equation for Q-values can be written as:

$$V_{\pi^\star}(b) = \max_{a\in\mathcal{A}} \left(u_{\mathcal{B}}(b, a) + \gamma\,\mathbb{E}\left[V_{\pi^\star}\left(B_{t+1}\right) \mid B_t = b, A_t = a\right]\right). \tag{2.19}$$

This equation shows that solving a POMDP in the belief MDP formulation is similar to solving a continuous state MDP, which is generally difficult. Fortunately, optimal methods for solving POMDPs with finite and small state, action, and observation spaces exist. Additionally, approximate methods have been developed for large POMDPs; refer to [29] for a comprehensive survey.

2.2.2 Model-free and Model-based Learning

RL methods can be broadly categorized into MFRL and MBRL approaches, as depicted in Figure 2.2. For a comprehensive introduction to MFRL, refer to [16]. For an extensive survey of MBRL, see [56]. MFRL methods learn the policy π, the value function $V_\pi(s)$, or the Q-value function $Q_\pi(s, a)$ by trial and error. MBRL methods learn the transition model $\mathcal{M}(s'|s, a)$ completely or partially through experience and plan over the learned model, as illustrated in Figure 2.3. MBRL improves sample efficiency, reducing interactions needed to find a good policy,

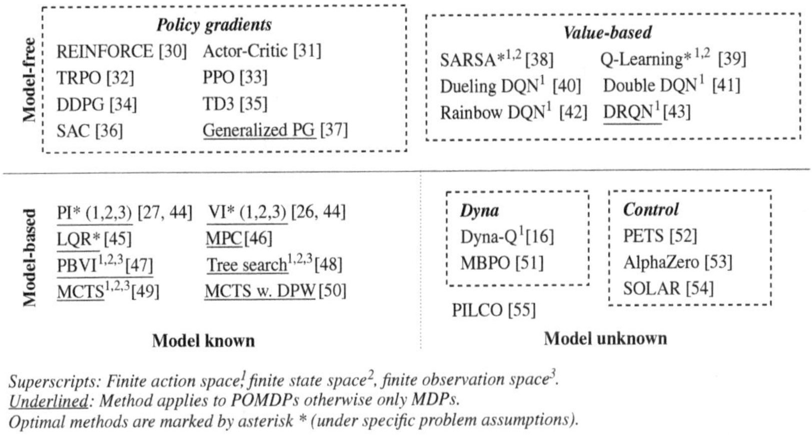

Figure 2.2 Difference between model-free and model-based RL methods. *Source*: [25] / with permission of Petteri Pulkkinen.

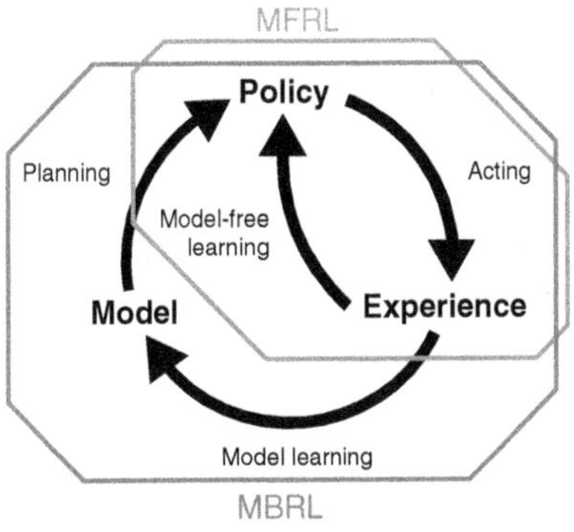

Figure 2.3 Overview of well-known and model-free and model-based RL methods. *Source*: [12] / with permission of Petteri Pulkkinen.

which is crucial in many real-world applications, including ISAC. In ISAC applications, MBRL can use existing broad knowledge of radio environments, communications, and radar systems and apply learning only where necessary. However, its performance depends on the quality of the learned model. Also, the computational complexity per real-world interaction in MBRL is typically higher than in MFRL.

2.2.2.1 Exploration-exploitation Tradeoff

The exploration and exploitation (E&E) tradeoff is a fundamental problem in RL. Consider MFRL with policy π_{ψ} or MBRL with model $\mathcal{M}_{\psi}$, where $\psi \in \mathcal{P}$ are the parameters to learn. Traditionally, E&E concerns the tradeoff in gathering more information about unknown parameters $\psi \in \mathcal{P}$ and maximizing current performance using existing knowledge of these parameters. E&E can also refer to obtaining more knowledge about the current state in POMDPs. By including ψ in the state, in some cases, POMDP solvers can be used to compute the optimal policies that directly balance E&E. However, this approach can be highly complicated and unfeasible for many problems. Therefore, more straightforward methods for balancing E&E have been developed.

Common E&E tradeoff strategy is the ϵ-*greedy* method, which selects a random action with probability ϵ and the best-known action otherwise [16, 39]. In continuous action spaces, *noise-based* exploration can also be used, where randomness is added to actions [34] or policy parameters [57]. *Explore-then-commit* strategy begins with completely random exploration before switching to exploitation [58]. Boltzmann exploration [59] selects actions probabilistically based on their estimated values. Thompson sampling (TS) samples model parameters or values from their posterior distributions to balance E&E. TS has been shown effective across multi-armed bandit (MAB), MDP, and POMDP settings, with strong theoretical guarantees in restless multi-armed bandit (RMAB) [60, 61] and MDP [62] problems.

In contrast to sampling-based exploration, another approach to balance E&E is to rely on *optimism in the face of uncertainty* principle, which encourages trying uncertain actions that may lead to higher rewards. For example, the widely used upper confidence bound (UCB) algorithm selects actions based on UCBs of their estimated value and has been successfully applied to MAB, MDP [63], and POMDP [49] settings. A related approach is recency-based exploration [64, 65], which favors exploring actions that have been used less recently.

2.2.2.2 Model-free Algorithms

MFRL algorithms can be broadly categorized into value-based [38–43] and policy gradient (PG) methods [30–37]. Advanced value-based methods can solve MDPs with continuous state spaces, but they are restricted to discrete action spaces.

In contrast, PG methods can be applied to problems with continuous action and state spaces.

2.2.2.2.1 Value-based Methods

Q-learning is a temporal difference (TD) learning algorithm combining ideas from Monte Carlo (MC) methods and dynamic programming [39]. In Q-learning, the optimal action-value is learned step-by-step using samples of state, action, reward, and next-state tuples obtained via interacting with the environment, thereby approximating the Bellman equation in (2.10). With finite action and state spaces (i.e., the tabular case), Q-value estimates $Q(s_t, a_t)$ are updated based on:

$$Q(s_t, a_t) \leftarrow Q(s_t, a_t) + \alpha_t \left[u_{t+1} + \gamma \max_{a_{t+1} \in \mathcal{A}} Q(s_{t+1}, a_{t+1}) - Q(s_t, a_t) \right]. \tag{2.20}$$

The Q-function $Q(s, a)$ is initialized arbitrarily for all state-action pairs (s, a). Then, actions are selected based on an E&E strategy (e.g., ϵ-greedy), and the Q-values are updated using (2.20) until convergence. To apply Q-learning to continuous or large state spaces, deep neural networks (DNNs) are often employed to approximate the action-value function [40–43, 66, 67].

2.2.2.2.2 Policy Gradient Methods

PG algorithms are a class of RL methods that optimize the policy directly rather than learning a value function to derive the policy [32–36]. These algorithms are particularly useful for problems with high-dimensional or continuous action spaces. A fundamental result for PG algorithms is that the gradient of the expected return with respect to the policy parameters is

$$\nabla_{\boldsymbol{\psi}} J(\boldsymbol{\psi}) = \mathop{\mathbb{E}}_{\pi_{\boldsymbol{\psi}}} \left[Q_{\pi_{\boldsymbol{\psi}}}(s, a) \nabla_{\boldsymbol{\psi}} \log \pi_{\boldsymbol{\psi}}(a|s) \right], \tag{2.21}$$

where $\boldsymbol{\psi}$ represents the policy parameters, $\pi_{\boldsymbol{\psi}}(a|s)$ is the policy PDF, and $Q_{\pi_{\boldsymbol{\psi}}}(s, a)$ is the action-value function.

One of the simplest PG algorithms is the *reward increment = nonnegative factor × offset reinforcement × characteristic eligibility (REINFORCE)* algorithm [30]. It uses the stochastic gradient descent (SGD) algorithm with an unbiased estimate of the gradient in (2.21) obtained by a MC trial to collect trajectories with state, action, and reward tuples. In practice, the vanilla REINFORCE is rarely used to solve RL problems, as it suffers from high gradient variance. The variance can be reduced by rewriting (2.21) as

$$\nabla_{\boldsymbol{\psi}} J(\boldsymbol{\psi}) = \mathop{\mathbb{E}}_{\pi_{\boldsymbol{\psi}}} \left[A_{\pi_{\boldsymbol{\psi}}}(s, a) \nabla_{\boldsymbol{\psi}} \log \pi_{\boldsymbol{\psi}}(a|s) \right], \tag{2.22}$$

where

$$A_{\pi_{\boldsymbol{\psi}}}(s, a) = Q_{\pi_{\boldsymbol{\psi}}}(s, a) - V_{\pi_{\boldsymbol{\psi}}}(s). \tag{2.23}$$

The gradient estimates obtained using (2.22) exhibit reduced variance, which improves the efficiency of the learning process. Methods using this are called actor-critic (AC) methods [31]. The "actor" updates the policy parameters using (2.21), while the "critic" estimates the value function (or advantage) to reduce the variance of the gradient estimate.

2.2.2.3 Model-based Algorithms

The methods and models in MBRL can be very application-specific. A learned model allows for making predictions about future states and rewards, hence avoiding unnecessary trial-and-error steps. The model simulates the environment and generates the synthetic data needed to plan and improve the policy. Application-dependent models may be specified using rigorous equations with possibly unknown parameters. One example is the linear state-space model as in Kalman filtering. However, model-based RL can also be applied when the functional form of the state transition model is entirely unknown by using function approximation techniques such as DNNs to estimate the parameters of the state transition distribution [52]. Generative models may also be employed to simulate the environment and predict future states [68]. Alternatively, Gaussian processes can be used to model the environment and perform probabilistic inference [55].

When the model is known, methods such as PI [27, 44], VI [26, 44, 47], or tree search-based methods [48, 49, 50] can be employed. However, when the model is unknown, planning in MBRL is typically based on the following approaches: (1) *Dyna architecture*, (2) *model-based PG*, and (3) *(approximate) optimal control*. The *Dyna architecture* is a hybrid approach that combines model-free and model-based methods [16, 51]. These methods learn a policy using MFRL algorithms by augmenting real-world data with data generated by the model. The model of the environment is learned and used to generate simulated experiences, which are then used to update the policy and value functions. *Model-based PG* methods, such as PILCO, use the learned model to evaluate the finite horizon cumulative reward gradient analytically to implement PG updates directly on the learned model [55]. Many practical MBRL methods are based on the *(approximate) optimal control* idea, where only the model is learned, and then an optimizer is employed to maximize the cumulative return. This approach considers system stability and constraints similarly to control theory, making it suitable for applications where safety and reliability are critical [69]. For example, model predictive control (MPC) uses a model of the system to predict future states and optimize control actions over a finite time window. Combining MPC with the learned models is known as the learning-based model predictive control (LMPC) approach [69]. In the famous AlphaZero, Monte Carlo tree search (MCTS) plans actions in simulated board games with self-play [53]. The PETS algorithm uses a gradient-free optimization method, namely cross-entropy method (CEM), to evaluate the expected return of different actions and select the best action based on these predictions [52].

The CEM method can also be applied for planning in the parameter space of the policy network [70]. A more computationally efficient version of the CEM for real-time planning was proposed in [71]. By learning latent state presentations of complex observations and enforcing linear dynamics in the latent space, a linear quadratic regulator (LQR) can be used to control a system [45, 54].

2.3 MBRL for ISAC Resource Allocation

This section introduces the C-POMDP framework for rigorously formulating ISAC resource allocation problems with constraints. An example of ISAC resource allocation formulations within the C-POMDP framework is given. We further describe key structural properties of the resulting C-POMDPs, which may be utilized in the development of MBRL algorithms.

2.3.1 C-POMDP Framework

The resource allocation problems in ISAC systems may be modeled as a C-POMDP with N_C constraints. A C-POMDP objective incorporates scalar random variables $U_{t+1}^{(i)}$ for all $i = 0, \dots, N_C$, referred to as utilities [72]. The utility $U_t^{(0)} \in \mathcal{U}_0$ corresponds to the regular POMDP reward to be maximized, and the utilities $U_t^{(i)} \in \mathcal{U}_i$ for all $i = 1, \dots, N_C$ are used to impose the constraints. For convenience, we define the expected utility as

$$u_i(s, a, s') := \mathbb{E}\left[U_{t+1}^{(i)} \mid S_t = s, A_t = a, S_{t+1} = s'\right] \tag{2.24}$$

as well as the marginal expectations of (2.24) as $u_i(s, a)$. In C-POMDPs, one is interested in expected rewards conditioned on the belief state

$$u_B^{(i)}(b, a, b') := \mathbb{E}[u_i(S_t, A_t, S_{t+1}) \mid B_t = b, A_t = a, B_{t+1} = b'], \tag{2.25}$$

and the marginal expectation $u_B^{(i)}(b_t, a_t)$. The objectives for sensing may be defined more directly by utilizing the form in (2.25), for example, when the reward is a function of the target state uncertainty covariance matrix.

The value function of each utility is the expected discounted sum of utilities starting from the belief $b \in \mathcal{B}$. It is defined as

$$V_\pi^{(i)}(b) = \mathbb{E}_\pi\left[\sum_{j=0}^{\infty} \gamma_i^j u_B^{(i)}(B_{t+j}, A_{t+j}) | B_t = b\right], \tag{2.26}$$

where $\gamma_i \in [0, 1)$ is the discount factor. The value-based objective can then be written as

$$\arg\max_\pi \quad V_\pi^{(0)}(b) \tag{2.27a}$$

$$\text{s.t.} \quad \mathrm{V}_\pi^{(i)}(\mathrm{b}) \geq 0 \quad \forall i = 1, \dots, N_C. \tag{2.27b}$$

The optimal policy π is the one that solves (2.27) for all $b \in \mathcal{B}$. This chapter concentrates on the objective of (2.27) which has practical applications in MBRL. However, interested readers may refer to the analytical results in [12] or MFRL algorithm in [11] utilizing an average reward objective.

2.3.2 Resource Allocation Using C-POMDPs

The C-POMDP is a suitable framework for modeling ISAC systems. It allows for capturing the trade-offs between communication and sensing subsystems under operational constraints. For instance, one may maximize the sensing performance under minimum tolerable communications quality of service (QoS) requirements or prioritize communications performance under sensing constraints. Alternatively, ISAC resource allocation can be formulated as a MOO by considering a weighted combination of utilities and finding Pareto-optimal solutions.

The *action space* in ISAC systems encompasses the available degrees of freedoms (DoFs), including allocations in centralized and distributed architectures, potentially parallel to different tasks. Resources such as transmitters (TXs), receivers, antennas, antenna subarrays, beams, beamformers, codes, and frequency-domain sub-channels may be dynamically allocated. Allocations may occur as a resource selection or power allocation under practical operational constraints, modeled by the action space $\mathcal{A}$. The action space $\mathcal{A}$ may be compatible with desirable waveform properties, e.g., by controlling the peak-to-average power ratio (PAPR), subcarrier power ratio (SPR), or peak-to-sidelobe ratio (PSR), or enforcing constraints such as similarity to a reference ambiguity function.

The MDP *state* may include various elements related to sensing, communications, and the operational environment:

1. *Target states*, comprising target kinematics and response characteristics. The accurate estimation of these states enables optimized waveform design and adaptation. Target response is essential for computing sensing signal-to-noise ratio (SNR) or signal-to-interference-plus-noise ratio (SINR) and contributes to tasks such as recognition and classification through features like the high range resolution profile (HRRP). Target responses vary based on target radar cross section (RCS) and its fluctuations as well as changes in sensing geometry (e.g., angle of arrival (AoA), angle of departure (AoD), distance, shadowing).
2. *Communication channel state* comprises channel gains, path loss, line-of-sight (LoS) and non-LoS propagation components, fast and slow fading, multipath, and experienced interference and noise levels at receivers. Optimizing communication performance relies on accurate channel state information at the transmitter (CSIT) and consequently receiver SINR levels.
3. *Spectrum state* may include information about spectrum users (e.g., primary, secondary, adversarial), their transmitted waveforms, RF fingerprints,

locations, and their usage patterns. It may also capture information on propagation environments (scatterers), interference, and clutter in sensing and communications. Receivers are always victims of such disturbances. Spectrum awareness (situational awareness [SA]) is essential at both transmit and receive ends to optimize performance while avoiding causing or suffering from interference. It also supports covert communications and coexistence with other systems.

The physically meaningful state variables mentioned above may not satisfy the Markov property. We refer to such variables as physical states:

Definition 2.1 (Physical state) A random variable $\widetilde{S}_t \in \widetilde{\mathcal{S}}$ is a physical state if it satisfies:

$$\begin{cases} \mathbb{P}\{O_t \mid S_t, A_{t-1}\} = \mathbb{P}\{O_t \mid \widetilde{S}_t, A_{t-1}\} \\ \mathbb{P}\{U_t \mid S_t, A_{t-1}\} = \mathbb{P}\{U_t \mid \widetilde{S}_t, A_{t-1}\}, \end{cases}$$

where $S_t \in \mathcal{S}$, $\widetilde{S}_t \in \widetilde{\mathcal{S}}$, $A_t \in \mathcal{A}$, and $O_t \in \mathcal{O}$. The physical state transitions are not required to satisfy the Markov property.

POMDPs naturally accommodate non-Markovian physical state processes by assuming an underlying latent state (e.g., a history or trajectory of physical states and actions). When necessary, we denote the physical state as $\tilde{s} \in \widetilde{\mathcal{S}}$ and the MDP state as $s \in \mathcal{S}$. Otherwise, one may assume $\widetilde{\mathcal{S}} = \mathcal{S}$.

Observations related to target states are derived from received sensing signals, while communication-related CSIT is obtained via estimation using uplink pilots, feedback, or relying on channel reciprocity. Sensing receivers capture primary and secondary (interference-only) data for target parameter estimation and SA. Communications receivers may also conduct spectrum monitoring. Dedicated RF sensors may be employed to enhance awareness. At a higher abstraction level, observations can include: (1) target detections and parameter estimates (range, Doppler, angle, response), (2) communication channel estimates, (3) AoA or location information, (4) estimated interference and noise statistics, (5) interference channel estimates, (6) sub-channel occupancy detections, and (7) shared information on radio environment from cooperative systems.

Sensing *utilities* are often derived from classical statistical and information theoretic criteria used in waveform optimization: mutual information (MI) between scattering coefficients and received signals [73–77], Cramér–Rao bounds (CRBs) for parameter estimation [78–87], mean square error (MSE) for beampattern shaping [88, 89], and target probability of detection (PD), which relates to received SINR [86, 90–93]. Communication utilities typically include MI, data rate, throughput, latency, outage probability, or symbol and bit error rates. In the

RL framework, utility functions can be formulated to reflect long-term system-level objectives. For comprehensive overviews of RL in communication systems, see [17, 18]. In the context of ISAC, relevant objectives include minimizing the time allocated to sensing tasks to free resources for other ISAC functionalities [19], as well as penalizing for interference and collisions among coexisting systems [9, 94, 95].

Belief state updates follow the equation (2.14) and may be implemented using techniques such as hidden Markov model (HMM), Kalman filter, or particle filter. In cases where model parameters are unknown or the MDP state is not the physical state, belief updates can be approximated using learned models. For example, in [43]:

$$B_t = \mathcal{G}_{\psi}(O_t, A_{t-1}, B_{t-1}), \tag{2.28}$$

where $\mathcal{G}_{\psi} : \mathcal{O} \times \mathcal{A} \times \mathcal{B} \rightarrow \mathcal{B}$ is implemented by an recursive neural network (RNN) with parameters $\boldsymbol{\psi}$. Alternatively, a nonparametric approach with no learning can be employed by recording a finite number of past observations and actions.

2.3.3 ISAC POMDP Structure and Assumptions

Section 2.3.2 provided a comprehensive overview of the possible actions, states, observations, and utilities in integrated sensing and communication systems. In practice, considering all these variables simultaneously when solving specific C-POMDP problems is often unnecessary. By isolating specific system components, it may be possible to design separate policies and belief propagation algorithms for the resulting subproblems. For instance, SA building or communication channel estimation tasks may be decoupled from downlink (DL) communication and radar sensing tasks. Such a hierarchical problem decomposition into more manageable subproblems with distinct structural properties facilitates the development of more efficient MBRL methods. Hence, the overall problem structure is simplified and made more tractable for performance analysis.

Assumption 2.1 The MDP state transitions are action-independent. Formally, this is expressed as $\mathbb{P}\{S_{t+1} \mid S_t, A_t\} = \mathbb{P}\{S_{t+1} \mid S_t\}$.

Assumption 2.1 is closely related to the transition dynamics in MABs [96]. It holds when interference patterns and channel coefficients evolve independently of the ISAC resource allocation strategy. Such an assumption is frequently adopted in cognitive radio literature, particularly in the context of opportunistic spectrum access [94, 97]. It is generally valid for modeling communication and sensing channel processes. However, in scenarios involving intelligent adversaries or reactive jammers, the interference may depend on past actions. To address this, the assumption can be relaxed as follows.

Assumption 2.2 The physical state transition is independent of the most recent action. Formally, $\mathbb{P}\{\widetilde{S}_{t+1} \mid A_t, S_t\} = \mathbb{P}\{\widetilde{S}_{t+1} \mid S_t\}$, which also implies $\mathbb{P}\{\widetilde{S}_{t+1} \mid A_t, B_t\} = \mathbb{P}\{\widetilde{S}_{t+1} \mid B_t\}$.

Assumption 2.2 is particularly reasonable when the timescale of the MDP is significantly shorter than the adaptation time of the interference sources. Although the dynamics are independent of the most recent action, they may still indirectly depend on earlier actions through the latent state $S_t \in \mathcal{S}$.

The observation model plays a crucial role in determining the complexity of solving a given C-POMDP problem. The simplest case arises when the physical state is fully observable.

Assumption 2.3 The physical state $\widetilde{S}_t \in \widetilde{\mathcal{S}}$ is fully observable at each time instance t, i.e., $O_t = \widetilde{S}_t$.

Assumption 2.3 is rarely valid in practical ISAC scenarios, where the physical state is inferred through noisy observations. If the observation noise is negligible or the physical state evolves slowly, sufficiently accurate estimates may be constructed to ignore observation uncertainty during resource allocation. For example, a point estimate derived from the underlying belief distribution may serve as an effective surrogate for the true state.[1] Nevertheless, Assumption 2.3 may not hold in general and can be relaxed under the following conditions.

Assumption 2.4 The observations are not controllable. Formally, $\mathbb{P}\{O_t \mid \widetilde{S}_t, A_{t-1}\} = \mathbb{P}\{O_t \mid \widetilde{S}_t\}$.

Assumption 2.4 is often justified in the context of communication channels, where pilot or feedback signals are defined in standards, thus decoupled from actions affecting DL communication or radar sensing. Similarly, spectrum state observations are typically independent of ISAC resource allocation, especially when dedicated sensing intervals or auxiliary monitoring spectrum sensors are used. In contrast, observations related to target tracking or radar sensing are generally affected by the choice of actions, violating Assumption 2.4. Indeed, the sensing waveform should be designed to optimize the detection performance or accuracy of the target state estimates.

2.3.4 MBRL Algorithms for ISAC

MBRL algorithms, which learn models of complex system dynamics to predict future behavior and optimize control actions accordingly, offer a compelling framework for ISAC resource allocation. This architecture, illustrated in Figure 2.4,

---

1 This belief distribution is computed using true observations, while the observation for the resource allocation problem is the resulting point estimate.

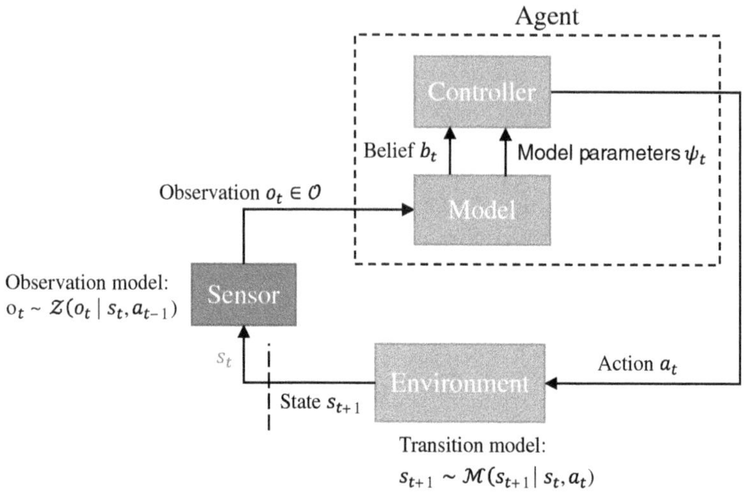

Figure 2.4 MBRL architecture based on approximate optimal control and learned models such as in LMPC. *Source*: [25] / with permission of Petteri Pulkkinen.

enables online decision-making and rapid adaptation to changes in the model or utility functions. A notable example of such an approach is LMPC, where decisions are made using MPC, a technique widely adopted in robotics and control applications [69].

2.3.4.1 Model Learning

Consider a model learning problem where the agent aims to find model parameters $\psi \in \mathcal{P}$ to minimize the following loss function:

$$\mathcal{L}(\psi) = \mathbb{E}\,[l_t(\psi)], \tag{2.29}$$

where $l_t(\psi)$ denotes the instantaneous loss incurred at time step t. For instance, when learning the transition dynamics of an MDP, a common choice of loss function is

$$l_t(\psi) = -\log \mathcal{M}_\psi(s_{t+1} \mid s_t, a_t), \tag{2.30}$$

which corresponds to minimizing the cross-entropy (CE), or equivalently, the Kullback–Leibler (KL) divergence between the true transition distribution and the model. In the context of POMDPs with fully observed physical states, i.e., $o_t = \tilde{s}_t$, an analogous loss function is given by

$$l_t(\psi) = -\log \mathcal{M}_\psi(\tilde{s}_{t+1} \mid b_t, a_t), \tag{2.31}$$

where b_t denotes the belief state at time t. For more general POMDPs, where only observations o_{t+1} are available, a forward-corrected loss function may be used [13]:

$$l_t(\psi) = -\log \int_{\tilde{s}_{t+1}\in\tilde{\mathcal{S}}} \mathcal{Z}(o_{t+1} \mid \tilde{s}_{t+1}, a_t)\mathcal{M}_\psi(\tilde{s}_{t+1} \mid b_t, a_t)\, d\tilde{s}_{t+1}. \tag{2.32}$$

Equation (2.32) is motivated by maximizing the likelihood of observations, a surrogate for learning the transition model. The forward correction method was initially proposed to address erroneous discrete labels in classification problems [98].

In online-to-batch learning, the agent observes a sequence of loss functions $l_1(\psi), \dots, l_T(\psi)$ and incrementally updates the parameters $\psi \in \mathcal{P}$ to minimize the objective in (2.29). The estimates of the parameters $\hat{\psi}_t \in \mathcal{P}$ can be updated at each step t as

$$\hat{\psi}_{t+1} = \mathcal{X}(\hat{\psi}_t, l_t), \tag{2.33}$$

where $\mathcal{X}$ denotes an online learning algorithm. Algorithms such as Online Gradient Descent (OGD) [99] can be employed:

$$\hat{\psi}_{t+1} = \mathrm{Proj}_{\mathcal{P}}\left(\hat{\psi}_t - \frac{D}{G\sqrt{t}}\nabla_\psi l_t(\psi)\mid_{\psi=\hat{\psi}_t}\right), \tag{2.34}$$

where D is an upper bound on the diameter of the parameter space $\mathcal{P}$, G bounds the norm of the gradient $\nabla_\psi l_t$, and $\mathrm{Proj}_{\mathcal{P}}(\cdot)$ denotes projection onto $\mathcal{P}$.

Another approach to learning the model parameters $\psi \in \mathcal{P}$ is to employ Bayes' rule. We can treat these unknown model parameters as state variables and form an augmented state space $\mathcal{S}' = \mathcal{S} \times \mathcal{P}$ when the belief update rule (2.14) could be used in this new state space [100]. While this approach is beneficial for identifying model parameter uncertainty, it can be challenging to implement in practice efficiently due to complicated relations between states, observations, and model parameters.

2.3.4.2 Myopic Control

While optimization of the constrained C-POMDP objective in (2.27) using approximate methods such as MPC, MCTS, and other online POMDP solvers may be tractable in specific cases, it remains computationally challenging for ISAC problems characterized by high-dimensional action and belief spaces and complex belief transition dynamics. To reduce the complexity, it is possible to consider a myopic (one-step) version of (2.27) as in [9, 10, 12, 13]:

$$\arg\max_{a_t\in\mathcal{A}} \quad u^{(0)}_{B,\psi}(b_t, a_t) \tag{2.35a}$$

$$\text{s.t.} \quad u^{(i)}_{B,\psi}(b_t, a_t) \geq 0 \quad \forall i \in [N_C]. \tag{2.35b}$$

Recall Assumptions 2.1 and 2.4, which together imply that the future evolution of the belief state is independent of the action selected at the current time step. Myopic policies can yield optimal performance in unconstrained problem formulations.

Corollary 2.1 *For $N_C = 0$ and under Assumptions 2.1 and 2.4, the myopic LMPC policy that optimizes (2.35), assuming the true model parameters $\psi^\star \in \mathcal{P}$, is optimal.*

Proof Starting from the Bellman optimality equation (2.19), and applying Assumptions 2.1 and 2.4, the optimal value function can be expressed as:

$$V^{(0)}_{\pi^\star,\psi^\star}(b_t) = \max_{a_t \in \mathcal{A}} \left(u^{(0)}_{B,\psi^\star}(b_t, a_t)\right) + \mathbb{E}\left[V^{(0)}_{\pi^\star,\psi^\star}(b_{t+1}) \mid b_t\right]. \tag{2.36}$$

Assumptions 2.1 and 2.4 ensure that b_{t+1} is independent of a_t. Consequently, maximizing the immediate utility $u^{(0)}_{B,\psi^\star}(b_t, a_t)$ is sufficient to achieve the optimal value.

The following result shows that the myopic policy satisfies the value constraints if the myopic problem (2.35) is always feasible.

Corollary 2.2 *The myopic policy that optimizes (2.35), assuming the true model parameters $\psi^\star \in \mathcal{P}$, satisfies the value constraints defined in (2.27b), provided the optimization problem (2.35) remains feasible for all $t \in \mathbb{N}$.*

Proof From the definition of the value function in (2.26), if the immediate constraint utility $u^{(i)}_{B,\psi^\star}(b_t, a_t)$ is nonnegative for all t, then $V_{B,\psi^\star}(b_t) \geq 0$ follows directly. Hence, the feasibility of the myopic problem guarantees that the value constraints are satisfied.

While Corollary 2.2 confirms that the value constraints are met, it is essential to note that the pointwise (myopic) constraints are generally more restrictive than the long-term value constraints. Therefore, even under Assumptions 2.1 and 2.4, the myopic controller may be suboptimal in constrained settings when $\gamma_i > 0$ for all $i \in [N_C]$. A more detailed analysis and performance bounds are provided in [12].

2.3.4.3 Pseudo-myopic Control

In general POMDP settings, both the state transition and observation processes may depend on the agent's actions, also making the belief state process action-dependent. Even when the state transitions are independent of actions—as specified by Assumption 2.1—the belief state process can still be influenced by actions if the observation model is action-dependent (i.e., Assumption 2.4 does not hold). In such cases, actions impact the quality and informativeness of observations, requiring the agent to consider how current decisions shape future beliefs. Greedy, myopic policies that focus solely on immediate utility may fail to gather critical information needed to improve long-term performance. Therefore, exploration is required to reduce uncertainty about the system state and unknown model parameters. Balancing this E&E trade-off in POMDPs is essential when maximizing immediate returns conflicts with the information acquisition process.

Solving the non-myopic planning problem in belief space to optimally balance E&E is computationally demanding. A practical alternative is to use *pseudo-myopic controllers*, which retain the computational simplicity of myopic control while incorporating mechanisms to promote exploration. This is achieved through regularization or reward augmentation, where the objective function is modified to include terms that reward information gain or penalize uncertainty. These E&E strategies have been extensively studied in RL, especially in the context of MABs, and have also been extended to general POMDPs with unknown parameters. Notable examples include entropy-regularized objectives [101] and methods based on optimism in the face of uncertainty and TS. Refer to Section 2.2.2.1 for more about the E&E algorithms in RL. Pseudo-myopic controllers for C-POMDPs can be expressed as:

$$\arg\max_{a_t \in \mathcal{A}} \quad \tilde{u}^{(0)}_{B,\psi}(b_t, a_t) \tag{2.37a}$$

$$\text{s.t.} \quad \tilde{u}^{(i)}_{B,\psi}(b_t, a_t) \geq 0 \quad \forall\, i \in [N_C], \tag{2.37b}$$

where the functions $\tilde{u}^{(i)}_{B,\psi}(b_t, a_t)$ for all $i = 0, \ldots, N_C$ serve as exploration-aware surrogates for the utilities $u^{(i)}_{B,\psi}(b_t, a_t)$.

Note: Although highly efficient E&E algorithms have been developed in settings such as MABs, they are generally not directly applicable to most ISAC problems. This limitation arises primarily from the high-dimensional and continuous action, state, and observation spaces and complicated state and belief transition dynamics encountered in ISAC resource allocation tasks.

2.4 Multicarrier ISAC Waveform Optimization

Multicarrier modulation, such as OFDM and its multiuser version OFDMA, is used in most current and emerging wireless communication and broadcast systems, including 4G, 5G, and 6G. Multicarrier waveforms are also used in radars. This section investigates frequency resource allocation for multicarrier ISAC systems operating in dynamic radio environments. Traditional batch-mode structured optimization approaches [73, 74, 79, 102–104] are effective in static or slowly varying operating environments but become inadequate in rapidly varying radio spectrum and target scenarios where accurate offline modeling is infeasible or of too high complexity. Moreover, these methods do not learn from their experiences. Online adaptation is essential in real-world, congested, and even contested environments with non-cooperative, cooperative, or adversarial systems.

This section explores online MBRL algorithms that learn directly from interactions with the environment and adapt policies in real time, proposed in [9–13]. These methods support intelligent and dynamic allocation of frequency and power resources, enhancing the robustness and performance of multicarrier ISAC systems in dynamic radio environments.

2.4.1 Multicarrier ISAC System Model

Consider a colocated and co-designed ISAC base station (BS) operating in a dynamic radio environment. The system model includes the ISAC system, a user equipment (UE) receiving data from the ISAC BS, and several external communications or sensing systems, which may be cooperative, non-cooperative, or even adversarial, as illustrated in Figure 2.5. A cooperative system shares awareness about the state of the spectrum with the ISAC system, such as channel state information and interference statistics. In contrast, a noncooperative system does not exchange any information. Moreover, an adversarial system may cause intentional interference. Unintentional interference between the co-designed sensing and communication functionalities may be significantly reduced through orthogonal design and SC or RB allocation to different users and tasks.

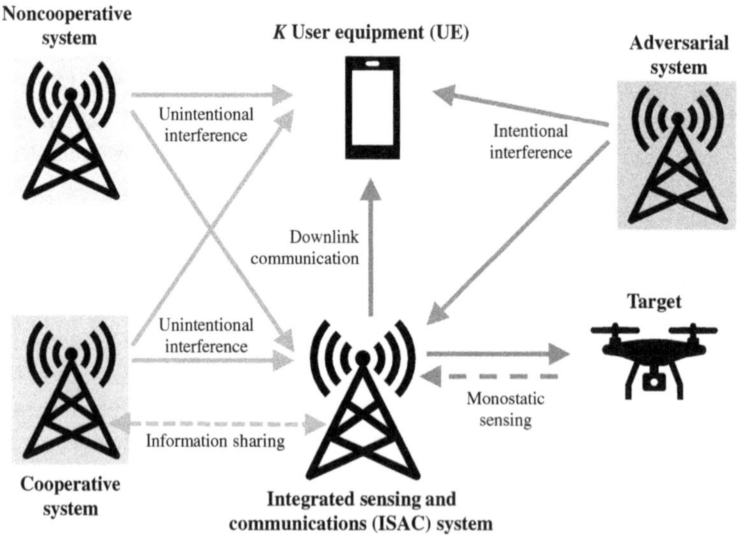

Figure 2.5 Model for ISAC operating in congested and contested radio spectrum. *Source*: [25] / with permission of Petteri Pulkkinen.

2.4.1.1 Transmitter and Receiver Signal Models

In one transmission frame corresponding to the timestep t of the MDP, the multicarrier system transmits a signal using N orthogonal SCs (e.g., OFDM signal) and M symbols. The SC spacing is f_Δ, and the symbol interval is T_o. For example, with OFDM $T_o = f_\Delta^{-1} + T_{cp}$, where T_{cp} is the cyclic prefix (CP) duration. On SC $n \in [N]$ and symbol $m \in [M]$, the TX signal can be written as

$$x_{n,m} = x_{n,m}^{(0)} + x_{n,m}^{(1)}, \tag{2.38}$$

where $x_{n,m}^{(0)}$ is the sensing waveform and $x_{n,m}^{(1)}$ is the communications signal. The power transmitted on SC n is defined as $\rho_n = \mathbb{E}\left[|x_{n,m}|^2\right]$. It is assumed to be constant over M symbols.

We consider a sensing receiver (RX) model for a single target that may extend in the delay domain. Also, we assume the target response in the frequency domain $\boldsymbol{\alpha} = [\alpha_1, \dots, \alpha_N]^{\mathrm{T}}$ (including RCS, path loss, and other losses) is approximately constant over the observation period of M OFDM symbols but is independent and identically distributed (IID) between the observation periods, corresponding to the well-known Swerling I model. Thus, the multicarrier sensing RX model on SC $n \in [N]$ and symbol $m \in [M]$ can be written as

$$y_{n,m}^{(0)} = \underbrace{\alpha_n e^{j2\pi(m f_d T_o - n f_\Delta \tau)}}_{:= h_{n,m}^{(0)}} x_{n,m} + \upsilon_{n,m}^{(0)}, \tag{2.39}$$

where τ is the two-way propagation delay (delay of the closest scatterer for extended targets) in the monostatic case, f_d is the Doppler frequency, and $\upsilon_{n,m}^{(0)}$ contains the noise and interference. A statistical model for the noise and interference term $\upsilon_{n,m}^{(0)}$ is provided in the sequel. The shorthand notation $h_{n,m}^{(0)} = \alpha_n e^{j2\pi(m f_d T_o - n f_\Delta \tau)}$ refers to the target channel, including the propagation delay and Doppler. Let $h_{n,m}^{(0)} \sim \mathcal{CN}(0, g_n^{(0)})$ where $g_n^{(0)}$ is the variance of the target channel $h_{n,m}^{(0)}$. The communications RX model can be written as

$$y_{n,m}^{(1)} = h_n^{(1)} x_{n,m} + \upsilon_{n,m}^{(1)}, \tag{2.40}$$

where $h_n^{(1)}$ is the communications channel, and $\upsilon_{n,m}^{(1)}$ comprises the noise and interference. Define shorthand notation $g_n^{(1)} := \left|h_n^{(1)}\right|^2$ for the channel gain.

2.4.1.2 Noise and Interference Statistics

The noise and interference are assumed to be quasi-stationary over time, i.e., wide-sense stationary (WSS) over a timespan of M multicarrier symbols. However, their statistics may change abruptly between observation periods to reflect nonstationary behavior. Under the WSS assumption, the samples $\upsilon_{n,m}^{(k)}$ and $\upsilon_{n',m'}^{(k)}$ are approximately uncorrelated for all $(n, m) \neq (n', m')$. This follows from the fact that the covariance matrix of a WSS process has a Toeplitz structure, which can be well-approximated by a circulant matrix when the dimensions are sufficiently large [105, 106]. The noise and interference are assumed to obey zero-mean circularly symmetric complex Gaussian distribution $v_{n,m}^{(k)} \sim \mathcal{CN}(0, \xi_n{}^{(k)})$, where $\xi_n{}^{(k)}$ denotes the total interference-plus-noise variance.

Although awareness on the state of the spectrum (e.g., channel state information or signal covariance matrices) is shared among cooperative systems, the actual TX signals may still cause unintentional interference. Hence, the total interference can be decomposed into contributions from cooperative, non-cooperative, and adversarial sources:

$$\xi_n{}^{(k)} = \underbrace{\xi_n^{\mathrm{co}\to k}}_{\text{cooperative systems}} + \overbrace{\xi_n^{\mathrm{nc}\to k}}^{\text{non-cooperative and adversarial systems}} + \underbrace{\sigma_k^2}_{\text{thermal noise}}. \tag{2.41}$$

2.4.2 C-POMDP Formulation

The ISAC resource allocation problem is formulated as a C-POMDP with the following components [10–13]:

- **Action $\boldsymbol{a}_t \in \boldsymbol{A}$**: The action consists of SC assignment and power allocation decisions. The binary vector $\boldsymbol{\omega}_t \in \{0,1\}^N := \mathcal{A}_\omega$ indicates the assignment to sensing and communications tasks, where $[\boldsymbol{\omega}_t]_n = \omega_n(t)$, and $\omega_n(t) = 1$ means that SC $n \in [N]$ is assigned to communications (and otherwise to sensing). The power allocation is defined by the vector $\boldsymbol{\rho}_t \in \mathbb{R}_+^N$, where $[\boldsymbol{\rho}_t]_n = \rho_n(t)$ denotes the TX power allocated to SC n. A total power constraint is imposed, $\mathbf{1}^{\mathrm{T}}\boldsymbol{\rho}_t \le P_{\mathrm{tot}}$, and a spectral mask constraint, $\boldsymbol{\rho}_t \le \boldsymbol{\rho}$, which ensures that interference caused to cooperative users remains at tolerated levels. These constraints define the feasible power allocation space $\boldsymbol{\rho}_t \in \mathcal{A}_p$, leading to a joint action space $\mathcal{A} = \mathcal{A}_p \times \mathcal{A}_\omega$. The overall action is then given by $a_t = \{\boldsymbol{\omega}_t, \boldsymbol{\rho}_t\}$.
- **Physical State $\tilde{\boldsymbol{s}}_t \in \widetilde{\boldsymbol{S}}$**: At observation period t, the physical state includes interference-plus-noise statistics and channel gains (including target response) for both sensing ($k = 0$) and communications ($k = 1$) functionalities. Let $\xi_n{}^{(k)}(t)$ and $g_n{}^{(k)}(t)$ denote the interference-plus-noise variance and channel gain for SC n, respectively. Let us define $\boldsymbol{\xi}_t{}^{(k)} \in \mathbb{R}_+^N$ and $\boldsymbol{g}_t{}^{(k)} \in \mathbb{R}_+^N$ as the vectors collecting these values across all SCs. The physical state is given by $\tilde{s}_t = \{\boldsymbol{g}_t{}^{(k)}, \boldsymbol{\xi}_t{}^{(k)}\}_{k\in\{0,1\}}$.
- **Observation $\boldsymbol{o}_t \in \boldsymbol{O}$**: An estimate of the interference-plus-noise power at SC n is found by

$$\hat{\xi}_n{}^{(k)}(t) = \frac{1}{N_{\mathrm{S}}} \sum_{i=1}^{N_{\mathrm{S}}} |\tilde{v}_{n,i}{}^{(k)}(t)|^2, \tag{2.42}$$

where N_s is the number of samples containing only interference and noise, and $\tilde{v}_{n,i}{}^{(k)}(t)$ denotes the ith sample at SC n collected during observation period t. Signal-free samples can be obtained either through dedicated silent transmissions or, for example, by extracting secondary data using guard and training cells as in constant false alarm rate (CFAR) processing [107]. The estimator $\hat{\xi}_n{}^{(k)}(t)$ obeys Gamma distribution with shape parameter N_s and rate $N_s/\xi_n{}^{(k)}(t)$. A separate belief tracker is used to estimate the channel gains independently of the C-POMDP resource allocation process. The tracker produces a posteriori estimates $\bar{\boldsymbol{g}}_t^{(k)}$, which approximate the true channel gains as $\boldsymbol{g}_t^{(k)} \approx \bar{\boldsymbol{g}}_t^{(k)}$. The obtained estimates $\bar{\boldsymbol{g}}_t^{(k)}$ are then used as observations in the resource allocation problem.

We consider the problem of maximizing the sensing MI, subject to MI (associated with data rate or capacity) constraints imposed for the communication function. The utility function for function $k \in \{0, 1\}$ is based on MI and is given by [12, 73]:

$$u_k(s_t, a_t, s_{t+1}) = \sum_{n=1}^{N} \mathbb{1}_{\{\omega_n(t)=k\}} \log\left(1 + \frac{g_n^{(k)}(t+1)}{\xi_n{}^{(k)}(t+1)} \rho_n(t)\right), \tag{2.43}$$

where $\mathbb{1}_{\{\cdot\}}$ is an indicator function. Equation (2.43) is only dependent on the physical state $\tilde{s}_{t+1}$ and action a_t.

2.4.3 MBRL Algorithm

This section introduces a MBRL algorithm proposed in [10–13] for solving the C-POMDP problem formulated in Section 2.4.2. The algorithm optimizes control actions based on the myopic controller introduced in Section 2.3.4.2. The SC channel quality is defined as:

$$\kappa_n^{(k)}(t) = \frac{g_n^{(k)}(t)}{\xi_n^{(k)}(t)}, \tag{2.44}$$

where $g_n^{(k)}(t)$ and $\xi_n^{(k)}(t)$ are the channel gain and interference-plus-noise power for SC n and function $k \in \{0, 1\}$ at time t, respectively. From the MI expression in (2.43) and under Assumption 2.2—which states that the physical state transition is independent of the most recent action—it is sufficient to learn the model

$$\mathcal{M}_{\psi,n}^{(k)}\left(\kappa_n^{(k)}(t+1) \mid b_t\right) := p(\kappa_n^{(k)}(t+1) \mid b_t), \tag{2.45}$$

for all SCs $n \in [N]$ and both sensing and communication functions $k \in \{0, 1\}$. Rather than learning the model at the SC domain, a number of adjacent SCs can be grouped into RBs, enabling learning in the RB domain to reduce dimensions. The RB size of 12 SCs is commonly used in wireless standards. Additionally, the model may be constructed separately for the channel gains $p(g_n^{(k)}(t+1) \mid b_t)$ and noise variances $p(\xi_n^{(k)}(t+1) \mid b_t)$ which can be advantageous when these components are not fully observable. This topic is discussed in more detail in [13].

2.4.3.1 Model Learning

The distributions $\mathcal{M}_{\psi,n}^{(k)}(\kappa_n^{(k)}(t+1)|b_t)$ are approximated with step approximations (histogram) such that the probability distribution is piecewise constant. In other words, a collection of disjoint sets $\{\mathcal{C}_i\}_{i=1}^{L_B}$ is defined where each set $\mathcal{C}_i \subset \mathbb{R}_+$ covers a specific range of channel qualities. To simplify the notation, assume that the collection $\{\mathcal{C}_i\}_{i=1}^{L_B}$ is the same for each $k \in \{0, 1\}$ and $n \in [N]$. Then, the channel-wise transition model is

$$\mathcal{M}_{\psi,n}^{(k)}\left(\kappa_n^{(k)}(t+1) \mid b_t\right) = \prod_{i=1}^{L_B}\left[\frac{\phi_{n,i}^{(k)}(b_t, \psi)}{\mathcal{V}_i}\right]^{\mathbb{1}_{\{\kappa_n^{(k)}(t+1)\in \mathcal{C}_i\}}}, \tag{2.46}$$

where $\mathcal{V}_i$ is the volume of the set $\mathcal{C}_i$, and $\phi_{n,i}^{(k)} : \mathcal{B} \times \mathcal{P} \to [0, 1]$ is the function with parameters $\psi \in \mathcal{P}$ to be learned for all $k \in \{0, 1\}$, $n \in [N]$, and $i \in [L_B]$. Intuitively, $\phi_{n,i}^{(k)}(\cdot, \cdot)$ is the probability that the state of the function k on channel n will fall into the histogram bin $\mathcal{C}_i$. The function $\phi_{n,i}^{(k)}(\cdot, \cdot)$ can be learned with any function approximation method using an output layer that satisfies the total probability constraint, e.g., a neural network with the SoftMax output layer. In this case, the model parameters are learned using the online learning scheme presented in Section 2.3.4.1 and the loss functions presented below.

CE Loss: Define the estimated channel quality as

$$\hat{\kappa}_n^{(k)}(t) = \frac{\bar{g}_n^{(k)}(t)}{\hat{\xi}_n^{(k)}(t)}, \tag{2.47}$$

where $\bar{\boldsymbol{g}}_t^{(k)}$ is a point estimate from the channel gain tracker, and $\hat{\xi}_n{}^{(k)}(t)$ is defined in (2.42). The CE loss in (2.31) can be directly used if the estimate $\hat{\kappa}_n{}^{(k)}(t) \approx \kappa_n{}^{(k)}(t)$ is accurate, for example, when N_s in (2.42) is large. Using the model defined above, the CE loss employed to learn the model can be written as:

$$\mathcal{L}(\psi) = -\sum_{k\in\{0,1\}}\sum_{n=1}^{N}\mathbb{E}\left[\sum_{i=1}^{L_B}\mathbb{1}_{\{\hat{\kappa}_n^{(k)}(t+1)\in \mathcal{C}_i\}}\log \phi_{n,i}^{(k)}(b_t, \psi)\right]. \tag{2.48}$$

Forward Correction Loss: The loss in (2.48) is not a feasible criterion when reliable estimates of $\hat{\kappa}_n{}^{(k)}(t)$ cannot be obtained, for example, when N_s in (2.42) is small. Fortunately, we can use the forward correction loss in equation (2.32) that maximizes the likelihood of observations, hence it is a surrogate for the model learning.[2] On SC $n \in [N]$ and function $k \in \{0, 1\}$, define $\mathcal{Z}_n{}^{(k)}\left(\hat{\kappa}_n{}^{(k)}(t) \mid \kappa_n{}^{(k)}(t)\right)$ to

2 A backward loss correction method was also proposed for model learning in [13], but it is not recommended due to unstable performance.

denote the likelihood of observation given the true channel quality. By employing the transition and observation models, equation (2.32) can be written as

$$l_t^{(k)}(\psi) = -\log \int_{\kappa \in \mathbb{R}_+} \mathcal{Z}_n^{(k)}\left(\hat{\kappa}_n^{(k)}(t+1) \mid q\right) \mathcal{M}_{\psi,n}^{(k)}(\kappa \mid b_t) \; d\kappa, \tag{2.49}$$

$$= -\log \sum_{i=1}^{L_B} \frac{1}{\mathcal{V}_i} \phi_{n,i}^{(k)}(b_t, \psi) \underbrace{\int_{\kappa \in C_i} \mathcal{Z}_n{}^{(k)}\left(\hat{\kappa}_n{}^{(k)}(t+1) \mid \kappa\right) d\kappa}_{\approx \mathcal{V}_i \mathcal{Z}_n{}^{(k)}\left(\hat{\kappa}_n{}^{(k)}(t+1) \mid \bar{\kappa}_i\right)} \tag{2.50}$$

$$\approx -\log \sum_{i=1}^{L_B} \phi_{n,i}^{(k)}(b_t, \psi) \mathcal{Z}_n^{(k)}\left(\hat{\kappa}_n^{(k)}(t+1) \mid \bar{\kappa}_i\right), \tag{2.51}$$

where (2.50) is obtained by the definition of the model (2.46), and the approximation (2.51) is obtained by assuming smoothness of the observation kernel with $\bar{\kappa}_i$ being the center point of the bin C_i. Thus, the forward corrected loss function can be written as

$$\mathcal{L}(\psi) = -\mathbb{E}\left[\sum_{n=1}^{N} \log \sum_{i=1}^{L_B} \mathcal{Z}_n^{(k)}\left(\hat{\kappa}_n^{(k)}(t+1) \mid \bar{\kappa}_i\right) \phi_{n,i}^{(k)}(b_t, \psi)\right], \tag{2.52}$$

where the predicted quality probabilities are weighted by the observation likelihoods.

2.4.3.2 Controller Algorithm

The controller optimizes the control actions for the myopic LMPC problem defined in (2.35). By using the model in (2.46), the expected utilities can be approximated by

$$u_{B,\psi}{}^{(k)}(b_t, a_t) \approx \sum_{n=1}^{N} \sum_{i=1}^{L_B} \mathbb{1}_{\{\omega_n(t)=k\}} \frac{\phi_{n,i}{}^{(k)}(b_t, \psi)}{2} \times [\log(1 + \check{\kappa}_i \, \rho_n(t)) + \log(1 + \hat{\kappa}_i \, \rho_n(t))], \tag{2.53}$$

where $\check{\kappa}_i$ and $\hat{\kappa}_i$ are the boundary values of the approximation bin C_i. The equation in (2.53) is a concave function of the power variables $\boldsymbol{\rho}_t$, which is advantageous for convex optimization techniques. However, the optimization problem in (2.35), combined with the utility functions defined in (2.53), results in a nonconvex mixed-integer problem. To address this, we first relax the problem by assuming all channels are used for communication and optimize the power allocation accordingly. Next, we determine the minimum number of SCs required to meet the communications demands. Subsequently, any excess power not needed for communications is reallocated for sensing by minimizing the power assigned to communications while meeting the communications MI constraint. Finally, the remaining power budget is optimized for sensing across the remaining SCs. The pseudo-code for the controller algorithm is detailed in Algorithm 2.1.

Algorithm 2.1: The controller algorithm.

Require: Model $\mathcal{M}_\psi$, belief state b_t

1: Solve $\boldsymbol{\rho}_{\text{comm}} \leftarrow \arg\max_{\boldsymbol{\rho}\in\mathcal{A}_p} u_\psi^{(1)}(b_t, \{\mathbf{1}, \boldsymbol{\rho}\})$
2: Solve $\boldsymbol{\omega}_t \leftarrow \arg\min_{\boldsymbol{\omega}\in\mathcal{A}_\omega} \mathbf{1}^\top\boldsymbol{\omega}$ s.t. $u_\psi^{(1)}(b_t, \{\boldsymbol{\omega}, \boldsymbol{\rho}_{\text{comm}}\}) \geq 0$
3: Solve $\boldsymbol{\rho}_{\text{comm}} \leftarrow \arg\min_{\boldsymbol{\rho}\in\mathcal{A}_p} \mathbf{1}^\top\boldsymbol{\rho}$ s.t. $u_\psi^{(1)}(b_t, \{\boldsymbol{\omega}_t, \boldsymbol{\rho}_{\text{comm}}\}) \geq 0$
4: Set $P_0 \leftarrow P_{\text{tot}} - \mathbf{1}^\top\boldsymbol{\rho}_{\text{comm}}$
5: Solve $\boldsymbol{\rho}_{\text{sensing}} \leftarrow \arg\max_{\boldsymbol{\rho}\in\mathcal{A}_p} u_\psi^{(0)}(b_t, \text{vec}(\boldsymbol{\omega}_t, \boldsymbol{\rho}))$ s.t. $\mathbf{1}^\top\boldsymbol{\rho} \leq P_0$
6: Set $\boldsymbol{\rho}_t = \boldsymbol{\rho}_{\text{comm}} + \boldsymbol{\rho}_{\text{sensing}}$
7: **return** $a_t = \{\boldsymbol{\omega}_t, \boldsymbol{\rho}_t\}$

2.4.4 Numerical Examples

In the numerical examples, spectrum dynamics are simulated using various models, as summarized in Table 2.1. For comprehensive simulation results and further analysis, see [9–13]. We evaluate the performance of the following methods:

- **MBRL (linear):** MBRL algorithm where the function $\phi_{n,i}(b_t, \psi)$ in (2.46) is linear before the SoftMax output layer.
- **MBRL (nonlinear):** MBRL algorithm where $\phi_{n,i}(b_t, \psi)$ is a nonlinear function implemented using a feedforward neural network (FNN).
- **Efficient MFRL:** A sample-efficient MFRL algorithm we proposed in [11], based on action-independent transition models. This approach has been developed only for the fully observable scenario.

Table 2.1 Summary of models used in simulations.

Model	Description	Reference
Markov chain	Uses discrete Markov chains to determine RX powers and which sub-channels are occupied	[9–13]
Frequency hopping	Interference frequency changes deterministically based on a finite history. A special case of the Markov model with an extended latent state space	[10–12]
Stationary	Channel qualities remain constant for the entire simulation period	[11, 12]
Matched	The employed randomly parameterized MBRL model is directly used for sampling state transitions	[11, 12]

- **Non-predictive:** A model-based optimization algorithm that does not use a learned model to predict future channel states.
- **Known Model:** A controller using the true model in the MBRL framework.
- **Uniform Allocation:** SCs are randomly assigned to sensing and communication tasks with equal probability, and power is allocated uniformly across SCs.

The memory used by learning-based methods is implemented using the recursive structure defined in (2.28). In Section 2.4.4.1, both MBRL and MFRL utilize a windowed memory of past observations. In contrast, the MBRL algorithms in Section 2.4.4.2 employ a recurrent memory structure based on a long short-term memory (LSTM) network.

2.4.4.1 Algorithm Comparison

We compare the performance of the proposed MBRL and efficient MFRL methods under various spectrum models, using both sensing and communication MIs. The evaluation considers four distinct simulation models: *matched*, *Markov chain*, *hopping*, and *stationary*. Figure 2.6a, b, respectively, shows the evolution of sensing and communications performance as a function of time. In this section, all MBRL methods use the CE model loss in (2.48) with $\hat{\kappa}_n^{(k)}(t) = \kappa_n^{(k)}(t)$.

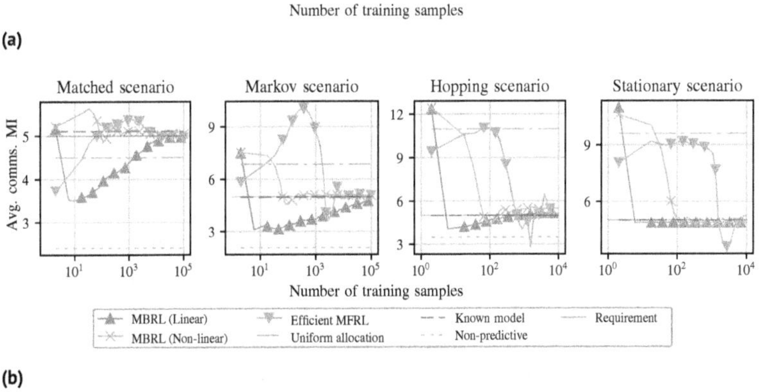

Figure 2.6 Performance comparison of MBRL and efficient MFRL with baseline methods. *Source*: [12] / IEEE / CC BY 4.0. (a) sensing performance and (b) communications performance with constraint $C = 5$.

Matched Model: The linear MBRL algorithm yields the best sensing performance in the matched model, approaching the known-model baseline. No significant difference exists in the sensing performance of the nonlinear MBRL and efficient MFRL, although they fall behind the linear MBRL. Regarding communications performance, the linear MBRL initially violates the constraint but recovers quickly, reaching near-feasible MI values after about 1,000 samples. In contrast, the nonlinear MBRL immediately satisfies the communication constraint. The efficient MFRL also converges quickly, within 50 samples, to feasible communications MI values.

Markov Chain Model: Under the Markov chain model, the performance trends are similar to those seen under the matched model. The linear MBRL improves sensing performance steadily, and its communication performance gradually approaches the constraint from below. The nonlinear MBRL again satisfies the communication requirement immediately and delivers sensing performance comparable to the linear model. The efficient MFRL initially prioritizes communications, resulting in lower sensing performance early in training. However, once the communications performance stabilizes after 1,000 samples, it reallocates resources and achieves excellent sensing results.

Hopping Model: Under the pseudo-deterministic hopping model, the convergence occurs generally faster due to a smaller variance in the state transitions. Both MBRL-based methods achieve good performance within 100 samples. However, the MFRL method takes over 1,000 samples to reach comparable levels. Additionally, its performance exhibits oscillations in both sensing and communications metrics.

Stationary Model: All methods converge fast in the stationary setting except MFRL. The linear MBRL achieves stability in just 10 learning steps, although its communications performance is slightly below the constraint due to approximation artifacts in the model. This gap could be closed by increasing the number of bins in the model. The nonlinear MBRL needs around 100 samples to converge and performs similarly to linear MBRL. In contrast, the efficient MFRL method converges approximately after 10,000 training samples. However, the performance of the MFRL after a large number of training samples is the best.

2.4.4.1.1 Discussion

Across all simulation models, the linear MBRL method consistently achieves strong sensing performance, often approaching the known-model baseline. Furthermore, the nonlinear MBRL variant better satisfies communication constraints while maintaining sensing performance close to the linear approach. Although the efficient MFRL method is substantially more sample-efficient than conventional MFRL algorithms, it still requires significantly more training samples than MBRL methods to achieve comparable performance. These observations highlight a fundamental trade-off: MBRL offers faster convergence and greater interpretability, whereas MFRL may yield improved performance in the asymptotic regime, provided that a large number of training samples are available.

2.4.4.2 Loss Correction for Model Learning

This section demonstrates the effectiveness of the forward correction loss introduced in Section 2.3.4.1, designed to improve model learning under partial observability. We evaluate three approaches:

- **Oracle:** Uses the model CE loss (2.48) but using $\kappa_n^{(k)}(t)$ instead of the estimate $\hat{\kappa}_n^{(k)}(t)$.
- **Forward Correction:** Applies the forward correction loss in (2.32) to account for observation noise.
- **Noisy:** Uses the CE loss in (2.48) using noisy estimates $\hat{\kappa}_n^{(k)}(t)$.

Figure 2.7 presents the sensing and communications MI for each method mentioned above. The forward correction method achieves performance close to the oracle despite relying only on noisy observations. In contrast, the noisy method shows that performance can deteriorate significantly without accounting for observation noise. These results highlight the benefit of applying forward correction when the system operates under partial observability. The learning algorithm can recover more accurate transition dynamics by accounting for uncertainty in the state estimates, leading to improved resource allocation performance.

Note: These experiments are conducted with $N_s = 1$ in (2.42). By increasing the number of samples $N_s \gg 1$, the performance gap between the forward correction and noisy method is expected to decrease.

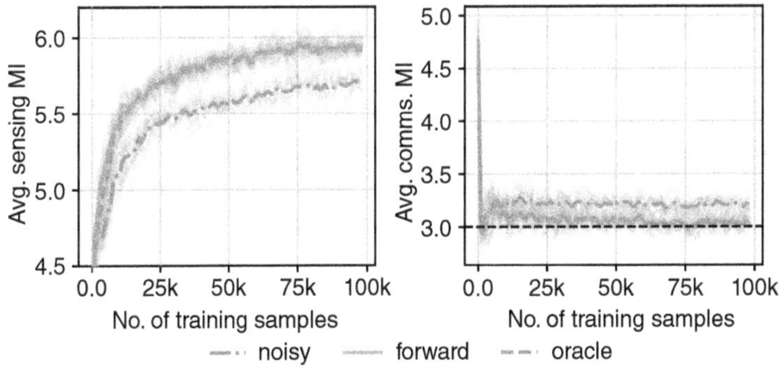

Figure 2.7 Comparison of MBRL performance with and without the forward correction method under a communications MI requirement of $C = 3$. *Source:* [13] / with permission of IEEE. Blurry lines indicate individual simulation runs, while solid lines represent the average among the runs.

2.5 Joint Beamspace and Frequency Resource Allocation

This section addresses spatial and frequency resource allocation in multicarrier MIMO ISAC systems. Spatial processing allows for more efficient use of the spectrum, management of interference, and focus of energy on desired users or targets. Specifically, we focus on learning in dynamic target environments, where the number of targets and their parameters are initially unknown. To adapt to such uncertainty, this chapter introduces online MBRL algorithms proposed in [14, 15] that enhance system performance by actively allocating beams in the beamspace domain and frequency resources in the SC or RB domain. Ideally, the frequency and spatial domain resources are orthogonal. Unlike the setting in Section 2.4, this section does not consider external interference from coexisting RF systems. However, the main challenge in the considered setting arises from the need to control the process of target observations and beliefs, which introduces an inherent E&E tradeoff.

2.5.1 Multicarrier MIMO ISAC System Model

The sensing sub-system is based on the monostatic OFDM model where the ISAC BS can transmit and receive signals simultaneously [108]. The ISAC BS is equipped with transmit and receive antenna arrays with L_{T} and L_{R} antenna elements, respectively. Steering vectors of the receive and transmit arrays in azimuth direction are denoted as $\boldsymbol{u}_{\mathrm{R}}(\theta)$ and $\boldsymbol{u}_{\mathrm{T}}(\theta)$, respectively.[3] The ISAC operates in the presence of Q radar targets with initially unknown locations. Simultaneously, the ISAC system serves K single antenna wireless UEs.

2.5.1.1 Transmit Signal Model

Denote the sub-carrier spacing of the OFDM waveform by f_Δ and the symbol interval by $T_{\mathrm{o}} = T_{\mathrm{cp}} + \frac{1}{f_\Delta}$, where T_{cp} is the CP duration. Each OFDM frame comprises N SCs and M OFDM symbols. In the frequency domain, the ISAC transmit signal can be written as

$$\boldsymbol{x}_{n,m} = \sum_{k=0}^{K} \boldsymbol{x}_{n,m}^{(k)}, \tag{2.54}$$

where $n = 1, \dots, N$ and $m = 1, \dots, M$ refer to the sub-carrier and the OFDM symbol indices, respectively. In addition, $\boldsymbol{x}_{n,m}^{(k)} \in \mathbb{C}^{L_{\mathrm{T}} \times 1}$ for $k \in [K]$ are the DL signals transmitted to UEs and $k = 0$ refers to the signal component dedicated to the sensing tasks. Let us denote $\boldsymbol{X} \in \mathbb{C}^{N \times M \times L_{\mathrm{T}}}$ as the tensor comprising all signal vectors of the form (2.54). An arbitrary codebook matrix is defined as $\Xi \in \mathbb{C}^{L_{\mathrm{T}} \times L}$, where the

3 Without loss of generality, the elevation angle is left out to simplify the notation.

columns are associated with L beams with $\|[\Xi]_{:,\ell}\|_2^2 = 1$. For example, for uniform linear array (ULA), Ξ can be the unitary discrete Fourier transform (DFT) matrix of size $L_\mathrm{T} \times L_\mathrm{T}$ such that we have $L = L_\mathrm{T}$ orthogonal beams. Using the beamspace basis, the signal components for $k = 0, \dots, K$ are defined as

$$\boldsymbol{x}_{n,m}^{(k)} = \Xi \boldsymbol{\Lambda}_n^{(k)} \boldsymbol{c}_{n,m}^{(k)}, \tag{2.55}$$

where ${\boldsymbol{\Lambda}_n}^{(k)} \in \mathbb{C}^{L\times L}$ is a diagonal matrix, and the vector $\boldsymbol{c}_{n,m}^{(k)} \in \mathbb{C}^{L\times 1}$ denotes the data symbols for L spatially multiplexed data streams. The covariance matrix of vector $\boldsymbol{c}_{n,m}^{(k)}$ is an identity matrix $\boldsymbol{I}_L$ of size L. The symbols are assumed to be independent among the indices n and m. The power of signal $\boldsymbol{x}_{n,m}^{(k)}$ is $\mathrm{trace}\left\{{\boldsymbol{\Lambda}_n}^{(k)}({\boldsymbol{\Lambda}_n}^{(k)})^\mathrm{H}\right\}$, where $(\cdot)^\mathrm{H}$ is the conjugate transpose operator. The power allocation is referred to as ${\rho_{n,\ell}}^{(k)} = [{\boldsymbol{\Lambda}_n}^{(k)}({\boldsymbol{\Lambda}_n}^{(k)})^\mathrm{H}]_{\ell,\ell}$, comprised in tensor $\boldsymbol{P} \in \mathbb{R}^{(K+1)\times N\times L}$ with total power:

$$P_\mathrm{tot} = \sum_{k=1}^{K}\sum_{n=1}^{N}\sum_{\ell=1}^{L} \rho_{n,\ell}^{(k)}, \tag{2.56}$$

where P_tot acts typically as a constraint for the designed waveforms.

In the case of single antenna UEs, only one spatial stream is transmitted to each user such that matrix $\boldsymbol{W}_n^{(k)} = \Xi\boldsymbol{\Lambda}_n^{(k)}$ for all $k \geq 1$ has at most one nonzero column.[4] However, we can transmit as many waveforms as needed for sensing-only purposes such that there are no constraints on the number of nonzero columns for the matrix $\boldsymbol{W}_n^{(0)} = \Xi\boldsymbol{\Lambda}_n^{(0)}$. The transmit signal model in (2.55) is useful for all k (even in the single stream) cases when optimizing the performance of the ISAC system.

2.5.1.2 Receive Signal Model

We consider a radar signal model for Q point targets with target scattering coefficient $\alpha_q \sim \mathcal{CN}(0, g_q^{(0)})$. The frequency domain received signal reflected off the targets can be written as

$$\boldsymbol{y}_{n,m}^{(0)} = \sum_{q=1}^{Q} \overbrace{\alpha_q e^{j2\pi(mf_{d,q}T_\mathrm{O} - nf_\Delta \tau_q)} \boldsymbol{u}_\mathrm{R}(\theta_q)\boldsymbol{u}_\mathrm{T}^\mathrm{H}(\theta_q)\boldsymbol{x}_{n,m}}^{\text{backscattered signal from target } q} + \boldsymbol{v}_{n,m}^{(0)}, \tag{2.57}$$

where τ_q is the two-way propagation delay (delay associated with the closest scatterer for extended targets), $f_{d,q}$ is the Doppler frequency, θ_q is the azimuth angle. In addition, $\boldsymbol{v}_{n,m}^{(0)} \sim \mathcal{CN}(\boldsymbol{0}, \sigma_0^2\boldsymbol{I})$ represents the RX noise. A block fading model is assumed for the communications such that the channel is quasistationary, i.e.,

4 This constraint comes from the method used to encode and decode the communications data.

constant over a single OFDM symbol. Therefore, the received communications signal at UE $k \in [K]$ can be written as

$$y_{n,m}^{(k)} = \left(\boldsymbol{h}_n^{(k)}\right)^{\mathrm{H}} \boldsymbol{x}_{n,m}^{(k)} + \overbrace{\sum_{k' \neq k} \left(\boldsymbol{h}_n^{(k)}\right)^{\mathrm{H}} \boldsymbol{x}_{n,m}^{(k')}}^{\text{inter-user interference}} + v_{n,m}^{(k)}, \tag{2.58}$$

where $\boldsymbol{h}_n^{(k)} \in \mathbb{C}^{L_T \times 1}$ is the communications channel to user k on sub-carrier n, and $\boldsymbol{v}_{n,m}^{(k)} \sim \mathcal{CN}(0, \sigma_k^2 \boldsymbol{I})$ is the RX noise.

2.5.1.3 GLRT-based Target Detection

The sensing subsystem performs a detection task using a generalized likelihood ratio test (GLRT) detector with nuisance parameters being the target scattering coefficient α, two-way propagation delay τ, Doppler shift f_d, and azimuth angle θ. Specifically, we use the maximum likelihood method to estimate α in a closed-form from (2.57) and plug it into the likelihood ratio test. This test statistic can be evaluated in a 3D grid of points for τ, f_d, and θ. Thus, the test statistic is

$$\eta(\tau, f_d, \theta) = \frac{1}{\sigma^2 C(\theta; \boldsymbol{X})} |S(\tau, f_d, \theta; \boldsymbol{X})|^2, \tag{2.59}$$

where

$$S(\tau, f_d, \theta; \boldsymbol{X}) = \sum_{n=1}^{N} \sum_{m=1}^{M} e^{j2\pi(m f_d T_\mathrm{O} - n f_\Delta \tau)} \boldsymbol{y}_{n,m}^{\mathrm{H}} \boldsymbol{u}_\mathrm{R}(\theta) \boldsymbol{u}_\mathrm{T}^{\mathrm{H}}(\theta) \boldsymbol{x}_{n,m} \tag{2.60}$$

can be interpreted as the matched filter, and

$$C(\theta; \boldsymbol{X}) = \sum_{n=1}^{N} \sum_{n=1}^{M} \left| \boldsymbol{u}_\mathrm{T}^{\mathrm{H}}(\theta) \boldsymbol{x}_{n,m} \right|^2 \tag{2.61}$$

is the beamshape of the transmit waveform. Due to the structure of the test statistic, it can be computed very efficiently using the fast Fourier transform [6].

For specific target parameters of τ, f_d and θ and probability of false alarm (PFA) P_{fa}, the detector makes a decision *target present* by

$$z(\tau, f_d, \theta) = \mathbb{1}_{\{\eta(\tau, f_d, \theta) \geq -\ln P_{\mathrm{fa}}\}}, \tag{2.62}$$

where $z(\tau, f_d, \theta) = 1$ refers to a decision that the target with the associated parameters is present. The PD of the GLRT detector is [15]

$$P_\mathrm{d}(\theta, \sigma_\alpha^2; \boldsymbol{X}) = P_{\mathrm{fa}}^{\left(C(\theta; \boldsymbol{X}) \frac{\sigma_\alpha^2}{\sigma^2} + 1\right)^{-1}}, \tag{2.63}$$

where σ_α^2 is the second order statistic of the scattering coefficient α. The PD conditional to the power allocation (i.e., marginal of [2.63]) can be approximated by

$$\hat{P}_{\rm d}(\theta, \sigma_\alpha^2; \boldsymbol{P}) = P_{\rm fa}^{\left(\mathbb{E}[C(\theta;\boldsymbol{X})|\boldsymbol{P}]\frac{\sigma_\alpha^2}{\sigma^2}+1\right)^{-1}}. \tag{2.64}$$

In (2.64), the expected beamshape can be written in a closed form

$$\mathbb{E}[C(\theta; \boldsymbol{X})] = M \sum_{\ell=1}^{L} \left|[\Xi^{\rm H} \boldsymbol{u}_{\rm T}(\theta)]_\ell\right|^2 \sum_{k=0}^{K} \sum_{n=1}^{N} \rho_{n,\ell}^{(k)}. \tag{2.65}$$

2.5.2 C-POMDP Formulation

The problem is formulated as C-POMDP as follows [15].

- **Action $\boldsymbol{a}_t \in \mathcal{A}$:** The actions are the powers $\rho_{n,\ell}^{(k)}$ for all SCs $n \in [N]$, beams $\ell \in [L]$, and function $k = 0, \dots, K$, represented as a tensor $\boldsymbol{P}$. The action space $\mathcal{A}$ takes into account constraints that ensure $\rho_{n,\ell}^{(k)} \geq 0$ and $\mathbf{1}^{\rm T}\text{vec}(\boldsymbol{P}) \leq P_{\rm tot}$. Also, we can have a constraint on the number of beams per UE written as $\sum_{\ell=1}^{L} \mathbb{1}_{\{\rho_{n,\ell}^{(k)}>0\}} \leq 1$ for all $k \in [K]$. Furthermore, we impose the unit SPR constraint:

$$\sum_{k=1}^{K} \rho_{n,\ell}^{(k)} \leq \frac{1}{N} \sum_{i=1}^{N} \sum_{k=1}^{K} \rho_{i,\ell}^{(k)}$$

  to manage the ambiguity function (AF) shape, particularly the range sidelobes and range resolution of the sensing waveform [74].[5]
- **State:** The state $s_t \in \mathcal{S}$ comprises the kinematic parameters of the targets, the target RCS responses, and communications channel state information. Also, the target RCS scattering second-order statistics $g_q^{(0)}(t)$ for all $q \in [Q]$ could be considered a part of the state. Notably, the state contains information on target angle $\theta_q(t)$, Doppler $f_{d,q}(t)$, and propagation delay $\tau_q(t)$ of all targets $q \in [Q]$.
- **Observations:** The observations of the sensing task utilize GLRT test statistics and detections found by using Equations (2.59) and (2.62). We do not assume an explicit form of the observations for communications functionality. However, we assume that there is a belief tracking algorithm (e.g., KF) that we can use to track and predict the communications channels.

5 Unit SPR constraint can be replaced by non-unit SPR constraint or other constraints to give more DoFs for frequency domain allocation.

The sensing task aims at maximizing the cumulative sum of PDs of the targets while simultaneously ensuring that UEs are served with acceptable data rates. As sum of PDs favors targets close by, other performance measures are also possible. For example, it is possible to generate tracks for the found targets and remove them from the search operation, which also enables detecting far-away targets as in [14]. For simplicity, only maximizing the cumulative sum of PDs is considered in this chapter. Therefore, the sensing utility function can be written as:

$$u_0(s_t, a_t, s_{t+1}) = \sum_{q=1}^{Q} \widehat{P}_{\mathrm{d}}(\theta_q(t+1), g_q^{(0)}(t+1); \boldsymbol{P}_t). \tag{2.66}$$

Let us define $g_{n,\ell}^{(k)} := [|\Xi^{\mathrm{H}} \boldsymbol{h}_n^{(k)}|^2]_\ell$ as the channel gain of user k on sub-carrier n, beam ℓ, represented by the tensor $\boldsymbol{G} \in \mathbb{R}^{K \times N \times L}$. The communications rate is defined as

$$R_k(\boldsymbol{G}, \boldsymbol{P}) = \frac{1}{T_\mathrm{o}} \sum_{n=1}^{N} \log_2\left(1 + \frac{\sum_{\ell \in [L]} g_{n,\ell}^{(k)} p_{n,\ell}^{(k)}}{\sum_{\substack{k' \neq k \\ \ell \in [L]}} g_{n,\ell}^{(k)} p_{n,\ell}^{(k')} + \sigma_k^2}\right), \tag{2.67}$$

where T_o is the OFDM symbol duration. Consequently, we can define the communications utility as

$$u_k(s_t, a_t, s_{t+1}) = R_k(\boldsymbol{G}_{t+1}, \boldsymbol{P}_t) - C_k, \tag{2.68}$$

where C_k is the constraint for the minimum tolerated rate for UEs.

2.5.3 MBRL Algorithm

Let the unknown parameters in the environment at time t be denoted by:

$$\Psi_t = \left\{\left(\Theta_q(t), g_q^{(0)}(t)\right)\right\}_{q \in [Q]},$$

where the uppercase notation emphasizes that $\Psi_t \in \mathcal{P}$ is treated as a random variable. We assume that these unknown parameters are functions of the underlying system state $S_t \in \mathcal{S}$. Since the target observations are action-dependent, the MBRL algorithm employed here utilizes the pseudo-myopic control scheme introduced in Section 2.3.4.3 to manage the E&E trade-off. The problem can be viewed as a variant of a stochastic MAB problem, where the true model parameters $\psi_t^\star \in \mathcal{P}$ are static or evolve slowly over time. However, the action space in ISAC is continuous, and the relationships between observations and rewards are considerably more complicated than in standard MAB formulations. As a result, conventional MAB algorithms are not directly applicable.

Instead, we adopt TS, a sampling-based E&E strategy that has proven to be effective in many RL domains [109].

The TS approach supports exploration by randomly sampling plausible environment parameters before solving the decision-making problem. The pseudo-myopic MBRL controller proceeds as follows [15]:

1. Compute the current belief distribution b_t over the unknown parameters, and draw a sample $\tilde{\psi}_{t+1|t} \sim \mathbb{P}\{\Psi_{t+1} \mid B_t = b_t\}$.
2. Solve the downstream optimization task in (2.37) using the sampled parameter realization $\tilde{\psi}_{t+1|t}$.

The surrogate sensing utility for the sampled parameters can be expressed as:

$$\tilde{u}_B^{(0)}(b_t, a_t) = \sum_{q=1}^{\tilde{Q}_{t+1}} \hat{P}_{\mathrm{d}}\left(\tilde{\theta}_q(t+1 \mid t), \tilde{g}_q^{(0)}(t+1 \mid t); \boldsymbol{P}_t\right), \tag{2.69}$$

where $\tilde{\theta}_q(t+1|t)$ and $\tilde{g}_q^{(0)}(t+1|t)$ correspond to the elements of $\tilde{\psi}_{t+1|t}$. Since the communications channel observations are assumed to be uncontrollable and the corresponding estimates reasonably accurate, no explicit exploration of the communications parameters is performed. Therefore, the surrogate communications utility is defined using the a priori point estimate $\bar{\boldsymbol{G}}_{t+1|t}$ of the channel gains:

$$\tilde{u}_B^{(k)}(b_t, a_t) = R_k(\bar{\boldsymbol{G}}_{t+1|t}, \boldsymbol{P}_t) - C_k, \tag{2.70}$$

which approximates the expected utility $u_B^{(k)}(b_t, a_t)$ for the communications task.

Given the above definitions, the controller $a_t = \pi(b_t)$ solves the following constrained optimization problem:

$$\underset{a_t \in \mathcal{A}}{\arg\max} \quad \sum_{q=1}^{\tilde{Q}_{t+1}} \hat{P}_{\mathrm{d}}\left(\tilde{\theta}_q(t+1|t), \tilde{g}_q^{(0)}(t+1|t); \boldsymbol{P}_t\right) \tag{2.71a}$$

$$\text{s.t.} \quad R_k(\bar{\boldsymbol{G}}_{t+1|t}, \boldsymbol{P}_t) \geq C_k. \tag{2.71b}$$

In the following, we present (1) a posterior tracking algorithm for updating the belief state over Ψ_t, and (2) an efficient algorithm for solving the optimization problem in (2.71).

2.5.3.1 Bayesian Learning of Unknown Parameters

This section addresses the belief state update from the perspective of estimating the parameter Φ_t based on a sequence of observations. Directly estimating the posterior distribution over the random set Φ_t is challenging due to its variable cardinality. To mitigate this, the estimation problem is reformulated using a discretized representation of the target kinematic space, encoded in a three-dimensional range-Doppler-azimuth (RDA) cube. Let g_i and η_i denote the gain associated with a hypothetical target and the GLRT test statistic for grid cell $i \in \mathcal{J}$, respectively. A zero value, $g_i = 0$, indicates no target in the ith cell. Accordingly, the belief update task reduces to estimating the gain posterior over a sparse representation of the RDA cube.

The GLRT statistics between RDA cells can ideally be assumed to be independent, allowing the posterior update to be parallelized among the cells [15]. We partition the RDA cells into two disjoint sets: the set of active cells $\mathcal{J}_\mathrm{A} \subseteq \mathcal{J}$, and the set of inactive cells $\mathcal{J} \setminus \mathcal{J}_\mathrm{A}$. A cell becomes active if a successful detection occurs within it.[6] Cells where no target present is decided remain inactive. For active cells $i \in \mathcal{J}_\mathrm{A}$ with a prior $p(g_i)$, Bayes' rule is used to update the posterior:

$$p(g_i \mid \eta_i) \propto p(\eta_i \mid g_i) p(g_i). \tag{2.72}$$

In (2.72), the observation likelihood can be written as:

$$p(\eta_i \mid g_i) = \left(C(\theta_{i_\theta}; \boldsymbol{X}) \frac{g_i}{\sigma_0^2} + 1 \right)^{-1} \exp\left(- \frac{\eta_i}{C(\theta_{i_\theta}; \boldsymbol{X}) \frac{g_i}{\sigma_0^2} + 1} \right), \tag{2.73}$$

where θ_{i_θ} is the azimuth angle of RDA cell $i \in \mathcal{J}$. The posterior of inactive cells $i \in \mathcal{J} \setminus \mathcal{J}_\mathrm{A}$ are updated efficiently as:

$$p(g_i \mid z_i = 0) \propto \left(1 - P_\mathrm{d}(\theta_{i_\theta}, g_i; \boldsymbol{X})\right) p(g_i), \tag{2.74}$$

where $z_i \in \{0, 1\}$ is the detection result in cell i. The computational benefits of this approach are realized when prior $p(g_i) = p(g_{i'})$ for all i and i' where $i_\theta = i'_\theta$ such that only a single posterior update per azimuth cell is required to cover all inactive range and Doppler cells. Furthermore, by discretizing the set of possible values of g_i, posterior updates in (2.72) and (2.74) can be done efficiently.

There are two approaches for specifying the prior distribution $p(g_i)$ of a newly activated cell $i \in \mathcal{J}_\mathrm{A}$. If the distance $d(i, i')$ (e.g., weighted Euclidean distance) between the newly activated cell i and an existing active cell $i' \in \mathcal{J}_\mathrm{A}$ is below a predefined threshold ϵ, the detection is associated with cell i', and the posterior distribution is transferred to cell i. This facilitates tracking hypothetical targets moving between cells in the RDA cube. Otherwise, the posterior $p(g_i)$ of any inactive cell $i \in \mathcal{J} \setminus \mathcal{J}_\mathrm{A}$ where $i_\theta = i'_\theta$ can be used as the prior. While activating too many cells can be computationally problematic, it can be avoided by monitoring the posterior distributions of active cells and deactivating those with low expected gain values.

6 Raw detections are pre-processed to suppress detections caused by a single target activating multiple adjacent RDA cells. For example, via using peak detection.

2.5.3.2 Controller Design

Consider the learned model in Section 2.5.3.1 used to sample $\tilde{g}_i(t+1 \mid t) \sim \mathbb{P}\{G_i(t+1) \mid B_t = b_t\}$ for all $i \in \mathcal{J}$ where $G_i(t)$ is a random realization of $g_i(t)$. Thus, we can rewrite the pseudo-myopic controller objective in (2.71) as:

$$\underset{\boldsymbol{P}}{\arg\max} \quad \sum_{i \in \mathcal{J}} \widehat{P}_{\mathrm{d}}(\theta_{i_\theta}, \tilde{g}_i(t+1 \mid t); \boldsymbol{P}) \tag{2.75a}$$

$$\text{s.t.} \quad \sum_{k=0}^{K} \sum_{n=1}^{N} \sum_{\ell=1}^{L} \rho_{n,\ell}^{(k)} \le P_{\mathrm{tot}} \tag{2.75b}$$

$$\rho_{n,\ell}^{(k)} \ge 0, \forall k = 0, \dots, K, n \in [N], \ell \in [L] \tag{2.75c}$$

$$\sum_{k=1}^{K} \rho_{n,\ell}^{(k)} = \frac{1}{N} \sum_{n'=1}^{N} \sum_{k=1}^{K} \rho_{n',\ell}^{(k)}, \ \forall\, n \in [N], \ell \in [L] \tag{2.75d}$$

$$\sum_{\ell=1}^{L} \mathbb{1}_{\{\rho_{n,\ell}^{(k)} > 0\}} \le 1, \ \forall\, k \in [K], n \in [N] \tag{2.75e}$$

$$R_k(\bar{\boldsymbol{G}}_{t+1|t}; \boldsymbol{P}) \ge C_k, \forall k \in [K], \tag{2.75f}$$

where (2.75b) is the total power constraint, (2.75c) ensures positivity, (2.75d) is unit SPR constraint (ratio between maximum power of a certain sub-carrier and average sub-carrier power) to control range ambiguities and promote sufficient range resolution, (2.75e) ensures at most a single data stream per user per sub-channel, and (2.75f) ensures the required data rate for each user.

To satisfy the constraints (2.75b)-(2.75d), we use the following transformation

$$\rho_{n,\ell}^{(k)} = \frac{1}{N} [\zeta(\boldsymbol{\Omega})]_{n,\ell}^{(k)} [\zeta(\boldsymbol{\omega})]_{\ell} P_{\mathrm{tot}}, \tag{2.76}$$

where $\boldsymbol{\Omega} \in \mathbb{R}^{K \times N \times L}$ and $\boldsymbol{\omega} \in \mathbb{R}^{L \times 1}$ are the parameters to be optimized, and $\zeta(\cdot)$ is a softmax function that operates on the first dimension of the input.[7] Let us define a shorthand notation for the conversion from the transformed variables $\boldsymbol{\omega}$ and $\boldsymbol{\Omega}$ to the power allocation as $\boldsymbol{P} = f(\boldsymbol{\omega}, \boldsymbol{\Omega})$. Additionally, we propose a surrogate rate equation

$$\widetilde{R}_k(\boldsymbol{G}; \boldsymbol{P}) = \frac{1}{T_{\mathrm{o}}} \sum_{n=1}^{N} \max_{\ell \in [L]} \log_2 \left(1 + \frac{g_{n,\ell}^{(k)} p_{n,\ell}^{(k)}}{\sum_{\ell' \ne \ell} g_{n,\ell'}^{(k)} p_{n,\ell'}^{(k)} + \sum_{k' \ne k} \sum_{\ell''=1}^{L} g_{n,\ell''}^{(k)} p_{n,\ell''}^{(k')} + \sigma_k^2}\right), \tag{2.77}$$

7 For example, $[\zeta(\boldsymbol{\Omega})]_{n,\ell}{}^{(k)} = \exp([\boldsymbol{\Omega}]_{n,\ell}{}^{(k)}) / \sum_{k'=1}^{K} \exp([\boldsymbol{\Omega}]_{n,\ell}^{(k')})$.

where the maximum operation takes the rate associated with the best beam. Hence, the power on the other beams may cause interference. The constraint (2.75e) is eliminated by replacing the rate equation in (2.75f) with (2.77). Using this reformulation, we can rewrite the optimization problem as

$$\underset{\omega,\Omega}{\arg\max} \quad \sum_{i\in\mathcal{J}} \hat{P}_{\mathrm{d}}(\theta_{i_\theta}, \tilde{g}_i(t+1\mid t); f(\omega,\Omega)) \tag{2.78a}$$

$$\text{s.t.} \quad \tilde{R}_k(\overline{G}_{t+1|t}; f(\omega,\Omega)) \geq C_k, \forall\, k \in [K]. \tag{2.78b}$$

Finally, after solving the problem, we can allocate all the excess power on each frequency and beam resource (in other beams than the best one) to the auxiliary sensing signal $k = 0$.

It is necessary to solve this problem efficiently using, for example, first-order optimization algorithms. We propose solving (2.75) as an unconstrained minimization problem where the loss function is written as

$$\begin{aligned} \mathcal{L}(\omega, \Omega; g, G) &= -\sum_{i\in\mathcal{J}} \hat{P}_{\mathrm{d}}(\theta_{i_\theta}, g_i; f(\omega,\Omega)) + \\ &\quad \sum_{k=1}^{K} \lambda_k \,\mathrm{ReLU}\big[C_k - \tilde{R}_k(G; f(\omega,\Omega))\big], \end{aligned} \tag{2.79}$$

where $\mathrm{ReLU}(x) = \max(0, x)$ is the ReLU function, the latter part of the right-hand side equation corresponds to the penalty of not satisfying the rate constraints, and $\lambda_k > 0$ for $k = 1, \dots, K$ is the weight for each constraint. The dimension of the sum over $i \in \mathcal{J}$ in (2.79) is not a problem since typically $g_i = 0$ for many i. Thus, the resulting unconstrained optimization problem can be solved efficiently using a first-order optimization algorithm such as AdaGrad [110]. In addition, computation time can be decreased by optimizing power allocation on RBs instead of SCs, where each RB combines multiple adjacent SCs. Typically, RBs of 12 SCs are used in wireless systems. All the proposed methods apply to RB-based power allocation by changing the dimensions and replacing the channels with average channels.

2.5.4 Numerical Examples

The performance of the proposed resource allocation algorithm is evaluated through simulations over a sequence of time steps. For more extensive simulation results and analysis, see [11, 14]. The performance indicators we compare at each time instance are (1) sum of PDs, and (2) violation of the rate constraint. We compare the MBRL method to three algorithms:

- **Sensing-only:** Optimizes (2.79) by setting weights $\lambda_k = 0$ for $k = 1, \dots, K$ to not consider the communications constraints.
- **Communications-only:** Excludes the sum of PDs of the loss in (2.79) only to find a feasible power allocation to satisfy communications constraints.

- **Random:** Algorithm randomly dividing the power into different resources. Especially, we sample $\boldsymbol{\omega}$ and $\boldsymbol{\Omega}$ from i.i.d. unit normal distribution, and use the transformation $\boldsymbol{P} = f(\boldsymbol{\omega}, \boldsymbol{\Omega})$ explained in Section 2.5.3.2 to obtain the power allocation.

For the MBRL method, the posterior of gains g_i for each cell $i \in \mathcal{J}$ is modeled using a discrete set of 16 values where one value $g_i = 0$ represents the absence of a target, and the remaining values are uniformly spaced on a decibel scale across a 35 dB dynamic range. The prior distribution, before any interaction with the environment, with $\mathbb{P}\{g_i = 0\}$ is computed based on the expected number of targets per beam being 4, ensuring a sparse representation in the RDA domain. The remaining probability mass is distributed uniformly over the rest of the discrete gain values.

The locations of the BS, UEs, and targets are shown in Figure 2.8. For clarity, UEs do not contribute as targets in the considered scenarios, meaning their RCS is zero. We define the target SNR as $MP_{\text{tot}}\,\sigma_\alpha^2/\sigma^2$, where integration is taken into account and σ^2 is the per SC noise power while total noise power is $N\sigma^2$. A standard monostatic radar equation is used for determining the SNR such that we define the target gain as

$$g_q^{(0)} = \frac{L_\text{T} L_\text{R} \sigma_\text{rcs}}{(4\pi)^3 \left(\frac{\tau_q}{2}\right)^4 c^2 f_\text{c}^2}, \tag{2.80}$$

where c is the speed of light, f_c is the carrier frequency, and σ_rcs is the mean RCS of the target (all targets have the same RCS). With the parameters defined in Table 2.2, the target SNR (with integration) varies from 19.93 dB to 32.48 dB when a single beam is pointed exactly at the target direction.

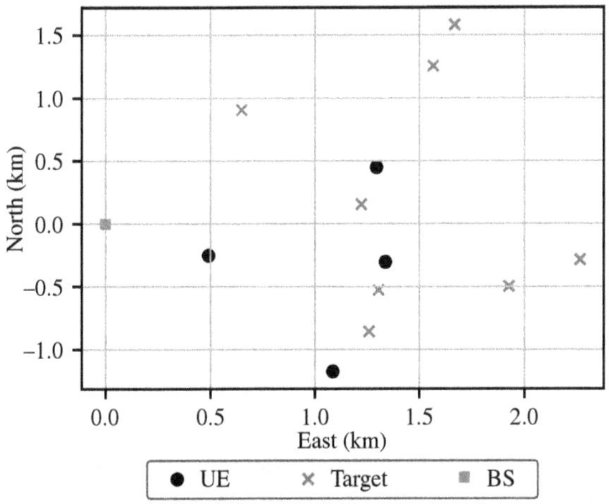

Figure 2.8 Base station (BS), UE, and target positions. *Source:* [15] / with permission of IEEE.

Table 2.2 Simulation parameters.

Description	Symbol	Value
# of SCs	N	1,200
# of symbols	M	128
Carrier frequency	f_c	6.0 GHz
Sub-carrier spacing	f_Δ	15.0 kHz
CP	T_{cp}	16.7 μs
Total power	P_{tot}	30.0 dBm
Noise power (per carrier)	$\sigma_k^2\ \forall\ k$	−132.2 dBm
# of TX/RX antennas	L_T, L_R	32
RB size	-	12
Rate requirement	$C_k\ \forall\ k \in [K]$	19.8 Mbps
# of targets	Q	8
Mean RCS	σ_{rcs}	1.0
# of users	K	4
# of MC iterations	N_{MC}	80
Prob. of False Alarm	P_{fa}	1.0e-06

For the communications, we use LoS channel model

$$\boldsymbol{h}_n^{(k)} = \beta_k \boldsymbol{u}_T(\phi_k), \tag{2.81}$$

where β_k is the channel complex gain, and ϕ_k is the LoS angle to the UE. The channel is Rayleigh fading channel such that β_k is sampled from zero-mean complex Gaussian distribution with variance $\sigma_{\beta_k}^2$. Using the free-space path loss model, the average gain of the channel is defined as

$$\sigma_{\beta_k}^2 = L_T \left(\frac{c}{4\pi f_c d_k} \right)^2, \tag{2.82}$$

where d_k is the user distance from the base station. Communications SNR $P_{tot}\,\sigma_{\beta_k}^2/(\sigma^2 N)$ varies from 34.39 dB to 43.63 dB. This high communications SNR means that the communications rate is primarily limited by mutual interference.

2.5.4.1 Results

Figure 2.9a shows the evolution of the sum of PDs over time. The proposed ISAC method consistently maintains high sensing performance, significantly outperforming the communications-only baseline and closely matching the performance of the sensing-only approach. These results indicate that incorporating communication constraints does not severely degrade sensing capabilities. Both the ISAC

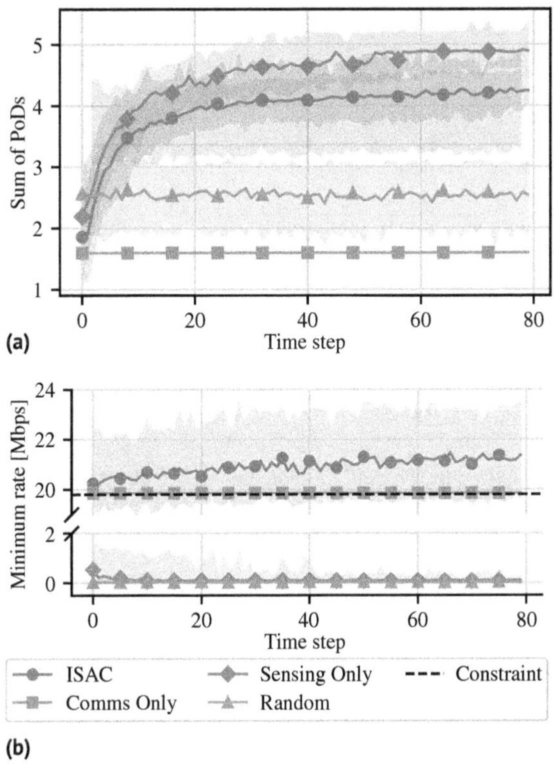

Figure 2.9 Comparison of sensing and communication performance over time. The solid lines show median values, while shaded regions represent the tenth and ninetieth percentiles across MC runs. *Source*: [15] / with permission of IEEE. (a) Sum probability of detection and (b) minimum communication rate.

and sensing-only methods benefit from closed-loop resource adaptation, which leads to substantial improvements in PD over time. In contrast, the random baseline's substantially lower PD performance emphasizes the importance of directing beams toward target locations after acquiring the model (or state) information.

Figure 2.9b shows the minimum user rate over time. Here, the ISAC and communications-only approaches consistently meet the required 19.8 Mbps constraint for all users. In contrast, the sensing-only and random methods fail to satisfy this rate requirement, highlighting the need to consider the communications constraints rigorously. The ISAC method demonstrates robust and balanced performance, maintaining high sensing quality while satisfying the communications rate constraints.

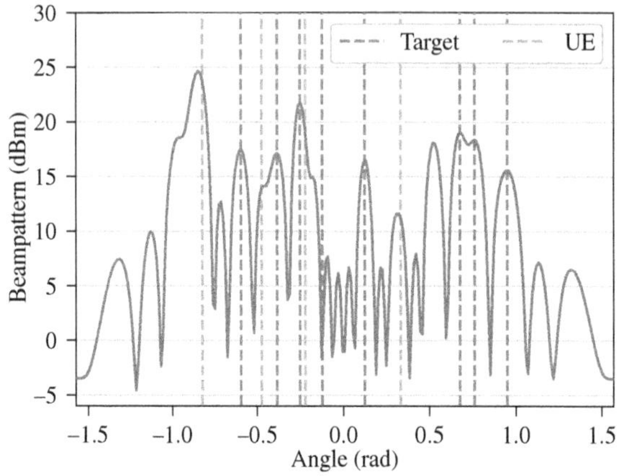

Figure 2.10 Final beampattern produced by the ISAC method. Beams are directed and power allocated toward UEs and detected targets. *Source*: [15] / with permission of IEEE.

The transmit beampattern in Figure 2.10 illustrates the spatial resource allocation at the final time step. Power is steered toward all UEs and most of the targets. Notably, one distant target is excluded from beamforming, as allocating power to it would reduce the overall sum of PDs. Creating tracks of the identified targets could help to detect low-gain targets, as targets with tracks can be removed from the search operation [14]. Additionally, a beam directed around 0.523 radians appears due to exploration promoted by the TS-based controller.

2.6 Conclusions and Future Directions

This chapter used RL to address waveform optimization and resource allocation in multicarrier and multiantenna ISAC systems. An overview of the theory and methods of MFRL and MBRL was provided. In particular, a C-POMDP framework and MBRL algorithms developed for joint optimization of sensing and communications were described in detail. Subsequently, we introduced a C-POMDP-based resource allocation framework paired with the large body of existing knowledge on properties of radio environments and propagation that can be exploited by the MBRL algorithms. Lastly, two illustrative examples were presented to demonstrate the application of MBRL techniques to practical ISAC resource allocation problems.

The first example focused on frequency-domain resource allocation in dynamic spectral environments. The method learns to allocate SCs or RBs and their powers to communications and sensing tasks to achieve the desired performance levels in ISAC. A MBRL algorithm was used to learn predictive models of SC channel qualities and consequently experienced SINR levels, enabling informed decision-making under uncertainty. This approach was benchmarked against a sample-efficient MFRL algorithm, and the results demonstrated the superior sample efficiency, flexibility, and interpretability of the MBRL method. Both fully and partially observable scenarios were considered. Partial observability is emphasized in practical settings where observations are noisy or incomplete due to short observation periods and coherence times, complexity of propagation phenomena, or dense spectrum use. To address this, we employed a loss correction mechanism that enhances model learning under observation uncertainty.

The second example considered joint spatial and frequency-domain resource allocation in multicarrier MIMO ISAC systems, emphasizing challenges related to unknown target parameters. Allocation of beamspace domain beams and frequency-domain SCs or RBs to wireless users and sensing targets and their power allocations were learned. In contrast to the first example, interference from external systems was not considered. This problem exhibited similarities with MAB formulations, and we applied algorithms that explicitly balance the E&E trade-off. The resulting policies achieved a high target detection performance while satisfying communication rate constraints in the considered scenarios.

Future research directions include the development of non-myopic LMPC policies for MBRL, which could further improve performance in adversarial and dynamic environments. Additionally, game-theoretic frameworks offer a promising path for modeling and solving multiagent problems involving cooperative, non-cooperative, and adversarial systems. Furthermore, the joint allocation of frequency and spatial resources, such as beams and SCs, in agile interference environments could be considered.

References

1 K. V. Mishra, M. R. B. Shankar, V. Koivunen, B. Ottersten, and S. A. Vorobyov. Toward millimeter-wave joint radar communications: A signal processing perspective. *IEEE Signal Processing Magazine*, 36(5):100–114, 2019.

2 F. Liu, Y. Cui, C. Masouros, J. Xu, T. X. Han, Y. C. Eldar, and S. Buzzi. Integrated sensing and communications: Toward dual-functional wireless networks for 6G and beyond. *IEEE Journal on Selected Areas in Communications*, 40(6): 1728–1767, 2022.

3 F. Liu, L. Zheng, Y. Cui, C. Masouros, A. P. Petropulu, H. Griffiths, and Y. C. Eldar. Seventy years of radar and communications: The road from separation to integration. *IEEE Signal Processing Magazine*, 40(5):106–121, 2023.

4 V. Koivunen, M. F. Keskin, H. Wymeersch, M. Valkama, and N. González-Prelcic. Multicarrier ISAC: Advances in waveform design, signal processing, and learning under nonidealities. *IEEE Signal Processing Magazine*, 41(5):17–30, 2024.

5 H. D. Griffiths and C. J. Baker. *An Introduction to Passive Radar*. Artech House, 2017.

6 K. Braun. OFDM radar algorithms in mobile communication networks. PhD thesis, 2014.

7 N. Levanon. Multifrequency radar signals. In *Record of the IEEE 2000 International Radar Conference [Cat. No. 00CH37037]*, RADAR-00, pages 683–688. IEEE, 2000.

8 P. Pulkkinen, T. Aittomäki, and V. Koivunen. Reinforcement learning based transmitter-receiver selection for distributed MIMO radars. In *2020 IEEE International Radar Conference (RADAR)*, pages 1040–1045, 2020.

9 P. Pulkkinen and V. Koivunen. Model-based online learning for resource allocation in joint radar-communication systems. In *IEEE International Conference on Acoustics, Speech, and Signal Processing (ICASSP22)*, pages 4103–4107, 2022.

10 P. Pulkkinen and V. Koivunen. Model-based online learning for joint radar-communication systems operating in dynamic interference. In *30th European Signal Processing Conference (EUSIPCO22)*, 2022.

11 P. Pulkkinen and V. Koivunen. Model-free online learning for waveform optimization in integrated sensing and communications. In *IEEE International Conference on Acoustics, Speech, and Signal Processing (ICASSP23)*, pages 1–5, 2023.

12 P. Pulkkinen and V. Koivunen. Model-based online learning for active ISAC waveform optimization. *IEEE Journal of Selected Topics in Signal Processing*, 18(5):737–751, 2024.

13 P. Pulkkinen and V. Koivunen. Partially observable model-based learning for ISAC resource allocation. In *IEEE International Conference on Acoustics, Speech and Signal Processing (ICASSP24)*, pages 12996–13000, 2024.

14 P. Pulkkinen, M. Esfandiari, and V. Koivunen. Cognitive beamspace algorithm for integrated sensing and communications. In *IEEE Radar Conference (RadarConf24)*, pages 1–6, 2024.

15 P. Pulkkinen, M. Esfandiari, and V. Koivunen. Beamspace and frequency domain ISAC resource allocation using reinforcement learning. In *58th Asilomar Conference on Signals, Systems, and Computers*, 2024.

16 R. S. Sutton and A. G. Barto. *Reinforcement Learning: An Introduction*. The MIT Press, Cambridge, MA, USA, 2nd edition, 2018.

17 N. C. Luong, D. T. Hoang, S. Gong, D. Niyato, P. Wang, Y. Liang, and D. I. Kim. Applications of deep reinforcement learning in communications and networking: A survey. *IEEE Communications Surveys Tutorials*, 21(4):3133–3174, 2019.

18 P. Mary, V. Koivunen, and C. Moy. Reinforcement learning for physical layer communications. In Y. C. Eldar, A. Goldsmith, D. Gündüz, and H. V. Poor,

editors, *Machine Learning and Wireless Communications*, pages 231–284. Cambridge University Press, Cambridge, UK, 2022.

19 P. Pulkkinen, T. Aittomäki, A. Ström, and V. Koivunen. Time budget management in multifunction radars using reinforcement learning. In *2021 IEEE Radar Conference (RadarConf21)*, pages 1–6, 2021.

20 B. Witherell, T. -Y Yu, D. Schvartzman, N. Goodman, A. Stringer, and G. Dolinger. Adaptive radar resource management for all-digital multi-function phased array radar using proximal policy optimization. In *Proceedings IEEE Radar Conference (RadarConf)*, pages 1–6, 2024.

21 S. Akbar, R. S. Adve, Z. Ding, and P. W. Moo. Task scheduling in cognitive multifunction radar using model-based DRL. *IEEE Transactions on Aerospace and Electronic Systems*, 61(2):2434–2449, 2025.

22 A. M. Ahmed, A. A. Ahmad, S. Fortunati, A. Sezgin, M. S. Greco, and F. Gini. A reinforcement learning based approach for multitarget detection in massive MIMO radar. *IEEE Transactions on Aerospace and Electronic Systems*, 57(5):2622–2636, 2021.

23 I. Bouhou, S. Fortunati, L. Gharsalli, and A. Renaux. POMDP-driven cognitive massive MIMO radar: Joint target detection-tracking in unknown disturbances. *arXiv preprint arXiv:2410.17967*, 2025.

24 C. E. Thornton, R. M. Buehrer, and A. F. Martone. Constrained contextual bandit learning for adaptive radar waveform selection. *IEEE Transactions on Aerospace and Electronic Systems*, 58(2):1133–1148, 2022.

25 P. Pulkkinen. Model-based reinforcement learning for radar and communications systems. Ph.d. dissertation, Aalto University, Espoo, Finland, 2025.

26 R. E. Bellman. *Dynamic Programming*. Princeton University Press, Cambridge, UK, 1957.

27 R. A. Howard. *Dynamic Programming and Markov Processes*. MIT Press, Cambridge, MA, 1960.

28 V. Krishnamurthy. *Partially Observed Markov Decision Processes: From Filtering to Controlled Sensing*. Cambridge University Press, Cambridge, UK, 2016.

29 M. Lauri, D. Hsu, and J. Pajarinen. Partially observable Markov decision processes in robotics: A survey. *IEEE Transactions on Robotics*, 39(1):21–40, 2022.

30 R. J. Williams. Simple statistical gradient-following algorithms for connectionist reinforcement learning. *Machine Learning*, 8(3-4):229–256, 1992.

31 V. R. Konda and J. N. Tsitsiklis. Actor-critic algorithms. In *Advances in Neural Information Processing Systems*, pages 1008–1014, 2000.

32 J. Schulman, S. Levine, P. Moritz, M. I. Jordan, and P. Abbeel. Trust region policy optimization. In *International Conference on Machine Learning*, pages 1889–1897. PMLR, 2015.

33 J. Schulman, F. Wolski, P. Dhariwal, A. Radford, and O. Klimov. Proximal policy optimization algorithms. *arXiv preprint arXiv:1707.06347*, 2017.

34 T. P. Lillicrap, J. J. Hunt, A. Pritzel, N. Heess, T. Erez, Y. Tassa, D. Silver, and D. Wierstra. Continuous control with deep reinforcement learning. *arXiv preprint arXiv:1509.02971*, 2015.

35 S. Fujimoto, H. van Hoof, and D. Meger. Addressing function approximation error in actor-critic methods. In *International Conference on Machine Learning*, pages 1587–1596. PMLR, 2018.

36 T. Haarnoja, A. Zhou, P. Abbeel, and S. Levine. Soft actor-critic: Off-policy maximum entropy deep reinforcement learning with a stochastic actor. In *International Conference on Machine Learning*, pages 1861–1870. PMLR, 2018.

37 K. Azizzadenesheli, Y. Yue, and A. Anandkumar. Policy gradient in partially observable environments: Approximation and convergence. *arXiv preprint arXiv:1810.07900*, 2020.

38 G. A. Rummery and M. Niranjan. On-line Q-learning using connectionist systems. Technical report, University of Cambridge, Department of Engineering, 1994.

39 C. J. C. H. Watkins and P. Dayan. Q-learning. *Machine Learning*, 8(3-4):279–292, 1992.

40 Z. Wang, T. Schaul, M. Hessel, H. van Hasselt, M. Lanctot, and N. De Freitas. Dueling network architectures for deep reinforcement learning. In *International Conference on Machine Learning*, pages 1995–2003. PMLR, 2016.

41 H. van Hasselt, A. Guez, and D. Silver. Deep reinforcement learning with double Q-learning. *arXiv preprint arXiv:1509.06461*, 2016.

42 M. Hessel, J. Modayil, H. van Hasselt, T. Schaul, G. Ostrovski, W. Dabney, D. Horgan, B. Piot, M. G. Azar, and D. Silver. Rainbow: Combining improvements in deep reinforcement learning. In *Thirty-second AAAI Conference on Artificial Intelligence*, 2018.

43 M. Hausknecht and P. Stone. Deep recurrent Q-learning for partially observable MDPs. *arXiv preprint arXiv:1507.06527*, 2015.

44 E. J. Sondik. The optimal control of partially observable Markov processes. PhD thesis, Stanford University, 1971.

45 B. D. O. Anderson and J. B. Moore. *Optimal Control: Linear Quadratic Methods*. Dover Publications, 2007.

46 J. B. Rawlings and D. Q. Mayne. *Model Predictive Control: Theory and Design*. Nob Hill Publishing, 2009.

47 J. Pineau, G. Gordon, and S. Thrun. Point-based value iteration: An anytime algorithm for POMDPs. In *Proceedings of the 18th International Joint Conference on Artificial Intelligence (IJCAI)*, pages 1025–1032, 2003.

48 S. Ross, J. Pineau, S. Paquet, and B. Chaib-draa. Online planning algorithms for POMDPs. *Journal of Artificial Intelligence Research*, 32:663–704, 2008.

49 D. Silver and J. Veness. Monte-Carlo planning in large POMDPs. In *Advances in Neural Information Processing Systems (NIPS)*, volume 23, pages 2164–2172, 2010.

50 Z. N. Sunberg and M. J. Kochenderfer. Online algorithms for POMDPs with continuous state, action, and observation spaces. *arXiv preprint arXiv:1709.06196*, 2018.

51 M. Janner, J. Fu, M. Zhang, and S. Levine. When to trust your model: Model-based policy optimization. In *33rd Conference on Neural Information Processing Systems (NeurIPS 2019)*, pages 12519–12530, 2019.

52 K. Chua, R. Calandra, R. McAllister, and S. Levine. Deep reinforcement learning in a handful of trials using probabilistic dynamics models. In *32nd Conference on Neural Information Processing Systems (NeurlPS 2018)*, 2018.

53 D. Silver, T. Hubert, J. Schrittwieser, I. Antonoglou, M. Lai, A. Guez, M. Lanctot, L. Sifre, D. Kumaran, T. Graepel, et al. Mastering Chess and Shogi by self-play with a general reinforcement learning algorithm. *Nature*, 588(7839):604–609, 2017.

54 M. Zhang, S. Vikram, L. Smith, P. Abbeel, M. J. Johnson, and S. Levine. Solar: Deep structured representations for model-based reinforcement learning. In *Proceedings of the 36th International Conference on Machine Learning (ICML)*, 2019.

55 M. P. Deisenroth and C. E. Rasmussen. PILCO: A Model-based and Data-Efficient Approach to Policy Search. In *Proceedings of the 28th International Conference on International Conference on Machine Learning*, ICML'11, pages 465–472, Omnipress, Madison, WI, USA, 2011.

56 T. M. Moerland, J. Broekens, P. Aske, and C. M. Jonker. Model-based reinforcement learning: A survey. *Foundations and Trends® in Machine Learning*, 16(1):1–118, 2023.

57 M. Plappert, R. Houthooft, P. Dhariwal, S. Sidor, R. Y. Chen, X. Chen, T. Asfour, P. Abbeel, and M. Andrychowicz. Parameter space noise for exploration. In *International Conference on Learning Representations (ICLR)*, 2018.

58 P. Auer, N. Cesa-Bianchi, and P. Fischer. Finite-time analysis of the multiarmed bandit problem. *Machine Learning*, 47:235–256, 05 2002.

59 N. Cesa-Bianchi, C. Gentile, G. Lugosi, and G. Neu. Boltzmann exploration done right. *Advances in Neural Information Processing Systems*, 30: 6284–6293, 2017.

60 S. Agrawal and N. Goyal. Analysis of Thompson sampling for the multi-armed bandit problem. In *Proceedings of the 25th Annual Conference on Learning Theory*, pages 39.1–39.26, 2012.

61 Y. H. Jung and A. Tewari. Regret bounds for thompson sampling in episodic restless bandit problems. *Advances in Neural Information Processing Systems*, 32, 9005–9014, 2019.

62 Y. Ouyang, M. Gagrani, A. Nayyar, and R. Jain. Learning unknown Markov decision processes: A Thompson sampling approach. *Advances in Neural Information Processing Systems*, 30, 1333–1342, 2017.

63 K. Dong, Y. Wang, X. Chen, and L. Wang. Q-learning with UCB exploration is sample efficient for infinite-horizon MDP. *arXiv preprint arXiv:1901.09311*, 2019.

64 J. Oksanen and V. Koivunen. An order optimal policy for exploiting idle spectrum in cognitive radio networks. *IEEE Transactions on Signal Process*, 63(5):1214–1227, 2015.

65 J. Oksanen and V. Koivunen. Learning spectrum opportunities in non-stationary radio environments. In *2017 IEEE International Conference on Acoustics, Speech and Signal Processing (ICASSP)*, pages 2447–2451, 2017.

66 V. Mnih, K. Kavukcuoglu, D. Silver, A. A. Rusu, J. Veness, M. G. Bellemare, A. Graves, M. Riedmiller, A. K. Fidjeland, G. Ostrovski, et al. Human-level control through deep reinforcement learning. *Nature*, 518(7540):529–533, 2015.

67 T. Schaul, J. Quan, I. Antonoglou, and D. Silver. Prioritized experience replay. In *International Conference on Learning Representations*, 2016.

68 D. Ha and J. Schmidhuber. Recurrent world models facilitate policy evolution. In S. Bengio, H. Wallach, H. Larochelle, K. Grauman, N. Cesa-Bianchi, and R. Garnett, editors, *Advances in Neural Information Processing Systems*, volume 31. Curran Associates, Inc., 2018.

69 L. Hewing, K. P. Wabersich, M. Menner, and M. N. Zeilinger. Learning-based model predictive control: Toward safe learning in control. *Annual Review of Control, Robotics, and Autonomous Systems*, 3(1):269–296, 2020.

70 T. Wang and J. Ba. Exploring model-based planning with policy networks. In *International Conference on Learning Representations (ICLR)*, 2020.

71 C. Pinneri, S. Sawant, S. Blaes, J. Achterhold, J. Stueckler, M. Rolinek, and G. Martius. Sample-efficient cross-entropy method for real-time planning. In Jens Kober, Fabio Ramos, and Claire Tomlin, editors, *Proceedings of the 2020 Conference on Robot Learning*, volume 155 of *Proceedings of Machine Learning Research*, pages 1049–1065. PMLR, 2021.

72 J. D. Isom, S. P. Meyn, and R. D. Braatz. Piecewise linear dynamic programming for constrained POMDPs. In *AAAI'08: Proceedings of the 23rd National Conference on Artificial Intelligence*, volume 1, pages 291–296, 2008.

73 M. Bică, K. -W. Huang, V. Koivunen, and U. Mitra. Mutual information based radar waveform design for joint radar and cellular communication systems. In *IEEE International Conference on Acoustics, Speech and Signal Processing (ICASSP16)*, pages 3671–3675, 2016.

74 M. Bica and V. Koivunen. Multicarrier radar-communications waveform design for RF convergence and coexistence. In *ICASSP 2019 - 2019 IEEE International Conference on Acoustics, Speech and Signal Processing (ICASSP)*, pages 7780–7784. IEEE, 2019.

75 C. Shi, F. Wang, S. Salous, and J. Zhou. Joint subcarrier assignment and power allocation strategy for integrated radar and communications system based on power minimization. *IEEE Sensors Journal*, 19(23):11167–11179, 2019.

76 X. Yuan, Z. Feng, W. Ni, Z. Wei, and R. P. Liu. Waveform optimization for MIMO joint communication and radio sensing systems with imperfect channel feedbacks. In *2020 IEEE International Conference on Communications Workshops (ICC Workshops)*, pages 1–6, 2020.

77 Z. Wei, J. Piao, X. Yuan, H. Wu, J. A. Zhang, Z. Feng, L. Wang, and P. Zhang. Waveform design for MIMO-OFDM integrated sensing and communication system: An information theoretical approach. *IEEE Transactions on Communications*, 72(1):496–509, 2024.

78 A. Ahmed, Y. D. Zhang, A. Hassanien, and B. Himed. OFDM-based joint radar-communication system: Optimal sub-carrier allocation and power distribution by exploiting mutual information. In *2019 53rd Asilomar Conference on Signals, Systems, and Computers*, pages 559–563. IEEE, 2019.

79 M. Bică and V. Koivunen. Radar waveform optimization for target parameter estimation in cooperative radar-communications systems. *IEEE Transactions on Aerospace and Electronic Systems*, 55(5):2314–2326, 2019.

80 S. D. Liyanaarachchi, T. Riihonen, C. B. Barneto, and M. Valkama. Optimized waveforms for 5G–6G communication with sensing: Theory, simulations and experiments. *IEEE Transactions on Wireless Communications*, 20(12):8301–8315, 2021.

81 S. Mura, D. Tagliaferri, M. Mizmizi, U. Spagnolini, and A. Petropulu. Waveform design for OFDM-based ISAC systems under resource occupancy constraint. In *2024 IEEE Radar Conference (RadarConf24)*, pages 1–6, 2024.

82 C. B. Barneto, S. D. Liyanaarachchi, T. Riihonen, M. Heino, L. Anttila, and M. Valkama. Beamforming and waveform optimization for OFDM-based joint communications and sensing at mm-waves. In *2020 54th Asilomar Conference on Signals, Systems, and Computers*, pages 895–899, 2020.

83 Z. Cheng, Z. He, and B. Liao. Hybrid beamforming design for OFDM dual-function radar-communication system. *IEEE Journal of Selected Topics in Signal Processing*, 15(6):1455–1467, 2021.

84 N. T. Nguyen, N. Shlezinger, K. -H. Ngo, V. -D. Nguyen, and M. Juntti. Joint communications and sensing design for multi-carrier MIMO systems. In *2023 IEEE Statistical Signal Processing Workshop (SSP)*, pages 110–114, 2023.

85 W. Wang, D. Qiao, L. Yang, Y. Zhan, and D. W. Kwan Ng. Optimal resource allocation design for wideband integrated sensing and communication systems. *IEEE Transactions on Wireless Communications*, 24(3):2140–2156, 2024.

86 C. Ravelo, D. Martín-Sacristán, and J. F. Monserrat. Exploiting zero forcing and block diagonalization precoding for multi-user MIMO ISAC: Bridging communications and sensing. *IEEE Access*, 12:98083–98095, 2024.

87 S. D. Liyanaarachchi, T. R., C. B. Barneto, and M. Valkama. Joint MIMO communications and sensing with hybrid beamforming architecture and OFDM waveform optimization. *IEEE Transactions on Wireless Communications*, 23(2):1565–1580, 2024.

88 Z. Cheng and B. Liao. QoS-aware hybrid beamforming and DOA estimation in multi-carrier dual-function radar-communication systems. *IEEE Journal of Selected Areas in Communications*, 40(6):1890–1905, 2022.

89 N. T. Nguyen, N. Shlezinger, Y. C. Eldar, and M. Juntti. Multiuser MIMO wideband joint communications and sensing system with subcarrier allocation. *IEEE Transactions on Signal Processing*, 71:2997–3013, 2023.

90 M. Temiz, E. Alsusa, and M. W. Baidas. Optimized precoders for massive MIMO OFDM dual radar-communication systems. *IEEE Transactions on Communications*, 69(7):4781–4794, 2021.

91 C. B. Barneto, T. Riihonen, S. D. Liyanaarachchi, M. Heino, N. González-Prelcic, and M. Valkama. Beamformer design and optimization for joint communication and full-duplex sensing at mm-waves. *IEEE Transactions on Communications*, 70(12):8298–8312, 2022.

92 F. Dong, F. Liu, Y. Cui, W. Wang, K. Han, and Z. Wang. Sensing as a service in 6G perceptive networks: A unified framework for ISAC resource allocation. *IEEE Transactions on Wireless Communications*, 22(5):3522–3536, 2023.

93 Y. Huo, C. Shan, Y. Ma, and H. Zhao. Waveform design with detection probability maximization for MIMO-OFDM integrated sensing and communication system. In *2024 IEEE 99th Vehicular Technology Conference (VTC2024-Spring)*, pages 1–5, 2024.

94 J. Lunden, V. Koivunen, and H. V. Poor. Spectrum exploration and exploitation for cognitive radio: Recent advances. *IEEE Signal Processing Magazine*, 32(3):123–140, 2015.

95 C. E. Thornton, R. M. Buehrer, and A. F. Martone. Constrained online learning to mitigate distortion effects in pulse-agile cognitive radar. In *IEEE Radar Conference (RadarConf21)*, pages 1–6, 2021.

96 T. Lattimore and C. Szepesvári. *Bandit Algorithms*. Cambridge University Press, Cambridge, UK, 2020.

97 J. Oksanen. Machine learning methods for spectrum exploration and exploitation. Ph.d. dissertation, Aalto University, Espoo, Finland, 2016.

98 G. Patrini, A. Rozza, A. K. Menon, R. Nock, and L. Qu. Making deep neural networks robust to label noise: A loss correction approach. In *2017 IEEE Conference on Computer Vision and Pattern Recognition (CVPR)*, pages 2233–2241, 2017.

99 E. Hazan. Introduction to online convex optimization. *Foundations and Trends® in Optimization*, 2(3-4):157–325, 2016.

100 H. Bai, D. Hsu, and W. S. Lee. Planning how to learn. In *2013 IEEE International Conference on Robotics and Automation*, pages 2853–2859, 2013.

101 H. Delecki, M. Vazquez-Chanlatte, E. Yel, K. Wray, T. Arnon, S. Witwicki, and M. J. Kochenderfer. Entropy-regularized point-based value iteration. *2025 11th International Conference on Control, Decision and Information Technologies (CoDIT)*, 2024.

102 C. Shi, Y. Wang, F. Wang, and H. Li. Joint optimization of subcarrier selection and power allocation for dual-functional radar-communications system. In *IEEE 11th Sensor Array and Multichannel Signal Process. Workshop (SAM20)*, pages 1–5, 2020.

103 A. Ahmed, Y. D. Zhang, and A. Hassanien. Joint radar-communications exploiting optimized OFDM waveforms. *Remote Sensing*, 13(21):4376, 2021.

104 S. D. Liyanaarachchi, C. B. Barneto, T. Riihonen, and M. Valkama. Joint OFDM waveform design for communications and sensing convergence. In *IEEE International Conference on Communications (ICC20)*, pages 1–6, 2020.

105 R. M. Gray. *Toeplitz and Circulant Matrices: A Review*. Now Foundations and Trends, 2006.

106 Z. Zhu and M. B. Wakin. On the asymptotic equivalence of circulant and Toeplitz matrices. *IEEE Transactions on Information Theory*, 63(5):2975–2992, 2017.

107 A. De Maio and M. Greco. *Modern Radar Detection Theory*. Scitech Publishing, Edison, NJ, 2016.

108 C. B. Barneto, T. Riihonen, M. Turunen, L. Anttila, M. Fleischer, K. Stadius, J. Ryynanen, and M. Valkama. Full-duplex OFDM radar with LTE and 5G NR waveforms: Challenges, solutions, and measurements. *IEEE Transactions on Microwave Theory and Techniques*, 67(10):4042–4054, 2019.

109 D. J. Russo, B. van Roy, A. Kazerouni, I. Osband, and Z. Wen. A tutorial on Thompson sampling. *Foundations and Trends in Machine Learning*, 11(1):1–96, 2018.

110 J. Duchi, E. Hazan, and Y. Singer. Adaptive subgradient methods for online learning and stochastic optimization. *Journal of Machine Learning Research*, 12:2121–2159, 2011.

3

Low-resolution Processing with Deep Learning

Farhang Yeganegi,[1,] Arian Eamaz,[1] Tara Esmaeilbeig,[2] and Mojtaba Soltanalian[1]*

[1] *Department of Electrical and Computer Engineering, University of Illinois Chicago, Chicago, IL, USA*
[2] *AI Research Lab, Nokia Bell Labs, Murray Hill, NJ, USA*

*Corresponding author: fyegan2@uic.edu.edu

3.1 Introduction

Quantization is a crucial step in digital-signal processing that converts continuous signals into discrete representations. However, achieving high-resolution quantization requires a large number of quantization levels, which can result in increased power consumption, higher manufacturing cost, and a reduced sampling rate in analog-to-digital converters (ADCs). To address these challenges, significant attention has been given to low-resolution quantization strategies, particularly the extreme case of one-bit quantization. In this setting, signals are compared with a fixed threshold, and only the sign of the result is retained, leading to a drastic simplification of ADC hardware. This one-bit approach allows for high-rate sampling with significantly lower power and cost, making it an attractive option for energy-constrained systems. One-bit ADCs have demonstrated remarkable utility in various domains, such as massive multiple-input and multiple-output (MIMO) systems [1] and array signal processing [2]. Furthermore, one-bit sigma-delta schemes have been extensively investigated, with notable research contributions available [3].

Covariance matrix recovery is a key task in many statistical signal processing applications, including direction-of-arrival (DoA) estimation, radar waveform design, target parameter estimation, communication channel estimation, and adaptive radar detection [4–7]. In the context of one-bit quantization, this task becomes more challenging due to the inherent loss of signal magnitude information. Prior works such as [8–10] address this issue by modeling the input signal as a stationary zero-mean Gaussian process. Under this assumption, the input covariance can be estimated by taking advantage of the *arcsine law,* which connects the covariance of an unquantized signal with that of its quantized counterpart [8]. Additionally, the *Bussgang law* [11] provides a similar relationship for the cross-correlation matrix between the input and quantized output, again under the stationary zero-mean Gaussian assumption. It is important to emphasize that these results hold in the absence of dithering, i.e., they assume a zero dither sequence.

Traditional one-bit quantization reconstructs signals by comparing their values to a fixed threshold, usually zero. While simple, this method presents inherent limitations in accurately estimating signal parameters. Specifically, it cannot distinguish between a signal $\boldsymbol{x}$ and any scaled version $\alpha\boldsymbol{x}$ for $\alpha > 0$, since both yield identical sign patterns. Consequently, recovering the signal's amplitude becomes infeasible, leading to the loss of critical information about the original signal.

Considerable research has been undertaken to examine the effects of adding dither to a signal before the quantization process, specifically on quantization error [12, 13]. In digital signal processing, the mean squared error (MSE) serves as a key metric for assessing quantization performance. Dithering provides a mechanism for controlling system behavior by enabling a trade-off between resolution and accuracy. Importantly, it has been shown that dithering effectively reduces the average quantization error [13]. A variety of dithering strategies have been proposed, including uniform, discrete, and Gaussian dithering schemes [14, 15].

Research has shown that introducing an appropriately designed dither signal can significantly improve the resolution and accuracy of digital measurement systems. In particular, within the framework of Cramér–Rao bound (CRB) analysis, Gaussian dithering demonstrates superior performance for input signals with bell-shaped distributions [16]. Although discrete dithering has also been explored, it typically yields higher quantization errors compared to uniform or Gaussian dithering [13]. However, discrete dithering remains appealing due to its simplicity and ease of implementation in hardware.

The design of a dither generator for producing Gaussian dithering in ADC quantization systems was presented in [17]. In this method, Gaussian dither is created using an analog random noise signal, typically sourced from a low-cost thermal noise diode. These diodes exploit quantum mechanical effects, specifically electron-hole pair generation, to produce truly random noise. Since the generated noise may have relatively low power, additional processing circuitry—such as gain control units, low-cost operational amplifiers, or similar components—can be incorporated to enhance the signal. Furthermore, the use of multiple random dithering was demonstrated in [18] as part of a 12-bit, 18-GS/s ADC implementation.

Recent studies have explored the use of random dithering as a means to enhance the estimation of signal characteristics [16]. Dithered one-bit quantization, in particular, has gained traction in a variety of signal processing tasks across the literature. These include sparse parameter estimation [19], compressed sensing (CS) [20–22], low-rank matrix sensing [23], matrix completion [15, 24–26], phase retrieval [27–29], and sampling theory [14, 30–32]. It is worth pointing out that the main difference between one-bit sampling with dithering discussed in this chapter and sigma-delta ($\Sigma\Delta$) sampling is that the one-bit $\Sigma\Delta$ ADC may not properly sample the input signals with large slopes, potentially leading to the loss of amplitude information.

Building on the effective use of dithering for signal amplitude recovery, recent works have extended this concept to covariance recovery from one-bit measurements. In particular, several studies have proposed the use of Gaussian dithering to aid in this process [33–35]. Notably, the works in [34, 35] achieved substantial improvements in estimating signal autocorrelation by integrating Gaussian dithering into the covariance matrix recovery framework through a modified arcsine law. The authors also introduced a modified Bussgang law, which enables the recovery of cross-correlation matrix elements between the input signal and its one-bit quantized measurements.

A key limitation of both the classical (non-dithered) and the initial version of the modified arcsine law is their dependence on the assumption that the input signal is stationary [8, 35]. However, in practical communication and signal processing scenarios, signals are typically non-stationary and exhibit time-varying variances. In these contexts, covariance recovery becomes even more vital for understanding the underlying dynamics of nonstationary systems—despite an expected reduction in accuracy. To address these limitations, the authors of [34] extended the modified arcsine law framework [35] to accommodate nonstationary input signals by incorporating Gaussian dithering. Their approach is grounded in an integral relationship that connects the autocorrelation of one-bit quantized signals to the entries of the original signal's non-Toeplitz covariance matrix. Furthermore, they provided a closed-form expression for the signal mean, which enabled the estimation of time-varying variances (i.e., volatility). Although the resulting integral appears analytically intractable, the authors addressed this challenge by employing the piecewise Padé approximation technique, which expresses the integrands as rational functions. This transformation enabled exact integration and led to the formulation of a nonconvex estimation criterion for autocorrelation recovery. The authors also proposed two numerical techniques—Gauss–Legendre quadrature and Monte Carlo integration—for recovering the input covariance matrix via a convex estimation criterion.

Beyond the modified arcsine law, the authors of [36] introduced a harmonic approximation to the one-bit Hermite law for modeling one-bit correlation. Their approach offers closed-form expressions with a finite number of terms and low computational complexity. Additionally, they derived approximation error bounds, providing theoretical insights into the signal statistics, quantization thresholds, and the parameters in the proposed approximation.

Building on these developments, the authors in [37–39] leveraged the fact that the effect of one-bit quantization can be neutralized in expectation when uniform dithering is applied. This holds under certain statistical assumptions on the dynamic range (or the scale parameter of the dithers). Using this idea, they demonstrated that the input covariance matrix can be accurately estimated from the sample covariance of one-bit data. Moreover, they established probabilistic bounds on the estimation error between the true and recovered covariance matrix, highlighting the robustness of this approach.

Covariance recovery is not only of theoretical interest but also serves as a fundamental building block for practical applications such as DoA estimation, radar, and wireless communication. In particular, DoA estimation relies heavily on accurate covariance information, making one-bit covariance recovery highly relevant in this context.

The problem of DoA estimation is of central importance in array signal processing, with critical applications spanning radar, sonar, and wireless communications systems [10]. Traditional DoA estimation methods typically assume that analog measurements from sensor arrays are digitized using high-resolution ADCs, allowing quantization errors to be neglected. However, the cost and power consumption of ADCs grow significantly with increased resolution and sampling rates. This becomes especially problematic in systems employing large-scale antenna arrays, where multiple ADCs are required [40]. A promising approach to mitigate these limitations is the use of one-bit quantization. Consequently, one-bit signal processing has garnered considerable attention in recent years for its potential to enable efficient DoA estimation under severe quantization constraints [41–43].

DoA estimation can be effectively framed as a sparse recovery problem, where the goal is to identify the angular positions of a small number of targets. In this context, deep unrolling methods have emerged as powerful tools for accurately recovering target locations [44, 45]. Recent advancements in the CS community have increasingly emphasized hybrid approaches that integrate model-based algorithms with data-driven learning. These techniques aim to solve sparse linear inverse problems—reconstructing sparse signals from underdetermined and noisy linear measurements—by combining the interpretability and theoretical guarantees of model-driven methods with the adaptability and performance benefits of deep learning [46, 47].

Another prominent example that highlights the importance of low-resolution processing in the literature is one-bit CS. The development of one-bit CS has sparked extensive research into quantized signal recovery, particularly emphasizing the role of dithering in enhancing reconstruction performance. One-bit CS was first introduced in [48], laying the groundwork for subsequent studies. A significant advancement came in [49], where the authors derived lower bounds on reconstruction error and introduced heuristic algorithms for signal recovery.

Much of the early literature focused on ditherless setups, leading to a variety of algorithmic approaches. Notably, binary iterative hard thresholding (BIHT) and its normalized variant, NBIHT, have emerged as prominent solutions in this space [49–51].

Several theoretical results have also been derived under ditherless assumptions. For instance, [49] established the number of measurements needed for robust signal recovery within a bounded error ball, focusing on Gaussian measurement ensembles. The work in [52] introduced the concept of random hyperplane tessellations to establish a probabilistic link between the Hamming distance and angular error in direction estimation. Further refinements by [53] analyzed the trade-off between distortion and sample complexity using Gaussian complexity, though their analysis was limited to Gaussian matrices and direction-only recovery.

Recent works have demonstrated that introducing dithering enables not only direction recovery but also full signal reconstruction under certain conditions [20, 54, 55]. For example, [56] employed scalar dithered quantization after random projection to construct quasi-isometric embeddings with bounded additive and multiplicative distortion in the ℓ_1-norm. In [54], a hybrid measurement strategy using fixed and zero dithering allowed simultaneous estimation of signal norm and direction. Similarly, [55] proposed injecting adaptive or random dither before binarization, particularly under Gaussian random projections. In a broader setting, [20] proposed a uniform dithering scheme compatible with the general restricted isometry property (RIP) matrices, extending theoretical guarantees beyond Gaussian systems and without relying on the high-resolution assumption. Additionally, [57] extended random hyperplane tessellation theory to uniform dithering, but restricted their analysis to sub-Gaussian matrices, lacking guarantees for deterministic sampling matrices, such as the Fourier sampling matrix. Lastly, [19] approached the dithered quantization problem from a statistical estimation standpoint by casting it as a generalized least absolute shrinkage and selection operator (LASSO) problem. This reformulation allowed them to inherit a wide range of theoretical results from the literature on regularized convex optimization.

In recent years, there has been a growing interest in data-driven approaches for one-bit CS, with several methods demonstrating extraordinary performance in signal reconstruction. These techniques—often based on deep neural networks or algorithm unrolling—have shown the ability to outperform traditional optimization-based methods by capturing complex signal structures and learning task-adapted inference strategies.

Traditional optimization methods, such as gradient descent, are built upon strong theoretical foundations. They are typically concise in formulation and come with performance guarantees under well-defined conditions. When addressing an optimization problem, one can select a method appropriate to the problem type and reasonably expect it to produce a solution consistent with its theoretical properties [58, 59].

Algorithm unrolling develops optimization methods by training on sample data. While it may not offer the same level of theoretical rigor as classical approaches, it often achieves superior performance when properly trained on given datasets. Although training is typically resource-intensive and conducted offline, a major advantage is the potential for faster online inference. In challenging scenarios such as nonconvex optimization—where traditional methods struggle to find accurate solutions—well-trained unrolled algorithms can outperform classical techniques, delivering higher-quality results within similar computational budgets [60, 61]. For example, algorithm unrolling methods have been shown to reconstruct sparse signals more accurately and efficiently than LASSO [62–64].

In algorithm unrolling, each iteration of an optimization algorithm is represented as a layer in a neural network. By stacking these layers, a deep network is formed in which the forward pass effectively simulates the iterative algorithm. Notably, unrolling methods such as the learned iterative shrinkage thresholding algorithm (LISTA) [62, 65, 66] have demonstrated significantly faster convergence and improved performance compared to traditional iterative shrinkage thresholding algorithm (ISTA) or fast ISTA (FISTA) optimizers. These approaches have been successfully applied in diverse domains, including image restoration and reconstruction, medical and biological imaging [47], and wireless communications [67]. A key advantage of unrolling methods lies in their ability to learn realistic signal priors from data while preserving the interpretability and structure of model-driven techniques. Moreover, by incorporating domain-specific knowledge into the network architecture, methods like LISTA can achieve competitive performance using smaller or shallower networks with fewer parameters than purely data-driven models [68].

The authors of [69] formulated the DoA estimation problem within the CS framework, specifically with regard to the data matrix. Their proposed algorithm was developed by unrolling the iterations of the fixed-point continuation algorithm, originally designed for one-bit CS problems. However, the approach in [69] presents two major limitations. First, the reliance on a ditherless quantization scheme is a significant drawback, as dithering is known to enhance recovery performance. Second, the absence of theoretical guarantees undermines the robustness and generalizability of the method. In contrast, [70] addressed these shortcomings by providing rigorous theoretical guarantees and reformulating the DoA estimation problem as a CS problem defined over the input covariance matrix, leading to a more principled and effective solution.

To address the one-bit CS problem using data-driven methods, algorithm unrolling has been adopted to transform traditional iterative solvers into trainable neural network architectures. For example, the BIHT algorithm has been unrolled into a network where each iteration corresponds to a layer, the sensing matrix and step sizes are treated as learnable weights, and the hard-thresholding operation serves as the activation function. This unrolled design has demonstrated strong empirical performance, highlighting the potential of model-aware deep learning in one-bit CS [47, 71]. Notably, [71] introduced dithering as learnable parameters. However, this approach requires not only the storage of quantized measurements but also the corresponding dither values for accurate signal reconstruction.

Unlike the classical CS setting—where unrolled solvers such as LISTA enjoy theoretical recovery guarantees under mild assumptions—no such guarantees currently exist for learned one-bit CS models. This limitation underscores a critical gap and presents an important direction for future research at the intersection of deep learning and quantized signal recovery. The approach proposed in [70], which leverages uniform dithering, addresses this gap by eliminating the need to store dither values during reconstruction and providing theoretical guarantees for recovery performance.

In this chapter, we provide an overview of key developments in one-bit covariance recovery, with particular emphasis on techniques that leverage dithering to mitigate the information loss caused by quantization. We review prominent approaches for recovering the input covariance matrix from coarsely quantized measurements, addressing both stationary and nonstationary signal models. The chapter also explores algorithm unrolling strategies—particularly deep unrolling architectures such as LISTA—that reinterpret classical iterative solvers as structured, trainable neural networks for efficient recovery. Building on these foundations, we formulate the DoA estimation problem within a CS framework using one-bit dithered measurements.

Furthermore, we expand the discussion beyond DoA estimation to highlight the growing importance of data-driven approaches in one-bit CS, a central topic in modern signal processing. A key distinction we emphasize is that, in the DoA estimation problem, the reconstructed covariance matrix from one-bit measurements serves as input to the deep unrolling network, while in the one-bit CS problem, the network directly processes the one-bit measurements to reconstruct the original signal. By drawing this comparison, we illustrate the effectiveness and versatility of data-driven methods in addressing various low-resolution inference scenarios.

***Notation*:** Throughout this chapter, bold lowercase letters (e.g., $\boldsymbol{x}$) denote vectors, and bold uppercase letters (e.g., $\boldsymbol{x}$) denote matrices. We represent a vector $\boldsymbol{x}$ and a matrix $\boldsymbol{X}$ in terms of their elements as $\boldsymbol{x} = [x_i]$ and $\boldsymbol{X} = [\boldsymbol{X}]_{i,j}$, respectively. The sets of real and complex numbers are represented by $\mathbb{R}$ and $\mathbb{C}$, respectively. The operator $\mathbb{E}[\cdot]$ denotes the mathematical expectation. The operators $(\cdot)^{\mathrm{T}}$, $(\cdot)^{\star}$, and $(\cdot)^{\mathrm{H}}$ denote the vector/matrix transpose, conjugate, and Hermitian transpose, respectively. The operator $\mathrm{diag}(\boldsymbol{b})$ creates a diagonal matrix with the entries of $\boldsymbol{b} = [b_i]$ along the diagonal. The Khatri–Rao product is denoted by $*$. The notation $[n]$ represents the index set $\{1, \ldots, n\}$. The ℓ_p-norm of a vector $\boldsymbol{x}$ is defined as $\|\boldsymbol{x}\|_p = \left(\sum_i x_i^p\right)^{1/p}$, and the ℓ_0-norm $\|\boldsymbol{x}\|_0$ counts the number of nonzero entries in $\boldsymbol{x}$. The Frobenius norm of a matrix $\boldsymbol{X} \in \mathbb{C}^{n_1 \times n_2}$ is given by $\|\boldsymbol{X}\|_{\mathrm{F}} = \sqrt{\sum_{r=1}^{n_1} \sum_{s=1}^{n_2} \left|[\boldsymbol{X}]_{r,s}\right|^2}$. For a square real matrix $\boldsymbol{X} \in \mathbb{R}^{p \times p}$, the operator norm (largest singular value) is denoted by $\|\boldsymbol{X}\| = \sup_{\boldsymbol{u} \in \mathbb{S}^{p-1}} \|\boldsymbol{X}\boldsymbol{u}\|_2$, the elementwise maximum norm by $\|\boldsymbol{X}\|_\infty = \max_{i,j}\left|X_{i,j}\right|$, and the maximum column norm by $\|\boldsymbol{X}\|_{1\to 2} = \max_{j \in [p]}\left\|\boldsymbol{x}_j\right\|_2$, where $\boldsymbol{x}_j$ is the jth column of $\boldsymbol{X}$. The vectorization of a matrix $\boldsymbol{X}$ is denoted by $\operatorname{vec}(\boldsymbol{X})$. We use $a \lesssim b$ to denote that there exists a constant $c > 0$ such that $a \leq cb$, and similarly, $a \gtrsim b$ if $a \geq cb$. For a complex number $a \in \mathbb{C}$, the operators $\Re(a)$ and $\Im(a)$ denote its real and imaginary parts, respectively. These operators are applied elementwise to vectors as well. The bounded sparse set $\mathcal{X}$ is defined as $\mathcal{X}(B, s) = \left\{\boldsymbol{x} \mid |x_i| \leq B, \forall i, \|\boldsymbol{x}\|_0 \leq s\right\}$. We write $\boldsymbol{x} \sim \mathcal{N}(\boldsymbol{d}, \boldsymbol{\Sigma})$ to indicate that $\boldsymbol{x}$ follows a multivariate normal distribution with mean $\boldsymbol{d}$ and covariance matrix $\boldsymbol{\Sigma}$. The elementwise maximum between two vectors $\boldsymbol{x}$ and $\boldsymbol{y}$ is denoted by $\max\{\boldsymbol{x}, \boldsymbol{y}\}$, and $\boldsymbol{x} \geq \boldsymbol{y}$ denotes component-wise inequality, i.e., $x_i \geq y_i$ for all i. The Q-function is defined as

$$Q(x) = \frac{1}{\sqrt{2\pi}} \int_x^\infty e^{-\frac{z^2}{2}}\, dz. \tag{3.1}$$

The notation $\boldsymbol{I}_M$ denotes an M by M identity matrix. Finally, the sub-Gaussian norm of a random variable X is defined as

$$\|X\|_{\psi_2} = \inf\left\{t > 0 : \mathbb{E}\left[e^{\frac{X^2}{t^2}}\right] \leq 2\right\}. \tag{3.2}$$

3.2 One-bit DoA Estimation Problem

In this section, we address the problem of DoA estimation under a highly quantized measurement model. Section 3.2.1 introduces the DoA estimation problem in the context of high-resolution measurements. In Section 3.2.2, we describe the dithered one-bit quantization model and highlight the benefits of dithering for covariance matrix estimation from one-bit data. Section 3.2.3 presents the deep unrolling-based solver tailored to the DoA estimation problem under dithered one-bit measurements. Finally, Section 3.2.4 provides numerical results demonstrating the effectiveness of the deep unrolling solver in this setting.

3.2.1 Problem Statement

Consider a uniform linear array (ULA) consisting of M sensors, which is illuminated by K narrowband sources. Let d denote the inter-element spacing, and λ the wavelength. The steering vector for a source located at angle θ_k is defined as:

$$[\boldsymbol{a}(\theta_k)]_m = e^{-\mathrm{j}\frac{2\pi(m-1)d\sin\theta_k}{\lambda}}, \quad m \in [M].$$

Stacking all K steering vectors forms the steering matrix:

$$\boldsymbol{A} = \begin{bmatrix} \boldsymbol{a}(\theta_1) & \boldsymbol{a}(\theta_2) & \cdots & \boldsymbol{a}(\theta_K) \end{bmatrix} \in \mathbb{C}^{M\times K}.$$

Let $s_k(n)$ denote the signal transmitted by the kth source at snapshot n. The array measurement at snapshot n is expressed as:

$$\begin{aligned} \boldsymbol{x}(n) &= \begin{bmatrix} x_1(n) & x_2(n) & \cdots & x_M(n) \end{bmatrix}^{\mathrm{T}} \in \mathbb{C}^M \\ &= \sum_{k=1}^{K} \boldsymbol{a}(\theta_k)s_k(n) + \boldsymbol{v}(n) \\ &= \boldsymbol{A}\boldsymbol{s}(n) + \boldsymbol{v}(n), \quad \text{for } n \in [N], \end{aligned} \tag{3.3}$$

where $\boldsymbol{s}(n) = [s_1(n), \ldots, s_K(n)]^{\mathrm{T}}$ and the noise vector $\boldsymbol{v}(n) \sim \mathcal{CN}(\boldsymbol{0}, \sigma^2\boldsymbol{I}_M)$ is assumed to be independent and identically distributed (i.i.d.) Complex Gaussian.

In practice, the DoA θ_k may not coincide exactly with a predefined overcomplete set of candidate DoAs $\{\bar{\theta}_1, \cdots, \bar{\theta}_L\}$. To accommodate this mismatch, we introduce an overcomplete formulation by defining zero-padded signals:

$$p_l(n) = \begin{cases} s_k(n), & \left|\bar{\theta}_l - \theta_k\right| \leq \frac{\Delta\theta}{2}, \\ 0, & \text{otherwise}, \end{cases} \quad l \in [L],\ n \in [N], \tag{3.4}$$

with $\Delta\theta$ denoting the interval between the elements in the overcomplete set. Using the extended steering matrix $\widehat{\boldsymbol{A}} = \begin{bmatrix} \boldsymbol{a}(\bar{\theta}_1) & \cdots & \boldsymbol{a}(\bar{\theta}_L) \end{bmatrix}$, we obtain the overcomplete form of (3.3) as:

$$\boldsymbol{x}(n) = \sum_{l=1}^{L} \boldsymbol{a}(\bar{\theta}_l)p_l(n) + \boldsymbol{v}(n), \tag{3.5}$$

$$= \widehat{\boldsymbol{A}}\boldsymbol{p}(n) + \boldsymbol{v}(n), \quad \text{for } n \in [N], \tag{3.6}$$

which yields the covariance matrix of $\boldsymbol{x}(n)$ as:

$$\begin{aligned} \boldsymbol{R} &= \mathbb{E}\left[\boldsymbol{x}(n)\boldsymbol{x}^{\mathrm{H}}(n)\right] \in \mathbb{C}^{M\times M} \\ &= \widehat{\boldsymbol{A}}\,\mathrm{diag}\left(\begin{bmatrix} \nu_1 & \nu_2 & \cdots & \nu_L \end{bmatrix}^{\mathrm{T}}\right)\widehat{\boldsymbol{A}}^{\mathrm{H}} + \sigma^2\boldsymbol{I}_M, \end{aligned} \tag{3.7}$$

with $\{\nu_l\}_{l=1}^{L}$ denoting variances of $\{p_l(n)\}_{l=1}^{L}$. We vectorize $\boldsymbol{R}$ to obtain:

$$\begin{aligned} \boldsymbol{h} &= \mathrm{vec}(\boldsymbol{R}) \in \mathbb{C}^{M^2} \\ &= \left(\widehat{\boldsymbol{A}}^{\star} \times \widehat{\boldsymbol{A}}\right)\begin{bmatrix} \nu_1 & \nu_2 & \cdots & \nu_L \end{bmatrix}^{\mathrm{T}} + \sigma^2\begin{bmatrix} \boldsymbol{e}_1^{\mathrm{T}} & \boldsymbol{e}_2^{\mathrm{T}} & \cdots & \boldsymbol{e}_M^{\mathrm{T}} \end{bmatrix}^{\mathrm{T}}, \end{aligned} \tag{3.8}$$

where $\boldsymbol{e}_m^{\mathrm{T}}$ is the mth column of the identity matrix. To work with realvalued representations, we separate the real and imaginary parts:

$$\begin{bmatrix} \Re(\boldsymbol{h}) \\ \Im(\boldsymbol{h}) \end{bmatrix} = \begin{bmatrix} \Re\left(\widehat{\boldsymbol{A}}^{\star} \times \widehat{\boldsymbol{A}}\right) \\ \Im\left(\widehat{\boldsymbol{A}}^{\star} \times \widehat{\boldsymbol{A}}\right) \end{bmatrix} \boldsymbol{\nu} + \begin{bmatrix} \Re\left(\sigma^2 \begin{bmatrix} \boldsymbol{e}_1^{\mathrm{T}} & \boldsymbol{e}_2^{\mathrm{T}} & \cdots & \boldsymbol{e}_M^{\mathrm{T}} \end{bmatrix}^{\mathrm{T}}\right) \\ \boldsymbol{0}_{M^2 \times 1} \end{bmatrix}, \tag{3.9}$$

where $\boldsymbol{\nu} = \begin{bmatrix} \nu_1 & \nu_2 & \cdots & \nu_L \end{bmatrix}^{\mathrm{T}} \in \mathbb{R}^L$. This representation forms the basis for estimating $\boldsymbol{\nu}$, and thus the DoAs, from the sample covariance of $\boldsymbol{x}(n)$. In the next section, we will examine how coarse (one-bit) quantization affects this process and how dithering can mitigate its negative impact on covariance estimation.

3.2.2 Dithered One-bit Quantization

Consider a zero-mean stationary Gaussian input signal, $\boldsymbol{x} \sim \mathcal{N}(\boldsymbol{0}, \boldsymbol{R})$, where $\boldsymbol{R}$ is a Toeplitz matrix determined by the autocorrelation function of $\boldsymbol{x}$, denoted as R. The input $\boldsymbol{x} \in \mathbb{R}^M$ represents a temporal sequence drawn from the ensemble $\{\boldsymbol{x}(n)\}$, where $n \in [N]$. Let x_i and x_j denote the ith and jth components of $\boldsymbol{x}$, respectively. Define the output $\boldsymbol{y} = \operatorname{sign}(\boldsymbol{x})$, where $\operatorname{sign}(\cdot)$ denotes the sign function applied elementwise. The autocorrelation function of the output, denoted by $R_o(l)$ for lag $l = |i - j|$, is related to the autocorrelation of $\boldsymbol{x}$ via the arcsine law [8]:

$$R_o(i, j) = R_o(l) = \mathbb{E}\left[y_i y_j\right] = \frac{2}{\pi} \sin^{-1}\left(\frac{R(l)}{R(0)}\right), \tag{3.10}$$

where y_i and y_j are the ith and jth entries of $\boldsymbol{y}$, and $R(l)$ denotes the input signal autocorrelation for lag l.

The arcsine law is a fundamental statistical result in one-bit quantization [9, 10], linking the covariance of a continuous-valued signal to that of its binary (quantized) version. However, a key limitation of this law is its assumption of a zero threshold during quantization, which leads to substantial information loss. This limitation is evident in (3.10), where it becomes impossible to uniquely recover both $R(0)$ and $R(l)$ from a single equation.

To mitigate this limitation, several works have proposed introducing a dithering sequence prior to quantization [34, 35, 38, 39]. Specifically, let $\boldsymbol{\tau}$ be a dithering sequence independent of the input signal, drawn from the distribution $\boldsymbol{\tau} \sim \mathcal{N}(\boldsymbol{d} = \boldsymbol{1}d, \boldsymbol{\Sigma})$. Define a new random process $\boldsymbol{w}$ such that $\boldsymbol{w} = \boldsymbol{x} - \boldsymbol{\tau}$. Since both $\boldsymbol{x}$ and $\boldsymbol{\tau}$ are Gaussian and independent, $\boldsymbol{w}$ is also Gaussian, with distribution $\boldsymbol{w} \sim \mathcal{N}(-\boldsymbol{d}, \boldsymbol{R} + \boldsymbol{\Sigma} = \boldsymbol{P})$. The autocorrelation function of the one-bit quantized output corresponding to this process, as a function of lag l is studied in the following:

Theorem 3.1 *Let p_l denote the autocorrelation of the random process $\boldsymbol{w}$ at lag $l \geq 1$, and let p_0 represent its variance* [35]. *Consider the one-bit quantized signal $\boldsymbol{y} = \operatorname{sign}(\boldsymbol{w})$. Then, the autocorrelation function of $\boldsymbol{y}$, denoted by $R_o(l)$, is given by:*

$$R_o(l) = \frac{e^{\frac{-d^2}{p_0+p_l}}}{\pi\sqrt{(p_0^2 - p_l^2)}} \left\{ \int_0^{\frac{\pi}{2}} \frac{1}{\beta_s} + \sqrt{\frac{\pi}{\beta_s}} \frac{\alpha_s}{2\beta_s} e^{\frac{\alpha_s^2}{4\beta_s}} - \sqrt{\frac{\pi}{\beta_s}} \frac{\alpha_s}{\beta_s} Q\left(\frac{\alpha_s}{\sqrt{2\beta_s}}\right) e^{\frac{\alpha_s^2}{4\beta_s}} \, d\theta \right\} - 1, \tag{3.11}$$

where α_s *and* β_s *are defined as:*

$$\alpha_s = \frac{d(\sin\theta + \cos\theta)}{p_0 + p_l}, \qquad \beta_s = \frac{p_0 - p_l \sin 2\theta}{2(p_0^2 - p_l^2)}. \tag{3.12}$$

Finding p_0 and $\{p_l\}$ in Theorem 3.1 results in input variance and autocorrelation recovery, which can be achieved by considering the relation:

$$R(i, j) = P(i, j) - \Sigma(i, j). \tag{3.13}$$

It is important to note that estimating p_0 and $\{p_l\}$ in the integral equation (3.11) is challenging, as only a single equation is available for two unknowns at each specific value of l [33]. To address this, the authors in [35] introduced a first-order statistical method that provides an additional equation, enabling the estimation of p_0 and $\{p_l\}$ efficiently:

Lemma 3.1 *The first moment (i.e., the mean) of the one-bit quantized signal, commonly estimated as* $\boldsymbol{\mu} \approx \frac{1}{N}\sum_{k=1}^{N} \boldsymbol{y}(k)$, *is determined by the statistical properties of the dithering sequence and the power of the underlying input signal* [35]. *Specifically, the expected value of the one-bit samples is given by:*

$$\boldsymbol{\mu} = \mathbb{E}[\boldsymbol{y}] = \mathbf{1}\mu = \mathbf{1}\left(2Q\left(\frac{d}{\sqrt{p_0}}\right) - 1\right). \tag{3.14}$$

As a result of this lemma, one can first estimate p_0 using (3.14), and subsequently estimate $\{p_l\}$ from the integral equation in (3.11). The authors of [35] empirically demonstrated the successful recovery of the input covariance matrix, either applying analytical approximation techniques to the integrands in expression (3.11) or employing numerical integration methods to solve the integral equation in (3.11).

A limitation of the approach in [35] is the assumption of a Toeplitz covariance matrix structure, which stems from the underlying stationarity assumption. This restriction was relaxed in [34], where the results were extended to accommodate nonstationary scenarios. In this setting, consider a dithering sequence $\boldsymbol{\tau}$ that is independent of the input signal and distributed as $\boldsymbol{\tau} \sim \mathcal{N}(\mathbf{d} = \mathbf{1}d, \boldsymbol{\Sigma})$. Define a new random process $\boldsymbol{w}$ such that $\boldsymbol{w} = \boldsymbol{x} - \boldsymbol{\tau}$. Since $\boldsymbol{x}$ and $\boldsymbol{\tau}$ are gaussian and independent, $\boldsymbol{w}$ is also Gaussian with distribution $\boldsymbol{w} \sim \mathcal{N}(-\mathbf{d}, \boldsymbol{R} + \boldsymbol{\Sigma} = \boldsymbol{P})$, where $\boldsymbol{P}$ is a non-Toeplitz covariance matrix due to the non-stationarity. The autocorrelation function of the one-bit quantized output corresponding to this process is formulated as follows:

Theorem 3.2 *Let* $\mathbf{P}$ *be the covariance matrix of the random vector* $\boldsymbol{W}$, *and define* $p_{0i} = \mathbf{P}(i, i)$, $p_{0j} = \mathbf{P}(j, j)$, *and* $p_{ij} = \mathbf{P}(i, j)$ [34]. *Consider the one-bit quantized random variable* $\mathbf{y} = \operatorname{sign}(\boldsymbol{w})$. *Then, the autocorrelation function of* $\mathbf{y}$ *is given by:*

$$R_o(i, j) = \frac{e^{\frac{-d^2(p_{0i}+p_{0j}-2p_{ij})}{2(p_{0i}p_{0j}-p_{ij}^2)}}}{\pi\sqrt{\left(p_{0i}p_{0j} - p_{ij}^2\right)}} \left\{\int_0^{\frac{\pi}{2}} \frac{1}{\beta_n} + \sqrt{\frac{\pi}{\beta_n}}\frac{\alpha_n}{2\beta_n} e^{\frac{\alpha_n^2}{4\beta_n}} - \sqrt{\frac{\pi}{\beta_n}}\frac{\alpha_n}{\beta_n} Q\left(\frac{\alpha_n}{\sqrt{2\beta_n}}\right) e^{\frac{\alpha_n^2}{4\beta_n}}\, d\theta\right\} - 1, \tag{3.15}$$

where the parameters α_n *and* β_n *are defined as:*

$$\begin{aligned} \alpha_n &= \frac{d\left(p_{0i}\sin\theta + p_{0j}\cos\theta - p_{ij}(\cos\theta + \sin\theta)\right)}{\left(p_{0i}p_{0j} - p_{ij}^2\right)}, \\ \beta_n &= \frac{p_{0j}\cos^2\theta + p_{0i}\sin^2\theta - p_{ij}\sin 2\theta}{2\left(p_{0i}p_{0j} - p_{ij}^2\right)}. \end{aligned} \tag{3.16}$$

Similarly to the stationary case, estimating the values of $\{p_{0i}\}$, $\{p_{0j}\}$, and $\{p_{ij}\}$ enables recovery of the input covariance matrix by (3.13). To facilitate efficient estimation, the authors of [34] proposed a first-order statistical method to compute the values of $\{p_{0i}\}$:

Lemma 3.2 *The mean of the one-bit quantized data* $\mathbf{y}$ *is determined by both the dithering distribution and the signal power* [34]. *Specifically, for each index* $i \in [M]$, *the expected value of the* i*th entry satisfies:*

$$\mathbb{E}[y_i] = 2Q\left(\frac{d}{\sqrt{p_{0i}}}\right) - 1, \tag{3.17}$$

where p_{0i} *is defined in the same manner as in Theorem 3.2.*

Based on this lemma, the values $\{p_{0i}\}$ can be estimated using (3.17), followed by the estimation of $\{p_{ij}\}$ using the integral equation in (3.15). In [34], the authors empirically validated the accurate recovery of the input covariance matrix with time-varying variance, using either analytical approximations of the integrands in (3.15) or numerical integration techniques to solve the integral equation in (3.15).

Although [34, 35] proposed accurate covariance matrix estimation schemes for both stationary and nonstationary scenarios, their approaches have two main limitations: (1) the mathematical formulations are valid only under the Gaussian assumption, and (2) the authors did not specify how many one-bit samples are required to achieve accurate covariance matrix estimation with high probability.

To extend the covariance matrix estimation framework to sub-Gaussian vectors $\boldsymbol{x}$, the authors of [38] introduced a two-bit dithered quantization scheme. In this approach, two bits are collected per entry of each sample vector using independent, uniformly distributed dithers during quantization. Specifically, the available data consists of the quantized sample vectors:

$$\begin{aligned} \boldsymbol{y}(n) &= \operatorname{sign}(\boldsymbol{x}(n) + \boldsymbol{\tau}(n)), \\ \bar{\boldsymbol{y}}(n) &= \operatorname{sign}(\boldsymbol{x}(n) + \bar{\boldsymbol{\tau}}(n)), \quad n \in [N], \end{aligned} \tag{3.18}$$

where the dithering vectors $\{\boldsymbol{\tau}(n), \bar{\boldsymbol{\tau}}(n)\}_{n=1}^{N}$ are all independent and uniformly distributed in $[-T, T]^{M}$. Using these quantized observations, we construct the estimator:

$$\widehat{\boldsymbol{R}} = \frac{1}{2}\left(\tilde{\boldsymbol{R}} + \tilde{\boldsymbol{R}}^{\mathrm{T}}\right), \tag{3.19}$$

where

$$\tilde{\boldsymbol{R}} = \frac{T^2}{N} \sum_{n=1}^{N} \boldsymbol{y}(n)\bar{\boldsymbol{y}}^{\mathrm{T}}(n). \tag{3.20}$$

The motivation for constructing such an estimator lies in the fact that the matrix $\tilde{\boldsymbol{R}}$ closely approximates the sample covariance matrix $\frac{1}{N}\sum_{n=1}^{N} \boldsymbol{x}(n)\boldsymbol{x}^{\mathrm{T}}(n)$ when the parameter T is appropriately chosen. This connection arises from a notable result in the quantization theory literature: "dithering effectively cancels quantization distortion in expectation." Mathematically, this result implies:

$$\mathbb{E}_{\boldsymbol{\tau}(n),\bar{\boldsymbol{\tau}}(n)}\left[T^2 \boldsymbol{y}(n)\bar{\boldsymbol{y}}^{\mathrm{T}}(n)\right] = \boldsymbol{x}(n)\boldsymbol{x}^{\mathrm{T}}(n), \tag{3.21}$$

where following the symmetrization, the covariance estimate $\widehat{\boldsymbol{R}}$ is constructed as (3.19). The following theorem provides an upper bound on the approximation error between the estimated covariance matrix $\widehat{\boldsymbol{R}}$ and the true covariance matrix $\boldsymbol{R}$:

Theorem 3.3 *Let $\boldsymbol{x}$ be a zero-mean, J-sub-Gaussian vector with covariance matrix $\boldsymbol{R}$. Let $\{\boldsymbol{x}(n)\}_{n=1}^{N} \overset{\text{i.i.d.}}{\sim} \boldsymbol{x}$ [38]. Let $\boldsymbol{M} \in [0, 1]^{M \times M}$ be a fixed symmetric mask.*

If $T^2 \geq C\log(N)\|\boldsymbol{R}\|_\infty$ for a universal constant C that only depends on J, then with probability at least $1-e^{-t}$,

$$\begin{aligned}\left\|\boldsymbol{M}\odot\widehat{\boldsymbol{R}}-\boldsymbol{M}\odot\boldsymbol{R}\right\| \leq & \, c\|M\|_{1\to 2}\left(T^2+T\|\boldsymbol{R}\|^{\frac{1}{2}}\right)\sqrt{\frac{\log(M)+t}{N}} \\ & + cT^2\|\boldsymbol{M}\|\frac{\log(M)+t}{N},\end{aligned} \tag{3.22}$$

where c only depends on J.

It is worth noting that the estimator $\boldsymbol{M}\odot\widehat{\boldsymbol{R}}$ is not guaranteed to be positive semidefinite. To address this, a more practical approach is to apply the projection onto the positive semidefinite cone, denoted by $P_{\mathrm{PSD}}(\boldsymbol{M}\odot\widehat{\boldsymbol{R}})$, with respect to the operator norm. This projection can be efficiently computed using singular value decomposition (SVD). Since convex projections are 1-Lipschitz, Theorem 3.3 provides the same error bound for the modified estimator, i.e., $\left\|P_{\mathrm{PSD}}(\boldsymbol{M}\odot\widehat{\boldsymbol{R}})-\boldsymbol{R}\right\|$.

Up to this point, we have considered the signal $\boldsymbol{x}$ to be real-valued. However, as shown in Section 3.2.1, the underlying signal model is inherently complex. Consequently, the covariance matrix estimation framework must be adapted to account for the complex-valued nature of $\boldsymbol{x}$. To this end, we assume that the real and imaginary components of each entry are quantized independently, each using two dither sequences. As a result, we obtain four dithered one-bit measurements:

$$\begin{aligned}\boldsymbol{y}_R(n) &= \mathrm{sign}(\Re(\boldsymbol{x}(n))+\boldsymbol{\tau}_R(n)),\\ \boldsymbol{y}_I(n) &= \mathrm{sign}(\Im(\boldsymbol{x}(n))+\boldsymbol{\tau}_I(n)),\\ \bar{\boldsymbol{y}}_R(n) &= \mathrm{sign}(\Re(\boldsymbol{x}(n))+\bar{\boldsymbol{\tau}}_R(n)),\\ \bar{\boldsymbol{y}}_I(n) &= \mathrm{sign}(\Im(\boldsymbol{x}(n))+\bar{\boldsymbol{\tau}}_I(n)),\ \ n\in[N],\end{aligned} \tag{3.23}$$

where the dithering vectors $\{\boldsymbol{\tau}_R(n),\boldsymbol{\tau}_I(n),\bar{\boldsymbol{\tau}}_R(n),\bar{\boldsymbol{\tau}}_I(n)\}_{n=1}^N$ are all independent and uniformly distributed in $[-T,T]^M$. We then construct the complex-valued binary vectors as follows:

$$\begin{aligned}\boldsymbol{y}(n) &= \boldsymbol{y}_R(n)+\mathrm{j}\boldsymbol{y}_I(n),\\ \bar{\boldsymbol{y}}(n) &= \bar{\boldsymbol{y}}_R(n)+\mathrm{j}\bar{\boldsymbol{y}}_I(n),\ \ n\in[N].\end{aligned} \tag{3.24}$$

Using these, we estimate $\boldsymbol{R}$ via:

$$\widehat{\boldsymbol{R}}=\frac{1}{2}\left(\tilde{\boldsymbol{R}}+\tilde{\boldsymbol{R}}^{\mathrm{H}}\right), \tag{3.25}$$

where

$$\tilde{\boldsymbol{R}}=\frac{T^2}{N}\sum_{n=1}^{N}\boldsymbol{y}(n)\bar{\boldsymbol{y}}^{\mathrm{H}}(n). \tag{3.26}$$

The following theorem characterizes the accuracy of $\widehat{\boldsymbol{R}}$ in approximating the true covariance matrix $\boldsymbol{R}$:

Theorem 3.4 *Let $\boldsymbol{x}$ be a zero-mean, J-sub-Gaussian vector with covariance matrix $\boldsymbol{R}$. Let $\{\boldsymbol{x}(n)\}_{n=1}^{N} \overset{\text{i.i.d.}}{\sim} \boldsymbol{x}$ [39]. Then there exists a constant $c > 0$ which only depends on J such that if $T^2 \gtrsim \log(N)\|\boldsymbol{R}\|_{\infty}$, the covariance estimator $\widehat{\mathbf{R}}$ fulfills, for any $t \geq 0$, with probability at least $1 - 8e^{-cNt}$:*

$$\| \widehat{\boldsymbol{R}} - \boldsymbol{R} \|_{\mathrm{F}} \lesssim T^2 \sqrt{\frac{M^2(\log(M) + t)}{N}}. \tag{3.27}$$

The result of Theorem 3.4 plays a key role in establishing the convergence bound of the deep unrolling solver presented in the following section.

3.2.3 DoA Estimation via Deep Algorithm Unrolling

In this section, our objective is to estimate the vector $\boldsymbol{\nu}$ in (3.9) using the dithered one-bit quantization model described in (3.23). Specifically, we assume that each complex-valued data sample $\boldsymbol{x}(n)$ is processed through the model in (3.23) for all $n \in [N]$. Using the resulting binary measurements, we construct an estimate of the true covariance matrix, denoted by $\widehat{\boldsymbol{R}}$, as given in (3.25). Substituting this estimate into (3.9), we obtain the following system of linear equations to recover $\boldsymbol{\nu}$:

$$\begin{bmatrix} \Re(\widehat{\boldsymbol{h}}) \\ \Im(\widehat{\boldsymbol{h}}) \end{bmatrix} = \begin{bmatrix} \Re\left(\widehat{\boldsymbol{A}}^{\star} \times \widehat{\boldsymbol{A}}\right) \\ \Im\left(\widehat{\boldsymbol{A}}^{\star} \times \widehat{\boldsymbol{A}}\right) \end{bmatrix} \boldsymbol{\nu} + \begin{bmatrix} \Re\left(\sigma^2 \begin{bmatrix} \boldsymbol{e}_1^{\mathrm{T}} & \boldsymbol{e}_2^{\mathrm{T}} & \cdots & \boldsymbol{e}_M^{\mathrm{T}} \end{bmatrix}^{\mathrm{T}}\right) \\ \boldsymbol{0}_{M^2 \times 1} \end{bmatrix}, \tag{3.28}$$

where $\widehat{\boldsymbol{h}} = \mathrm{vec}(\widehat{\boldsymbol{R}})$. To simplify the notation, we rewrite (3.28) as:

$$\boldsymbol{b} = \boldsymbol{\Phi}\boldsymbol{\nu} + \boldsymbol{z}, \tag{3.29}$$

where

$$\begin{aligned} \boldsymbol{\Phi} &= \begin{bmatrix} \Re\left(\widehat{\boldsymbol{A}}^{\star} \times \widehat{\boldsymbol{A}}\right) \\ \Im\left(\widehat{\boldsymbol{A}}^{\star} \times \widehat{\boldsymbol{A}}\right) \end{bmatrix} \in \mathbb{R}^{2M^2 \times L}, \\ \boldsymbol{b} &= \begin{bmatrix} \Re(\widehat{\boldsymbol{h}}) \\ \Im(\widehat{\boldsymbol{h}}) \end{bmatrix} \in \mathbb{R}^{2M^2}, \\ \boldsymbol{z} &= \begin{bmatrix} \Re\left(\sigma^2 \begin{bmatrix} \boldsymbol{e}_1^{\mathrm{T}} & \boldsymbol{e}_2^{\mathrm{T}} & \cdots & \boldsymbol{e}_M^{\mathrm{T}} \end{bmatrix}^{\mathrm{T}}\right) \\ \boldsymbol{0}_{M^2 \times 1} \end{bmatrix} \in \mathbb{R}^{2M^2 \times L}. \end{aligned} \tag{3.30}$$

As discussed in Section 3.2.1, the transmitted signal is zero-padded, which implies that the vector $\boldsymbol{\nu}$ admits a sparse representation. In the inverse problems literature, the formulation in (3.29) with a sparse vector $\boldsymbol{\nu}$ is recognized as a CS problem [72, 73]. A commonly used approach for recovering sparse vectors in such problems is the ℓ_1-relaxation method, formulated as follows:

$$\mathcal{P}: \min_{\boldsymbol{\nu}} \frac{1}{2}\|\tilde{\boldsymbol{b}} - \boldsymbol{\Phi}\boldsymbol{\nu}\|_2^2 + \gamma\|\boldsymbol{\nu}\|_1, \tag{3.31}$$

where $\tilde{\boldsymbol{b}} = \boldsymbol{b} - \boldsymbol{z}$. To solve the optimization problem in (3.31), one widely adopted algorithm is ISTA, which updates the estimate of $\boldsymbol{\nu}$ using the following iteration rule [74]:

$$\boldsymbol{\nu}^{(k+1)} = \mathrm{ST}_{\gamma/L}\left(\boldsymbol{\nu}^{(k)} + \frac{1}{L}\boldsymbol{\Phi}^{\mathrm{T}}\left(\tilde{\boldsymbol{b}} - \boldsymbol{\Phi}\boldsymbol{\nu}^{(k)}\right)\right), \tag{3.32}$$

where $\mathrm{ST}_\eta(\cdot) = \mathrm{sign}(\cdot)\max(0, |\cdot| - \eta)$ denotes the soft-thresholding operator with threshold $\frac{\gamma}{L}$, and L is typically chosen as the largest eigenvalue of the Gram matrix $\boldsymbol{\Phi}^{\mathrm{T}}\boldsymbol{\Phi}$. As highlighted in the literature, selecting an appropriate value for γ is crucial for ensuring the effective performance of ISTA in recovering the unknown sparse vector. This task becomes especially challenging when the observation vector $\boldsymbol{b}$ is corrupted by noise. By comparing (3.9) with (3.28), it is clear that we are dealing with a noisy system of equations in (3.28) due to the use of the estimated covariance matrix $\widehat{\boldsymbol{R}}$. Furthermore, determining a suitable value of γ that consistently yields good performance across multiple instances of the sparse vector $\boldsymbol{\nu}$ adds another layer of difficulty. To address these challenges, recent works have proposed integrating learning into classical iterative algorithms [58, 62, 64]. A key approach involves constructing a neural network that mirrors the structure of the iterative algorithm, where each layer corresponds to a single iteration. Within this framework—known as *algorithm unrolling*—the activation function in each layer is designed to mimic the proximal operator used in the original update process. Building on this idea, to unroll ISTA in (3.32), LISTA has been proposed with the following update process [62]:

$$\boldsymbol{\nu}^{(k+1)} = \mathrm{ST}_{\eta^{(k)}}\left(\boldsymbol{W}_1^{(k)}\boldsymbol{\nu}^{(k)} + \boldsymbol{W}_2^{(k)}\tilde{\boldsymbol{b}}\right), \; k \in [I], \tag{3.33}$$

where the learnable parameters are denoted by $\boldsymbol{\zeta}_I \triangleq \left\{\eta^{(k)}, \boldsymbol{W}_1^{(k)}, \boldsymbol{W}_2^{(k)}\right\}_{k=1}^{I}$. It is straightforward that with $\boldsymbol{W}_1 = \boldsymbol{I}_L - \frac{1}{L}\boldsymbol{\Phi}^{\mathrm{T}}\boldsymbol{\Phi}$, $\boldsymbol{W}_2 = \frac{1}{L}\boldsymbol{\Phi}^{\mathrm{T}}$ and $\eta = \frac{\gamma}{L}$, LISTA boils down to ISTA. Given a set of S training samples $\{\boldsymbol{\nu}_j, \tilde{\boldsymbol{b}}_j\}_{j=1}^{S}$, the parameters $\boldsymbol{\zeta}_I$ are learned through the following training process:

$$\min_{\boldsymbol{\zeta}_I} \quad \mathbb{P}_S \ell_{\boldsymbol{\zeta}_I}, \tag{3.34}$$

where

$$\mathbb{P}_S \ell_{\boldsymbol{\zeta}_I} = \frac{1}{S}\sum_{i=1}^{S} \| \boldsymbol{\nu}^{(I+1)}(\boldsymbol{\zeta}_I, \tilde{\boldsymbol{b}}_i, \boldsymbol{\nu}^{(1)}) - \boldsymbol{\nu}_i \|_2^2. \tag{3.35}$$

In [62], it was demonstrated that the update process in (3.33) must satisfy certain necessary conditions to ensure convergence:

Remark 3.1 For LISTA to successfully recover the target parameter $\boldsymbol{\nu}$, meaning that $\boldsymbol{\nu}^{(k)} \to \boldsymbol{\nu}$ as $k \to \infty$, the following conditions must be satisfied [62]:

$$\boldsymbol{W}_1^{(k)} - \left(\boldsymbol{I}_L - \boldsymbol{W}_2^{(k)}\boldsymbol{\Phi}\right) \to \mathbf{0}, \text{ as } k \to \infty, \tag{3.36}$$

$$\eta^{(k)} \to 0, \text{ as } k \to \infty. \tag{3.37}$$

Under conditions of Remark 3.1, the update process in (3.33) reduces to

$$\boldsymbol{\nu}^{(k+1)} = \mathrm{ST}_{\eta^{(k)}}\left(\boldsymbol{\nu}^{(k)} + \boldsymbol{W}^{(k)\mathrm{T}}(\tilde{\boldsymbol{b}} - \boldsymbol{\Phi}\boldsymbol{\nu}^{(k)})\right), \; k \in [I]. \tag{3.38}$$

Interestingly, in the update process shown in (3.38), the learnable parameters reduce to $\boldsymbol{\zeta}_I = \left\{\boldsymbol{W}^{(k)}, \eta^{(k)}\right\}_{k=1}^{I}$, which is advantageous in terms of computational complexity during training. To summarize the methodology, Figure 3.1 illustrates the overall framework for recovering the sparse vector $\boldsymbol{\nu}$ from one-bit dithered measurements. We conclude this section with the following theorem, which establishes the convergence of LISTA with one-bit data:

Theorem 3.5 *Let $\{\boldsymbol{x}(n)\}_{n=1}^{N}$ be a sequence of i.i.d J-subGaussian random variables, and assume the uniform dither scale parameter T satisfies the condition stated in Theorem 3.4 [70]. Given initialization $\boldsymbol{\nu}^{(1)} = \mathbf{0}$ and parameters $\left\{\boldsymbol{W}^{(k)}, \eta^{(k)}\right\}_{k=1}^{\infty}$, let the sequence $\left\{\boldsymbol{\nu}^{(k)}\right\}_{k=2}^{\infty}$ be generated by (3.38). If $\boldsymbol{\nu} \in \mathcal{X}(B, s)$ for sufficiently small s, then there exists a choice of parameters $\left\{\boldsymbol{W}^{(k)}, \eta^{(k)}\right\}$ such that, for all $\boldsymbol{\nu} \in \mathcal{X}(B, s)$ and constants $c_1, c_2, C, t \geq 0$, the following holds with probability at least $1 - 8e^{-c_1 N t}$:*

$$\| \boldsymbol{\nu}^{(k)} - \boldsymbol{\nu} \|_2 \leq sBe^{-c_2 k} + CT^2M^2\sqrt{\frac{\log(M) + t}{N}}. \tag{3.39}$$

3.2.4 Numerical Illustrations

This section presents a comprehensive evaluation of the DoA estimation framework under severely quantized measurement constraints, specifically using

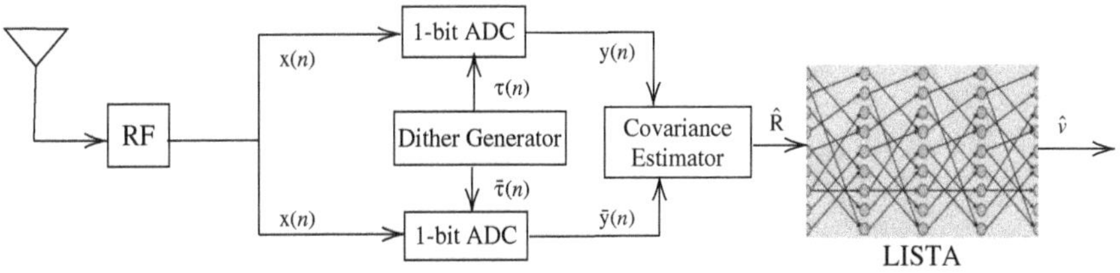

Figure 3.1 A simplified illustration of the overall system for DoA estimation from one-bit dithered data.

one-bit quantization with dithering. The experimental setup is designed to assess the estimation accuracy of LISTA.

While the formulations developed in Section 3.2.1 are applicable to arbitrary array geometries, our empirical investigations focus on the widely adopted ULA

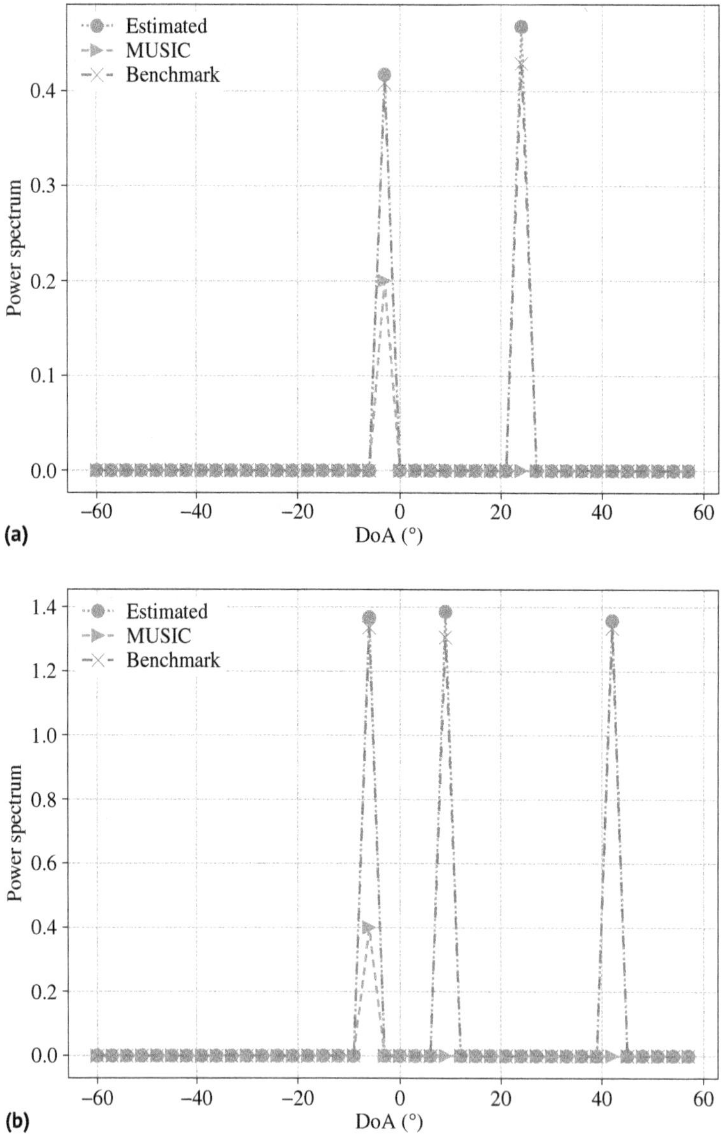

Figure 3.2 The estimated DoAs by LISTA from one-bit data for (a) two targets and (b) three targets, compared with the MUSIC algorithm.

configuration. In particular, we consider arrays with $M = \{8, 16\}$ antenna elements, spaced at half the wavelength, i.e., $d = \lambda/2$. Each antenna is assumed to be equipped with a dithered one-bit ADC. The angular domain is discretized with a resolution of $\Delta\theta = 1°$, and two experimental scenarios are examined: one with $K = 2$ and another with $K = 3$ targets. In both cases, the true DoAs are drawn uniformly at random from the interval $[-60°, 60°]$. To construct the training dataset, we simulate $N = 10^4$ snapshots of received signals based on the overcomplete signal model presented in (3.5). These signals are then subjected to one-bit quantization with dithering, as described in Section 3.2.2. From the quantized observations, the covariance matrices are estimated according to the formulation in (3.25).

The training dataset consists of $S = 1600$ labeled examples, with an additional 400 samples reserved for validation. The LISTA-based recovery algorithm is trained to infer target DoAs from the one-bit covariance estimates. Figure 3.2a, b display the performance of LISTA for the aforementioned two configurations. In both cases, LISTA successfully recovers all target directions, whereas the classical MUSIC algorithm—operating on the same quantized data—fails to resolve more than one target. This clearly demonstrates the robustness of LISTA in processing severely quantized measurements.

3.3 One-bit CS Problem

In this section, we address the problem of dithered one-bit CS based on the framework proposed in [71]. Section 3.3.1 introduces the problem formulation along with well-known classical solvers. In Section 3.3.2, we present model-based deep learning approaches for these classical methods. In particular, we introduce a model-driven one-bit CS autoencoder (AE) architecture that enables joint optimization of the encoder (sensing parameters) and decoder (parameters of the classical solvers) when one-bit quantization with dithering is used during data acquisition. Finally, Section 3.3.3 provides numerical results that demonstrate the effectiveness of the learned one-bit compressive AE in reconstructing signals from dithered one-bit measurements.

3.3.1 Problem Formulation and Classic Solvers

In this section, we focus on the one-bit CS measurement model, i.e., the encoder module, whose dynamics can be described as follows:

$$\boldsymbol{r} = \text{sign}(\boldsymbol{\Phi}\boldsymbol{x} - \boldsymbol{b}), \tag{3.40}$$

where $\boldsymbol{\Phi} \in \mathbb{R}^{m \times n}$ is the sensing matrix, $\boldsymbol{b} \in \mathbb{R}^m$ represents the dithering sequence, and $\boldsymbol{x} \in \mathbb{R}^n$ is a K-sparse signal. Interestingly, the measurement model in (3.40) can be expressed as a system of linear inequalities:

$$\boldsymbol{r} \odot (\boldsymbol{\Phi}\boldsymbol{x} - \boldsymbol{b}) \geq \boldsymbol{0}, \tag{3.41}$$

or equivalently

$$\boldsymbol{R}(\boldsymbol{\Phi x} - \boldsymbol{b}) \geq \mathbf{0}, \tag{3.42}$$

where $\boldsymbol{R} \triangleq \text{diag}(\boldsymbol{r})$. Before introducing potential optimization strategies to address the feasibility problem in (3.42), we first present the following definition:

Definition 3.1 Consistent Reconstruction Property (CRP) Let $\boldsymbol{x}$ be a K-sparse signal, and let $\widehat{\boldsymbol{x}}$ denote the output of an arbitrary reconstruction algorithm solving the one-bit feasibility problem defined in (3.42). Then, such a reconstruction algorithm is said to be consistent when:

$$\text{sign}(\boldsymbol{\Phi x} - \boldsymbol{b}) = \text{sign}(\boldsymbol{\Phi}\widehat{\boldsymbol{x}} - \boldsymbol{b}). \tag{3.43}$$

The concept of CRP has been instrumental in establishing theoretical guarantees in the domain of one-bit sensing [15, 20, 22, 49]. To incorporate CRP, one can formulate the ℓ_0 minimization problem to address the feasibility in (3.42):

$$\mathcal{P}_0 : \quad \min_{\boldsymbol{x} \in \mathbb{R}^n} \quad \|\boldsymbol{x}\|_0, \quad \text{s.t.} \quad \boldsymbol{R}(\boldsymbol{\Phi x} - \boldsymbol{b}) \geq \mathbf{0}. \tag{3.44}$$

The ℓ_0 minimization problem is nonconvex and known to be NP-hard. To address this, it is commonly relaxed to the following convex ℓ_1 minimization problem:

$$\mathcal{P}_1 : \quad \min_{\boldsymbol{x} \in \mathbb{R}^n} \quad \|\boldsymbol{x}\|_1, \quad \text{s.t.} \quad \boldsymbol{R}(\boldsymbol{\Phi x} - \boldsymbol{b}) \geq \mathbf{0}. \tag{3.45}$$

Building on the regularization-based relaxation method introduced in [48], we reformulate the above problem as follows:

$$\mathcal{P}_2 : \quad \min_{\boldsymbol{x} \in \mathbb{R}^n} \quad \|\boldsymbol{x}\|_1 + \frac{1}{2}\|\rho(\boldsymbol{R}(\boldsymbol{\Phi x} - \boldsymbol{b}))\|_2^2, \tag{3.46}$$

where $\rho(\boldsymbol{c}) \triangleq \max\{-\boldsymbol{c}, \mathbf{0}\}$. Note that the function $\rho(\cdot)$ can be expressed using the well-known rectified linear unit (ReLU) function as $\rho(\boldsymbol{c}) = \text{ReLU}(-\boldsymbol{c})$. The second term in the objective imposes a quadratic penalty on the negative components of the vector $\boldsymbol{R}(\boldsymbol{\Phi x} - \boldsymbol{b})$, thereby promoting the CRP. Since the objective function of $\mathcal{P}_2$ is a composite convex function, it can be efficiently addressed using a proximal algorithm. It is straightforward to verify that the corresponding proximal updates for solving $\mathcal{P}_2$ are given by:

$$\begin{aligned} \tilde{\boldsymbol{d}}_k &= -(\boldsymbol{R\Phi})^{\mathrm{T}}\rho(\boldsymbol{R}(\boldsymbol{\Phi x}_{k-1} - \boldsymbol{b})), \\ \boldsymbol{x}_k &= \text{ST}_\tau\left(\boldsymbol{x}_{k-1} - \delta\tilde{\boldsymbol{d}}_k\right). \end{aligned} \tag{3.47}$$

In the remainder of this chapter, we refer to the iterations defined in (3.47) as the *generalized renormalized fixed-point iteration (G-RFPI).*

Another widely used approach for addressing the problem in (3.44) is the *BIHT* algorithm [20, 55]. BIHT is a simple yet effective first-order iterative reconstruction technique for one-bit CS, operating under the assumption that the sparsity level K is known in advance. It is specifically tailored to address the following formulation of $\mathcal{P}_0$:

$$\mathcal{P}_3: \quad \min_{\boldsymbol{x}\in\mathbb{R}^n} \quad \|\rho(\boldsymbol{R}(\boldsymbol{\Phi}\boldsymbol{x}-\boldsymbol{b}))\|_1, \quad \text{s.t.} \quad \|\boldsymbol{x}\|_0 = K, \tag{3.48}$$

where $\rho(\cdot)$ is defined as previously described. By computing the sub-gradients of the objective function in (3.48), the following update steps are obtained:

$$\begin{aligned} \boldsymbol{u}_k &= \boldsymbol{x}_{k-1} + \frac{\delta}{2}\boldsymbol{\Phi}^{\mathrm{T}}(\boldsymbol{r} - \mathrm{sign}(\boldsymbol{\Phi}\boldsymbol{x}_{k-1} - \boldsymbol{b})), \\ \boldsymbol{x}_k &= \mathcal{H}_K(\boldsymbol{u}_k), \end{aligned} \tag{3.49}$$

where $\mathcal{H}_K(\cdot)$ is a sparsity projection operator that preserves the K largest-magnitude components of its input while setting all others to zero, thereby enforcing the K-sparsity constraint. Throughout the remainder of this chapter, we refer to the above iterative scheme as the *generalized BIHT* (G-BIHT) algorithm.

In general, the performance and convergence of G-RFPI and G-BIHT are highly sensitive to the selection of the parameters $\{\delta, \tau\}$ and δ, respectively. It is important to note that there is no straightforward method for properly selecting these parameters in either algorithm. Moreover, both algorithms rely on fixed parameters throughout all iterations. This motivates the development of a methodology for designing a decoder function that leverages adaptive parameters, potentially leading to significant improvements in the performance of both algorithms.

In the following section, we present an over-parameterization of the iterative steps in the G-RFPI and G-BIHT algorithms, which serves as the foundation for constructing a one-bit compressive AE. This framework facilitates the joint optimization of the encoder—defined in (3.40) and parameterized by the sensing matrix $\boldsymbol{\Phi}$ and the dithering vector $\boldsymbol{b}$—together with a family of decoder functions derived from the previously discussed iterative recovery algorithms.

3.3.2 Model-based Deep Learning Models for One-bit CS

In this section, we focus on designing a *model-driven one-bit CS-based AE* architecture that enables joint optimization of both encoder and decoder parameters under one-bit quantization with dithering in the data acquisition process (i.e., the encoder) for a K-sparse input signal. Specifically, the decoder is implemented as a model-based deep network, constructed by unrolling the iterations of the earlier introduced first-order optimization algorithms into the layers of a deep neural network.

Generally, an arbitrary AE is characterized by two main functions: (1) an encoder function $f_{\zeta_e}^{\text{Encoder}} : \mathbb{R}^n \mapsto \mathbb{R}^m$ parameterized by a set of variables ζ_e, which maps the input signal into a transformed signal space, and (2) a decoder function $f_{\zeta_d}^{\text{Decoder}} : \mathbb{R}^m \mapsto \mathbb{R}^n$ parameterized by ζ_d, which reconstructs the original signal from the encoded measurements. Thus, the overall dynamics of an AE can be described by the composition $\widehat{\boldsymbol{x}} = f_{\zeta_d}^{\text{Decoder}} \circ f_{\zeta_e}^{\text{Encoder}}(\boldsymbol{x})$, where $\widehat{\boldsymbol{x}}$ represents the reconstructed signal. Motivated by this structure, we seek to interpret a one-bit CS system within the AE framework. This perspective enables a unified approach for jointly designing the sensing matrix $\boldsymbol{\Phi}$ and the dithering vector $\boldsymbol{b}$ to effectively capture the information of a K-sparse signal under one-bit quantization. In addition, it facilitates learning the parameters of an iterative signal recovery algorithm. To this end, we modify and unroll the iterations of the G-RFPI algorithm, defined in (3.47), and the G-BIHT method, defined in (3.49), into the layers of a deep neural network. We then leverage deep learning tools to jointly tune the parameters of the one-bit compressive AE.

3.3.2.1 Structure of the Encoding Module

Based on the dithered one-bit quantization data acquisition model defined in (3.40), we define the encoder module of the one-bit compressive AE as follows:

$$f_{\zeta_e}^{\text{Encoder}}(\boldsymbol{x}) = \widetilde{\text{sign}}(\boldsymbol{\Phi}\boldsymbol{x} - \boldsymbol{b}), \tag{3.50}$$

where $\zeta_e = \{\boldsymbol{\Phi}, \boldsymbol{b}\}$ represents the set of learnable encoder parameters, and $\widetilde{\text{sign}}(\boldsymbol{x}) = \tanh(t \cdot \boldsymbol{x})$ with a large constant $t > 0$ (t was set to 50 in numerical investigations). The non-differentiable sign function $\text{sign}(\cdot)$ is replaced by the smooth surrogate $\widetilde{\text{sign}}(\cdot)$ to facilitate effective training with standard learning algorithms, such as stochastic gradient-based optimization methods.

3.3.2.2 Structure of the Decoding Module

In this section, we present the architecture of the decoder function $f_{\zeta_d}^{\text{Decoder}}$ for the one-bit compressive AE. Specifically, the decoder is constructed by unrolling the iterations of the G-RFPI algorithm, defined in (3.47), and the G-BIHT method, defined in (3.49), onto the layers of a deep neural network as follows:

- **Learned G-RFPI (LG-RFPI)**: We over-parameterize the iterations of the G-RFPI algorithm in (3.47) by introducing learnable parameters for each iteration. Specifically, we parameterize the decoder function by the set

$\zeta_d = \{\delta_i, \tau_i\}_{i=0}^{L-1}$. Given the encoder parameters $\zeta_e = \{\mathbf{\Phi}, \boldsymbol{b}\}$, the decoder function is expressed as follows:

$$f_{\zeta_d}^{\text{Decoder}}(\boldsymbol{z}_0) = g_{\phi_{L-1}} \circ \cdots \circ g_{\phi_1} \circ g_{\phi_0}(\boldsymbol{z}_0; \mathbf{\Phi}, \boldsymbol{R}, \boldsymbol{b}), \tag{3.51}$$

with $\phi_i = \{\delta_i, \tau_i\}$ and

$$g_{\phi_i}(\boldsymbol{z}; \mathbf{\Phi}, \boldsymbol{R}, \boldsymbol{b}) = \text{ST}_{\tau_i}\left(\boldsymbol{z} + \delta_i(\boldsymbol{R}\mathbf{\Phi})^{\text{T}}\rho(\boldsymbol{R}(\mathbf{\Phi}\boldsymbol{z} - \mathbf{b}))\right). \tag{3.52}$$

Figure 3.3 illustrates the computational dynamics within the ith layer of the LG-RFPI deep architecture.

- **Learned G-BIHT (LG-BIHT)**: We similarly unroll the G-BIHT iterations defined in (3.49) by parameterizing the decoder function with $\zeta_d = \{\delta_i\}_{i=0}^{L-1}$. Given the encoder parameters $\zeta_e = \{\mathbf{\Phi}, \boldsymbol{b}\}$, the decoder function in this case is expressed as follows:

$$f_{\zeta_d}^{\text{Decoder}}(\boldsymbol{z}_0) = g_{\phi_{L-1}} \circ \cdots \circ g_{\phi_1} \circ g_{\phi_0}(\boldsymbol{z}_0; \mathbf{\Phi}, \boldsymbol{r}, \boldsymbol{b}, K), \tag{3.53}$$

with $\phi_i = \{\delta_i\}$ and

$$g_{\phi_i}(\boldsymbol{z}; \mathbf{\Phi}, \boldsymbol{r}, \boldsymbol{b}, K) = \mathcal{H}_K(\boldsymbol{z} + \delta_i\mathbf{\Phi}^{\text{T}}(\boldsymbol{r} - \text{sign}\,(\mathbf{\Phi}\boldsymbol{z} - \mathbf{b}))). \tag{3.54}$$

Figure 3.4 depicts the computational flow of the ith layer in the LG-BIHT deep architecture.

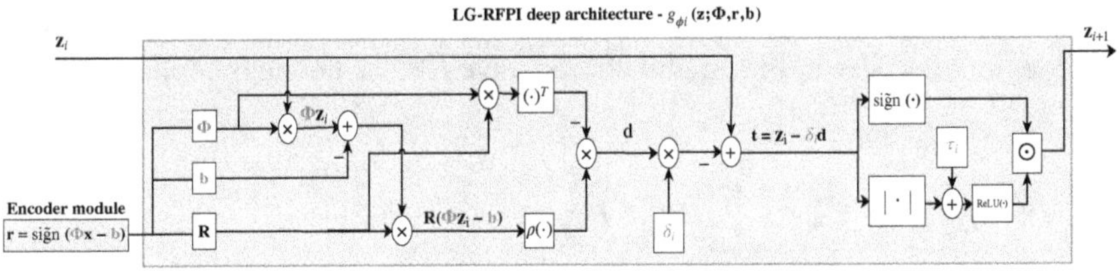

Figure 3.3 Illustration of the computational flow in the ith layer of the LG-RFPI deep architecture, with trainable parameters $\mathbf{\Phi}$, b, $\{\delta_i\}$, and $\{\tau_i\}$.

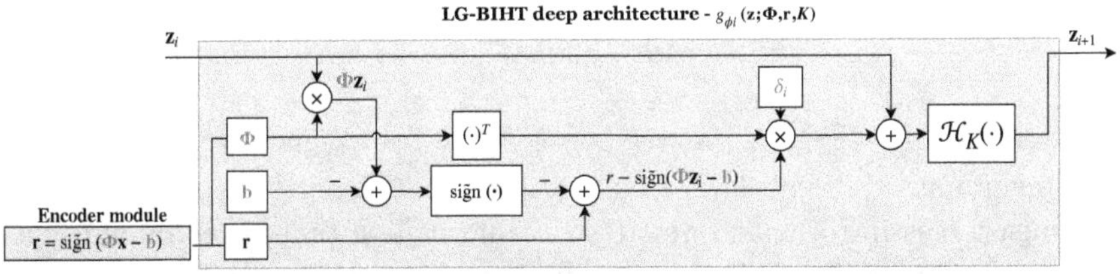

Figure 3.4 Illustration of the computational flow in the ith layer of the LG-BIHT deep architecture, with trainable parameters $\mathbf{\Phi}$, b, and $\{\delta_i\}$.

In the next section, building on the established structures of the encoder and decoder modules, we present the training framework for the one-bit compressive AE. Specifically, we define an appropriate loss function to facilitate the training of the unrolled deep architectures, aiming to jointly optimize all network parameters—both in the encoder and decoder—in an end-to-end fashion using standard deep learning techniques.

3.3.2.3 Loss Function Characterization and Training Method

The training of an AE requires the definition of a suitable loss function that quantifies the similarity between the input signal $\boldsymbol{x}$ and its reconstructed counterpart $\hat{\boldsymbol{x}}$. As previously discussed, the output of the one-bit compressive AE is given by $\hat{\boldsymbol{x}} = f^{\text{Decoder}}_{\zeta_d} \circ f^{\text{Encoder}}_{\zeta_e}(\boldsymbol{x})$. A commonly adopted choice for the loss function in this context is the MSE loss, defined as:

$$\mathbb{P}_S \ell_{\zeta_e,\zeta_d} = \frac{1}{S}\sum_{i=1}^{S}\left\|\boldsymbol{x}_i - f^{\text{Decoder}}_{\zeta_d} \circ f^{\text{Encoder}}_{\zeta_e}(\boldsymbol{x}_i)\right\|_2^2, \tag{3.55}$$

where the objective is to minimize this loss over the parameters of both the encoder $\boldsymbol{\zeta}_e$ and the decoder $\boldsymbol{\zeta}_d$ across the set of training samples $\{\boldsymbol{x}_i\}_{i=1}^{S}$. Nevertheless, in deep architectures with many layers and parameters, relying solely on a simple output loss such as MSE can hinder effective training due to the vanishing gradient problem during backpropagation. To mitigate this issue, a more effective strategy for training the one-bit compressive AE is to adopt a cumulative MSE loss that incorporates reconstruction errors from intermediate layers. Let $\tilde{g}_i = g_{\phi_i} \circ g_{\phi_{i-1}} \circ \cdots \circ g_{\phi_0} \circ f^{\text{Encoder}}_{\zeta_e}(\boldsymbol{x})$. To jointly train the parameters of both the encoder and decoder in the one-bit compressive AE, we define the following loss function:

$$\begin{aligned}\mathbb{P}_S \ell_{\zeta_e,\zeta_d}[L] &= \underbrace{\sum_{i=0}^{L-1} w_i \mathbb{P}_S \ell_{\zeta_e,\zeta_d(i)}}_{\text{accumulated MSE loss of all layers}} \\ &+ \underbrace{\lambda_r \sum_{i=0}^{L-1} \text{ReLU}(-\delta_i) + \lambda_r \sum_{i=0}^{L-1} \text{ReLU}(-\tau_i)}_{\text{regularization term for the step-sizes and shrinkage thresholds}},\end{aligned} \tag{3.56}$$

where $\mathbb{P}_S \ell_{\zeta_e,\zeta_d(i)} = \frac{1}{S}\sum_{j=1}^{S} \| \boldsymbol{x}_j - \tilde{g}_i(\boldsymbol{x}_j) \|_2^2$, w_i is the assigned weight reflecting the importance of the ith layer's output, and $\lambda_r > 0$ denotes the regularization parameter. It is important to note that as information propagates through the network layers, one expects the quality of reconstruction to improve progressively. Therefore, a natural approach for assigning the importance weights w_i is to increase them with the layer index, giving more emphasis to the outputs of deeper layers. In this chapter, we adopt a logarithmic weighting scheme, where the weights are defined as $w_i = \log(i+2)$ for $i \in \{0, \dots, L-1\}$. Moreover, for the LG-BIHT-based decoder, we omit the last term in (3.56) since the associated parameters $\{\tau_i\}$ are not present in its formulation.

For the training procedure, our numerical investigations indicate that an incremental learning strategy is particularly effective for jointly training the encoder and decoder parameters. The adopted approach operates as follows: during the lth incremental training stage (for $l \in [L]$), we optimize the loss function $\mathbb{P}_S \ell_{\zeta_e, \zeta_d}[l]$ by learning the parameter set $\boldsymbol{\zeta}_l = \boldsymbol{\zeta}_e \cup \{\phi_i\}_{i=0}^{l}$, which includes the encoder parameters and the decoder parameters up to the lth layer. Training in each stage is performed using mini-batches of size B. Upon completing the lth training round, the $(l+1)$th layer is appended to the network, and the corresponding loss function is updated to $\mathbb{P}_S \ell_{\zeta_e, \zeta_d}[l+1]$. A new training phase is then conducted for the expanded network. Notably, the parameters learned in stage l are used to initialize the same parameters in stage $(l+1)$.

3.3.3 Numerical Illustrations

In this section, we present various simulation results to evaluate the performance of the one-bit compressive AEs. For training, we randomly generate K-sparse signals of dimension $n = 128$, i.e., $\boldsymbol{x} \in \mathbb{R}^{128}$, where the nonzero entries are drawn from the standard normal distribution. The decoder is implemented with a fixed number of $L = 30$ layers, which corresponds to unrolling 30 iterations of the optimization updates given in (3.52) and (3.54). The learnable sensing matrix is assumed to be $\boldsymbol{\Phi} \in \mathbb{R}^{512\times 128}$. Each result is averaged over 128 independent realizations of the system parameters. We consider the case that $m > n$, which aligns with practical one-bit quantization settings—wherein a large number of low-precision measurements are typically available, in contrast to conventional CS scenarios.

For training, we employ the Adam optimizer with a learning rate of 10^{-3} to update the parameters of the deep architectures. As detailed in Section 3.3.2.3, training is performed using the incremental batch-learning strategy with mini-batches of size 64 across each round $l \in [30]$, and a total of 200 training epochs per round. For the LG-RFPI deep architecture, the sparsity level K of each training signal is uniformly sampled from the set $\{16, 24, 32\}$ within each mini-batch. We evaluate the trained model on target signals with $K \in \{16, 32\}$ (seen during training), as well as on $K \in \{8, 40\}$ (unseen during training) to assess generalization. In contrast, the LG-BIHT deep architecture assumes prior knowledge of the signal sparsity level. As such, it does not require training over different values of K. Instead, we train this architecture specifically for signals with $K = 16$ and evaluate its performance on $K \in \{16, 24\}$.

Performance of the LG-RFPI AE :

Figure 3.5 shows the MSE between the source signal $\boldsymbol{x}$ and the recovered signal $\hat{\boldsymbol{x}}_i$ as a function of the total number of optimization iterations i, for $i \in \{0, \ldots, 29\}$, and for sparsity levels (a) $K = 8$, (b) $K = 16$, (c) $K = 32$, and (d) $K = 40$. This figure compares the following scenarios:

- *Case 1*: The G-RFPI algorithm with a randomly generated sensing matrix $\boldsymbol{\Phi}$ and dithering vector $\boldsymbol{b}$, whose elements are i.i.d. samples from the standard normal distribution, using fixed values for δ and $\alpha = \dfrac{\delta}{\tau}$.
- *Case 2*: The G-RFPI algorithm using a learned sensing matrix $\boldsymbol{\Phi}$ and dithering vector $\boldsymbol{b}$, with fixed values for δ and α, as in Case 1.
- *Case 3*: The LG-RFPI AE method, corresponding to the iterative updates in (3.52), using learned parameters: $\boldsymbol{\Phi}$, $\boldsymbol{b}$, $\{\delta_i\}_{i=0}^{L-1}$, and $\{\tau_i\}_{i=0}^{L-1}$.

Note that the focus of this section is on recovering the amplitude information of the underlying K-sparse signal through the use of the dithering vector.

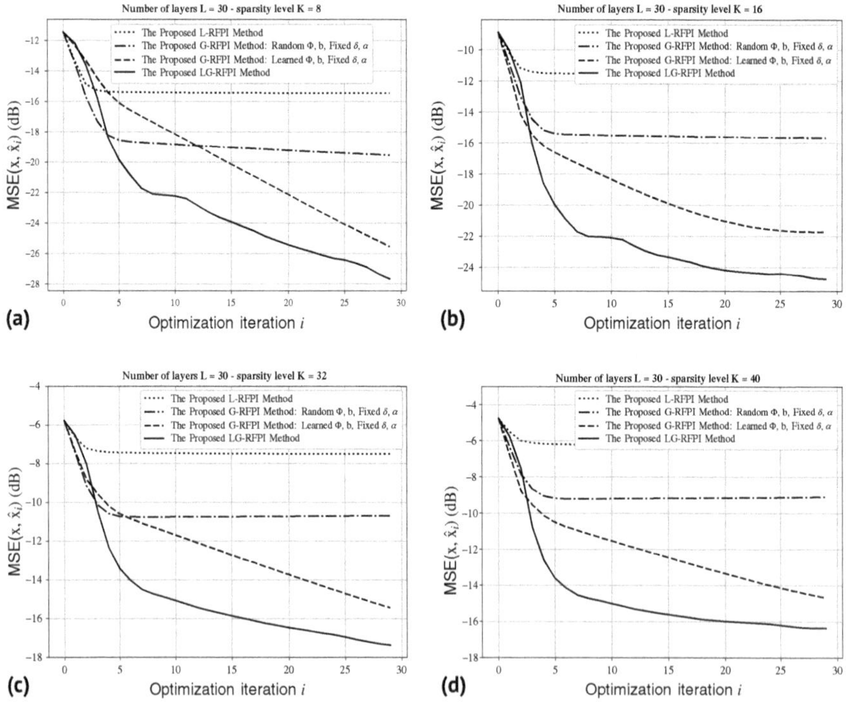

Figure 3.5 The performance of the LG-RFPI AE and the G-RFPI method in recovering the amplitude information of the K-sparse signals for sparsity levels: (a) $K = 8$, (b) $K = 16$, (c) $K = 32$, and (d) $K = 40$.

While the traditional RFPI method is limited to recovering the normalized signal (i.e., its direction only), we also include the performance of the Learned RFPI (L-RFPI) method—which significantly outperforms the standard RFPI—in recovering amplitude information, for comparison purposes. As shown in Figure 3.5, the G-RFPI algorithm with a randomly generated sensing matrix and dithering vector (Case 1) achieves accurate recovery of the signal's amplitude information across all tested sparsity levels $K \in \{8, 16, 32, 40\}$. In contrast, both the standard RFPI and its learned variant (L-RFPI) fail to recover amplitude information due to the absence of dithering. Next, we examine the impact of learning a distribution-specific (data-driven) sensing matrix and dithering vector (Case 2). As shown in Figure 3.5, this approach achieves noticeably lower MSE in recovering amplitude information compared to the vanilla G-RFPI method by learning a proper sensing matrix and dithering vector and utilizing them during the data-acquisition process. Finally, Figure 3.5 shows that the LG-RFPI method (Case 3) significantly outperforms all other approaches, achieving much lower MSE within only a few iterations. In addition, LG-RFPI demonstrates strong generalization to unseen sparsity levels $K \in \{8, 40\}$ (see Figure 3.5a, d). This generalization capability can be attributed to the model-driven nature of the deep network architecture.

Performance of the LG-BIHT AE:

Figure 3.6 illustrates the MSE between the true signal $\boldsymbol{x}$ and the recovered signal $\hat{\boldsymbol{x}}_i$ as a function of the optimization iteration i, for sparsity levels (a) $K = 16$ and (b) $K = 24$. This figure compares the following scenarios:

- *Case 1*: The G-BIHT algorithm with a randomly generated sensing matrix $\boldsymbol{\Phi}$ and dithering vector $\boldsymbol{b}$, where the elements of both are i.i.d. samples from the standard normal distribution, and a fixed set of step sizes $\{\delta_i\}_{i=0}^{L-1}$.

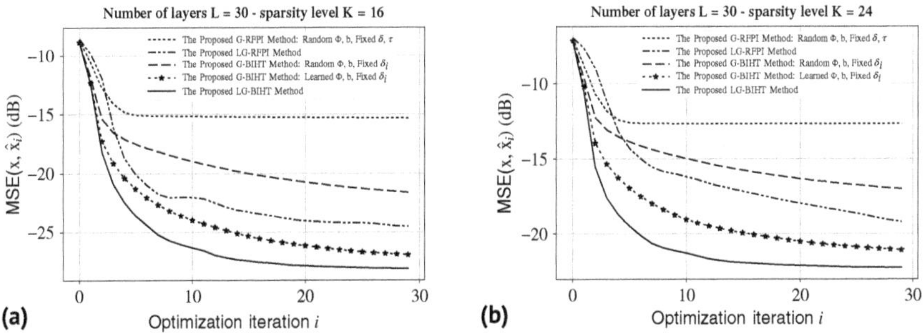

Figure 3.6 The performance of the G-BIHT and the corresponding LG-BIHT AE in recovering the amplitude information of the signal for sparsity levels: (a) $K = 16$ and (b) $K = 24$.

- *Case 2*: The G-BIHT algorithm using a learned sensing matrix $\mathbf{\Phi}$ and dithering vector $\boldsymbol{b}$, with the step sizes $\{\delta_i\}_{i=0}^{L-1}$ fixed as in Case 1.
- *Case 3*: The LG-BIHT AE method, corresponding to the iterative updates in (3.54), using learned $\mathbf{\Phi}$, $\boldsymbol{b}$, and $\{\delta_i\}_{i=0}^{L-1}$.

For comparison, we also include the numerical results of the LG-RFPI AE and G-RFPI methods. As shown in Figure 3.6, the G-BIHT algorithm with randomly generated $\mathbf{\Phi}$ and $\boldsymbol{b}$ (Case 1) significantly outperforms its G-RFPI counterpart, achieving high accuracy in just a few iterations. Meanwhile, the LG-RFPI method still yields lower MSE than the vanilla G-RFPI, highlighting the benefit of learning in that context. Furthermore, a comparison between Case 2 of the G-BIHT algorithm (with learned $\mathbf{\Phi}$ and $\boldsymbol{b}$) and both the LG-RFPI and vanilla G-BIHT (Case 1) demonstrates the strength of the learned sensing setup and the underlying G-BIHT framework. Remarkably, by learning only $\mathbf{\Phi}$ and $\boldsymbol{b}$—while keeping the step sizes fixed—G-BIHT achieves better performance than LG-RFPI, despite the latter relying on fully learned variables. Finally, Figure 3.6a, b shows that the LG-BIHT algorithm (Case 3) significantly outperforms all other methods, achieving much lower MSE in fewer iterations—particularly when compared to the LG-RFPI AE. This superior performance stems from the explicit incorporation of the known sparsity level K into the recovery process. As previously discussed, when the sparsity level is known a priori, it is advantageous to use either the G-BIHT algorithm (when avoiding learning) or the LG-BIHT approach. Notably, similar to the earlier methods, LG-BIHT also generalizes well to unseen sparsity levels such as $K = 24$ (see Figure 3.6b), despite the network not being exposed to this value during training.

3.4 Conclusion

This chapter presents a comprehensive overview of low-resolution signal processing techniques with a focus on dithered one-bit quantization. Key results on covariance matrix recovery under both stationary and nonstationary signal models were discussed, highlighting the benefits of dithered quantization in mitigating the loss of information caused by extreme quantization. Several theoretical results were provided for covariance estimation, including performance bounds under sub-Gaussian models. The chapter also introduced algorithm unrolling as a model-based deep learning approach for solving inverse problems in this context. In particular, the LISTA architecture was applied to the DoA estimation problem using covariance estimates from one-bit data. The framework achieves accurate recovery while maintaining computational efficiency and interpretability. Finally, the discussion extended to the one-bit CS problem, where model-driven AEs were designed to jointly optimize sensing and recovery. These architectures enable end-to-end learning and offer practical advantages in scenarios where low-bit data acquisition is necessary due to hardware constraints. The combination of dithering, one-bit quantization, and deep unrolling provides a robust and scalable solution for modern sensing systems. This chapter serves as a foundation for further developments in low-resolution signal processing using learning-based methods.

References

1 A. Mezghani and A. L. Swindlehurst. Blind estimation of sparse broadband massive MIMO channels with ideal and one-bit ADCs. *IEEE Transactions on Signal Processing*, 66(11):2972–2983, 2018.

2 C. Liu and P. Vaidyanathan. One-bit sparse array DoA estimation. In *IEEE International Conference on Acoustics, Speech and Signal Processing*, pages 3126–3130, 2017.

3 P. Deift, F. Krahmer, and C. Güntürk. An optimal family of exponentially accurate one-bit sigma-delta quantization schemes. *Communications on Pure and Applied Mathematics*, 64(7):883–919, 2011.

4 R. Maronna. Robust M-estimators of multivariate location and scatter. *The Annals of Statistics*, 4(1):51–67, 1976.

5 A. Weiss. Blind direction-of-arrival estimation in acoustic vector-sensor arrays via tensor decomposition and Kullback-Leibler divergence covariance fitting. *IEEE Transactions on Signal Processing*, 69:531–545, 2020.

6 L. Xu, J. Li, and P. Stoica. Target detection and parameter estimation for MIMO radar systems. *IEEE Transactions on Aerospace and Electronic Systems*, 44(3):927–939, 2008.

7 H. Djelouat, M. Leinonen, L. Ribeiro, and M. Juntti. Joint user identification and channel estimation via exploiting spatial channel covariance in mMTC. *IEEE Wireless Communications Letters*, 10(4):887–891, 2021.

8 J. H. Van Vleck and D. Middleton. The spectrum of clipped noise. *Proceedings of the IEEE*, 54(1):2–19, 1966.

9 G. Jacovitti and A. Neri. Estimation of the autocorrelation function of complex Gaussian stationary processes by amplitude clipped signals. *IEEE Transactions on Information Theory*, 40(1):239–245, 1994.

10 O. Bar-Shalom and A. Weiss. DoA estimation using one-bit quantized measurements. *IEEE Transactions on Aerospace and Electronic Systems*, 38(3):868–884, 2002.

11 J. J. Bussgang. Crosscorrelation functions of amplitude-distorted gaussian signals. Research Laboratory of Electronics, MIT, Cambridge, MA, USA, Rep. 216, 1952.

12 L. Schuchman. Dither signals and their effect on quantization noise. *IEEE Transactions on Communication Technology*, 12(4):162–165, 1964.

13 P. Carbone. Quantitative criteria for the design of dither-based quantizing systems. *IEEE Transactions on Instrumentation and Measurement*, 46(3):656–659, 1997.

14 A. Eamaz, K. Mishra, F. Yeganegi, and M. Soltanalian. UNO: Unlimited sampling meets one-bit quantization. *IEEE Transactions on Signal Processing*, 72:997–1014, 2024.

15 A. Eamaz, F. Yeganegi, and M. Soltanalian. Matrix completion from one-bit dither samples. *IEEE Transactions on Signal Processing*, 74:1616–1630, 2024.

16 F. Xi, Y. Xiang, S. Chen, and A. Nehorai. Gridless parameter estimation for one-bit MIMO radar with time-varying thresholds. *IEEE Transactions on Signal Processing*, 68:1048–1063, 2020.

17 I. S. Robinson, J. Toplicar, and J. G. Heston. Analog to digital conversion using differential dither. US Patent 10,298,256, May 21, 2019.

18 A. M. A. Ali and P. Gulati. Background calibration of reference, DAC, and quantization non-linearity in ADCs. US Patent 10,547,319,B2, May 2, 2020.

19 C. Thrampoulidis and A. S. Rawat. The generalized LASSO for sub-Gaussian measurements with dithered quantization. *IEEE Transactions on Information Theory*, 66(4):2487–2500, 2020.

20 C. Xu and L. Jacques. Quantized compressive sensing with RIP matrices: The benefit of dithering. *Information and Inference: A Journal of the IMA*, 9(3):543–586, 2020.

21 S. Dirksen and S. Mendelson. Non-Gaussian hyperplane tessellations and robust one-bit compressed sensing. *Journal of the European Mathematical Society*, 23(9):2913–2947, 2021.

22 A. Eamaz, F. Yeganegi, D. Needell, and M. Soltanalian. Harnessing the power of sample abundance: Theoretical guarantees and algorithms for accelerated one-bit sensing. *IEEE Transactions on Information Theory*, 70(9):6690–6713, 2024.

23 F. Yeganegi, A. Eamaz, and M. Soltanalian. Low-rank matrix sensing with dithered one-bit quantization. In *2024 IEEE International Symposium on Information Theory (ISIT)*, pages 527–532. IEEE, 2024.

24 A. Eamaz, F. Yeganegi, and M. Soltanalian. One-bit matrix completion with time-varying sampling thresholds. In *2023 International Conference on Sampling Theory and Applications (SampTA)*, pages 1–5. IEEE, 2023.

25 A. Eamaz, F. Yeganegi, Y. Hu, S. Sun, and M. Soltanalian. Automotive radar sensing with sparse linear arrays using one-bit Hankel matrix completion. In *2024 IEEE Radar Conference (RadarConf24)*, pages 1–6. IEEE, 2024.

26 A. Eamaz, F. Yeganegi, Y. Hu, M. Soltanalian, and S. Sun. Collaborative automotive radar sensing via mixed-precision distributed array completion. *ICASSP 2025-2025 IEEE International Conference on Acoustics, Speech and Signal Processing (ICASSP)*, 2025.

27 A. Eamaz, F. Yeganegi, and M. Soltanalian. One-bit phase retrieval: More samples means less complexity? *IEEE Transactions on Signal Processing*, 70:4618–4632, 2022.

28 A. Eamaz, F. Yeganegi, and M. Soltanalian. OPeRA: Leveraging the sample size and complexity trade-off towards efficient one-bit phase retrieval. In *2022 56th Asilomar Conference on Signals, Systems, and Computers*, pages 81–85. IEEE, 2022.

29 A. Eamaz, F. Yeganegi, and M. Soltanalian. Global one-bit phase retrieval via sample abundance-including an application to STFT measurements. In *2023 57th Asilomar Conference on Signals, Systems, and Computers*, pages 925–931. IEEE, 2023.

30 A. Eamaz, K. Mishra, F. Yeganegi, and M. Soltanalian. Unlimited sampling via one-bit quantization. In *2023 International Conference on Sampling Theory and Applications (SampTA)*, pages 1–5. IEEE, 2023.

31 A. Eamaz, F. Yeganegi, and M. Soltanalian. HDR imaging with one-bit quantization. In *2024 IEEE 13rd Sensor Array and Multichannel Signal Processing Workshop (SAM)*, pages 1–5. IEEE, 2024.

32 A. Eamaz, F. Yeganegi, and M. Soltanalian. Streamlining UNO: A generalized sampling approach to optimal one-bit modulo sensing. In *ICASSP 2025-2025 IEEE International Conference on Acoustics, Speech and Signal Processing (ICASSP)*, pages 1–5. IEEE, 2025.

33 A. Eamaz, F. Yeganegi, and M. Soltanalian. Modified arcsine law for one-bit sampled stationary signals with time-varying thresholds. In *ICASSP 2021-2021 IEEE International Conference on Acoustics, Speech and Signal Processing (ICASSP)*, pages 5459–5463. IEEE, 2021.

34 A. Eamaz, F. Yeganegi, and M. Soltanalian. Covariance recovery for one-bit sampled non-stationary signals with time-varying sampling thresholds. *IEEE Transactions on Signal Processing*, 70:5222–5236, 2022.

35 A. Eamaz, F. Yeganegi, and M. Soltanalian. Covariance recovery for one-bit sampled stationary signals with time-varying sampling thresholds. *Signal Processing*, 206:108899, 2023.

36 C. Liu and Y. Chou. Approximation and analysis of the one-bit Hermite law. In *ICASSP 2025-2025 IEEE International Conference on Acoustics, Speech and Signal Processing (ICASSP)*, pages 1–5. IEEE, 2025.

37 S. Dirksen and J. Maly. Tuning-free one-bit covariance estimation using data-driven dithering. *IEEE Transactions on Information Theory*, 70(7):5228–5247, 2024.

38 S. Dirksen, J. Maly, and H. Rauhut. Covariance estimation under one-bit quantization. *The Annals of Statistics*, 50(6):3538–3562, 2022.

39 T. Yang, J. Maly, S. Dirksen, and G. Caire. Plug-in channel estimation with dithered quantized signals in spatially non-stationary massive MIMO systems. *IEEE Transactions on Communications*, 72(1):387–402, 2024.

40 D. Ho and B. Rao. Antithetic dithered 1-bit massive MIMO architecture: Efficient channel estimation via parameter expansion and PML. *IEEE Transactions on Signal Processing*, 67(9):2291–2303, 2019.

41 S. Sedighi, M. Soltanalian, and B. Ottersten. On the performance of one-bit DoA estimation via sparse linear arrays. *IEEE Transactions on Signal Processing*, 69:6165–6182, 2021.

42 X. Huang and B. Liao. One-bit MUSIC. *IEEE Signal Processing Letters*, 26(7):961–965, 2019.

43 K. Yu, Y. Zhang, M. Bao, Y. Hu, and Z. Wang. DoA estimation from one-bit compressed array data via joint sparse representation. *IEEE Signal Processing Letters*, 23(9):1279–1283, 2016.

44 X. Su, Z. Liu, J. Shi, P. Hu, T. Liu, and X. Li. Real-valued deep unfolded networks for off-grid DoA estimation via nested array. *IEEE Transactions on Aerospace and Electronic Systems*, 59(4):4049–4062, 2023.

45 L. Wu, Z. Liu, and J. Liao. DoA estimation using an unfolded deep network in the presence of array imperfections. In *International Conference on Signal and Image Processing*, pages 182–187, 2022.

46 R. Liu, S. Cheng, L. Ma, X. Fan, and Z. Luo. Deep proximal unrolling: Algorithmic framework, convergence analysis and applications. *IEEE Transactions on Image Processing*, 28(10):5013–5026, 2019.

47 J. Zhang, B. Chen, R. Xiong, and Y. Zhang. Physics-inspired compressive sensing: Beyond deep unrolling. *IEEE Signal Processing Magazine*, 40(1):58–72, 2023.

48 P. T. Boufounos and R. G. Baraniuk. 1-bit compressive sensing. In *2008 42nd Annual Conference on Information Sciences and Systems*, pages 16–21. IEEE, 2008.

49 L. Jacques, J. N. Laska, P. T. Boufounos, and R. G. Baraniuk. Robust 1-bit compressive sensing via binary stable embeddings of sparse vectors. *IEEE Transactions on Information Theory*, 59(4):2082–2102, 2013.

50 Y. Plan and R. Vershynin. One-bit compressed sensing by linear programming. *Communications on Pure and Applied Mathematics*, 66(8):1275–1297, 2013.

51 M.P. Friedlander, H. Jeong, Y. Plan, and Ö. Yılmaz. NBIHT: An efficient algorithm for 1-bit compressed sensing with optimal error decay rate. *IEEE Transactions on Information Theory*, 68(2):1157–1177, 2021.

52 Y. Plan and R. Vershynin. Dimension reduction by random hyperplane tessellations. *Discrete & Computational Geometry*, 51(2):438–461, 2014.

53 S. Oymak and B. Recht. Near-optimal bounds for binary embeddings of arbitrary sets. *arXiv preprint arXiv:1512.04433*, 2015.

54 K. Knudson, R. Saab, and R. Ward. One-bit compressive sensing with norm estimation. *IEEE Transactions on Information Theory*, 62(5):2748–2758, 2016.

55 R.G. Baraniuk, S. Foucart, D. Needell, Y. Plan, and M. Wootters. Exponential decay of reconstruction error from binary measurements of sparse signals. *IEEE Transactions on Information Theory*, 63(6):3368–3385, 2017.

56 L. Jacques. Small width, low distortions: Quantized random embeddings of low-complexity sets. *IEEE Transactions on Information Theory*, 63(9):5477–5495, 2017.

57 M. Dereziński and E. Rebrova. Sharp analysis of sketch-and-project methods via a connection to randomized singular value decomposition. *arXiv preprint arXiv:2208.09585*, 2022.

58 T. Chen, X. Chen, W. Chen, H. Heaton, J. Liu, Z. Wang, and W. Yin. Learning to optimize: A primer and a benchmark. *Journal of Machine Learning Research*, 23(189):1–59, 2022.

59 D. Gilton, G. Ongie, and R. Willett. Deep equilibrium architectures for inverse problems in imaging. *IEEE Transactions on Computational Imaging*, 7:1123–1133, 2021.

60 X. Chen, Y. Zhang, C. Reisinger, and L. Song. Understanding deep architecture with reasoning layer. *Advances in Neural Information Processing Systems*, 33:1240–1252, 2020.

61 C. Yang, Y. Gu, B. Chen, H. Ma, and H. So. Learning proximal operator methods for nonconvex sparse recovery with theoretical guarantee. *IEEE Transactions on Signal Processing*, 68:5244–5259, 2020.

62 X. Chen, J. Liu, Z. Wang, and W. Yin. Theoretical linear convergence of unfolded ISTA and its practical weights and thresholds. *Advances in Neural Information Processing Systems*, 31: 9061–9071, 2018.

63 X. Chen, J. Liu, Z. Wang, and W. Yin. Hyperparameter tuning is all you need for LISTA. *Advances in Neural Information Processing Systems*, 34:11678–11689, 2021.

64 A. Aberdam, A. Golts, and M. Elad. Ada-LISTA: Learned solvers adaptive to varying models. *IEEE Transactions on Pattern Analysis and Machine Intelligence*, 44(12):9222–9235, 2021.

65 K. Gregor and Y. LeCun. Learning fast approximations of sparse coding. In *Proceedings of the 27th International Conference on International Conference on Machine Learning*, pages 399–406, 2010.

66 J. Liu and X. Chen. ALISTA: Analytic weights are as good as learned weights in LISTA. In *International Conference on Learning Representations (ICLR)*, 2019.

67 A. Balatsoukas-Stimming and C. Studer. Deep unfolding for communications systems: A survey and some new directions. In *2019 IEEE International Workshop on Signal Processing Systems (SiPS)*, pages 266–271. IEEE, 2019.

68 A. Eamaz, F. Yeganegi, and M. Soltanalian. Physics-Inspired Binary Neural Networks: Interpretable Compression with Theoretical Guarantees. *arXiv:2502.01908v3*, 2025.

69 P. Xiao, B. Liao, and N. Deligiannis. DeepFPC: A deep unfolded network for sparse signal recovery from 1-bit measurements with application to DoA estimation. *Signal Processing*, 176:107699, 2020.

70 F. Yeganegi, A. Eamaz, T. Esmaeilbeig, and M. Soltanalian. Deep learning-enabled one-bit DoA estimation. In *2024 IEEE 13rd Sensor Array and Multichannel Signal Processing Workshop (SAM)*, pages 1–5. IEEE, 2024.

71 S. Khobahi and M. Soltanalian. Model-based deep learning for one-bit compressive sensing. *IEEE Transactions on Signal Processing*, 68:5292–5307, 2020.

72 E. J. Candès, J. Romberg, and T. Tao. Robust uncertainty principles: Exact signal reconstruction from highly incomplete frequency information. *IEEE Transactions on Information Theory*, 52(2):489–509, 2006.

73 E. J. Candes and T. Tao. Decoding by linear programming. *IEEE Transactions on Information Theory*, 51(12):4203–4215, 2005.

74 A. Beck and M. Teboulle. A fast iterative shrinkage-thresholding algorithm for linear inverse problems. *SIAM Journal on Imaging Sciences*, 2(1):183–202, 2009.

4

Reinforcement Learning for Intelligent Radar Target Detection

Adam Umra,[1,] Aya Mostafa Ahmed,[2] Aydin Sezgin,[1] Maria Sabrina Greco,[3] and Fulvio Gini[3]*

[1] *Faculty of Electrical Engineering and Information Technology, Ruhr University Bochum, Bochum, Germany*
[2] *Corporate Research, Robert Bosch GmbH, Renningen, Germany*
[3] *Department of Information Engineering, University of Pisa, Pisa, Italy*

*Corresponding author: adam.umra@rub.de

4.1 Introduction

Radar target detection is rapidly advancing toward intelligent and adaptive approaches due to increasingly demanding applications in automotive safety, autonomous navigation, and remote sensing. Traditional radar systems struggle in dynamic environments, prompting the adoption of cognitive radar concepts, multiple-input and multiple-output (MIMO) architectures, and reinforcement learning (RL) to dynamically optimize performance. This chapter introduces a unified framework that integrates RL-driven waveform selection with colocated MIMO radar and statistical detection techniques, thereby enhancing detection performance in complex and uncertain scenarios. The following sections provide background context and motivation for the framework, outline the scope of the chapter, and conclude by describing its organization.

4.1.1 Background and Motivation

In recent decades, radar technology has undergone significant advances, particularly in the development of intelligent and adaptive radar systems [1, 2]. These advances have been driven by increasingly demanding applications across a wide range of critical domains [3], where radar systems are required to operate with high accuracy, reliability, and robustness under highly dynamic and unpredictable environmental conditions. Conventional radar systems, however, are typically based on fixed transmission strategies and simplified assumptions regarding noise, clutter, and interference statistics. Such model-driven approaches rely on predefined waveforms and static system configurations, which inherently limit their ability to adapt to nonstationary environments. As a result, their performance degrades in the presence of rapidly changing clutter dynamics, evolving interference, or mismatches between assumed and actual statistical models. These limitations highlight the inadequacy of traditional radar methodologies in complex operational scenarios and underscore the need for more flexible and adaptive solutions. In response, the research community has increasingly shifted toward data-driven and intelligent radar systems that can dynamically adjust their operational strategies based on real-time environmental interactions and feedback.

A significant development gaining attention recently is the concept of *cognitive radar* [4, 5]. Cognitive radar systems explicitly aim to overcome the rigidity and limitations inherent in traditional radar by incorporating principles from cognitive science and intelligent control. Such systems achieve adaptive capabilities through a real-time perception-action loop, enabling continuous sensing, learning, and dynamic decision-making based on instantaneous environmental assessments [6]. More specifically, cognitive radar dynamically adjusts critical parameters such as transmit waveform, power allocation, modulation schemes, and beamforming characteristics in response to the sensed radar environment. Through this feedback mechanism, cognitive radar achieves more effective resource management, thus significantly improving target detection, classification accuracy, and tracking performance, particularly in cluttered or interference-rich operational scenarios. However, as these adaptive capabilities become more sophisticated, cognitive radar increasingly requires mechanisms that support flexible and context-aware operation across multiple dimensions. One such dimension is spectrum awareness, which has motivated recent efforts on cognitive radar coexistence and operation in congested frequency bands. For example, [7] and [8] investigate spectrum sensing, collaborative sharing, and operation in spectrally dense environments, reinforcing the role of environmental awareness within the cognitive radar loop. More broadly, effective cognitive radar operation relies on access to adaptable transmission degrees of freedom to support real-time waveform selection and control. In this context, the inherent waveform diversity offered by MIMO radar systems provides a natural architectural foundation for cognitive radar implementation.

MIMO radar systems leverage multiple transmitting and receiving antenna elements, thereby harnessing spatial diversity, coherent processing gains, and advanced adaptive beamforming methodologies. The foundational principles of MIMO radar were initially demonstrated by research highlighting significant detection capability improvements obtained by exploiting spatially distributed antenna configurations [9]. Subsequent investigations into closely spaced, colocated antenna array configurations demonstrated additional performance benefits, including coherent array gain, enhanced spatial resolution, improved angular resolution, and notably reduced sidelobe levels [10, 11]. Furthermore, the waveform diversity inherent in MIMO radar systems facilitates more effective illumination and identification of targets, enhanced multipath mitigation, and superior clutter suppression capabilities. Together, these features make MIMO radar robust and reliable, even in rapidly changing and highly unpredictable environments. Building upon the foundational principles of MIMO radar systems, the massive MIMO (MMIMO) radar leverages large-scale antenna arrays to significantly enhance spatial processing capabilities [12]. This paradigm shift enables the system to achieve unprecedented spatial resolution and superior target detection performance, particularly in complex or cluttered environments. A key innovation in MMIMO radar is the incorporation of a robust Wald-type hypothesis test [13], specifically designed to operate effectively under the high-dimensional settings characteristic of massive antenna configurations.

Complementing transmitter waveform diversity, receiver cognition is equally critical, as it governs radar decision-making and directs waveform selection. To this end, RL is adopted to enable cognitive radar operation [14–16]. Traditional radar systems, as outlined above, typically require explicit statistical modeling of noise, clutter, and interference. In contrast, RL approaches avoid these strict modeling requirements by relying instead on iterative policy refinement driven by continuous, reward-based interactions with the radar environment [17, 18]. This capability allows RL algorithms to adaptively optimize radar operational parameters, such as waveform selection and beamforming strategies, by directly linking performance metrics, including detection probability and false alarm rate, to the policy updating mechanism. Such direct feedback and adaptability empower RL-based radar methods to excel in dynamically changing environments, significantly enhancing their suitability for integration into cognitive radar architectures.

Overall, recent innovations in cognitive radar technology—including advanced MIMO radar architectures and RL methodologies—signal a decisive move toward intelligent, adaptive, and robust radar systems capable of meeting the stringent performance demands of modern operational scenarios.

4.1.2 Scope

As previously discussed, real-world radar environments are inherently complex and pose significant challenges to conventional radar signal processing techniques. To tackle these challenges, this chapter introduces a framework that effectively combines a robust Wald-type statistical detection technique [13, 19] with RL-driven adaptive waveform optimization [16]. Wald-type detection methods are particularly appealing due to their inherent capability to maintain a constant false alarm rate (CFAR) even in the presence of significant environmental uncertainties. Recent research has highlighted the effectiveness of these methodologies, particularly in MMIMO radar systems. The approach leverages the high dimensionality of large antenna arrays to enhance detection robustness, delivering reliable performance even in the absence of accurate prior knowledge of disturbance statistics. In [13], the authors provide analytical expressions for false alarm and detection probabilities, with numerical simulations reinforcing the theoretical findings. These results underscore the viability and practical advantages of MMIMO systems in complex radar environments.

While robust detection mechanisms form the cornerstone of reliable radar operation, waveform optimization plays an equally critical role in enhancing radar performance. Optimal waveform design significantly improves radar resource management and maximizes the signal-to-clutter-plus-noise ratio (SCNR) under diverse operational conditions [20]. Building on this observation, [16] proposed an RL-based adaptive waveform optimization framework for multitarget detection in cognitive MMIMO radar systems. By formulating radar operation as a perception-action cycle, this method enables autonomous sensing, continuous environmental learning, and real-time adaptation of transmitted waveforms. The RL framework optimizes beamforming patterns dynamically, yielding substantial performance improvements compared to conventional omnidirectional and non-RL adaptive techniques. Simulations under challenging conditions—including heavy-tailed clutter, low signal-to-noise ratios (SNRs), and rapidly changing scenarios—demonstrated the superiority of the RL-based algorithm in multitarget detection. This approach complements the robust Wald-type detection framework developed in [13], providing a solution for reliable radar performance in unpredictable environments.

While in [16] the authors emphasize how RL-inspired algorithms enable autonomous decision-making under uncertainty, learning-based methods have, in parallel, driven advances in waveform optimization, yielding both data-driven and model-based strategies for enhanced radar operation. For example, model-based learning has been applied to MIMO radar waveform design in multitarget scenarios [21], while algorithm unrolling has been used for MIMO radar beampattern synthesis [22]. Deep learning has also been leveraged for automotive radar waveform optimization [23]. Collectively, these contributions illustrate that waveform optimization may be pursued independently or in concert with detection, with the two tasks forming a mutually reinforcing loop. Fundamentally, their objectives differ: detection seeks reliable target identification under uncertainty, whereas waveform design focuses on shaping the transmitted signal for improved sensing performance. Complementary to this radar-centric perspective, Chapter 2 of this book addresses integrated sensing and communications (ISAC) systems, introducing a model-based RL framework for joint spatial and spectral resource allocation that enhances learning efficiency and interpretability in dynamic 6G environments.

Building upon the RL-based multitarget detection framework in [16], the authors in [24] introduced refinements aimed at enhancing multitarget detection capabilities in MMIMO radar systems. Their work incorporated modifications to the traditional ϵ-greedy RL policy, specifically designed to facilitate rapid reacquisition of missed targets. Additionally, they proposed a novel adaptive strategy for autonomously tuning RL hyperparameters, significantly reducing dependence on manual adjustments. Through extensive simulations in adverse environments characterized by strong clutter and low SNRs, their approach demonstrated substantial performance gains, further solidifying the potential of RL-based detection and optimization in complex radar applications. To enable simultaneous radar detection and communication, [25] extended the radar framework to an RL-based dual-functional MIMO system that uses a shared transmit waveform. The approach enhances weak-target detection through a refined reward mechanism, while a low-complexity two-step beamforming design efficiently embeds communication data using convex optimization and rotation-based precoding [26]. However, this topic will not be within the scope of this chapter. Extending the practical applicability of the integrated radar framework, [27] expanded its scope to encompass realistic two-dimensional (2D) clutter scenarios. By leveraging planar MMIMO array configurations, this extension effectively addresses the challenges posed by complex 2D clutter environments, reinforcing the robustness and adaptability of the combined Wald-type detection and RL-driven waveform optimization methodologies.

In particular, this work builds on the earlier contributions [16, 24, 27], which provide the foundational basis for the algorithms and evaluations presented here. The developments in this chapter should be understood as a continuation of that line of research, extending it with additional background and offering new insights, especially in the context of dynamic environments. Specifically, we present the RL-based waveform selection algorithm that works hand in hand with robust asymptotic statistical techniques to formulate a fully adaptive, data-driven detection framework for MMIMO radar systems. A key advantage of this approach is that it achieves high detection performance without requiring any prior knowledge of the environment. In addition, we provide a discussion of the MMIMO system's signal and disturbance models alongside the robust Wald-type test. As we will demonstrate, asymptotic statistical analysis is a powerful method for describing the statistical behavior of the derived test, especially as the number of virtual antennas tends toward infinity. To provide a well-rounded and complete understanding of the topic, we also introduce the fundamental concepts of RL, Markov decision processes (MDPs), and relevant learning algorithms. The core contribution of this chapter lies in the seamless integration of RL into the MMIMO detection problem, resulting in an adaptive RL-based waveform selection algorithm designed to maximize the system's probability of detection (P_D), particularly in challenging 2D clutter environments. By leveraging the strengths of both RL and robust statistical methodologies, this approach represents a significant step toward autonomous, intelligent radar systems capable of operating effectively in highly dynamic and uncertain environments.

4.1.3 Organization of the Chapter

The remainder of this chapter is structured as follows. Section 4.2 provides an overview of the underlying framework, establishing the foundational concepts necessary for the subsequent discussions. In Section 4.3, we introduce the MIMO radar system model, detailing its core components, including the array architecture, transmitted waveforms, signal propagation characteristics, and the impact of noise on received signals. Additionally, we formulate the target detection problem, laying the groundwork for the integration of advanced signal processing and learning-based techniques. Section 4.4 explores the application of RL within cognitive radar systems. A particular emphasis is placed on the definition of states, actions, and rewards, demonstrating how these elements are designed to optimize detection performance. Furthermore, a comprehensive discussion of the *SARSA* algorithm is provided, highlighting its role in refining the decision-making process and its seamless integration into the radar signal processing chain. In Section 4.5, we conduct a comparative analysis of the RL-based approach against non-RL techniques. The evaluation includes key performance metrics such as detection probability and false alarm rate, offering insights into the practical advantages of adopting a learning-driven methodology in MMIMO radar systems. Finally, Section 4.6 summarizes the key findings of this work and discusses potential avenues for future research.

4.2 Key Concepts of the Framework

This section presents the fundamental principles underpinning the proposed cognitive radar framework, integrating colocated MMIMO radar, robust statistical detection methods (Wald-type detector), and RL. Leveraging synergy among these components, we develop a data-driven, adaptive detection system optimized for MMIMO radar applications. This section outlines the principal components of the proposed framework and clarifies their respective roles within the overall detection architecture.

4.2.1 Colocated MMIMO Radar: Principles and Advantages

As discussed earlier, this framework employs colocated MMIMO radar, a technology that has significantly advanced modern radar systems by leveraging the coordinated use of multiple closely spaced transmit and receive antennas. Unlike conventional radar configurations that are constrained by limited antenna arrays, colocated MMIMO radar capitalizes on sophisticated signal processing techniques. This approach substantially enhances radar capabilities, facilitating superior detection, precise localization, and effective tracking of targets, even in environments characterized by high interference or considerable clutter [10].

A distinguishing feature of colocated MMIMO radar is its simultaneous signal transmission and reception across multiple antennas, which can be configured as a uniform planar array (UPA). This configuration provides extended spatial coverage in azimuth and elevation dimensions, allowing superior resolution between closely spaced targets. Consequently, MMIMO radar significantly improves situational awareness and effectively detects subtle target movements critical for applications like autonomous navigation and ISAC [26, 28]. Furthermore, MMIMO radar employs orthogonal waveforms, where each transmit antenna emits unique signals. This waveform diversity facilitates efficient signal separation, enabling concurrent detection of multiple objects with high positional accuracy. Additionally, the use of a virtual array extends the effective aperture, improving angular resolution and detecting low-observable targets that traditional systems might miss. MMIMO radar's unique waveform independence enhances parameter identifiability, enabling the distinction of significantly more targets compared to conventional phased-array radars. Moreover, sophisticated optimization of transmitted waveforms and adaptive beampattern design, achieved via convex optimization methods, provides adaptive signal enhancement and real-time suppression of sidelobes.

Overall, the high angular resolution, flexible waveform design, and advanced signal processing capabilities of colocated MMIMO radar make it a powerful and versatile foundation for next-generation radar systems.

4.2.2 Wald-type Detector for Robust Target Detection

Despite the advanced spatial capabilities of MMIMO radar, conventional detection methods typically struggle with unknown or non-Gaussian disturbances. Addressing this challenge, the Wald-type detector [13] offers a robust statistical approach, ensuring reliable detection under diverse disturbance conditions without prior distributional knowledge. By providing mathematically rigorous closed-form expressions for detection and false alarm probabilities, the Wald-type detector guarantees asymptotic convergence to chi-squared distributions, an effect enhanced as the antenna count increases. This property ensures robust detection performance even without secondary data for disturbance estimation, which is crucial in unpredictable real-world scenarios. The detector is computationally efficient and adaptable to various noise environments, proving especially beneficial in applications demanding real-time adaptability. This robust and mathematically rigorous detection method effectively complements the spatial diversity advantages offered by MMIMO radar.

4.2.3 RL for Adaptive Beamforming

RL operates on a reward–action cycle, in which an intelligent agent observes the current state of its environment, takes an action based on its policy, and then receives feedback in the form of a reward signal [17]. The policy in RL is essentially the strategy or mapping that tells the agent which action to take in a given state. It can be deterministic—always choosing the same action for a given state—or stochastic—choosing an action according to a probability distribution. Over time, through repeated interaction with its environment, the agent refines this policy to maximize cumulative rewards. By directly interacting with the operational environment and continuously updating its behavior based on observed rewards, RL avoids reliance on predefined models or assumptions. Instead, it exploits ongoing experience to maximize long-term cumulative rewards, leading to robust and adaptive decision-making.

Within this framework, RL is employed to adaptively control transmit beamforming based on detection feedback. This adaptability proves particularly effective under challenging conditions, such as low SNR, heavy-tailed disturbances, or rapidly changing operational scenarios [16]. By relying on data-driven feedback and reward-driven strategies, RL continuously adapts to environmental changes, thereby sustaining high detection accuracy under dynamic and unpredictable conditions. This underscores the transformative potential of RL for multitarget detection in MMIMO radar, where its reward–action cycle enables continual learning and near-optimal performance.

4.2.4 Integration into a Cognitive Radar Framework

Figure 4.1 illustrates the closed-loop interaction between MMIMO signal reception, Wald-type detection, and RL-based beamforming adaptation. At its foundation, this integrated framework exploits the inherent spatial diversity and high-resolution capabilities of colocated MIMO radar, significantly improving detection accuracy and enabling precise target localization. The utilization of Wald-type detection introduces rigorous statistical robustness into the system, enhancing its capability to maintain performance integrity even when faced with unpredictable and complex disturbances, such as clutter or interference. Furthermore, the incorporation of RL ensures continual system adaptation and optimization in real time. As operational environments evolve, RL dynamically updates the beamforming strategy, leveraging past experiences and adapting its policies to optimize system performance continuously. As depicted in Figure 4.1, this integrated process begins with signal reception, which the Wald-type detector analyzes to identify potential targets. The detector outcome informs the RL module, prompting policy adjustments that directly influence the subsequent beamforming strategy. Following these updates, the system transmits new signals, initiating a continuously adapting cycle. By seamlessly combining advanced radar technology, robust statistical methods, and adaptive machine learning, this framework ensures superior performance in dynamically changing and complex operational conditions.

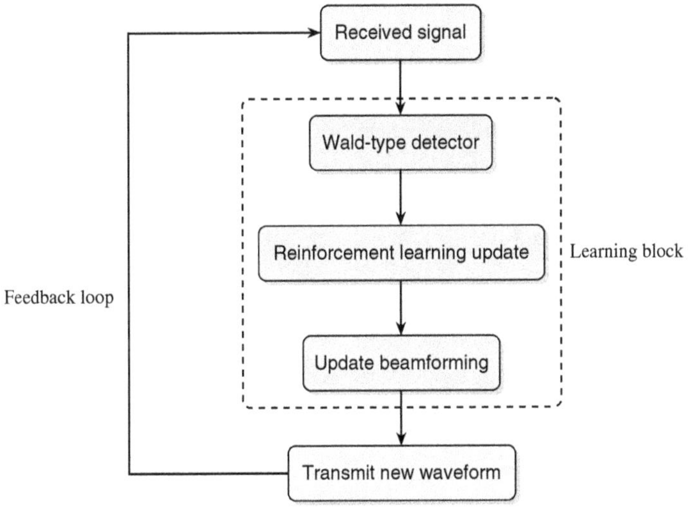

Figure 4.1 RL-driven adaptive beamforming for target detection.

Having introduced the key conceptual components of the proposed framework, we now turn to a precise mathematical formulation of the radar system. This formal signal model serves as the foundation upon which both the statistical detection strategy and the learning-based waveform adaptation are developed.

4.3 MIMO Radar Model

Notation: Bold lowercase letters (e.g., $\boldsymbol{x}$) denote column vectors, and bold uppercase letters (e.g., $\boldsymbol{A}$) denote matrices. The superscript $(\cdot)^{\mathrm{T}}$ indicates transpose, $(\cdot)^{\mathrm{H}}$ denotes Hermitian transpose, $(\cdot)^{*}$ the complex conjugate and $\mathrm{vec}(\cdot)$ is the vectorization operator. The identity matrix is represented by $\boldsymbol{I}$. The trace operator is $\mathrm{Tr}\{\cdot\}$, and $\otimes$ denotes the Kronecker product.

4.3.1 System Setup

Consider a colocated MIMO radar system equipped with M_T transmit (TX) antennas and M_R receive (RX) antennas. When the numbers of transmit and receive antennas are large, this configuration corresponds to a MMIMO system. In what follows, we refer to the system simply as MIMO, noting that MMIMO represents a particular regime within the broader MIMO framework. For ease of notation and analytical modeling, these antennas are assumed to be arranged on the xy-plane as UPAs. The arrangement on a planar grid allows for 2D angular resolution in azimuth and elevation. We define $N = M_T M_R$ as the total number of spatial

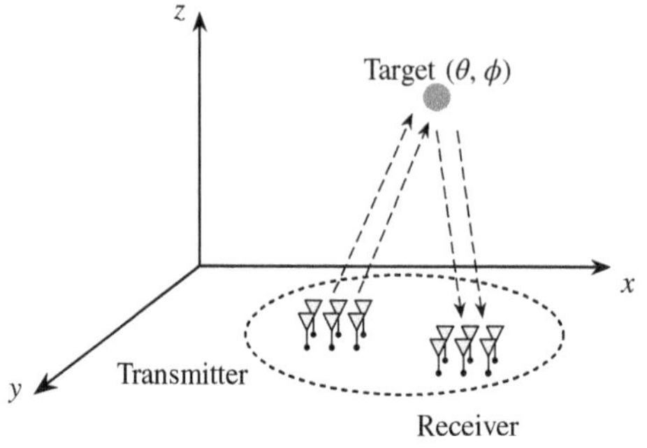

Figure 4.2 Colocated MIMO radar.

channels, sometimes called virtual array elements, because each transmitter-receiver pair is associated with one such channel. Consider a target scene where the azimuth and elevation angles are, respectively, denoted by θ and ϕ, measured with respect to the radar boresight on the xy-plane as shown in Figure 4.2. It is standard practice to position antenna elements at half-wavelength spacing to minimize spatial aliasing [29]. Therefore, if the wavelength is λ, the inter-element spacing in both x and y directions is $d_x = d_y = \lambda/2$.

4.3.2 Transmitted Waveform

A distinctive feature of colocated MIMO radar is the use of multiple, mutually orthonormal waveforms. By transmitting waveforms that are orthogonal in time or frequency (e.g., via code or frequency division), the radar system is able to separate returns from different transmit elements at the receiver with minimal interference. Let $\mathbf{s}(t) \in \mathbb{C}^{M_T}$ be the composite transmit signal at time t, expressed as:

$$\mathbf{s}(t) = \boldsymbol{W}\, \boldsymbol{\Psi}(t), \tag{4.1}$$

where $\boldsymbol{\Psi}(t) \in \mathbb{C}^{M_T}$ is a vector containing M_T orthonormal waveforms, and $\boldsymbol{W} \in \mathbb{C}^{M_T \times M_T}$ is a beamforming (or weighting) matrix. The orthonormal waveforms, satisfy:

$$\int_0^T \boldsymbol{\Psi}(t)\, \boldsymbol{\Psi}^{\mathrm{H}}(t)\, \mathrm{d}t = \boldsymbol{I}_{M_T}, \tag{4.2}$$

where T is the pulse duration. The integral implies that the cross-correlation among different waveforms is zero, thereby allowing them to be distinguished at the receiver.

Each column of $\boldsymbol{W}$, denoted $\boldsymbol{w}_i$, acts as a spatial steering vector associated with one of the orthonormal waveforms. The matrix $\boldsymbol{W}$ can be written as:

$$\boldsymbol{W} = \left[\boldsymbol{w}_1, \ldots, \boldsymbol{w}_{M_T}\right]^{\mathrm{T}}, \tag{4.3}$$

where $\boldsymbol{w}_i \in \mathbb{C}^{M_T}$ defines the amplitudes and phases for the ith transmitted signal across the M_T antennas. Varying $\boldsymbol{W}$ results in different radar beamformers. For instance, if the radar system wishes to direct more energy in a certain angular sector, the corresponding elements in $\boldsymbol{w}_i$ can be set to induce constructive interference in that direction.

A crucial operational constraint in radar systems is the total transmit power. The transmit power is controlled by:

$$\mathrm{Tr}\left\{\boldsymbol{W}\boldsymbol{W}^{\mathrm{H}}\right\} = P_T, \tag{4.4}$$

where P_T is the total available power.

4.3.3 Received Signal Model

We now focus on how the transmitted signals propagate through the medium and how they are captured at the receive array. Consider a point-like target located at azimuth θ and elevation ϕ. In a colocated MIMO radar, each transmit antenna emits its respective waveform, and the backscattered signals from the target are captured by each of the M_R receive antennas. Assuming a baseband representation [10], the received signal vector $\hat{\boldsymbol{r}}(t) \in \mathbb{C}^{M_R}$ can be written as:

$$\hat{\boldsymbol{r}}(t) = \alpha\, \boldsymbol{a}_R(\theta, \phi)\, \boldsymbol{a}_T^{\mathrm{T}}(\theta, \phi)\, \boldsymbol{s}(t-\tau) + \hat{\boldsymbol{c}}(t), \tag{4.5}$$

where τ is the round-trip delay for the signal to travel from the transmit array to the target and back to the receive array. The parameter $\alpha \in \mathbb{C}$ models the target's reflectivity, including radar cross-section and propagation losses. Under simple models such as Swerling 0 [30], α is taken as a deterministic complex constant. The vectors $\boldsymbol{a}_T(\theta, \phi) \in \mathbb{C}^{M_T}$ and $\boldsymbol{a}_R(\theta, \phi) \in \mathbb{C}^{M_R}$ are the transmit and receive steering vectors, respectively. They contain the phases induced by wave propagation across each antenna element, which depend on the angles of departure and arrival. The term $\hat{\boldsymbol{c}}(t)$ represents the aggregated disturbance. In our model, the disturbance originates from a 2D field of interference sources—such as thermal noise and clutter reflections—and is then vectorized. For UPAs of size $\sqrt{M_T} \times \sqrt{M_T}$, one convenient way to express the steering vector is via a Kronecker product:

$$\boldsymbol{a}_T(\theta, \phi) = \boldsymbol{a}_{T,x}(\theta, \phi) \otimes \boldsymbol{a}_{T,y}(\theta, \phi), \tag{4.6}$$

where $\boldsymbol{a}_{T,x}(\theta, \phi) \in \mathbb{C}^{\sqrt{M_T}}$ and $\boldsymbol{a}_{T,y}(\theta, \phi) \in \mathbb{C}^{\sqrt{M_T}}$ account for the phase progression along x and y axes, respectively [29]. By using $d_x = d_y = \frac{\lambda}{2}$, the spatial frequencies associated with azimuth and elevation are:

$$f_x = \frac{1}{2}\sin(\theta)\cos(\phi), \quad f_y = \frac{1}{2}\sin(\theta)\sin(\phi).$$

Each one-dimensional steering vector is then written as:

$$\boldsymbol{a}_{T,x}(f_x) = \left[1, e^{j2\pi f_x}, \dots, e^{j2\pi(\sqrt{M_\mathrm{T}}-1)f_x}\right]^\mathrm{T}, \tag{4.7}$$

$$\boldsymbol{a}_{\mathrm{T},y}(f_y) = \left[1, e^{j2\pi f_y}, \dots, e^{j2\pi(\sqrt{M_\mathrm{T}}-1)f_y}\right]^\mathrm{T}. \tag{4.8}$$

This factorization reduces the complexity of dealing with 2D array responses by separating them into independent factors along each dimension. Such a property is particularly useful for fast or low-complexity algorithmic implementations of angle estimation. Naturally, the preceding definitions also apply to the receive steering vector $\boldsymbol{a}_R(\theta, \phi)$. Employing spatial frequencies simplifies the practical implementation of the algorithm by allowing us to efficiently handle the steering vectors in a factorized manner. However, for clarity and consistency in the theoretical analysis, we will continue to represent the array response explicitly in terms of the azimuth and elevation angles.

Let us define τ' as an estimate of the true delay τ. A bank of matched filters, each tuned to one of the orthonormal waveforms, is applied to the received signal:

$$\boldsymbol{R}(\tau') = \int_0^T \hat{\boldsymbol{r}}(t)\, \boldsymbol{\Psi}_{\mathrm{MF}}^{\mathrm{H}}\left(t - \tau'\right)\mathrm{d}t, \tag{4.9}$$

where $\boldsymbol{\Psi}_{\mathrm{MF}}(t)$ matches the waveforms in $\boldsymbol{\Psi}(t)$ and thus $\boldsymbol{R} \in \mathbb{C}^{M_R \times M_T}$. By substituting the received signal in (4.5) into (4.9) and assuming perfect time of arrival (ToA) estimation (i.e., $\tau' = \tau$), the cross-correlation of orthonormal waveforms becomes $\boldsymbol{I}_{M_T}$. This simplifies the expression to:

$$\boldsymbol{R}(\tau) = \alpha\, \boldsymbol{a}_R(\theta, \phi)\, \boldsymbol{a}_T^{\mathrm{T}}(\theta, \phi)\, \boldsymbol{W} + \boldsymbol{C}, \tag{4.10}$$

where $\boldsymbol{C}$ represents the filtered disturbance and is defined as:

$$\boldsymbol{C} = \int_0^T \hat{\boldsymbol{c}}(t)\, \boldsymbol{\Psi}_{\mathrm{MF}}^{\mathrm{H}}\left(t - \tau'\right) \mathrm{d}t \tag{4.11}$$

In subsequent stages of signal processing, it is quite common to transform the matrix-valued output of the matched filter, denoted by $\boldsymbol{R}(\tau)$, into its vectorized counterpart. This reorganization from a matrix format into a vector format proves especially advantageous for leveraging various tools in linear algebra, such as vector-based decompositions and optimizations. Concretely, applying the vectorization operator to $\boldsymbol{R}(\tau)$ yields:

$$\mathrm{vec}\left(\boldsymbol{R}(\tau)\right) = \alpha \left[\left(\boldsymbol{W}^{\mathrm{T}}\, \boldsymbol{a}_T(\theta, \phi)\right) \otimes \boldsymbol{a}_R(\theta, \phi)\right] + \boldsymbol{c} \in \mathbb{C}^{M_R M_T}, \tag{4.12}$$

where $\boldsymbol{c} = \mathrm{vec}(\boldsymbol{C}) \in \mathbb{C}^{M_R M_T}$ represents the noise contribution, now formatted as a vector.

To further streamline the notation and capture this structure more compactly, the following vector is defined:

$$\boldsymbol{h}(\theta, \phi) \triangleq \left(\boldsymbol{W}^{\mathrm{T}}\, \boldsymbol{a}_T(\theta, \phi)\right) \otimes \boldsymbol{a}_R(\theta, \phi). \tag{4.13}$$

Hence, the vectorized form of the matched filter output becomes:

$$\boldsymbol{r} = \mathrm{vec}\left(\boldsymbol{R}(\tau)\right) = \alpha\, \boldsymbol{h}(\theta, \phi) + \boldsymbol{c}. \tag{4.14}$$

When K pulses are transmitted and subsequently received, one can refer to the measured vector at the kth pulse as:

$$\boldsymbol{r}^{(k)} = \alpha^{(k)}\, \boldsymbol{h}^{(k)}(\theta, \phi) + \boldsymbol{c}^{(k)}, \quad k = 1, \dots, K, \tag{4.15}$$

where each $\boldsymbol{r}^{(k)}$, $\alpha^{(k)}$, $\boldsymbol{h}^{(k)}(\theta, \phi)$, and $\boldsymbol{c}^{(k)}$ represents the corresponding quantity for the kth pulse.

4.3.4 Detection Problem

Having introduced the signal model for a MIMO radar configuration, we now turn our attention to the problem of detecting targets. Importantly, as the following discussion will show, working within a large-scale regime enables us to avoid many of the restrictive assumptions typically made in radar detection research. In a radar system, multiple pulses are typically transmitted over a coherent processing interval, and spatial filtering is used at the receiver to partition the angle space into discrete bins. These bins aim to cover the antenna field of view with a grid of azimuth and elevation angles. Suppose there are L discrete azimuth bins and I discrete elevation bins, forming an $L \times I$ grid. Each bin (l, i), with $l \in \{1, \dots, L\}$ and $i \in \{1, \dots, I\}$, corresponds to a particular azimuth angle θ_l and elevation angle ϕ_i [27]. Let $\boldsymbol{r}_{l,i}^{(k)}$ denote the measurement vector obtained from the kth transmitted pulse associated with bin (l, i). The single-snapshot detection problem for a MIMO system can be formulated as

$$\begin{aligned} \mathcal{H}_0 &: \ \boldsymbol{r}_{l,i}^{(k)} = \boldsymbol{c}_{l,i}^{(k)}, \quad k = 1, \dots, K, \\ \mathcal{H}_1 &: \ \boldsymbol{r}_{l,i}^{(k)} = \alpha_{l,i}^{(k)}\, \boldsymbol{h}_{l,i}^{(k)} + \boldsymbol{c}_{l,i}^{(k)}, \quad k = 1, \dots, K, \end{aligned} \tag{4.16}$$

where the null hypothesis $\mathcal{H}_0$ defines the absence of a target and $\mathcal{H}_1$ its presence. A considerable number of studies investigate the formulation of decision statistics capable of distinguishing the null hypothesis $\mathcal{H}_0$ from its alternative $\mathcal{H}_1$ [31, 32]. Many of these methods depend on simplifying assumptions, such as the availability of a sufficiently large set of independent and identically distributed measurement vectors, the modeling of disturbance as Gaussian with a diagonal or block-diagonal covariance structure, and the maintenance of perfect orthogonality in the waveform matrix. While these assumptions facilitate analytical tractability, they are often too restrictive and commonly fail to hold in realistic operating conditions [33]. In [13], a method was introduced to relax the first two assumptions by exploiting the extensive number of virtual antenna channels. This approach concentrates on a single-snapshot scenario and enforces only the following assumption [13]:

Assumption 4.1 Let c represent a discrete-time, complex circular, possibly non-Gaussian random process $\{c_n, \forall n\}$ whose autocorrelation function decays polynomially. Formally:

$$r_c[m] \triangleq \mathbb{E}\{c_n\, c^*_{n-m}\} = \mathcal{O}\left(|m|^{-\gamma}\right), \quad m \in \mathbb{Z}, \quad \gamma > \frac{\epsilon}{\epsilon - 1}, \; \epsilon > 1. \quad (4.17)$$

This assumption is quite general, capturing a broad range of realistic disturbance models, including all (possibly non-Gaussian) autoregressive–moving-average (ARMA) processes and compound Gaussian processes. Moreover, it naturally extends to higher-dimensional cases—for instance, 2D autoregressive (AR) fields—so long as stationarity and polynomial correlation-decay are maintained and the process is appropriately vectorized. Under these conditions, one can reliably estimate the disturbance covariance from limited data and still obtain the same asymptotic distribution for the detection statistic, thereby ensuring both CFAR and robust detection performance. Leveraging this flexible disturbance model, which accurately captures the spatial correlation of observed samples, the robust Wald-type detector was proposed in [13]. In the limit as $N \to \infty$, this detector follows a χ^2 distribution under both $\mathcal{H}_0$ and $\mathcal{H}_1$, irrespective of the actual disturbance probability density function (pdf). Concretely, to form the Wald statistic, one only needs a $\sqrt{N}$-consistent estimator of α and its associated asymptotic error covariance. Under Assumption 4.1, the detector is given by [13]:

$$\Lambda^{(k)}_{l,i} = \frac{2\left|\left(\boldsymbol{h}^{(k)}_{l,i}\right)^{\mathrm{H}} \boldsymbol{r}^{(k)}_{l,i}\right|^2}{\left(\boldsymbol{h}^{(k)}_{l,i}\right)^{\mathrm{H}} \hat{\boldsymbol{\Gamma}} \left(\boldsymbol{h}^{(k)}_{l,i}\right)}, \quad (4.18)$$

where $\hat{\boldsymbol{\Gamma}}$ denotes the estimate of the unknown disturbance covariance matrix:

$$\boldsymbol{\Gamma} = \mathbb{E}\left\{\boldsymbol{c}^{(k)}_{l,i} \left(\boldsymbol{c}^{(k)}_{l,i}\right)^{\mathrm{H}}\right\}. \quad (4.19)$$

Further details on the derivation of $\hat{\boldsymbol{\Gamma}}$ and on the asymptotic distribution of $\Lambda^{(k)}_{l,i}$ can be found in Appendix A.

To discriminate between $\mathcal{H}_0$ and $\mathcal{H}_1$, a threshold is introduced:

$$\Lambda^{(k)}_{l,i} \underset{\mathcal{H}_0}{\overset{\mathcal{H}_1}{\gtrless}} \eta, \quad (4.20)$$

where the threshold η must be chosen to control the false alarm probability P_{FA}, which is crucial in radar applications. Hence, η is selected to satisfy:

$$\Pr\left(\Lambda_{l,i}^{(k)} > \eta \,\middle|\, \mathcal{H}_0\right) = \int_{\eta}^{\infty} f_{\Lambda|\mathcal{H}_0}(a \,|\, \mathcal{H}_0)\, \mathrm{d}a = P_{\mathrm{FA}}, \tag{4.21}$$

where $f_{\Lambda|\mathcal{H}_0}$ is the pdf of $\Lambda_{l,i}^{(k)}$ under $\mathcal{H}_0$. The threshold η that enforces any desired nominal P_{FA}, regardless of the disturbance distribution, is given by [13]:

$$\eta = -2\ \ln(P_{\mathrm{FA}}). \tag{4.22}$$

Notably, the above results for the Wald-type detector remain valid for any array geometry, since no specific structure of the steering vector is assumed. It has been shown that an asymptotic regime is achieved for $N \geq 10^4$ when $P_{\mathrm{FA}} = 10^{-4}$ [16]. However, providing a more general and rigorous criterion for this requirement remains an open challenge. Equation (4.22) ensures that the robust Wald-type detector in (4.18) enjoys the CFAR property. Nevertheless, CFAR is only one facet of an effective radar detection procedure. A second key goal is ensuring good detection performance in terms of P_D. In the next section, we introduce an RL-based algorithm that improves detection capabilities while preserving the CFAR property.

4.4 RL for Cognitive Radar Sensing

This section formulates the radar detection problem within an RL framework. A comprehensive discussion of RL fundamentals is provided in Chapter 2. Here, we offer only a brief overview to establish the context for the proposed approach.

In an RL framework, an agent interacts with its environment by observing a state, taking an action, and receiving a reward that measures the quality of that action. Over multiple time steps, the agent refines its *policy*—that is, the mapping from states to actions—so as to maximize the cumulative reward. Using this paradigm, the radar iteratively adjusts angular bin selection and beamforming weights based on observed detection outcomes. Figure 4.3 illustrates a high-level view of the proposed RL-driven cognitive radar. In this architecture, the MIMO radar acts as the agent, issuing actions (e.g., beamforming commands, angle-bin priorities), while the environment furnishes states and reward signals. This closed-loop mechanism enables the radar to optimize its transmission strategy over time.

In the sections that follow, we present the RL formulation for our target detection scenario. We begin by reviewing some fundamentals of RL and the MDP. Next, we define the specific states, actions, and rewards tailored to this radar application. Finally, we discuss how beamforming weights are updated and describe how

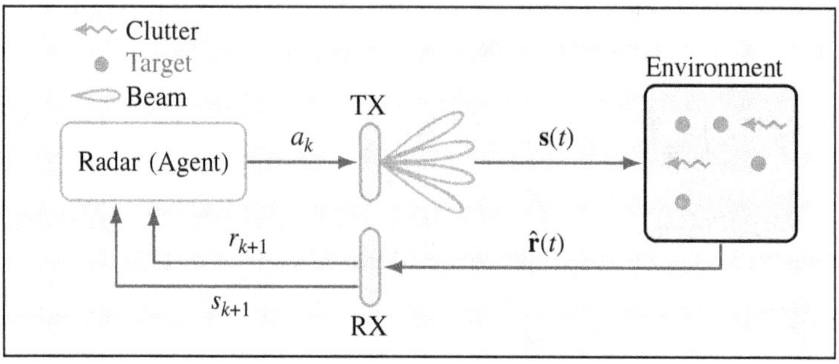

Figure 4.3 Proposed RL-based MIMO radar system forming beams for multitarget tracking.

a single-step on-policy algorithm, SARSA, can be applied to address the detection problem.

4.4.1 Fundamentals

As described previously, in RL, an agent learns how to make decisions by interacting with an environment [17]. At each time step k, the agent:

1. Observes the current *state* s_k,
2. Chooses and performs an *action* a_k,
3. Receives a *reward* r_{k+1},
4. Moves to a *new state* s_{k+1} based on the environment's rules.

If the *Markov property* holds—meaning the probability of moving to the next state s_{k+1} is conditionally independent of all previous states and actions given the current state s_k and action a_k—we can use a MDP [17, 34] to mathematically model this scenario. Although real-world radar or control systems may not perfectly satisfy this property, RL techniques still often perform effectively or can be adapted for environments with limited observability.

A *policy* π guides how the agent selects actions based on its current state. The primary aim in RL is to improve this policy over time, so the agent maximizes its cumulative reward (often measured with a discount factor or an average over many steps). Many RL algorithms exist—such as Q-learning, SARSA, or actor-critic—that specify different ways of updating the agent's policy from experience [35].

Broadly, RL algorithms can be grouped into two families: *on-policy* and *off-policy* methods. The distinction lies in how the agent learns from its interactions with the environment. In *on-policy* algorithms, such as SARSA, the agent learns using the very policy it is currently following. In other words, the agent evaluates and improves the same policy it uses to make decisions. For example, consider a robot navigating through a maze: if it takes a slightly random path to reach the goal, on-policy learning will use the outcomes of that exact path—including the randomness—to update its knowledge. By contrast, *off-policy* algorithms, such as Q-learning, allow the agent to learn about the best possible policy while still behaving according to another, possibly more exploratory, policy. Continuing the maze example, the robot might wander randomly to gather data, but in the background it learns how it would have behaved if it had followed a more optimal route. This separation can make off-policy methods converge more quickly to high-quality solutions, but it also comes with drawbacks: the learning process can become unstable or highly variable. On the other hand, on-policy methods tend to be more stable and reliable in practice, especially when there is a delicate balance between trying new actions (exploration) and leveraging known good actions (exploitation) [17].

Another important aspect of RL is the set of general challenges these algorithms face. One of the most fundamental is the *exploration–exploitation trade-off*. An agent must constantly decide whether to use its current knowledge to achieve good results (*exploit*) or to try actions it has not fully tested (*explore*), which might lead to even better outcomes in the future. Returning to the maze analogy, the robot faces this dilemma at every junction: it can follow a path it already knows leads closer to the goal (exploit), or it can venture down an unfamiliar corridor that might reveal a shortcut—or a dead end (explore). A second key challenge is convergence, which refers to whether the algorithm eventually settles on a stable and effective policy. Poorly chosen hyperparameters, such as the learning rate (step size) or the discount factor, can cause learning to stall or oscillate indefinitely. These challenges are particularly critical in radar applications: exploration may correspond to trying unconventional beampatterns, which can be risky if done too often, while convergence problems may lead directly to unreliable detection in environments that are already dynamic and uncertain.

Here, we focus on the SARSA algorithm, an on-policy method that updates its action-values using actual transitions $(s_k, a_k, r_{k+1}, s_{k+1}, a_{k+1})$ sampled from its own behavior. The name refers to the quintuple *state-action-reward-state-action*. Nevertheless, the overall RL-based radar framework described can naturally incorporate other learning algorithms if needed. Building on this foundation, we now examine how to cast the MIMO radar detection problem within an RL framework by defining the state space, action space, and reward function. Specifically, our objective is to develop a policy that adaptively determines the beamforming weights of the radar [16, 27]. By leveraging RL, the policy can enhance detection performance in environments where disturbance processes are uncertain or dynamically evolving over time.

4.4.2 State-space Definition

At every pulse index k, a detection statistic $\Lambda_{l,i}^{(k)}$ is computed to assess whether bin (l, i) might contain a target, using robust detection methods [27]. A straightforward way to capture this detection result is through a binary variable:

$$\overline{\Lambda}_{(l,i)}^{k} = \begin{cases} 1, & \Lambda_{(l,i)}^{k} > \eta, \\ 0, & \text{otherwise.} \end{cases} \tag{4.23}$$

If $\Lambda_{(l,i)}^{k}$ exceeds η, then bin (l, i) is tentatively marked as containing a target; otherwise, it is deemed empty. To define the RL state, let us consider how many angle bins are labeled as having targets at time k. Concretely,

$$s_k = \sum_{l=1}^{L} \sum_{i=1}^{I} \overline{\Lambda}_{(l,i)}^{k}, \tag{4.24}$$

which counts the total number of bins exceeding the detection threshold. The integer s_k can range from 0 to M, where M represents the maximum number of resolvable targets. Hence, the RL state space is $\mathcal{S} = \{0, 1, 2, \dots, M\}$.

4.4.3 Action Space Definition

The radar agent must decide at each time step k, how to transmit to maximize the detection probability. Since we have M_T transmit antennas, we can shape the radar illumination pattern by optimizing the beamforming matrix $\boldsymbol{W}_k$, defined at time step k. Moreover, we typically do not want to form beams for *all* angle bins simultaneously if only a subset of them is likely to contain targets. Instead, we choose a subset of angle bins that appear most promising based on the detection statistics and optimize for these bins. An action can be viewed as comprising two key tasks: first, selecting the candidate angle bins most likely to contain targets based on the current environmental state; and second, optimizing the beamforming matrix $\boldsymbol{W}_k$ to direct transmit power toward those selected bins. Hence, the *action* at time step k can be defined as:

$$a_k \in \mathcal{A} = \{\Theta_m \mid m \in \{0, 1, \dots, M\}\}, \tag{4.25}$$

where each Θ_m represents m candidate positions $(\hat{\theta}_1, \hat{\phi}_1), \dots, (\hat{\theta}_m, \hat{\phi}_m)$ that most probably contain targets. Typically, m is chosen based on the estimate of s_k (the number of bins exceeding the threshold). For instance, if $s_k = 2$, the radar might select Θ_2 to form beams toward the two most likely target bins. The cardinality of Θ_m can be from 0 to M. The beamforming matrix $\boldsymbol{W}_k$ is optimized to direct as much power as possible toward those m bins. This can be formulated with the following optimization problem:

$$\begin{aligned} \max_{\boldsymbol{W}_k} \quad \min_{j\in\{1,\dots,m\}} \quad & \boldsymbol{a}_T^{\mathrm{T}}(\hat{\theta}_j,\hat{\phi}_j)\,\boldsymbol{W}_k\,\boldsymbol{W}_k^{\mathrm{H}}\,\boldsymbol{a}_T^{*}(\hat{\theta}_j,\hat{\phi}_j), \\ \text{s.t.} \quad & \mathrm{Tr}(\boldsymbol{W}_k\boldsymbol{W}_k^{\mathrm{H}}) = P_T, \end{aligned} \tag{4.26}$$

where $\boldsymbol{a}_T(\hat{\theta}_j,\hat{\phi}_j)$ is the steering vector for the jth selected position, and P_T is the total transmit power constraint. The optimization problem can be solved using maximum power design [36]. If $m = 0$ (i.e., no bins appear to have a target), we may choose $\boldsymbol{W}_k$ to be an identity-like matrix.

4.4.4 Reward Function Design

Having defined the radar agent's action space, we now turn our attention to designing a suitable reward function to guide its behavior toward effective target detection. Reward functions are critical in shaping the agent's behavior. For radar detection, a natural performance objective is the P_D, offset by the cost of false alarms, P_{FA}. More precisely, let $\mathcal{T}$ denote the set of bins predicted to contain a target, and let $\mathcal{N}$ denote its complement, consisting of bins predicted not to contain a target. We define the detection performance at bin (l, i), via an asymptotic detection probability $\widehat{P}_{D_{l,i}}^{(k)}$ derived in [13], which can be approximated by a first-order Marcum Q function [37]:

$$\widehat{P}_{D_{l,i}}^{(k)} \overset{N\to\infty}{\doteq} Q_1\left(\sqrt{\rho_{l,i}^{(k)}},\ \sqrt{\eta}\right), \tag{4.27}$$

where η is the detection threshold from (4.23), and $\rho_{l,i}^{(k)}$ is:

$$\rho_{l,i}^{(k)} = \frac{2\left|\widehat{\alpha}_{l,i}^{(k)}\right|^2 \left\|\boldsymbol{h}_{l,i}^{(k)}\right\|^4}{\left(\boldsymbol{h}_{l,i}^{(k)}\right)^{\mathrm{H}} \widehat{\boldsymbol{\Gamma}}_{l,i} \boldsymbol{h}_{l,i}^{(k)}}, \qquad \widehat{\alpha}_{l,i}^{(k)} = \frac{\left(\boldsymbol{h}_{l,i}^{(k)}\right)^{\mathrm{H}} \boldsymbol{r}_{l,i}^{(k)}}{\left\|\boldsymbol{h}_{l,i}^{(k)}\right\|^2}. \tag{4.28}$$

Using these, one can define a reward function that sums the detection probabilities in the bins that contain targets, and perhaps subtracts the detection probabilities (i.e., false alarm probabilities) for bins that are empty. Thus:

$$r_{k+1} = \sum_{(l,i)\in\mathcal{T}} \widehat{P}_{D_{l,i}}^{(k)} - \sum_{(l,i)\in\mathcal{N}} \widehat{P}_{D_{l,i}}^{(k)}. \tag{4.29}$$

This form encourages the radar to place energy where it will improve target detection while incurring penalties for mistakenly elevating the detection statistic in bins without actual targets.

4.4.5 SARSA Algorithm for Target Detection

In the previous sections, we defined the key components of our RL framework, including states, actions, and the reward function. Building upon these foundations, we now introduce the SARSA algorithm. Specifically, this chapter proposes a SARSA-based RL framework designed to address the MIMO target detection problem. SARSA is a model-free, on-policy RL method known for inherently balancing exploration and exploitation, making it particularly advantageous in real-time scenarios where environmental conditions are often nonstationary [17]. Moreover, its simplicity in implementation and adaptive nature contribute to robust performance even under dynamic conditions. Central to the SARSA algorithm is the Q-function, denoted as $Q(s, a)$. This function approximates the long-term return, defined as the expected cumulative discounted reward, that an agent accrues by taking action a in state s and subsequently following the current policy. The learning mechanism involves iteratively updating the estimate $Q(s, a)$ using the observed tuple $\{s_k, a_k, r_{k+1}, s_{k+1}, a_{k+1}\}$ [17]. Specifically, the Q-function is represented as a matrix $\boldsymbol{Q} \in \mathbb{R}^{(M+1)\times(M+1)}$, where each element is updated according to:

$$Q(s_k, a_k) \leftarrow Q(s_k, a_k) + \alpha_{\text{lr}}\left[r_{k+1} + \gamma\, Q(s_{k+1}, a_{k+1}) - Q(s_k, a_k)\right], \tag{4.30}$$

where $\alpha_{\text{lr}} \in (0, 1]$ is the learning rate controlling the update magnitude, and $\gamma \in (0, 1]$ is the discount factor that determines the relative weighting of future rewards with respect to immediate ones.

An essential characteristic of the SARSA algorithm is its on-policy nature, meaning that the updates to the Q-function directly depend on actions taken according to the current policy. This feature ensures that the learning process closely mirrors the agent's actual behavior. However, the theoretical RL framework typically assumes a stationary MDP, where environmental conditions remain constant over time. In contrast, radar applications inherently involve nonstationary environments due to variations in target positions, SNR, and disturbance statistics. Such nonstationarity undermines the concept of a globally optimal policy, as optimality becomes dependent on continuously evolving environmental dynamics. Consequently, the effectiveness of SARSA-based strategies tailored to radar applications should be carefully validated before being generalized to other contexts. A practical solution to address environmental nonstationarity involves approximating the nonstationary MDP as a sequence of stationary sub-MDPs, enabling the use of stationary-based learning methods. Nonetheless, effectively managing transitions between these stationary sub-MDPs remains critical. The objective here is to improve the robustness of the SARSA algorithm in response to environmental variations, including fluctuations in target count, SNR levels, and angular positions. To achieve this goal, we adopt the quasi ϵ-greedy policy with target recovery introduced in [24]. The policy is formally defined as:

Algorithm 4.1: RL-based algorithm for target detection.

Initialize $k = 0$, $\boldsymbol{Q} = \boldsymbol{0}_{(M+1)\times(M+1)}$, initial state $s_0 = 1$, initial action $a_0 = 1$, and beamforming matrix $\boldsymbol{W}_0 = I$.

repeat:

Execute action a_k by transmitting the waveform using the current beamforming matrix $\boldsymbol{W}_k$.

Acquire the received signals $\boldsymbol{r}_{l,i}^{(k)}$ for all l and i.

Compute the Wald-type test statistics $\Lambda_{l,i}^{(k)}$ for all l and i, as defined in (4.18).

Determine the threshold-adjusted statistics $\overline{\Lambda}_{l,i}^{(k)}$ for all l and i according to (4.23).

Derive the subsequent state s_{k+1} based on the relation in (4.24).

Evaluate the reward r_{k+1} in accordance with (4.29).

Select the next action a_{k+1} using the quasi ϵ-greedy policy described in (4.31).

Identify the set Θ_m as determined by s_{k+1}.

Update $Q(s_k, a_k)$ using the update rule in (4.30).

Set $s_k \leftarrow s_{k+1}$ and $a_k \leftarrow a_{k+1}$.

if $s_{k+1} \neq 0$ **then**

  Compute the optimized beamforming matrix $\boldsymbol{W}_{k+1}$ as specified in (4.26).

else

  Reset the beamforming matrix by setting $\boldsymbol{W}_{k+1} \leftarrow \boldsymbol{I}$.

end if

until the observation period K terminates.

$$a^\star = \begin{cases} \arg\max_{a\in\mathcal{A}} Q(s_k, a), & m_{k+1} < m_k, \\ \begin{cases} \text{a random action from } \mathcal{A}', & \text{w.p. } \epsilon, \\ \arg\max_{a\in\mathcal{A}} Q(s_{k+1}, a), & \text{w.p. } 1-\epsilon, \end{cases} & m_{k+1} \geq m_k, \end{cases} \tag{4.31}$$

where $\mathcal{A}'$ denotes the set of actions with a higher number of angular bins compared to the current state. According to this policy, if at time step $k+1$ the radar detects an equal or greater number of targets compared to step k, it employs the strategy defined for the condition $m_{k+1} \geq m_k$. This prevents the radar, during the exploration phase, from concentrating its power on fewer angular bins than those corresponding to the number of potentially detected targets. Conversely, if the number of detected targets decreases, the algorithm prioritizes recovery by selecting the greedy action based on the previous state s_k. Further details on this policy are available in [24].

Building on these foundational principles, Algorithm 4.1 details the complete RL framework for MIMO target detection. The algorithm integrates the SARSA methodology to iteratively optimize the beamforming matrix and facilitate robust detection decisions.

4.5 Results

In this section, we present comprehensive simulation results to evaluate the effectiveness of the proposed RL-based beamforming approach compared to non-RL methods. The section is structured as follows: First, we specify the essential simulation parameters and the disturbance model utilized throughout the simulations. Next, we detail and analyze multiple simulation scenarios, starting with stationary targets to establish baseline performance and continuing with increasingly complex scenarios involving dynamic behaviors, such as the appearance and disappearance of targets and target movement.

4.5.1 System Parameters

The simulations consider $L = I = 20$ azimuth/elevation bins, for a total of 400 bins. The grid is expressed in terms of the spatial frequency $\boldsymbol{f}_x = \boldsymbol{f}_y = [-0.5 : 0.45]$. Other important parameters can be found in Table 4.1. All parameters remain the same for the following simulations, unless otherwise stated.

The disturbance vector $\boldsymbol{c}_{l,i}^{(k)}$ is modeled as a 2D circular complex AR process [38]. The mathematical formulation of this 2D AR process is given by

$$c_{n_x,n_y} = \sum_{i=1}^{p} \sum_{j=1}^{q} \rho_{i,j} c_{n_x-i,n_y-j} + w_{n_x,n_y}, \tag{4.32}$$

Table 4.1 Simulation parameters.

Parameter	Value
Probability of false alarm P_{FA}	10^{-4}
Receive antennas M_R	100
Transmit antennas M_T	100
Power P_T	1
Heaviness of the tails μ	2
Noise variance σ_w^2	1
Number of states	11
Number of actions	11
Discount factor γ	0.8

where c_{n_x,n_y} represents the disturbance at the spatial index (n_x, n_y) in the 2D spatial domain. The coefficients $\rho_{i,j}$ define the dependency structure of the AR process, capturing spatial correlations. The term w_{n_x,n_y} denotes the driving noise at each spatial location (n_x, n_y), which serves as the stochastic excitation of the system.

To model the statistical characteristics of the disturbance more accurately, the driving noise w_{n_x,n_y} follows a heavy-tailed t-distribution [13]. This distribution accounts for the non-Gaussian behavior often observed in cluttered radar environments and impulsive noise scenarios:

$$p_w(w_{n_x,n_y}) = \frac{\mu}{\sigma_w^2}\left(\frac{\mu}{\xi}\right)^{\mu}\left(\frac{\mu}{\xi} + \frac{|w_{n_x,n_y}|^2}{\sigma_w^2}\right)^{-(\mu+1)}, \tag{4.33}$$

where the parameter μ controls the heaviness of the tails, governing the likelihood of extreme noise values, and σ_w^2 represents the variance of the noise. The scale parameter is defined by $\xi = \mu/(\sigma_w^2(\mu - 1))$. This formulation ensures robustness in high-clutter scenarios, providing a more realistic representation of radar disturbances in high-resolution radar systems. For these simulations, these two parameters are selected as shown in Table 4.1.

The power spectral density (PSD) in the 2D spatial frequency domain characterizes the distribution of disturbance power across spatial frequencies and is expressed as:

$$S(f_x, f_y) = \sigma_w^2 \left| 1 - \sum_{n=1}^{p}\sum_{l=1}^{q} \rho_{n,l} e^{-j2\pi(nf_x + lf_y)} \right|^{-2}. \tag{4.34}$$

This equation describes how the AR coefficients influence the spectral properties of the disturbance process, shaping its frequency-domain behavior. The specific AR coefficients used in this model are defined by the following coefficient matrix:

$$\rho = \begin{bmatrix} 0 & 0.1 & 0.1 \\ 0.1 & 0 & 0 \\ 0.05 & 0 & 0 \end{bmatrix}. \tag{4.35}$$

These coefficients govern the spatial correlations in the disturbance process and lead to the PSD distribution shown in Figure 4.4.

We now focus on the key hyperparameters of the RL algorithm, ϵ and α_{lr}. Their selection is crucial, as they significantly impact performance. Typically, practitioners tune these parameters heuristically based on experience, but in dynamic or nonstandard environments, fixed values may quickly become suboptimal, necessitating an adaptive approach. An adaptive, data-driven strategy introduced in [24] automates their selection, reducing manual tuning and improving performance via numerical simulations.

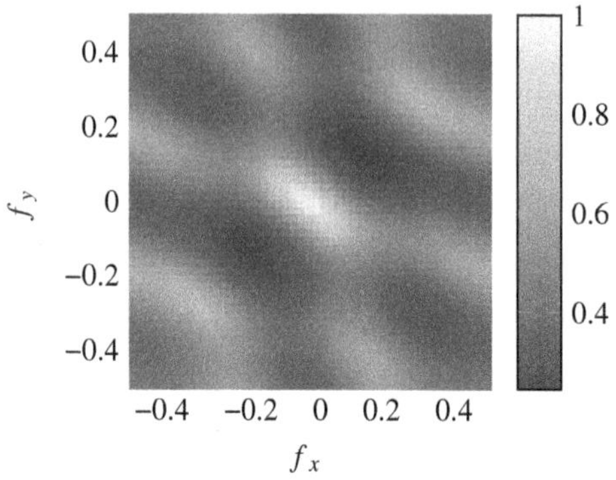

Figure 4.4 PSD of the modeled disturbance process.

To track reward changes, d_k is defined as:

$$d_k = \begin{cases} r_1, & k = 1, \\ r_k - r_{k-1}, & k > 1, \end{cases} \tag{4.36}$$

where r_k is the observed reward at time k. This detects environmental changes based on reward fluctuations.

Let x_k be a hyperparameter at time k, representing either ϵ_k or $\alpha_{\text{lr},k}$. It updates as:

$$x_{k+1} = \begin{cases} \max\{c_1 x_k, x_{\min}\}, & |d_k| < \delta_1, \\ \min\{c_2 x_k, x_{\max}\}, & \delta_1 \leq |d_k| < \delta_2, \\ x_{\max}, & |d_k| \geq \delta_2, \end{cases} \tag{4.37}$$

where $c_1 \in (0, 1)$ and $c_2 \in (1, \infty)$, with bounds $x_{\min}$ and $x_{\max}$. The initial value $x_0 = x_{\max}$ ensures maximum exploration or learning rate at the start. Hyperparameter updates are paused when two consecutive exploratory actions occur, preventing spurious adjustments due to transient reward dips. Thresholds δ_1 and δ_2 must be carefully set based on reward variability. For radar detection, $\delta_1 \approx 0.5$ distinguishes minor fluctuations, while $\delta_2 \geq 1.8$ ensures updates only for significant environmental shifts [24]. This adaptive mechanism enables SARSA to dynamically adjust ϵ and α_{lr} in response to system changes, minimizing manual intervention while maintaining close to optimal performance. The parameters used for the following simulations are shown in Table 4.2 and are adopted from [24].

Table 4.2 Parameters for adaptive selection of ϵ and $\alpha_{\rm lr}$.

x	$x_{\min}$	$x_{\max}$	c_1	c_2	δ_1	δ_2
ϵ	0.1	0.8	0.8	2	0.5	1.8
$\alpha_{\rm lr}$	0.2	0.6	0.9	2.5	0.5	1.8

4.5.2 Simulations

In the following simulations, we extensively analyze various scenarios to evaluate and benchmark system performance under diverse and complex operational conditions. These scenarios are meticulously crafted to thoroughly investigate and elucidate the radar's target detection and tracking capabilities, rigorously addressing both stationary and dynamically moving targets. The selected scenarios provide a comprehensive representation of real-world radar operational contexts, thereby ensuring the generalizability and robustness of our analysis.

Our comparative analysis involves a detailed assessment of the performance of an RL-based detection algorithm against two distinct non-RL-based methodologies:

1. **Orthogonal Waveform Selection:**
   In this non-RL-based method, the radar system selects an orthogonal beamformer at each iteration without referencing historical data or previous decision outcomes. The simplicity of this method, characterized by minimal computational overhead and ease of practical implementation, makes it attractive for systems requiring rapid decision-making without significant computational resources. Nonetheless, its major drawback lies in the limited exploitation of the radar's inherent high-resolution focusing capabilities, potentially compromising target detection performance due to its inability to adapt based on prior information.
2. **Non-RL (NRL) Beamforming Technique:**
   In contrast to the orthogonal waveform, the NRL beamforming technique proactively incorporates decision statistics directly into its transmission power allocation strategy. This method represents an advancement over purely random waveform selection, as it actively utilizes available statistical insights to dynamically enhance the allocation of resources, thereby improving overall detection efficiency. Despite this sophistication, it fundamentally diverges from RL-based methods, which typically adopt a structured state-action-reward learning paradigm. Instead, NRL beamforming relies explicitly on instantaneous values of test statistics to guide decision-making, without iterative learning from historical states.

Through this comprehensive comparative evaluation, the objective is to clearly delineate the respective strengths, operational applicability, and limitations of RL-based approaches relative to these established non-RL strategies. Particular attention is given to scenarios involving highly dynamic target conditions.

4.5.2.1 Stationary Targets

To thoroughly evaluate and leverage the benefits of RL within our system, we start our analysis with a controlled scenario involving stationary targets. In this setup, three stationary targets are positioned at predefined spatial coordinates and remain fixed throughout the simulation. Each target maintains a constant SNR, thus providing a stable and predictable environment conducive to accurately assessing baseline detection performance. Detailed specifications for these targets, including their precise locations and corresponding SNR levels, are summarized in Table 4.3. This carefully structured scenario is particularly valuable for examining the system's response to noise and interference under static conditions.

Figure 4.5 provides a comparative analysis of detection performance between our proposed RL-based beamforming method and the two alternative methods. As depicted in Figure 4.5a, the RL-based algorithm consistently outperforms the orthogonal and the NRL approach, demonstrating significantly enhanced detection accuracy across all target scenarios, even under challenging low-SNR conditions. A critical advantage of our RL method lies in its inherent capability to learn from the environment and adapt dynamically over time. In the initial stages (first ten time steps), the RL agent prioritizes exploring and characterizing the surrounding interference environment. Through continuous interaction and iterative learning, the agent progressively refines and optimizes its beamforming strategy, thereby achieving notable improvements in detection performance. In contrast, the other two approaches, illustrated in Figure 4.5b, c, demonstrate substantial limitations, particularly noticeable in scenarios involving targets with

Table 4.3 Stationary targets.

Target	Time interval
	[1,50]
1	(−0.3, −0.3)[1]
	−5 dB[2]
2	(0, 0)
	−15 dB
3	(0.4, 0.4)
	−10 dB

1 Target's position.
2 Target's SNR.

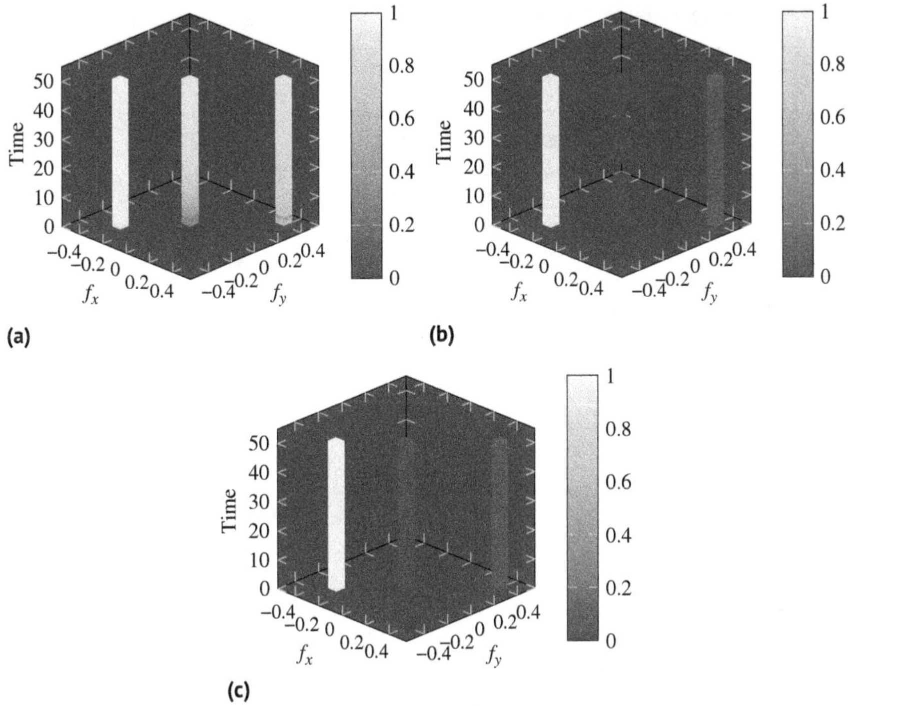

Figure 4.5 Detection performance of three static targets (a) RL-based beamforming, (b) equal power allocation, and (c) NRL-based beamforming.

lower SNR levels, which often remain undetectable. The NRL approach shows a slightly higher P_D for Target 2 compared to the orthogonal approach, which is due to the fact that it also operates beamforming but still rarely directs the beam in this direction. This contrast emphasizes the significant benefits of employing an RL-based approach, where intelligent adaptation and learned decision-making processes lead to superior target identification performance.

4.5.2.2 Dynamic Targets: Appearance and Disappearance

We now examine a dynamic scenario involving two targets that appear and disappear at different intervals. Specifically, as summarized in Table 4.4, Target 1 is present from the start of the observation period until discrete-time 60, after which it disappears at discrete-time 61. Conversely, Target 2 is initially absent and appears at time step 31. This setup introduces nonstationary behavior into the environment, creating a challenging scenario to evaluate the robustness and adaptability of the considered RL-based algorithm.

Once again, we compare our algorithm against the two alternative methods. Figure 4.6 illustrates the time evolution of theP_D for each target. Unlike Figure 4.5 presented in Section 4.5.2.1, we now adopt a different visualization strategy by

Table 4.4 Dynamic targets scenario.

Target	Time interval		
	[1,30]	**[31,60]**	**[61,90]**
1	(−0.4, −0.3)[1]	(−0.4, −0.3)	Absent
	−10 dB[2]	−10 dB	"
2	Absent	(0.5, −0.5)	(0.5, −0.5)
	"	−8 dB	−8 dB

1 Target's Position
2 Target's SNR

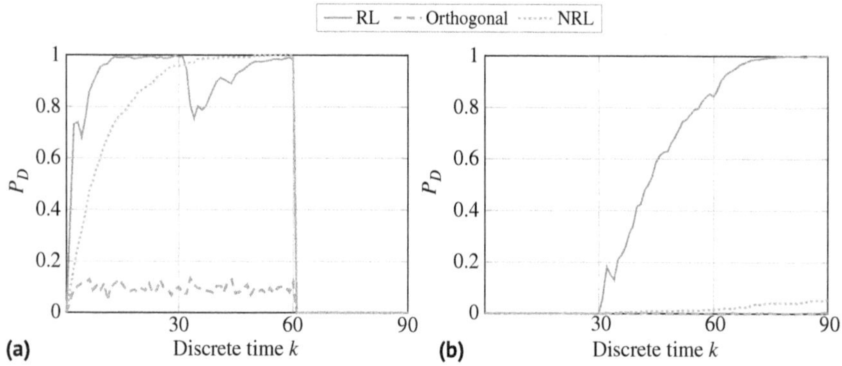

Figure 4.6 P_D for two dynamic targets (a) target 1 and (b) target 2.

plotting the detection performance of all three methods on the same graph for the target's position.

By aligning the P_D curves with the target appearance and disappearance schedule presented in Table 4.4, we clearly observe the responsiveness and temporal dynamics of each method. The results distinctly emphasize the superior adaptability of the RL-based algorithm in handling sudden environmental changes. Compared to the other two algorithms, the RL-based approach demonstrates heightened sensitivity to dynamic target conditions, maintaining stable detection performance even during abrupt scenario transitions. As illustrated in Figure 4.6a, the P_D for Target 1 steadily increases until time step 30, reaching one before decreasing after the emergence of Target 2. Subsequently, P_D increases again until time step 61 and then drops down to zero, coinciding with the disappearance of the target. Notably, the NRL approach also achieves a P_D of one for Target 1, though more slowly than the RL-based approach. However, after Target 2 appears, the NRL algorithm maintains a P_D of one for Target 1. While this might initially suggest a performance advantage, closer examination reveals that this is due to inadequate detection of Target 2, as depicted in Figure 4.6b. Overall, the RL-based beamforming approach offers superior adaptability in responding to new targets. This adaptability emerges naturally from the underlying SARSA learning mechanism, further enhanced by the adaptive hyperparameter selection strategy. In contrast to the RL and NRL approaches, the orthogonal approach exhibits the weakest performance overall, showing marginally better detection for Target 1, where the SNR is higher. The observed performance supports the conclusion that such methods offer a promising avenue for real-time adaptive signal processing in complex and evolving environments.

4.5.2.3 Moving Target

Finally, we investigate the system's performance in a dynamic scenario involving a moving target, as illustrated in Figure 4.7. This figure provides a detailed visualization of the target's trajectory within the spatial frequency domain, characterized by coordinates f_x and f_y. Notably, the target remains stationary for two consecutive time steps at each position.

The trajectory consists of three distinct segments: initially, the target moves diagonally upward from coordinates $(-0.5, -0.5)$ to $(0, 0)$. Then it progresses vertically upward to $(0, 0.25)$. Finally, it moves horizontally to the right, concluding at $(0.45, 0.25)$. Throughout the scenario, the SNR remains fixed at -12 dB. Unlike the dynamic scenarios discussed previously in Section 4.5.2.2, where targets appeared and disappeared at static positions, here we have a single target continuously changing its location.

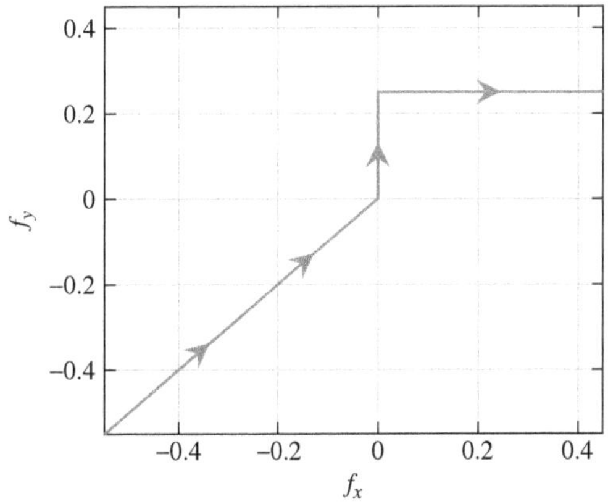

Figure 4.7 Target movement.

The detection performance of the three evaluated methods under these dynamic conditions is depicted in Figure 4.8. The plots display every second time step, as the target changes position every two time steps. The results demonstrate that the RL-based approach, illustrated in Figure 4.8a, consistently outperforms the alternative techniques depicted in Figure 4.8b, c, exhibiting greater adaptability and robustness in tracking moving targets. Specifically, the RL-based method gradually improves over time, eventually achieving a detection probability (P_D) of one. This performance can be attributed to the RL method's beam coverage, which encompasses not only the targeted angle bin but also neighboring bins, enabling detection even when the target changes position.

Examining the reward progression over time in Figure 4.9 reveals that the reward initially decreases following each positional change but then quickly recovers and continues to increase, demonstrating the RL system's ability to adapt dynamically. While the NRL beamforming method also shows a gradual improvement in detection performance, it fails to match the high detection probability achieved by the RL approach. Conversely, the orthogonal approach

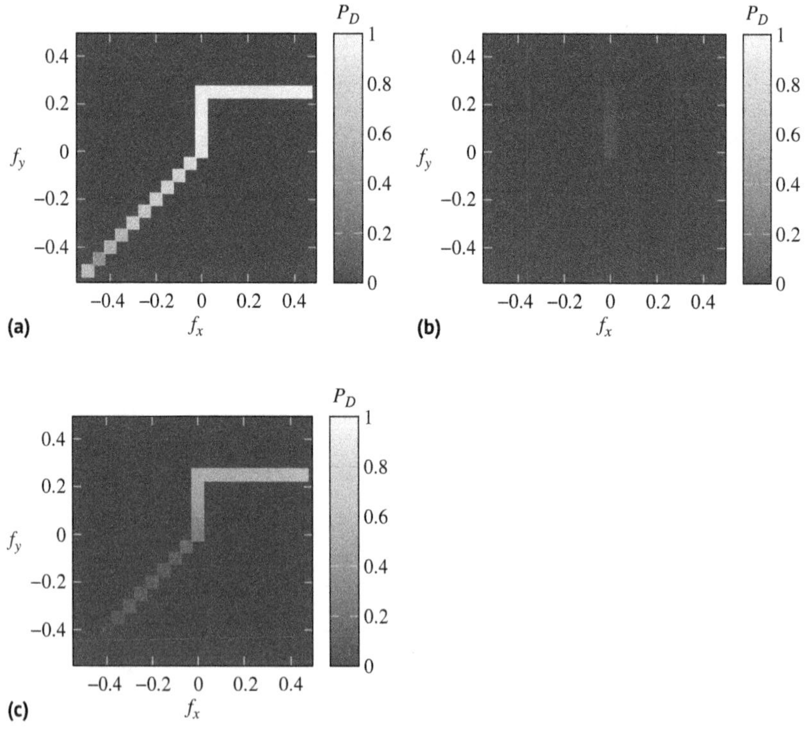

Figure 4.8 Detection performance of RL beamforming, vs. Orthogonal and NRL for a moving target (a) RL-based beamforming, (b) orthogonal , and (c) NRL-based beamforming.

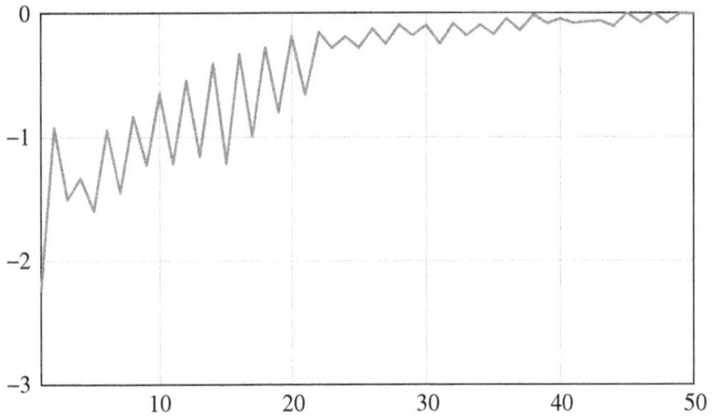

Figure 4.9 RL reward for a moving target.

completely fails in this scenario, which aligns with expectations based on its limited adaptability.

4.6 Conclusions

In this chapter, we have explored significant limitations inherent in conventional radar detection systems, emphasizing their vulnerability to dynamically evolving operational environments and complex clutter dynamics. Traditional methods frequently encounter substantial challenges, particularly under complex, nonstationary conditions, significantly limiting their effectiveness in practical, real-world deployments.

To effectively address these limitations, we introduced an integrated RL-based adaptive detection framework specifically designed for MMIMO radar systems. This innovative approach uniquely leverages the spatial diversity benefits inherent in colocated MIMO radar systems, the robustness provided by Wald-type statistical detectors, and the dynamic adaptability afforded by RL algorithms. The integration of these elements empowers radar systems to iteratively refine their operational strategies autonomously through continuous real-time feedback, resulting in substantial enhancements to detection accuracy, robustness, and reliability.

A particularly significant advantage of the proposed adaptive framework is its capability to operate effectively without relying on explicit prior knowledge of environmental statistical properties. Unlike traditional radar detection methodologies, which heavily depend on detailed, predefined statistical models of disturbances and clutter, the RL-driven system dynamically evolves its detection and waveform selection strategies through continuous interaction and learning from its operational environment. This inherently data-driven approach ensures persistent improvement and adaptability, making it especially effective in uncertain and dynamically changing operational scenarios.

Comprehensive simulation studies conducted throughout this chapter robustly validated the efficacy and superior performance of the RL-based framework across a diverse spectrum of scenarios. These scenarios encompassed stationary targets, targets exhibiting dynamic appearances and disappearances, and those characterized by complex motion trajectories. Consistently, the adaptive system demonstrated superior performance compared to traditional detection techniques, such as orthogonal beamforming and NRL-based waveform selection methods. Notably, the RL-driven framework exhibited exceptional resilience under highly challenging conditions, including heavy-tailed clutter distributions and low-SNR environments—situations in which the other methods struggled.

Looking ahead, several promising areas for further investigation are emerging. Future research could focus on extending the adaptive detection framework to handle even more complex and dynamic environmental conditions. Additionally, there is considerable potential in advancing RL algorithm development, specifically enhancing convergence speed, stability, and adaptive capabilities. Another exciting research direction involves exploring hybrid methodologies that effectively integrate machine learning approaches with classical statistical detection methods, potentially achieving even greater performance improvements in radar systems.

While the advantages of RL in radar applications are compelling, it is also important to acknowledge certain limitations and practical challenges [39]. RL approaches often require large volumes of training data, which may not always be feasible to obtain in radar contexts where scenarios are highly variable. Moreover, the computational overhead associated with training and online adaptation can pose barriers to real-time implementation, particularly in resource-constrained systems [40]. The robustness of RL methods may also be sensitive to reward function design, as misspecified or overly simplified reward structures could misguide the learning process. These considerations highlight the importance of carefully balancing algorithmic sophistication with operational feasibility when deploying RL-based radar detection systems.

In conclusion, the insights and developments presented in this chapter highlight the potential and considerable advantages of integrating RL into radar detection technologies. RL-based adaptive methods represent a compelling new frontier with the potential to significantly enhance radar system capabilities, marking this research domain as critical for ongoing innovation, exploration, and technological advancement.

Appendix A–Estimation of the Disturbance Statistics

In general, the disturbance covariance matrix $\boldsymbol{\Gamma}$ is unknown. However, in the context of MMIMO radar (see [13]), it is possible to estimate the entries of $\boldsymbol{\Gamma}$ directly from the *single* snapshot under test by exploiting robust methods for dependent data. Specifically, under Assumption 4.1 we can form a consistent sample estimate $\hat{\boldsymbol{\Gamma}}_l$ by truncating the empirical covariance at lag l. This leads to the following entry-wise definition ([13, Remark 1]):

$$\left[\hat{\Gamma}_l\right]_{i,j} = \begin{cases} \hat{c}_i\,\hat{c}_j^*, & j-i \le l, \\ \hat{c}_i^*\,\hat{c}_j, & i-j \le l, \\ 0, & |i-j| > l, \end{cases} \tag{A.1}$$

where $\hat{c}_n = y_n - \hat{\alpha}\,h_n$ is the residual disturbance after subtracting the fitted signal, and l is the *truncation lag* (see [41] and [13, Theorem 2]) which must grow more slowly than $N^{1/3}$ to ensure consistency. Here, $N = M_T\,M_R$ is the total number of *virtual* spatial antenna channels.

In particular, if Assumption 4.1 in [13] holds, Theorem 3 of [13] guarantees the following asymptotic distributional results:

i. Under $\mathcal{H}_0$:

$$\Lambda_{l,i}^{(k)}\left(\boldsymbol{r}_{l,i}^{(k)} \mid \mathcal{H}_0\right) \underset{N\to\infty}{\overset{d}{\sim}} \chi_2^2(0). \tag{A.2}$$

Hence, the false-alarm behavior becomes *CFAR* (constant false-alarm rate) in the large-system regime, *without* needing secondary data.

ii. Under $\mathcal{H}_1$:

$$\Lambda_{l,i}^{(k)}\left(\boldsymbol{r}_{l,i}^{(k)} \mid \mathcal{H}_1\right) \underset{N\to\infty}{\overset{d}{\sim}} \chi_2^2\left(\zeta\right), \tag{A.3}$$

where

$$\zeta = 2\,|\alpha|^2\,\frac{||\boldsymbol{h}||^4}{\boldsymbol{h}^{\mathrm{H}}\,\boldsymbol{\Gamma}\,\boldsymbol{h}}. \tag{A.4}$$

Thus, in the massive-antenna regime $N \to \infty$, one obtains a *closed-form* P_D at a given threshold η, via standard properties of the noncentral χ^2 distribution

$$P_D(\eta) \underset{\mathrm{N}\to\infty}{\to} Q_1\left(\sqrt{\zeta}, \sqrt{\eta}\right), \tag{A.5}$$

where $Q_1(\cdot, \cdot)$ is the first-order Marcum Q-function [37]. Hence, once the (nominal) false-alarm probability P_{FA} is set, one obtains η accordingly, and the detection probability converges to a simple closed-form function of ζ. Notably, ζ depends on the disturbance only through the ratio $||\boldsymbol{h}||^4 \,/\, (\boldsymbol{h}^{\mathrm{H}}\, \boldsymbol{\Gamma}\, \boldsymbol{h})$, so even if the exact disturbance distribution is unknown, it suffices that $\hat{\Gamma}_l$ consistently estimates Γ in that ratio.

In short, by exploiting robust estimation theory for dependent observations [42] and the large system dimension ($N \to \infty$), the proposed Wald-type detector exhibits an *invariant* behavior to the true disturbance statistics and thus attains a powerful CFAR property, *without* requiring secondary (training) snapshots from homogeneous ranges.

References

1 A. M. Ahmed, U. S. K. P. M. Thanthrige, A. Sezgin, and F. Gini. Resilient sparse array radar with the aid of deep learning. In *2023 IEEE 97th Vehicular Technology Conference (VTC2023-Spring)*, pages 1–5, 2023.

2 E. Čišija, A. M. Ahmed, A. Sezgin, and H. Wymeersch. RIS-aided mmWave mimo radar system for adaptive multi-target localization. In *2021 IEEE Statistical Signal Processing Workshop (SSP)*, pages 196–200, 2021.

3 S. Haykin. Cognitive radar: A way of the future. *IEEE Signal Processing Magazine*, 23(1):30–40, 2006.

4 S. Z. Gurbuz, H. D. Griffiths, A. Charlish, M. Rangaswamy, M. S. Greco, and K. Bell. An overview of cognitive radar: Past, present, and future. *IEEE Aerospace and Electronic Systems Magazine*, 34(12):6–18, 2019.

5 M. S. Greco, F. Gini, P. Stinco, and K. Bell. Cognitive radars: On the road to reality: Progress thus far and possibilities for the future. *IEEE Signal Processing Magazine*, 35(4):112–125, 2018.

6 S. Haykin, Y. Xue, and P. Setoodeh. Cognitive radar: Step toward bridging the gap between neuroscience and engineering. *Proceedings of the IEEE*, 100(11):3102–3130, 2012.

7 P. Stinco, M. S. Greco, and F. Gini. Spectrum sensing and sharing for cognitive radars. *IET Radar, Sonar & Navigation*, 10:595–602, 2016.

8 P. Stinco, M. Greco, F. Gini, and B. Himed. Cognitive radars in spectrally dense environments. *IEEE Aerospace and Electronic Systems Magazine*, 31(10):20–27, 2016.

9 A. M. Haimovich, R. S. Blum, and L. J. Cimini. MIMO radar with widely separated antennas. *IEEE Signal Processing Magazine*, 25(1):116–129, 2008.

10 J. Li and P. Stoica. MIMO radar with colocated antennas. *IEEE Signal Processing Magazine*, 24(5):106–114, 2007.

11 D. Bliss and K. Forsythe. Multiple-input multiple-output (MIMO) radar and imaging: degrees of freedom and resolution. In *The Thirty-Seventh Asilomar Conference on Signals, Systems & Computers, 2003*, volume 1, pages 54–59, 2003.

12 A. M. Ahmed, U. S. K. P. M. Thanthrige, A. E. Gamal, and A. Sezgin. Deep learning for doa estimation in mimo radar systems via emulation of large antenna arrays. *IEEE Communications Letters*, 25(5):1559–1563, 2021.

13 S. Fortunati, L. Sanguinetti, F. Gini, M. S. Greco, and B. Himed. Massive MIMO radar for target detection. *IEEE Transactions on Signal Processing*, 68:859–871, 2020.

14 F. Zhou, D. Zhou, and G. Yu. Target tracking in interference environments reinforcement learning and design for cognitive radar soft processing. In *2008 Congress on Image and Signal Processing*, volume 4, pages 73–77, 2008.

15 P. Liu, Y. Liu, T. Huang, Y. Lu, and X. Wang. Decentralized automotive radar spectrum allocation to avoid mutual interference using reinforcement learning. *IEEE Transactions on Aerospace and Electronic Systems*, 57(1):190–205, 2021.

16 A. M. Ahmed, A. A. Ahmad, S. Fortunati, A. Sezgin, M. S. Greco, and F. Gini. A reinforcement learning based approach for multitarget detection in massive MIMO radar. *IEEE Transactions on Aerospace and Electronic Systems*, 57(5):2622–2636, 2021.

17 R. S. Sutton and A. G. Barto. *Reinforcement Learning: An Introduction*. A Bradford Book, Cambridge, MA, USA, 2018.

18 S. A. Flandermeyer, R. G. Mattingly, and J. G. Metcalf. Deep reinforcement learning for cognitive radar spectrum sharing: A continuous control approach. *IEEE Transactions on Radar Systems*, 2:125–137, 2024.

19 A. M. Ahmed, S. Fortunati, A. Sezgin, M. S. Greco, and F. Gini. Robust reinforcement learning-based wald-type detector for massive MIMO radar. In *2021 29th European Signal Processing Conference (EUSIPCO)*, pages 846–850, 2021.

20 F. Gini, A. De Maio, and L. Patton. *Waveform Design and Diversity for Advanced Radar Systems*. Institution of Engineering and Technology, 2012.

21 J. Kweon, F. Gini, M. S. Greco, M. Rangaswamy, and V. Monga. Model-based learning for mimo radar waveform design in the presence of multiple targets. *IEEE Transactions on Aerospace and Electronic Systems*, 61(3):7165–7178, 2025.

22 K. Metwaly, J. Kweon, K. Alhujaili, F. Gini, M. S. Greco, M. Rangaswamy, and V. Monga. MIMO radar beampattern design via algorithm unrolling. *IEEE Transactions on Aerospace and Electronic Systems*, 60(6):9204–9220, 2024.

23 S. Khobahi, A. Bose, and M. Soltanalian. Deep radar waveform design for efficient automotive radar sensing. In *2020 IEEE 11th Sensor Array and Multichannel Signal Processing Workshop (SAM)*, pages 1–5, 2020.

24 F. Lisi, S. Fortunati, M. S. Greco, and F. Gini. Enhancement of a state-of-the-art RL-based detection algorithm for massive MIMO radars. *IEEE Transactions on Aerospace and Electronic Systems*, 58(6):5925–5931, 2022.

25 W. Zhai, X. Wang, X. Cao, M. S. Greco, and F. Gini. Reinforcement learning based dual-functional massive MIMO systems for multi-target detection and communications. *IEEE Transactions on Signal Processing*, 71:741–755, 2023.

26 A. M. Ahmed, L. Gharsalli, S. Fortunati, and A. Sezgin. Reinforcement learning for cognitive integrated communication and sensing systems. In *2023 20th European Radar Conference (EuRAD)*, pages 395–398, 2023.

27 A. Umra, A. M. Ahmed, and A. Sezgin. Towards smarter sensing: 2D clutter mitigation in RL-driven cognitive MIMO radar. In *2025 Joint European Conference on Networks and Communications & 6G Summit (EuCNC/6G Summit)*, pages 1–6, 2025.

28 S. Buzzi, C. D'Andrea, and M. Lops. Using massive MIMO arrays for joint communication and sensing. In *2019 53rd Asilomar Conference on Signals, Systems, and Computers*, pages 5–9, 2019.

29 W. Tan, S. D. Assimonis, M. Matthaiou, Y. Han, X. Li, and S. Jin. Analysis of different planar antenna arrays for mmWave massive MIMO systems. In *2017 IEEE 85th Vehicular Technology Conference (VTC Spring)*, pages 1–5, 2017.

30 P. Swerling. Probability of detection for fluctuating targets. *IRE Transactions on Information Theory*, 6(2):269–308, 1960.

31 J. Li, L. Xu, P. Stoica, K. W. Forsythe, and D. W. Bliss. Range compression and waveform optimization for MIMO radar: A Cramer–Rao bound based study. *IEEE Transactions on Signal Processing*, 56(1):218–232, 2008.

32 I. Bekkerman and J. Tabrikian. Target detection and localization using MIMO radars and sonars. *IEEE Transactions on Signal Processing*, 54(10):3873–3883, 2006.

33 B. Friedlander. On signal models for MIMO radar. *IEEE Transactions on Aerospace and Electronic Systems*, 48(4):3655–3660, 2012.

34 M. L. Puterman. *Markov Decision Processes: Discrete Stochastic Dynamic Programming*. John Wiley & Sons, 2014.

35 C. Szepesvari. *Algorithms for Reinforcement Learning*. Morgan and Claypool Publishers, 2010.

36 P. Stoica, J. Li, and Y. Xie. On probing signal design for MIMO radar. *IEEE Transactions on Signal Processing*, 55(8):4151–4161, 2007.

37 A. H. Nuttall. Some integrals involving the (q_m)-function. Technical Report 4297. Naval Underwater Systems Center, New London, CT, 1974.

38 D. E. Dudgeon and R. M. Mersereau. *Multidimensional Digital Signal Processing*. Prentice Hall Professional Technical Reference, 1990.

39 G. Dulac-Arnold, N. Levine, D. J. Mankowitz, J. Li, C. Paduraru, S. Gowal, and T. Hester. Challenges of real-world reinforcement learning: Definitions, benchmarks and analysis. *Machine Learning*, 110:2419–2468, 2021.

40 C. Thornton and R. Buehrer. When is cognitive radar beneficial? insights from dynamic spectrum access. In *2023 IEEE Radar Conference (RadarConf23)*, pages 1–6, 2023.

41 H. White and I. Domowitz. Nonlinear regression with dependent observations. *Econometrica*, 52(1):143–161, 1984.

42 S. Fortunati, F. Lisi, A. M. I. Ahmed, A. Sezgin, M. S. Greco, and F. Gini. *Fundamental Limits for ISAC—Asymptotics in Massive MIMO Sensing Systems*, pages 119–147. Springer Nature Singapore, Singapore, 2023.

5

Learning Approaches for Detection and Mapping in Autonomous Agents' Networks

Anna Guerra,[1,] Siwei Zhang,[2] Gianluca Fontanesi,[3] Francesco Guidi,[4] Davide Dardari,[1] and Petar M. Djurić[5]*

[1] *Department of Electrical, Electronic, and Information Engineering "Guglielmo Marconi," University of Bologna, Cesena, Italy*
[2] *Institute of Communications and Navigation, German Aerospace Center, Wessling, Germany*
[3] *Radio Systems Research, Nokia Bell Labs, Stuttgart, Germany*
[4] *Institute of Electronics, Information Engineering and Telecommunications, National Research Council of Italy, Bologna, Italy*
[5] *Department of Computer and Electrical Engineering, Stony Brook University, Stony Brook, NY, United States*

*Corresponding author: anna.guerra3@unibo.it

5.1 Introduction

This chapter investigates the role of learning in enhancing the sensing and communication performance of autonomous agents (AAs). Here, we define an "autonomous agent" as any device that operates independently in an environment to achieve specific goals without direct human control, such as an on-ground robot, unmanned aerial vehicle (UAV), or autonomous underwater vehicle (AUV).

Before exploring the core content, we clarify the intended meaning of three foundational terms used throughout this chapter: *sensing*, *communication*, and *learning*. *Sensing* refers to the capability of AAs to detect targets using wireless signals, to accurately localize and track them (i.e., *detection and localization*), and to reconstruct the environment electromagnetically (i.e., *mapping*). These functionalities are essential for supporting context-aware autonomous systems. The term *communication* denotes the exchange of information between agents and higher-level infrastructure, such as edge devices or a base station (BS). *Learning* refers to the agents' ability, either individually or collectively, to (1) infer the system's state from sensed data and network interactions and (2) make decisions that enhance inference accuracy—for example, navigation or adaptive sensor configurations such as beamforming patterns.

This leads to the central question motivating this chapter: *Why use AAs for sensing?* Multi-agent systems have attracted increasing attention in sensing and exploration applications, e.g., for search-and-rescue [1, 2], environmental monitoring [3–5], and extraterrestrial missions [6, 7]. For such applications, having a network of AAs cooperating to achieve a common goal brings several advantages. First, the exploration efficiency improves due to the possibility of exploring the environment from different points of view determined by the AAs' positions. Second, the risk of mission failure is reduced when multiple agents are used rather than a single one, and the overall learning capability is increased thanks to the sharing of knowledge among AAs. Furthermore, the possibility of using networks of AAs is increasingly attractive because of their flexibility and reconfigurability.

The capability of AAs to organize into well-structured formations and maintain them during navigation is an essential feature for sensing and localization applications. Localization based on AAs extends and improves the anchor-based localization problem. In fact, the concept of anchors or reference nodes is transformed into a mobile sensor, which can adapt its position in order to enhance its sensing capabilities. For such applications, AAs form a dynamic network of sensors acting as anchors that collectively detect and estimate the position of targets [8].

A fundamental challenge in AA operations is optimizing trajectories to maximize sensing efficiency while preserving resources like battery life. For instance, in post-disaster scenarios, networks of AAs are deployed to detect and localize victims, supporting rescue operations [9]. Effective trajectory planning ensures timely responses, and recent studies have explored this optimization in the contexts of wireless communication, localization, and energy efficiency [10–15]. Specifically, optimizing the trajectory can lead to a better channel [13] or energy use [15].

Traditionally, model-based optimization techniques, such as nonlinear and dynamic programming, have been employed to improve sensing capabilities. By systematically minimizing sensing errors and efficiently dealing with safety constraints such as obstacle avoidance and energy conservation, dynamic programming proves to be an invaluable tool in optimizing sensing capabilities [16].

However, when accurate system models are unavailable or unreliable, machine learning (ML), particularly reinforcement learning (RL), has emerged as a powerful technique for optimizing AAs' policies in dynamic environments [17–19]. RL enables AAs to learn optimal strategies through real-time interaction with their environment, adapting based on feedback. Techniques such as Q-learning and deep Q-learning are commonly used for trajectory optimization in small-scale and large-scale state spaces. Deep RL helps mitigate the curse of dimensionality through neural network (NN)-based function approximation, yet interpretability remains a challenge [20, 21]. However, deep RL is limited because these solutions are often tailored to specific locations or maps. A practical solution is sharing prior model knowledge to improve the convergence speed of RL training when the UAV faces a new environment [14].

In the following, we consider three use cases, summarized in Figure 5.1 and detailed below:

- **Scenario I—Dynamic Radar Network (DRN):** A multi-agent system forms a dynamic localization infrastructure, where agents with known positions (anchors) collaboratively track moving targets using onboard radars [8, 22]. For applications like surveillance or disaster response, AAs with multi-antenna radars can generate probabilistic occupancy maps via Bayesian filtering [23]. The environment state includes both target positions and the estimated map.
- **Scenario II—Infrastructure-free Relative Localization:** A swarm of agents operates in GPS-denied environments (e.g., space, underground), where self- and target-localization rely on inter-agent measurements [7]. Here, the environmental state includes both target and agent positions.
- **Scenario III—Integrated Sensing and Communication (ISAC):** A single agent with a massive MIMO array performs beamforming to simultaneously sense the environment and communicate with a terrestrial BS, fully integrating both functions [24].

Across all considered scenarios, we assume AAs as comprising three functional blocks, as illustrated in Figure 5.2:

1. **Sensing and Inference:** The first stage involves acquiring informative data through onboard sensors. For instance, when equipped with radar systems, an AA can extract information from signals backscattered by the environment. These measurements provide localization information about environmental targets. Based on the acquired data and the information exchanged within the

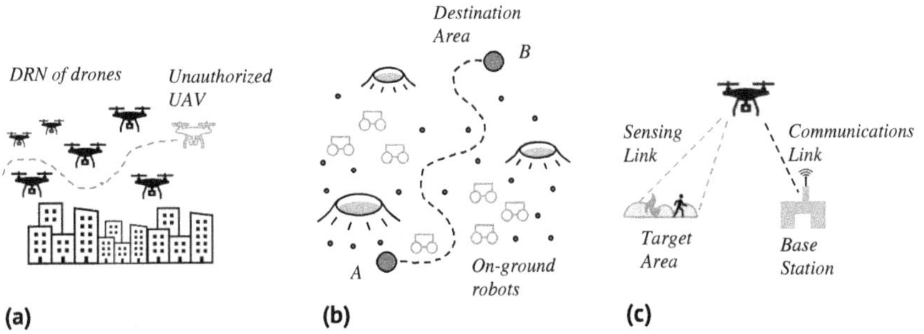

Figure 5.1 Case studies: (a) Target tracking by a DRN, (b) relative localization by a swarm of on-ground robots, (c): ISAC using a single UAV equipped with multiple antennas.

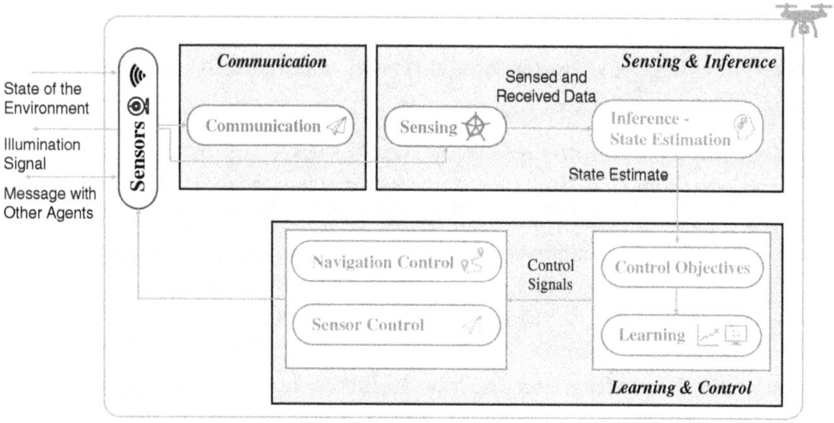

Figure 5.2 Functional architecture of an AA.

network, each agent estimates environmental features of interest (namely, the *state of the environment*). Bayesian estimators can be employed to compute the posterior probability distribution of the state, conditioned on the history of observations [25].

2. **Communication:** As part of a network, AAs share their information, be it raw measurements, processed estimates, or learned models, with neighboring agents or with a centralized entity (e.g., the cloud). Similarly, they receive data from nearby AAs through multi-hop communication protocols.
3. **Learning and Control:** Based on learned and sensed information, agents make decisions that improve sensing and communication performance. The strategies and policies that lead to these actions will be categorized into two major families: (1) model-based learning and (2) data-driven approaches. Each AA updates its position and sensor configuration (e.g., the phase profile of an antenna array) according to a control policy adapted to the assigned task.

Building on the introduction of key capabilities and architectural paradigms, the remainder of the chapter is organized as follows. In Section 5.2, we define how sensing, representing the foundation for learning, can be conceived and implemented in AA networks. Then, Section 5.3 describes how navigation decisions can be formulated as the analytical solution of an optimization problem. Section 5.4 discusses action selection based on received rewards using RL strategies [23], or using an NN and a beamformer optimizer [24]. Section 5.5 will explore the potential of combining the benefits of model-based and data-driven approaches. Finally, Section 5.6 concludes the chapter and outlines future research directions.

5.2 Sensing with Autonomous Agents

Recent research has highlighted the growing importance of AAs in enhancing sensing capabilities, particularly in dynamic environments [10, 26]. Sensing is at the core of AA operations across terrestrial, underwater, aerial, and space domains, enabling critical tasks such as object detection, localization, and environmental mapping. Indeed, AA-based networks significantly improve situational awareness and mission planning, critical for time-sensitive applications such as disaster response. Their flexibility allows rapid deployment and reconfiguration to meet specific operational needs and provides a clear advantage over static and fixed sensor infrastructures [27]. Beyond sensing, these networks also serve as communication backbones that allow localization, information exchange, and coordination in distributed environments [28–30].

A significant challenge in AA-based sensing networks is ensuring accurate, timely, and synchronized sensor readings across multiple agents. Spatiotemporal coordination is crucial for maintaining data integrity, and ISAC strategies can enhance information sharing, localization, and decision-making. In addition, the overall network architecture significantly influences sensing performance.

This section establishes foundational definitions for the subsequent discussions on AAs' decision-making for improving sensing capabilities. Specifically, we outline architectures for multi-AA sensing and explore key sensing techniques for detection, localization, and mapping. Finally, we discuss integrating sensing and learning approaches to support adaptive agent behaviors.

5.2.1 Architectures for Multi-agent Sensing

Autonomous sensing systems can be structured in different ways depending on their environment, mission objectives, and communication constraints. The architecture of these systems significantly impacts their efficiency, scalability, and robustness. We now categorize key architectural frameworks for multi-agent sensing and discuss their advantages, challenges, and potential applications [26, 31].

Centralized multi-agent sensing relies on a fusion center, such as a ground station or cloud server, where sensor data from multiple AAs is collected, processed, and analyzed. The fusion center then makes informed decisions and communicates instructions to the agents. This approach ensures high accuracy since it processes global information and maintains a consistent decision-making model. However, it introduces significant communication overhead, which may be problematic in bandwidth-limited environments. Additionally, centralized systems are vulnerable to single points of failure, making them less robust when the fusion center becomes inaccessible or compromised.

Distributed multi-agent sensing, in contrast, allows each agent to process its sensor data locally while exchanging relevant information with nearby agents. This approach enhances resilience and improves scalability by eliminating dependence on a centralized node, as agents can operate independently with minimal coordination overhead. Nonetheless, the distributed nature of this architecture presents challenges, such as inconsistencies in local decisions, which may lead to suboptimal global performance. Moreover, the onboard processing capabilities of individual agents are typically more limited than those of centralized systems. For example, in networks of UAVs, onboard processing is constrained by the energy required for flight, allowing only simple and computationally lightweight algorithms.

Hybrid multi-agent sensing architectures attempt to balance the trade-offs between centralized and distributed approaches by incorporating elements of both. In such systems, groups of agents may perform local data processing in clusters before transmitting processed information to a centralized fusion center. This structure improves fault tolerance by combining localized decision-making with global oversight, allowing the system to adapt dynamically to mission requirements. However, designing efficient coordination protocols that regulate data exchange between agents and fusion centers remains a complex challenge, especially in highly dynamic networks.

Following the aforementioned classification, networks can also be categorized according to whether they rely on external infrastructure. In this regard, **infrastructure-assisted** networks leverage external infrastructure, such as 6G BSs or satellite networks, to enhance sensing capabilities and communication. By exploiting the advanced computational resources available at the network edge, these systems significantly reduce onboard processing demands while enabling high-precision sensing. However, this approach comes with certain limitations. Its reliance on external communication infrastructure makes it vulnerable in remote or adversarial environments where connectivity may be unreliable or unavailable. Additionally, offloading data processing to an external system can introduce latency, potentially impacting real-time decision-making and responsiveness.

In contrast, for applications such as deep-sea exploration, space, or underground operations, **infrastructure-free** sensing is very important. In these GPS-denied environments, agents must operate autonomously without relying on external positioning systems. Several techniques enable effective operation under these conditions. Radio-based localization methods, including ultra-wideband (UWB) and acoustic signaling, allow agents to estimate their relative positions through inter-agent distance measurements. Simultaneous localization and mapping (SLAM) techniques enable agents to construct a map of their surroundings while concurrently determining their locations. Additionally, inter-agent communication is fundamental in improving collective localization accuracy, as agents can share position estimates and collaboratively refine their understanding of the environment. The primary challenges in infrastructure-free sensing lie in maintaining accurate self-localization without absolute reference points and ensuring proper synchronization of sensing data among agents without external timing signals. Additionally, it is essential to note that managing a communication network without infrastructure is highly complex, involving challenges such as resource allocation, medium access control, dynamic network topology, limited bandwidth, security vulnerabilities, and the need for self-organizing routing protocols.

Overall, selecting the appropriate architecture for multi-agent sensing depends on the application's requirements and constraints. While centralized architectures provide high accuracy and consistency, distributed and hybrid approaches offer increased resilience and scalability. Infrastructure-assisted sensing enhances precision and offloads computational effort, whereas infrastructure-free sensing is indispensable in environments where external support is unavailable.

5.2.2 Integrated Sensing and Communication

AAs can use existing communication hardware for simultaneous sensing, optimizing resource utilization and operational efficiency. By integrating sensing and communication, AAs can enhance situational awareness, improve decision-making, and enable more efficient coordination within a networked environment [32]. For instance, AVs, drones, and robotic platforms with multi-antenna arrays can detect and track objects while maintaining seamless communication with infrastructure nodes and other agents. In this dual-role configuration, AAs facilitate real-time data exchange and environmental mapping while acting as radars.

Advanced signal processing and deep-learning-based beamforming techniques enable dynamic antenna reconfiguration, optimizing communication quality and sensing accuracy. By continuously adapting their sensing and transmission strategies, AAs can refine localization accuracy, detect and classify targets more effectively, and maintain robust connectivity in dynamic and complex environments.

Furthermore, ISAC systems empower AAs to operate more intelligently by integrating sensing data with communication protocols. This fusion allows for cooperative sensing, where multiple agents share detected information, leading to enhanced environmental perception and improved decision-making in applications such as autonomous driving, surveillance, search-and-rescue, and industrial automation.

While sensing provides raw data and environmental cues, agents can transform this information into actionable decisions through learning mechanisms.

In the following sections, we review model-based, data-driven, and hybrid learning approaches for decision-making. These methods enable AAs to operate effectively in unknown environments by improving their sensing, inference, and decision-making capabilities through advanced learning techniques. Model-based approaches rely on analytical representations of the environment to optimize navigation and sensing strategies, offering structured and interpretable decision-making frameworks. In contrast, data-driven methods, such as RL, allow agents to learn directly from interactions with the environment, progressively enhancing performance without requiring prior models. Hybrid approaches combine these paradigms by integrating analytical structure with adaptive learning, thereby balancing efficiency and flexibility. To highlight the benefits and limitations of these approaches, we present representative case studies that illustrate how different learning strategies support reliable detection and mapping under dynamic and uncertain conditions.

5.3 Model-based Learning

The development of a network comprising AAs involves the crucial task of designing optimized trajectories to ensure that mission objectives are effectively achieved. We refer to this task as "agents' state optimization." The design of the pathways followed by these agents has been extensively explored in a plethora of studies [33–45]. In control engineering, significant attention has been given to the optimal placement of sensors for localization applications [33, 45], while other studies have tackled the problem through optimal control methods [44].

Methods such as information-seeking optimal control and utilizing metrics like Shannon or Fisher information (FI) have been thoroughly investigated for applications in localization and tracking [7, 36–38, 40–43, 45]. The authors in [45] studied the optimal network configurations using a FI metric to enhance the collaborative tracking of ground targets by UAVs. This analysis accounts for a non-Bayesian Fisher information matrix (FIM), without considering prior knowledge or motion models. Additionally, these methodologies often overlook the dynamic nature of the environment, neglecting the influence of non-line-of-sight (NLOS) biases and the target's radar cross section (RCS) on the measurement models.

As the name suggests, model-based learning for decision-making uses explicit mathematical models of the system dynamics and observation processes to:

- estimate the state of agents in a network and the state of the environment (e.g., state inference);
- optimize the state of agents to proactively achieve some tasks (e.g., navigation).

Figure 5.3 illustrates the concept of *self-aware swarm navigation*, which integrates estimation and control in a closed-loop framework. On the left, the *swarm localization* module seeks to minimize the expected squared error between the estimated position $\hat{\boldsymbol{p}}_k$ and the true position $\boldsymbol{p}_k$, using local observations and

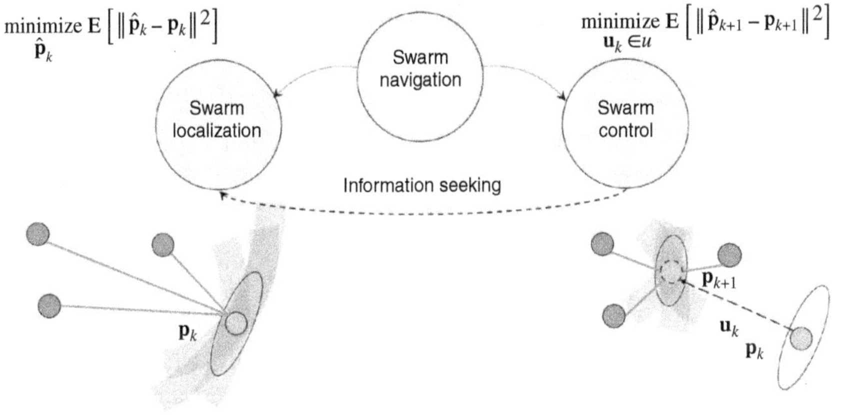

Figure 5.3 Self-aware swarm navigation: estimation and control.

inter-agent interactions. On the right, the *swarm control* module determines a control input $\boldsymbol{u}_k \in \mathcal{U}$ that minimizes the expected localization error at the next time step $k + 1$, i.e., between $\hat{\boldsymbol{p}}_{k+1}$ and $\boldsymbol{p}_{k+1}$. The central block, *swarm navigation*, coordinates the flow between estimation and control by guiding agents toward regions of higher expected information gain. This *information-seeking behavior* is represented by the dashed feedback loop from control back to localization, encouraging movement strategies that improve future estimation accuracy. The ellipses around the positions $\boldsymbol{p}_k$ and $\boldsymbol{p}_{k+1}$ represent the agents' uncertainty.

These navigation policies are obtained by solving an optimization problem to improve a mission-related performance metric. In state estimation and optimization, the AAs network requires a certain level of system model awareness. We distinguish three levels of model awareness in network navigation.

- At the first level, the network exploits the observation models in state estimation algorithms, e.g., weighted least squares (WLS), extended Kalman filters (EKFs), or particle filters (PFs). The state optimization uses only the "hard output" of state estimates. Any information about estimation uncertainty, which the localization algorithm might provide, is decoupled from state optimization. This level of information usage corresponds to a traditional navigation system.
- At the second level, the state uncertainty provided by the model is loosely considered for state optimization. Hence, the optimizer exploits the soft outputs of state estimates. This estimation uncertainty can be obtained by Cramér–Rao bound (CRB), posterior Cramér–Rao bound (PCRB), the posterior covariance matrix from an EKF, the *a posteriori* probability density function (pdf) from a PF, or the belief from a message passing (MP) algorithm. The state controller is designed with the knowledge of the state uncertainty, which is

referred to as stochastic control [46]. At levels one and two, the state optimization objectives are defined externally to the estimation tasks, where minimizing the state uncertainty is not an optimization objective.

- At the third level of model awareness, information usage is fundamentally different. The agents close the estimation-optimization loop by understanding how to change their state so that the uncertainty is actively minimized.

In the following, we refer to the state optimization using the third type of model awareness as the *information-seeking optimization*, which fully exploits the system model for state optimization decisions. Our information-seeking state optimization concept is based on the projected steepest gradient descent (PSGD) algorithm. Although agents' control is often a multi-objective problem, we focus on a single objective of information-seeking with collision avoidance constraints in this part. In contrast, the multi-objective nature of a swarm navigation system will be discussed in Section 5.5.

For the sake of generalizability, in the following, we assume a distributed multi-agent network where AAs communicate through multiple hops. We assume that each communication hop incurs a delay of one time slot. Note that the centralized multi-agent architecture can be considered a specific instance of the distributed architecture where all nodes are connected.

For this purpose, let us define:

- $\boldsymbol{s}_k$: the state vector at time k;
- $\hat{\boldsymbol{s}}_{i,k}$: the estimated state vector by the ith agent at time k;
- $\boldsymbol{p}_{i,k}$: the ith agent position at time instant k;
- $\boldsymbol{u}_{i,k}$: the ith agent position navigation command at time instant k;
- $\boldsymbol{\tau}_{i,k} = [\dots, h_{ij,k}, \dots]$: the vector containing the number of hops present between agent i and agent j at time instant k;
- $\boldsymbol{\ell}_{i,k} = \left[\dots, \boldsymbol{p}_{j,l_k}, \dots\right]$, $l_k = k - h_{ij,k} + 1$, $j \in \mathcal{N}_i$: the vector of agents' positions available at agent i at time k delayed by the number of communication hops $h_{ij,k} \in \boldsymbol{\tau}_{i,k}$ present between the agent i and its neighboring agents in $\mathcal{N}_i$;
- $\boldsymbol{\Sigma}_{i,k}$: the state covariance matrix at the ith agent at time step k;
- $\boldsymbol{I}_{i,k}$: the FIM, i.e., the inverse of the state covariance matrix $\boldsymbol{I}_{i,k} = \left(\boldsymbol{\Sigma}_{i,k}\right)^{-1}$, quantifying the information content of observations.

The primary objective of navigation is to optimize the formation of the AAs to minimize errors related to sensing and estimation, enhancing the overall efficacy and reliability of the aerial network in dynamic conditions. More specifically, position information-seeking aims to control the swarm to a favorable position, where the predicted measurements provide the richest additional information on the state of interest. This problem can be formulated as:

$$(\boldsymbol{\ell}_{i,k+1})^{\star} = \underset{\boldsymbol{\ell}_{i,k+1}}{\operatorname{argmin}} \quad \mathcal{C}(\boldsymbol{I}_{i,k}(\hat{\boldsymbol{s}}_{i,k+1|k}, \boldsymbol{\ell}_{i,k+1})), \tag{5.1a}$$

$$\text{s.t.} \quad \boldsymbol{h}(\boldsymbol{\ell}_{i,k+1}) \geq \mathbf{0}, \tag{5.1b}$$

where $\mathcal{C}(\cdot)$ is an optimal experimental design (OED) function and $\boldsymbol{h}(\boldsymbol{\ell}_{i,k+1})$ is a function accounting for the constraints. For a localization problem, $\boldsymbol{I}_{i,k}$ is an information metric that quantifies the accuracy of the localization and estimation processes, and it depends on the positions of the UAVs and the state of the environment (e.g., comprising the position coordinates of the target).

The control signal of the ith AA is given by:

$$\boldsymbol{u}_{i,k} = \left[(\boldsymbol{\ell}_{i,k+1})^{\star}\right]_i - \boldsymbol{p}_{i,k},$$

where $[\cdot]_i$ is an operator that picks the ith entry of the optimal formation in (5.1).

According to the optimality criteria outlined in [40], the cost function $\mathcal{C}(\boldsymbol{I}_{i,k})$ can be written as:

$$\mathcal{C}(\boldsymbol{I}_{i,k}) = \begin{cases} \operatorname{tr}[\boldsymbol{\Sigma}_{i,k}], & A\text{–optimality} \\ -\ln\det[\boldsymbol{I}_{i,k}], & D\text{–optimality}. \end{cases} \tag{5.2}$$

A-optimality (A stands for "average") focuses on minimizing the trace of the covariance matrix of the estimators. The trace of a matrix, denoted by $\operatorname{tr}[\cdot]$ in (5.2), is the sum of its diagonal elements and, in this context, represents the sum of the variances of the parameter estimates. By minimizing the trace, an A-optimal design seeks to minimize the average variance of the parameter estimates. D-optimality (D for "determinant") aims to minimize the determinant of the covariance matrix of the estimators, or equivalently, maximize the determinant of the information matrix. The determinant of the covariance matrix, denoted with $\det[\cdot]$ in (5.2), can be thought of as a measure of the volume of the confidence ellipsoid around the parameter estimates in the parameter space.

The optimization problem in (5.1) provides a principled way to guide AAs toward positions that maximize the information gained about a target or system state, thereby improving localization and sensing performance. The approach formulates trajectory planning as a constrained optimization problem, where the goal is to minimize estimation uncertainty using information-theoretic criteria such as A-optimality or D-optimality. By solving this problem iteratively via projected gradient descent, agents adapt their movements while satisfying safety constraints, enabling efficient and reliable operation in dynamic, multi-agent environments.

Traditionally, the approach to solving this issue involves model-based optimization techniques, such as nonlinear programming. This can be done either through a greedy approach or by deriving closed-form solutions that facilitate rapid decision-making. For instance, the work presented in [22] frames the UAV navigation challenge as minimizing the posterior covariance matrix related to target estimation while also considering UAV anti-collision and mobility constraints. Note that minimizing the posterior covariance matrix reduces the uncertainty of the state estimation by accounting for both the state transition and observation models.

The control problem formulated in (5.1) is a high-dimensional nonconvex optimization problem for the considered large-scale network of AAs. Instead of finding the optimal solution in one step, we adopt the PSGD method [47, Ch. 5] to design a low-complexity swarm controller.

PSGD is an iterative method to find a locally optimal solution to a constrained optimization problem. The general idea of PSGD is to iteratively apply the following steps [7]:

i. Analytically calculate the gradient of the cost function with respect to the AAs' positions.
The gradient of the objective function is written as:

$$\boldsymbol{c}_{i,k+1} = \text{vec}\{\boldsymbol{c}_{j,k+1} : j \in \mathcal{N}_i\} = \nabla_{\boldsymbol{\ell}_{i,k+1}} f_{\text{p}}(\boldsymbol{\ell}_{i,k+1}), \tag{5.3}$$

which points to the direction where the value of:

$$f_{\text{p}}(\boldsymbol{\ell}_{i,k+1}) \triangleq C(\boldsymbol{I}_{i,k}(\hat{\mathbf{s}}_{i,k+1|k}, \boldsymbol{\ell}_{i,k+1})),$$

ascends most steeply. Here j indicates the jth agent of the neighboring set $\mathcal{N}_i$ of the agent i at time instant k.

ii. Compute an initial solution using the steepest descent of the cost function.
An unconstrained solution $\tilde{\boldsymbol{\ell}}_{i,k}$ can first be found by the steepest gradient descent method, similar to step (i) of PSGD as:

$$\tilde{\boldsymbol{\ell}}_{i,k+1} = -\mu \frac{\boldsymbol{c}_{i,k+1}}{||\boldsymbol{c}_{i,k+1}||}, \tag{5.4}$$

where μ is the chosen step size of the gradient descent.

iii. Calculate the projection matrix to restrict the initial solution to the tangent space of any potentially violated constraints.
The activated constraint vector can be denoted as:

$$\boldsymbol{h}_{\text{a}}(\boldsymbol{\ell}_{i,k+1}) = \text{vec}\{h_{\text{c},ij}(\boldsymbol{\ell}_{i,k+1}) : h_{\text{c},ij}(\boldsymbol{\ell}_{i,k+1}) \leq 0\},$$

with length L_a, i.e., the constraints that either have been violated or are at the boundary of violation. The constraint gradient matrix $\boldsymbol{N}_i \in \mathbb{R}^{2|\mathcal{N}_i| \times L_a}$ is a $2|\mathcal{N}_i| \times L_a$ real-valued matrix defined as:

$$\boldsymbol{N}_i = \nabla_{\boldsymbol{\ell}_{i,k+1}} \boldsymbol{h}_a^{\mathrm{T}}(\boldsymbol{\ell}_{i,k+1}),$$

and the projection matrix $\boldsymbol{P} \in \mathbb{R}^{2|\mathcal{N}_i| \times 2|\mathcal{N}_i|}$ is defined as:

$$\boldsymbol{P}_i = \boldsymbol{I} - \boldsymbol{N}_i(\boldsymbol{N}_i^{\mathrm{T}}\boldsymbol{N}_i)^{-1}\boldsymbol{N}_i^{\mathrm{T}}, \tag{5.5}$$

with $\boldsymbol{I}$ being the identity matrix and where $[\cdot]^{\mathrm{T}}$ denotes the transpose operator. The projection matrix projects the unconstrained solution $\tilde{\boldsymbol{\ell}}_{i,k+1}$ onto the tangent space of the activated constraints.

iv. Apply corrections to address the nonlinearity effects of the constraints (termed "nonlinearity correction") and derive the projected control for AAs.
The control $\boldsymbol{\delta}_i$ after the projection is expressed as $\boldsymbol{\delta}_i = \boldsymbol{P}_i \tilde{\boldsymbol{\ell}}_{i,k+1}$. In addition, the quantity of constraint violations needs to be compensated, leading to the final solution:

$$(\boldsymbol{\ell}_{i,k+1})^{\star} = \boldsymbol{\delta}_i - \boldsymbol{N}_i(\boldsymbol{N}_i^{\mathrm{T}}\boldsymbol{N}_i)^{-1}\boldsymbol{h}_a(\boldsymbol{0}). \tag{5.6}$$

Again, the control signal of the ith AA is given by $\boldsymbol{u}_{i,k} = \left[(\boldsymbol{\ell}_{i,k+1})^{\star}\right]_i - \boldsymbol{p}_{i,k}$.

We now present two case studies illustrating the application of the information-seeking framework. The first case addresses a relative localization problem, where agents operate under a centralized architecture, and their positions are unknown and subject to estimation. The second case focuses on a distributed formation of agents collaboratively tracking a moving target.

5.3.1 Case Study 1: Swarm Control and Relative Localization

We now consider a network of AAs that operates to localize a source during an extraterrestrial mission in the absence of global navigation satellite systems (GNSS). The agents form a centralized network, and the state of the AAs (e.g., their positions) is unknown. Therefore, the mission goal of the network is to estimate a state vector including the positions of all the agents, i.e., $\boldsymbol{s}_k = \mathrm{vec}\{\boldsymbol{p}_{i,k}, i \in \mathcal{N}\}$. The layout is depicted in Figure 5.4 (left).

We consider the following movement model for the agents:

$$\boldsymbol{p}_{i,k+1} = \boldsymbol{p}_{i,k} + \boldsymbol{u}_{i,k} + \omega_{i,k}, \quad \forall i \in \mathcal{N}, \tag{5.7}$$

i.e., the control command $\boldsymbol{u}_{i,k}$ is directly applied to the positions of the agents and $\omega_{i,k} \sim \mathcal{N}(\boldsymbol{0}, \boldsymbol{Q}_i)$ is the additive Gaussian noise with zero mean and covariance $\boldsymbol{Q}_i$.

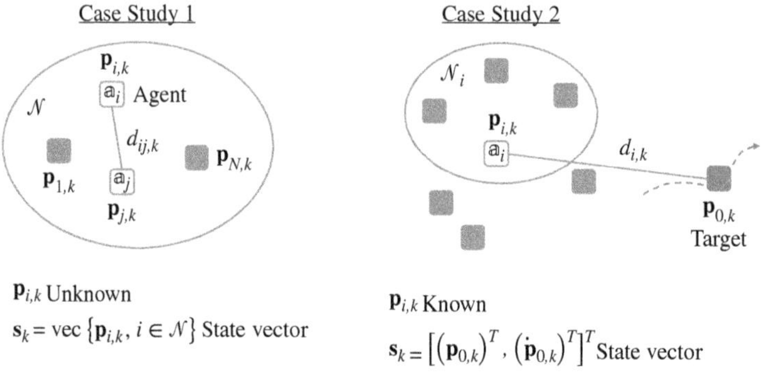

Figure 5.4 Layout of the network configuration employed for the case studies.

The trace of the state-estimation covariance matrix, calculated from $p(\tilde{\mathbf{z}}_{k+1}; \mathbf{s}_{k+1})$, being the distribution of the predicted ranging measurements $\tilde{\mathbf{z}}_{k+1}$ related to the state $\mathbf{s}_{k+1}$, is a suitable figure of merit for swarm optimization. However, the covariance matrix is an empirical metric, which is difficult to be directly exploited as an objective function. Instead, an FI-seeking objective can be formulated with the predicted FIM, denoted as $\boldsymbol{I}_{\mathbf{s}_{k+1}}$, given a snapshot of predicted measurements $\tilde{\mathbf{z}}_{k+1}$. The predicted CRB, denoted as $\text{CRB}[\mathbf{s}_{k+1}]$, is a lower bound of the predicted estimation covariance $\text{cov}_{\tilde{\mathbf{z}}_{k+1}}[\hat{\mathbf{s}}_{k+1}]$, i.e.,

$$\text{cov}_{\tilde{\mathbf{z}}_{k+1}}[\hat{\mathbf{s}}_{k+1}] \succeq \text{CRB}[\mathbf{s}_{k+1}], \tag{5.8}$$

where $\succeq$ denotes the matrix inequality in the semidefinite sense, i.e., $\boldsymbol{A} \succeq \boldsymbol{B}$ means $\boldsymbol{A} - \boldsymbol{B}$ is positive semidefinite.

For an ordinary network geometry and a high signal-to-noise ratio (SNR), the CRB is a tight bound suitable for modeling the estimation of covariance [48]. Then, the FI-seeking can be written with the following cost function

$$\mathcal{C}(\boldsymbol{I}_{i,k}) \triangleq \text{tr}[\boldsymbol{\Lambda}_k \text{CRB}[\mathbf{s}_{k+1}]], \tag{5.9}$$

where the diagonal weighting matrix

$$\boldsymbol{\Lambda}_k = \text{diag}\{\lambda_{i,k} \otimes [1, 1] : i = 1, \dots, \mathcal{N}\},$$

indicates the objective emphasis on the specific agents. The entities $\lambda_{i,k}$ of the weighting matrix can be either binary valued from $\{0, 1\}$ to select certain dimensions of the state for optimization, or arbitrary nonnegative real values, which assign different weights to specific dimensions. By minimizing the cost function in (5.9), the swarm will constantly optimize its position so that the value of this cost function gradually decreases.

Collision avoidance is another crucial objective for swarm control. Due to the imperfection of the controller and the position uncertainty, an agent $i \in \mathcal{N}$ needs to keep its distance d_{ij} to another node $j \in \mathcal{N}/\{i\}$ larger than a minimal tolerated distance $d_{\min}$.

The gradients of the cost function and the constraints with respect to the control command are essential for the PSGD algorithm. The gradient of CRB weighted by a diagonal matrix $\boldsymbol{\Lambda}$, i.e., $\nabla_{\boldsymbol{u}_k} \mathrm{tr}[\boldsymbol{\Lambda}\ \mathrm{CRB}[\boldsymbol{s}_{k+1}]]$ is essential for FI seeking, which represents the knowledge of the relationship between the swarm positions and the state estimation uncertainty, corresponding to the third level of the model awareness. Next, we will derive the closed-form expression of this gradient,

$$\boldsymbol{c}_k = \nabla_{\boldsymbol{u}_k} \mathrm{tr}\Big[\boldsymbol{\Lambda}\ \mathrm{CRB}[\boldsymbol{s}_{k+1}]\Big]. \tag{5.10}$$

If the FIM has full rank, the CRB is expressed as the inverse of the FIM. Subsequently, a closed-form expression of the derivative c_l of the trace of the weighted CRB, $\mathrm{tr}[\boldsymbol{\Lambda I}^{-1}_{(\boldsymbol{s}_{k+1}}]$ with respect to the lth element of $\boldsymbol{u}_k$, i.e., $u_l = [\boldsymbol{u}_k]_l$ is derived. Utilizing the derivative chain rule:

$$\partial \boldsymbol{A}^{-1} = -\boldsymbol{A}^{-1} \partial \boldsymbol{A} \boldsymbol{A}^{-1},$$

and the property of the trace $\mathrm{tr}[\boldsymbol{AB}] = \mathrm{tr}[\boldsymbol{BA}]$, the derivative can be rewritten in the form:

$$c_l = \frac{\partial \mathrm{tr}[\boldsymbol{\Lambda I}^{-1}_{\boldsymbol{s}_{k+1}}]}{\partial u_l} = -\mathrm{tr}\Bigg[\underbrace{\boldsymbol{I}^{-1}_{\boldsymbol{s}_{k+1}} \boldsymbol{\Lambda I}^{-1}_{\boldsymbol{s}_{k+1}}}_{\triangleq \boldsymbol{A}} \frac{\partial \boldsymbol{I}_{\boldsymbol{s}_{k+1}}}{\partial u_l}\Bigg]. \tag{5.11}$$

The expanded closed-form expression of (5.11) is lengthy, and therefore is omitted here and can be found in the Appendix of [7]. The derivative $\boldsymbol{c}_k$ of $\mathrm{tr}[\boldsymbol{\Lambda I}^{-1}_{\boldsymbol{s}_{k+1}}]$ with respect to $\boldsymbol{u}_k$ is obtained by stacking all c_l, i.e., $\boldsymbol{c}_k = \mathrm{vec}\{c_l : l = 1, \dots, 2|\mathcal{N}|\}$. In practice, the initial control can be set to zero i.e., $\boldsymbol{u}_k = \boldsymbol{0}$, and the current agents' positions are replaced by their estimates. Hence, the gradient is evaluated at the current agents' estimated positions.

Resulting formations from PSGD-based multi-agent navigation with two different information-seeking weight strategies, namely homogeneous FI seeking and heterogeneous FI seeking, are shown in Figure 5.5. The setup includes three anchors shown as square markers with limited coverages shown with circles, one stationary agent, or a target, shown as a diamond marker with the index 1 and $|\mathcal{N}| - 1 = 49$ movable agents shown as round markers with the indices $i = 2, \dots, |\mathcal{N}|$. In the homogeneous setting shown in Figure 5.5a, the FI seeking weights for all agents are set equally as $\lambda_{i,k} = 1,\ i = 1, \dots, |\mathcal{N}|$. In the heterogeneous setting shown in Figure 5.5b, the emphasis is on the stationary agent with a higher weight $\lambda_{1,k} = 1$. The movable agents have lower FI seeking weights $\lambda_{i,k} = 0.01,\ i = 2, \dots, |\mathcal{N}|$. The resulting formations are notably different. In the

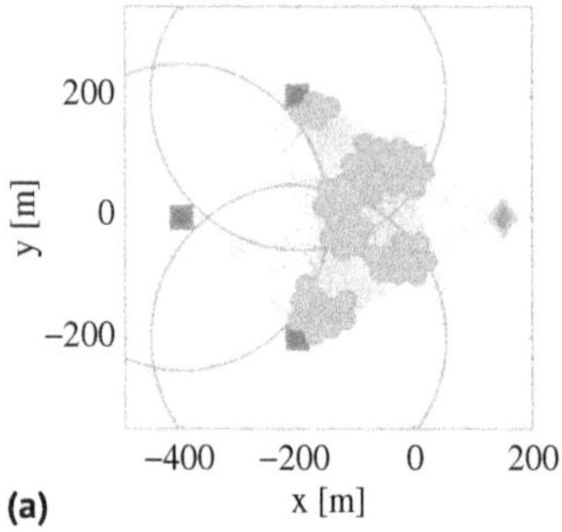

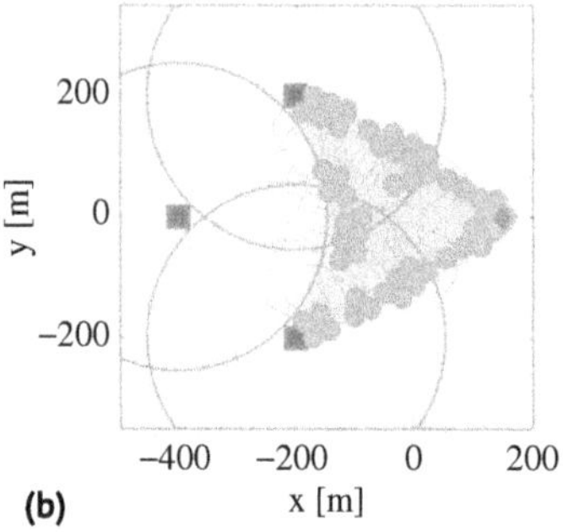

Figure 5.5 (a) Homogeneous FI seeking weights: $\lambda_{i,k} = 1,\ i = 1, \dots, |\mathcal{N}|$. (b) Heterogeneous FI seeking weights: $\lambda_{1,k} = 1, \lambda_{i,k} = 0.01,\ i = 2, \dots, |\mathcal{N}|$. Resulting formations from PSGD-based multi-agent navigation with different information-seeking weights λ_i. The setup includes three anchors shown as square markers with limited coverages shown with circles, one stationary agent, or a target, shown as a diamond marker with the index 1 and $|\mathcal{N}| - 1 = 49$ movable agents shown as round markers with the indices $i = 2, \dots, |\mathcal{N}|$.

heterogeneous setting, movable agents move more proactively to approach the stationary agent, which is beneficial for minimizing the position uncertainty of the stationary agent at the cost of increasing their position uncertainty. In the homogeneous setting, the movable agents are closer to the anchors so that the average position uncertainty of all agents is minimized.

5.3.2 Case Study 2: Dynamic Radar Network

We now present an example of how the navigation problem in (5.1) can be applied to a DRN of UAVs. Unlike Case study 1, we consider a distributed formation and a Bayesian problem in which we aim to track a moving target.

A DRN of UAVs is a group of collaborative UAVs operating in outdoor settings and equipped with onboard radars for target detection and environmental sensing. In this setup, we assume that the UAVs' positions (denoted by $\boldsymbol{p}_{i,k}$) are known, typically through the use of enhanced GNSS sensors. We consider a distributed network formation where each UAV operates autonomously as a central unit with surveillance capabilities like detecting and tracking unauthorized targets. We recall that such architecture enhances the reliability of the network against UAV failures or external threats. The layout is depicted in Figure 5.4 (right).

The goal of the DRN navigation is to optimize the accuracy in tracking the position, $\boldsymbol{p}_{0,k}$, and the velocity, $\dot{\boldsymbol{p}}_{0,k}$, of a moving noncooperative target. Hence, the state vector at instant k is $\boldsymbol{s}_k = \left[(\boldsymbol{p}_{0,k})^{\mathrm{T}}, (\dot{\boldsymbol{p}}_{0,k})^{\mathrm{T}}\right]^{\mathrm{T}}$. The state evolves according to the following dynamic model:

$$\boldsymbol{s}_{k+1} = \boldsymbol{A}\, \boldsymbol{s}_k + \boldsymbol{q}_k, \tag{5.12}$$

where $\boldsymbol{A}$ is the transition matrix, assumed to be known, and $\boldsymbol{q}_k \sim \mathcal{N}(\boldsymbol{0}, \boldsymbol{Q}_k)$ is the process noise.

All UAVs collect radar measurements, providing information on the target state. Indeed, starting from the received signal echoes, each UAV estimates information about the target state, e.g., the distance and angle from the target or the Doppler shift [49]. Subsequently, such information, together with UAV positions, is exchanged between UAVs via multi-hop links. At the end of this communication step, each UAV combines the gathered information, exploitable for target tracking in a vector. Let $\boldsymbol{z}_{i,k} = [\ldots, \boldsymbol{z}_{j,l_k}, \ldots]^{\mathrm{T}}$ be the information available to the ith UAV at time instant k, where the generic element $\boldsymbol{z}_{j,l_k} = \left[\boldsymbol{o}_{j,l_k}, \boldsymbol{p}_{j,l_k}\right]^{\mathrm{T}}$, with $j \in \mathcal{N}_{i,k}$, contains the radar estimates and the position of the jth UAV belonging to the neighboring set of the ith UAV (denoted by $\mathcal{N}_{i,k}$) delayed by l_k time steps due to the multi-hop connection with the ith agent.

Here, we consider only ranging information and line-of-sight (LOS) propagation conditions in the sequel. We refer the interested reader to [22] for NLOS scenarios and other types of measurements. In such a case, the observation model related to the generic radar measurement can be written as:

$$o_{i,k} = r_{i,k} = h(\boldsymbol{p}_{i,k}, \boldsymbol{p}_{0,k}) + n_{i,k}, \tag{5.13}$$

where $h(\boldsymbol{p}_{i,k}, \boldsymbol{p}_{0,k})$ is a function that relates the data to the target state and, in case of ranging measurements, it corresponds to the UAV-target distance: $d_{i,k} = h(\boldsymbol{p}_{i,k}, \boldsymbol{p}_{0,k}) = \|\boldsymbol{p}_{i,k} - \boldsymbol{p}_{0,k}\|$.

Starting from the transition and measurement model previously described, each UAV can perform tracking to estimate the state of the target. As described in Appendix A, a Bayesian estimator can be used to compute the a posteriori probability distribution of the target state given the acquired information. In our case, we adopt an EKF algorithm to compute the Gaussian belief of the state as: $b_i^{(k)}(\boldsymbol{s}_k) = f(\boldsymbol{s}_k|\boldsymbol{o}_{i,1:k}) \overset{\mathrm{EKF}}{=} \mathcal{N}(\boldsymbol{m}_{i,k}, \boldsymbol{\Sigma}_{i,k})$, where $\boldsymbol{m}_{i,k}$ and $\boldsymbol{\Sigma}_{i,k}$ are the conditional mean vector and the covariance matrix of the state and $\boldsymbol{o}_{i,\,1:k}$ is the acquired information by the ith UAV up to time instant k. The EKF filter algorithm generates estimates that minimize the mean-squared error, conditioned on the history of acquired information. Thus, the state estimate at time k, $\hat{\boldsymbol{s}}_{i,k}$, is given by the conditional mean, $\hat{\boldsymbol{s}}_{i,k} = \boldsymbol{m}_{i,k}$.

The autonomous control in (5.1) is designed to predict each UAV's next location to optimize its target tracking performance, taking into account the positions and estimates of neighboring UAVs. Starting from the information model described in

(5.13) and from the output of the EKF, it is possible to write the information matrix for the dynamic scenario as [33]:

$$\boldsymbol{I}_{i,k} = \left[\boldsymbol{\Sigma}_i^- - \boldsymbol{\Sigma}_i^- \boldsymbol{J}_{i,k}^{\mathrm{T}} \left(\boldsymbol{S}_{i,k}\right)^{-1} \boldsymbol{J}_{i,k} \boldsymbol{\Sigma}_i^-\right]_{11}^{-1}, \tag{5.14}$$

where $\boldsymbol{\Sigma}_i^- = \boldsymbol{\Sigma}_{i,k|k-1}$ is the predictive covariance, $\boldsymbol{J}_{i,k}$ is the Jacobian matrix, $\boldsymbol{S}_{i,k} = \boldsymbol{J}_{i,k} \boldsymbol{\Sigma}_i^- \boldsymbol{J}_{i,k}^{\mathrm{T}} + \boldsymbol{R}_{i,k}$, and $\boldsymbol{R}_{i,k}$ is the covariance matrix that depends on the statistical characterization of the measurement noise. Then, according to the matrix inversion lemma [50], (5.14) can be reformulated in a more convenient form as:

$$\boldsymbol{I}_{i,k} = \left[(\boldsymbol{\Sigma}_i^-)^{-1} + \boldsymbol{J}_{i,k}^{\mathrm{T}} \left(\boldsymbol{R}_{i,k}\right)^{-1} \boldsymbol{J}_{i,k}\right]_{11}, \tag{5.15}$$

where $(\boldsymbol{\Sigma}_i^-)^{-1}$ corresponds to the predictive information matrix of the target position, while the second term $\boldsymbol{J}_{i,k}^{\mathrm{T}} \left(\boldsymbol{R}_{i,k}\right)^{-1} \boldsymbol{J}_{i,k}$ corresponds to the FIM for nonrandom parameters. Such a term highlights the effect of the information model (encapsulated in $\boldsymbol{R}_{i,k}$) and of the UAV-target geometric configuration (in the Jacobian matrix, $\boldsymbol{J}_{i,k}$) on localization performance. In this setup, the Jacobian matrix is given by:

$$\boldsymbol{J}_{i,k} = \nabla_{\boldsymbol{s}_k} h(\boldsymbol{p}_{i,k}, \boldsymbol{p}_{0,k}) = \begin{bmatrix} \vdots & \vdots & \vdots & \vdots \\ \cos(\alpha_{j,l_k}) & \sin(\alpha_{j,l_k}) & 0 & 0 \\ \vdots & \vdots & \vdots & \vdots \end{bmatrix}, \tag{5.16}$$

with α_{j,l_k} being the angle between the jth UAV and the target computed with the information available at the ith UAV at time instant k. The expanded closed-form solution of (5.1) is omitted here and can be found in [22].

Figure 5.6 (left) shows a simulated outdoor scenario where we considered a group of five UAVs for tracking a moving target. The UAVs' positions, depicted with square markers, were initially uniformly distributed in the area with $x_{i,0} \sim \mathcal{U}(-0.2, 1)$ km, $y_{i,0} \sim \mathcal{U}(-1.4, 0.4)$ km, and $z_{i,0} \sim \mathcal{U}(0.6, 1.5)$ km. The initial position of the target, shown as a triangle, was $[0, 10, 30]$ m, and its velocity $[0.3, 0.4, 0]$ m/step. The tracking estimator considered a linear transition model. The target's RCS was 0.1 m^2. We assumed that UAVs exchange ranging measurements. The standard deviation of the ranging error was set to 10^{-4} m considering a target with a RCS of 1 m^2 at a distance of 1 m. Unless otherwise specified, we set the maximum number of hops to $h = 2$, the communication range per hop to $r = 0.5$ km, and the probability of losing a packet was set to 1/5. The temporal horizon was 3,000 steps, and the number of Monte Carlo iterations was set to 500. Finally, the UAVs were programmed to keep a safe distance of 5 m from obstacles and a minimum distance of 1 m from each other and from the target.

Figure 5.6 (right) illustrates the empirical cumulative density function (ECDF) of the localization error. For a given threshold, the ECDF indicates the proportion of times the observed localization error falls below that threshold.

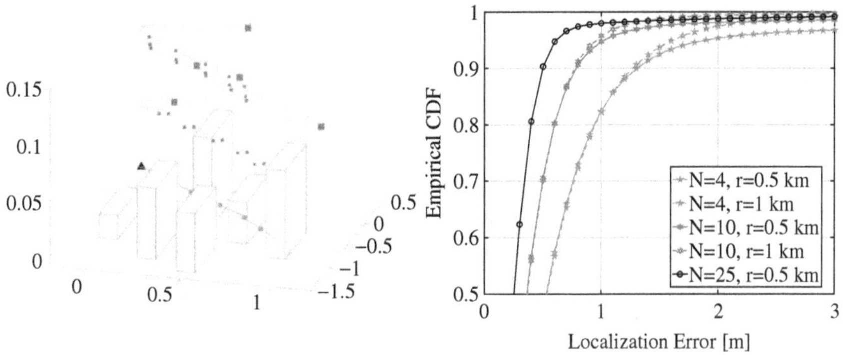

Figure 5.6 Left: simulated scenario with distances indicated in km. Right: empirical CDF of the target-state estimation error as a function of the number of UAVs present in the network and the maximum communication distances.

The performance evaluation considers various configurations, including different network sizes and communication ranges, while fixing the number of communication hops to one. As anticipated, increasing the number of UAVs, from 4 to 25, significantly improves localization accuracy, highlighting the benefits of denser agent deployments. On the other hand, variations in the communication range show a marginal effect in the simulated environment. This can be attributed to the AAs' tendency to converge rapidly near the target and remain within mutual proximity.

Model-based techniques are especially suitable when accurate agent dynamics, sensor noise, and environment structure models are available. They are computationally efficient and interpretable, making them attractive for safety-critical missions. However, their performance may degrade in highly uncertain or rapidly changing environments, motivating the need for data-driven alternatives.

5.4 Data-driven Learning

5.4.1 Background on MDPs and RL

In model-free learning, the interaction between an AA and the environment can be described under the umbrella of the Markov decision process (MDP) or as a partially observable MDP if the state is unknown [51]. An MDP is defined by a tuple $(\mathcal{S}, \mathcal{A}, \mathcal{R}, p, \gamma)$ containing the state space ($\mathcal{S}$), the action space ($\mathcal{A}$), the reward space ($\mathcal{R}$), the probability of state transition (i.e., $p(s'|s,\ a)$ denotes the probability of transitioning to state $s' \in \mathcal{S}$ from state $s \in \mathcal{S}$ after taking action $a \in \mathcal{A}$), and the discount factor γ, determining the importance of future rewards relative to immediate ones. A policy $\pi(a|s) = p(A_t = a|S_t = s)$ is a function that maps the observations or the states to the actions and, hence, *learning* is the process that allows inferring such a policy [52].

The goal in an MDP is to find an optimal policy $\pi^\star : \mathcal{S} \rightarrow \mathcal{A}$ that maximizes the expected cumulative reward, known as the *return* function, which is inherently linked to the *Q-values* used in RL. The action-value function, or *Q-function,* denoted by $Q^\pi(s, a)$, represents the expected return starting from state s, taking action a, and thereafter following policy π. The Bellman equation further relates this to future states where the optimal policy is obtained by selecting the action that maximizes $Q^\star(s, a) = \mathbb{E}[R(s, a) + \gamma \max_{a'} Q^\star(s', a')]$. Here, $R(s, a)$ is the immediate reward. This formulation provides a recursive structure that enables learning through dynamic programming or model-free methods, such as Q-learning [53].

RL is a method to solve MDPs when the transition probabilities and rewards are unknown [54]. In RL, the learning is carried out through interactions ("trial-and-error"), and actions are selected to maximize the sum of the discounted rewards over the future ("delayed rewards"). In autonomous navigation, RL enables agents to learn navigation strategies by interacting with their environment and adjusting their behavior based on the received rewards [23, 55].

The policy can be estimated through a policy-search or value-based learning approach. In policy-search learning, the policy is represented by a NN with observations as inputs and actions as outputs, so that the policy is directly inferred by approximations [52]. Thus, learning consists of adjusting the parameters of the NN to find the optimal input/output relationship. Possible algorithms to perform such updates are represented by *policy gradient methods.* Unfortunately, such methods might have slow convergence and can get stuck on local maxima.

In contrast, value-based learning maps a state-action pair into a *value* function, defined as the expected return, which can be a discounted reward sequence over a certain time horizon. Since the policy should return action for the UAV, it assesses the values for every possible action and decides either on the optimal action (*exploitation*) or a random one (*exploration*). Implementations of this policy include approaches based on dynamic programming, Monte Carlo methods, and temporal-difference learning. In high-dimensional problems, deep RL represents an interesting tool, as the adoption of deep neural networks (DNNs) reduces a high-dimensional problem to one of very low dimension.

5.4.2 Case Study 3: Target Detection and Environment Mapping by UAV Networks

We consider a set of $\mathcal{N} = \{1, \dots, i, \dots, N\}$ AAs employed in the environment, and a set of $\mathcal{T} = \{1, \dots, j, \dots, N_{\text{targets}}\}$ targets to be detected with certain reliability (see Appendix A). In this scenario, the state vector for each AA $\boldsymbol{s}_{i,k}$ at discrete time k contains the AA location, the map of the environment, and a detection variable, i.e., $\boldsymbol{s}_{i,k} = [\boldsymbol{p}_{i,k}, \boldsymbol{m}_k, t_{i,k}]^{\mathrm{T}}$, where $\boldsymbol{p}_{i,k} = [x_{i,k}, y_{i,k}, h]^{\mathrm{T}} \in \mathbb{R}^3$ is the true AA position with h being a constant altitude, $\boldsymbol{m}_k \in \mathbb{B}^{N_{\text{cell}}}$ is the true map at time k described as a vector of N_{cell} cells in which the map is discretized, each cell has a binary status (1 if occupied and 0 if free), and $t_{i,k}$ contains the target ID associated with the ith AA, and which can be empty in case no target is associated to the considered AA. In the navigation algorithm, we consider that the state coincides with the AA positions, that is, $\boldsymbol{s}_{i,k} = \left[\boldsymbol{p}_{i,k}\right]^{\mathrm{T}} \in \mathbb{R}^3$. We also assume that the AAs move in a grid so that $\boldsymbol{s}_{i,k} = \left[\boldsymbol{p}_{i,k} \bmod \Delta\right]^{\mathrm{T}}$, where Δ is the grid step and mod is the modulo operator. Note that, differently from Section 5.3, the state here represents the agents' positions so it also includes the agents' index i. Consequently, for a single-agent case, the state space considered for RL navigation purposes is $|\mathcal{S}| = N_{\text{cell}}$. Similarly, the AA navigation actions can be defined as $\boldsymbol{a}_{i,k} = \Delta\boldsymbol{p}_{i,k} = \left[\Delta x_{i,k}, \Delta y_{i,k}, 0\right]^{\mathrm{T}} \in \mathbb{R}^3$, where $\Delta\boldsymbol{p}_{i,k}$ is a position displacement in a continuous space. As before, since the AAs are constrained to move on a grid with only four available actions, we have $\boldsymbol{a}_{i,k} = \left(\Delta\boldsymbol{p}_{i,k} \bmod \Delta\right)$ and the action space is defined as $\mathcal{A} = \left\{[\Delta,\ 0],\ [-\Delta,\ 0],\ [0,\ \Delta],\ [0,\ -\Delta]\right\}$ corresponding to the right, left, up, and down directions.

In our case study, the map is estimated using a Bayesian filtering approach, namely a Bayesian occupancy grid (BOG). The goal of the estimation is to infer the a posteriori probability mass function of the occupancy of each cell based on the history of radar measurements with the algorithm described in Appendix B. Hence, at each instant k, each AA initially senses the environment to collect a vector with the received signal strength indicators (RSSIs) from which it can estimate the presence of the target and a vector of backscattered energy bins to infer the map.

Successively, we consider a Q-learning algorithm where the policy is learned at run-time while the AAs navigate the environment. For discrete states and actions, the Q-value can be represented by a Q-table that, at each time instant and for each agent, is updated by [52]:

$$Q_i(\boldsymbol{s}_{i,k}, \boldsymbol{a}_{i,k}) \leftarrow Q_i(\boldsymbol{s}_{i,k}, \boldsymbol{a}_{i,k}) + \alpha\left[R_{i,k+1} + \gamma \max_{\boldsymbol{a}} Q_i(\boldsymbol{s}_{i,k+1}, \boldsymbol{a}) - Q_i(\boldsymbol{s}_{i,k}, \boldsymbol{a}_{i,k})\right], \tag{5.17}$$

where α is the learning rate, and the max operator is used to have a greedy policy. A critical aspect in the application of RL is the design of the reward function, which shapes agent behavior toward the desired objectives. To this end, we define two distinct types of rewards: a primary (extrinsic) reward and a secondary (intrinsic) reward. Extrinsic rewards are typically task-specific and assign a real-valued signal to each state-action pair. In contrast, intrinsic rewards are only indirectly dependent on the actual state of the environment, as they are computed from the AA's belief estimates over the state space [53]. As in [55], we can write the reward as a weighted sum of extrinsic and intrinsic rewards as:

$$r_{i,k+1} = \omega_{\text{ext}}\, r_{\text{ext};i,k+1} + \omega_{\text{int}}\, r_{\text{int};i,k+1}, \tag{5.18}$$

where the extrinsic reward (denoted by $r_{\text{ext};i,k+1}$) relates to target detection, whereas the intrinsic reward (denoted by $r_{\text{int};i,k+1}$) to mapping coverage and accuracy. In our scenario, the detection reward is expressed as a function of the RSSI measured from the assigned target as:

$$r_{\text{ext};i,k+1} \triangleq \frac{\mathsf{RSSI}_{n_i,i,k+1}}{\mathsf{RSSI}_{\max}}, \tag{5.19}$$

where $\mathsf{RSSI}_{n_i,i,k+1}$ is the measured RSSI from the $n_{i,k+1}$th target associated with the ith UAV at time $k+1$, and $\mathsf{RSSI}_{\max}$ is the RSSI an AA would experience at a distance of a single cell (i.e., the minimum possible distance) from a target.

The rewards for mapping coverage and accuracy are given by:

$$r^{(c)}_{\text{int};i,k+1} \triangleq \frac{\sum_{j\in\mathcal{I}_{i,k+1}} \mathbb{I}(j \in \mathcal{D}_{i,k+1})}{N_{\text{cell}}}, \tag{5.20}$$

$$r^{(a)}_{\text{int};i,k+1} = -\sum_{j\in\mathcal{I}_{i,k+1}} b_{i,k+1}(m_j)\log_2\big(b_{i,k+1}(m_j)\big), \tag{5.21}$$

where $\mathbb{I}$ is the indicator function, i.e., $\mathbb{I}(x) = 1$ if x is true and 0 otherwise, $\mathcal{D}_{i,k+1} \subseteq \mathcal{I}_{i,k+1}$ indicates the cells visited for the first time, at time $k+1$, whereas $\mathcal{I}_{i,k+1}$ represents the set of indices of all the cells illuminated by the ith AA at the same instant $k+1$. The symbol $b_{i,k+1}(m_j)$ is the belief of the occupancy state of the jth cell as predicted by the ith agent at time slot $k+1$.

With Eq. (5.21), the agent is rewarded for exploring regions with higher uncertainty. This choice is motivated by the fact that areas already mapped with high accuracy are no longer of interest to the agent, whose objective is to cover and map as much of the environment as possible. As a result, the reward encourages navigation toward regions that have not yet been accurately mapped.

An illustrative example of the resulting map is shown in Figure 5.7, based on the simulation scenario introduced in [53]. In this setup, agents were equipped with sub-THz radar systems for mapping and radio-frequency receivers for active detection. Specifically, the radar operated at 140 GHz using a planar square array of 100 antenna elements with a peak gain of 26 dBi. On the other hand, the RF receiver functioned at a central frequency of 2.4 GHz with a 120 MHz bandwidth. Each target transmitted at 10 dBm, while the receiver noise floor was set to −90 dBm.

Navigation was addressed through a tabular Q-learning algorithm in a multi-agent configuration. The learning parameters were fixed at $\alpha = 0.99$ and $\gamma = 0.9$, while exploration was governed by a decaying ϵ-greedy strategy to gradually shift from exploration to exploitation. Each training episode lasted 150 time steps, and the training phase included 100 episodes.

Figure 5.7 presents the occupancy maps and the agents' trajectories obtained in the final episode of a single Monte Carlo run. The reward function was designed to promote joint optimization of mapping and detection tasks, with weights $\omega_{\text{ext}} = 1$

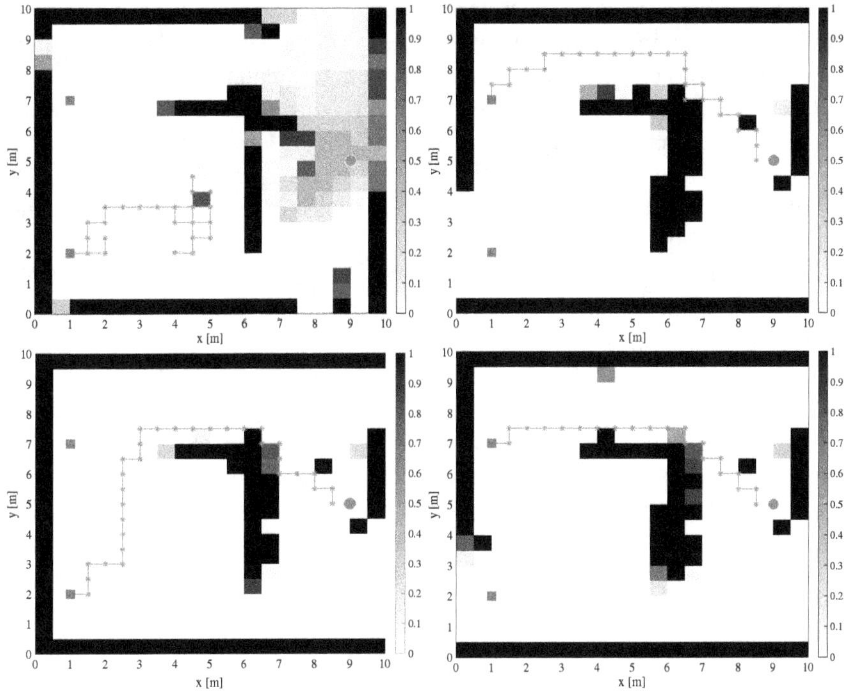

Figure 5.7 Example of trajectories for the last episode, i.e., $e = 100$, and two agents, and for a single Monte Carlo iteration. Top: independent learning, Bottom: centralized cooperative learning. *Source:* [53] / IEEE / CC BY 4.0.

and $\omega_{\text{int}} = 0.3$. Two learning strategies were compared: (1) independent learning, where each AA maintains and updates its own Q-table (top row); and (2) centralized learning, where a shared Q-table is updated by all agents (bottom row). The results indicate that centralized learning enables agents to complete the mission effectively, consistently reaching the target source. In contrast, the independent approach may fail in scenarios with NLOS conditions, as illustrated in Figure 5.7 (top, left).

5.4.3 DNN-based Solutions for Sensing

In the previously introduced MDP framework, AAs use Q-learning to enhance target detection and mapping via optimized online navigation. However, due to the limited energy and computational resources onboard UAVs, not all learning can occur online.

This section explores how offline-trained deep learning (DL) can support sensing by enabling adaptive beamforming systems. As a powerful subfield of ML,

DL excels at handling large datasets and making autonomous decisions without explicit programming. Its core innovation lies in DNNs, which use layered nonlinear computations to extract features directly from raw data.

Recent advances, such as improved backpropagation and the widespread availability of GPUs, have driven the resurgence of DNNs. However, a key challenge in applying DL to mobile communication systems is designing effective DNN architectures tailored to specific scenarios for robust offline training and online performance [56].

We now present an in-depth analysis of DNN-based beamforming in ISAC UAV systems, including architectural design, implementation constraints, and performance optimization. A major advantage of this method is its ability to model complex, nonlinear mappings between input signals and optimal beam-patterns, thereby complementing the Q-learning framework by optimizing antenna array configurations at the physical layer. While Q-learning handles real-time, high-level navigation in a discretized grid of the environment, DNN-based beamforming is trained offline and deployed online to support UAV operation across continuous positions. This dual-layer learning strategy, combining navigation policy with beamforming weight optimization, provides a robust framework for ISAC system implementation.

At the physical layer, the DNN receives raw antenna data as input, extracts spatial features through convolutional layers, and utilizes fully connected layers to generate beamforming weights that serve both sensing and communication objectives. This approach enables joint radar and communication performance optimization, outperforming traditional beamforming techniques [57].

Here, we examine a single UAV performing concurrent monostatic radar sensing and cellular communication. The goal is to enhance network performance by selecting the optimal serving ground base station (GBS) and configuring the UAV's beam-pattern parameters to meet both sensing and communication requirements. The model includes practical hardware constraints and realistic antenna specifications, assuming M patch antenna elements with cardioid radiation patterns.

To reduce cellular network interference, the beam-pattern optimization integrates null-steering, placing radiation nulls toward non-serving GBSs. This maintains signal strength toward the serving GBS and radar target while minimizing co-channel interference.

Our objective is to minimize a transmit beam-pattern gain error, optimizing the antenna beamforming weights $\boldsymbol{W}$ together with the optimal GBS, denoted by the index $\hat{k}_n$ at time instant n, subject to a connectivity constraint. The transmit beam-pattern characterizes the power of the transmitted signal directed toward the target of interest. The selection of the optimal GBS can be performed according to different association rules as described in [24].

We assume that the target is stationary and its position is known beforehand. We denote the estimated position of the target as $\hat{\boldsymbol{u}}_0$. Let $B^{\star}(\boldsymbol{q}_{\ell}(n),\ \boldsymbol{u}_i)$ denote the desired beam-pattern, which specifies the desired beamforming weights and directions for the N UAV positions along the ℓth trajectory (denoted by $\boldsymbol{q}_{\ell}(n)$ at time instant n) toward a generic destination denoted by $\boldsymbol{u}_i$. The transmit beam-pattern gain toward the target location $\boldsymbol{u}_0$ can be expressed as:

$$B(\boldsymbol{q}(n),\ \boldsymbol{u}_0) = \left|\boldsymbol{a}^{\mathrm{H}}(\boldsymbol{q}(n),\ \boldsymbol{u}_0)\,\boldsymbol{w}_0\right|^2 + \left|\boldsymbol{a}^{\mathrm{H}}(\boldsymbol{q}(n),\ \boldsymbol{u}_0)\,\boldsymbol{w}_{\hat{k}}\right|^2, \tag{5.22}$$

where $\boldsymbol{a}(\boldsymbol{q}(n),\ \boldsymbol{u}_0)$ is the array steering vector toward $\boldsymbol{u}_0$ when the UAV is located at $\boldsymbol{q}(n)$, $\boldsymbol{w}_0$, and $\boldsymbol{w}_{\hat{k}}$ are, respectively, the beamforming vectors for sensing and communication with the $\hat{k}$ BS. We can then define a cost function C that quantifies the beam-pattern matching error along the ℓth trajectory as:

$$C(\boldsymbol{q}_{\ell}(n),\ \boldsymbol{W}_{\ell}(n)) = \sum_{\forall \ell,\, n=1}^{N} |B^{\star}(\boldsymbol{q}_{\ell}(n),\ \boldsymbol{u}_0) - B(\boldsymbol{q}_{\ell}(n),\ \boldsymbol{u}_0)|^2, \tag{5.23}$$

where $\boldsymbol{W}_{\ell}(n)$ denotes the beamforming matrix associated with the ℓ trajectory and given by:

$$\boldsymbol{W}_{\ell}(n) = [\underbrace{\boldsymbol{w}_0(n)}_{\text{Sensing}},\ \underbrace{\boldsymbol{w}_{\hat{k}_n}(n)}_{\text{Comm.}}] \in \mathbb{C}^{M\times 2}. \tag{5.24}$$

To enhance the communications between the UAV and the selected GBS, we consider the signal-to-noise-plus-interference ratio (SINR) at the $\hat{k}$th GBS formulated as:

$$\mathrm{SINR}(\boldsymbol{q},\ \boldsymbol{u}_{\hat{k}}) = \frac{|\boldsymbol{h}_{\hat{k}}^{\mathrm{H}}(\boldsymbol{q},\ \boldsymbol{u}_{\hat{k}})\boldsymbol{w}_{\hat{k}}|^2}{\sigma^2 + |\boldsymbol{h}_0^{\mathrm{H}}(\boldsymbol{q},\ \boldsymbol{u}_0)\boldsymbol{w}_0|^2}, \tag{5.25}$$

where the channel vector between the UAV and the ith destination can be written as the following $M \times 1$ vector:

$$\boldsymbol{h}_i(\boldsymbol{q}(n), \boldsymbol{u}_i) = \sqrt{\mathrm{PL}_i}\, e^{j2\pi f_c \frac{d_i}{c}}\, \boldsymbol{a}(\boldsymbol{q}(n), \boldsymbol{u}_i), \tag{5.26}$$

where PL_i is a path-loss term accounting for LOS/NLOS propagation, and d_i is the distance to the ith destination/target. We have considered the worst-case scenario where the receiver cannot cancel the interference from the radar signals before decoding its desired information signal [58].

Given this setup, the optimization problem can be formulated as follows:

$$\mathcal{P}: \min_{\boldsymbol{W}, \hat{k}} \quad C(\boldsymbol{q}_{\ell}, \boldsymbol{W}_{\ell}(n)), \tag{5.27a}$$

$$\text{s.t.} \quad z(n) > z_{\hat{k}} = h_{\text{BS}}, \tag{5.27b}$$

$$\boldsymbol{q}(1) = \boldsymbol{q}_{\text{I}}, \ \boldsymbol{q}(N) = \boldsymbol{q}_{\text{F}}, \tag{5.27c}$$

$$\| \boldsymbol{w}_0 \|^2 + \| \boldsymbol{w}_{\hat{k}} \|^2 \leq P_{\max}, \tag{5.27d}$$

$$\text{SINR}(\boldsymbol{q}(n), \ \boldsymbol{u}_{\hat{k}}) \geq \gamma_{\text{SINR}}. \tag{5.27e}$$

In this formulation, (5.27b) ensures that the UAV height complies with local height regulations, with h_{BS} indicating the minimum required height. Condition (5.27c) guarantees that the UAV starts and ends its trajectory at the specified locations. Constraint (5.27d) imposes a power limit on the UAV, with $P_{\max}$ representing the maximum allowable transmitted power. Finally, (5.27e) ensures the minimum connectivity requirement with the GBS.

We note that, according to (5.22), we would like to find $\boldsymbol{W}$ under the total transmit power constraint such that the available transmit power is used to maximize the signal power at the locations of interest.

5.4.4 Case Study 4: Single-agent Autonomous Systems & Navigation

We now consider a heterogeneous cellular infrastructure comprising a single UAV and terahertz (THz) terrestrial GBSs. The UAV is required to function as a mobile dual-function radar-communication (DFRC) system, simultaneously probing its surrounding environment using a monostatic radar while maintaining communication with its associated terrestrial GBSs (Figure 5.8). Our objective is to enhance the operation of such networks by strategically selecting the GBS and optimizing

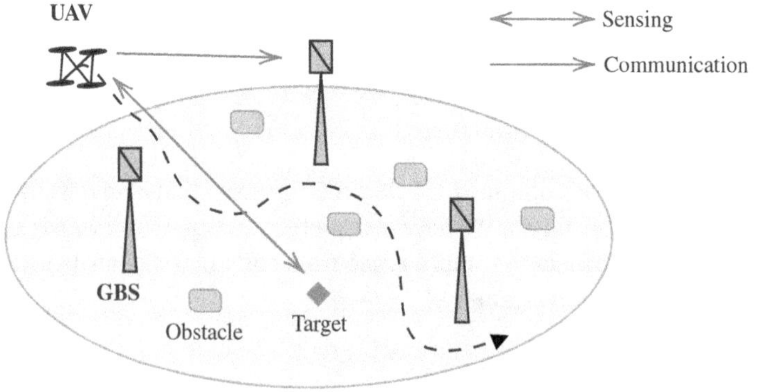

Figure 5.8 Considered scenario, where a DFRC UAV needs to autonomously navigate the environment while guaranteeing reliable communication and sensing performance.

the UAV's orientation and beam-pattern configuration. This ad-hoc beam-pattern optimization technique ensures that both communication and sensing requirements are met effectively.

Problem (5.27) is highly nonconvex. Currently available ad hoc tools, such as classical approaches like long short-term memory (LSTM) networks or beam-pattern optimizers [57], address the various requirements of DFRC-UAV beamforming for a single point at a time. However, these tools do not include the GBS association procedure. Additionally, the UAV may follow multiple trajectories and scan different GBS associations. During the training process and the various GBS associations along its trajectories, conditions (5.27d) and (5.27e) might be satisfied for individual communication or sensing tasks but fail when considered jointly. Consequently, recalculating weights via a beam optimizer for each UAV position and GBS association would result in lengthy procedures with high computational resources and power consumption requirements, which is unsuitable due to the need for a fast response from the UAV. A DNN-based beamformer can help by quickly producing the required beamforming matrix for any point within the UAV's flying area, significantly reducing the time compared to traditional optimizers.

Problem (5.27) can be viewed as a beamforming design problem for each UAV's position along the trajectory $\boldsymbol{q}_\ell$. In our approach, we use two conventional optimizers to generate the training data for the DNN, accounting for both the sensing and communication performance, like LSTM networks and a beam-pattern optimizer proposed in [57]. The learning process involves two main phases: training and inference. During the training phase, we first generate a comprehensive dataset by simulating various UAV flight trajectories within the operational area. For each position along these trajectories, the chosen antenna pattern optimizer calculates the optimal beamforming weights that balance communication and sensing performance. The antenna optimizer needs the azimuth and elevation coordinates of the beam center, the tolerable side lobe level (SLL) range bounded by the minimum values (SLL_{min}), the required effective isotropic radiated power (EIRP), and the null position so that the condition (5.27d) is satisfied. These optimized weights serve as the ground truth for training the DNN (see Figure 5.9). Once this training phase is completed, the UAV is able to generate, in the inference step, a beamforming matrix for any point in the flight area under consideration in a reduced computational time. To satisfy condition (5.27d), as shown in Figure 5.9, another DNN provides the most appropriate UAV-GBS association.

We employ an FFNN implementation in which the inputs for the communication beamforming weights are for each UAV position and for all the trajectories, the azimuth, and elevation toward the associated GBS, the azimuth and elevation of the interfering GBS based on the selected GBS association rule, and the EIRP. To generate the sensing weights, we provide the FFNN with the azimuth and elevation toward the target position and the associated EIRP. The system model

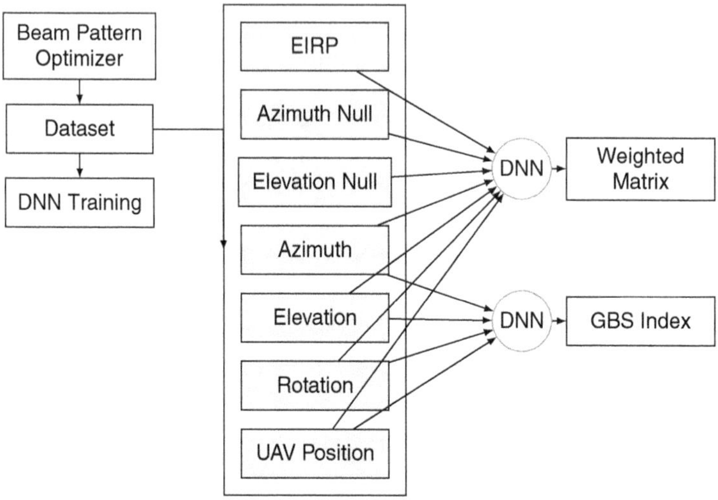

Figure 5.9 Architecture of the proposed NN solution.

considers an area of 2.25 square kilometers with a total of 5 GBSs. The UAV operates at a height of 100 meters and travels at a speed of 10 meters per second. The time interval for updates, denoted as δ_t, is set to 1 second. The SINR threshold is set to 0.3 dB. The height of each GBS is 2 meters, and the maximum EIRP is 37 dBm. The system utilizes a bandwidth of 100 MHz and operates at a frequency of 0.3 THz. The noise power is specified as −110 dBm. Additionally, the UAV is trained over 100 trajectories. We consider a patch antenna that generates a cardioid radiation pattern with a maximum of $M = 100$ elements. The direction angles, measured in azimuth and elevation, are defined geometrically by the positions of the UAV, the associated GBS, and the target. Based on these considerations, the nulling operation must be performed in the direction of the non-associated GBSs.

5.4.4.1 Implementation Considerations for UAVs

5.4.4.1.1 Computational Efficiency

The computational complexity of the proposed radiation pattern optimization algorithm is primarily influenced by the number of antenna elements provided as input. This algorithm uses the array factor (AF) formula to estimate the approximate radiation pattern for subsequent calculations of beamwidth, SLL, nulls, and EIRP. As the total number of antenna elements increases, the algorithm's runtime also increases due to the need for larger matrices and more multiplications involved in the AF calculation.

The computational complexity of the proposed DNN algorithm can be analyzed in terms of the number of multiplications in its main components. The computational complexity of a fully connected network is typically given by multiplying the parameters of each fully connected layer and depends on the number of features in the input vector and the number of neurons in the layer [59], $O(\zeta_i \zeta_n)$, where ζ_i is the number of features and ζ_n the number of neurons in the layer. Finally, the overall complexity is given by the sum of all layers of the DNN.

5.4.4.1.2 Numerical Results and Real-time Adaptation

The radiation pattern comparison graph in Figure 5.10 illustrates the performance characteristics of a uniform linear array (ULA) receiver antenna system, presenting both conventional (obtained as an analytical solution of (5.27)) and NN-based implementations. The analysis reveals two distinct patterns: the original Rx radiation pattern (indicated by a dashed line) and its NN implementation (shown as a black solid line). The signal environment encompasses three critical angular positions: a primary target at approximately 66.24° and two interferers located at 21.24° and 126.24°, respectively. The gain characteristics extend from approximately −40 to +10 dB across the angular spectrum.

The NN implementation closely mirrors the original pattern while maintaining essential characteristics for interference suppression. Minor variations in sidelobe structure are observed but do not significantly impact overall system performance.

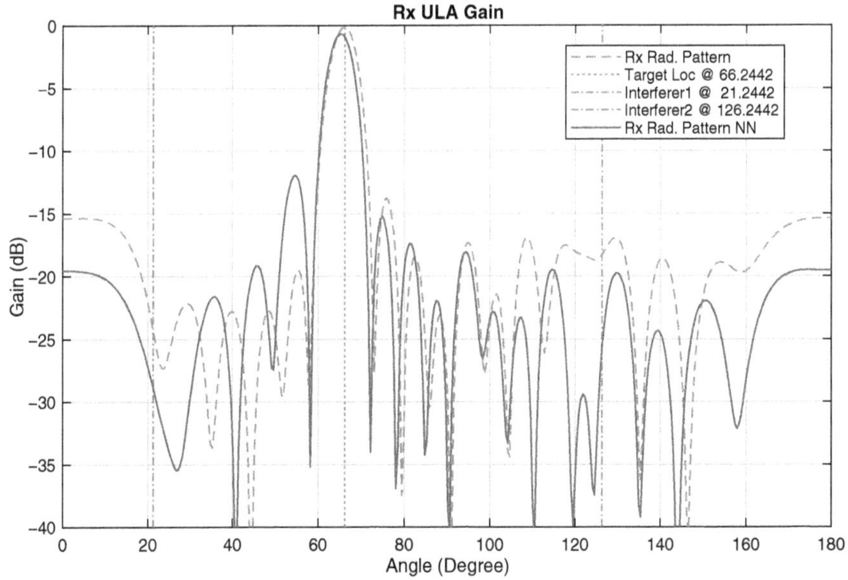

Figure 5.10 Radiation pattern comparison.

This comparison validates the effectiveness of NN-based approaches in maintaining critical radiation pattern characteristics while achieving desired interference suppression objectives.

In UAV systems, a real-time response is of vital importance. We have calculated and compared the beamforming updates of DNN inference with the beamforming optimizer response in the case of a new UAV position. The proposed DNN solution can predict beamforming weights with significantly reduced time compared to running the beam-pattern optimizer for any point of the possible UAV trajectories. The training duration and inference time for the proposed model are reported in Figure 5.11. It can be seen that the fast inference of the DNN ensures timely beamforming updates and adapts to changing conditions (e.g., UAV location or GBS connection) while maintaining stable performance. Figure 5.12 examines the three-dimensional radiation pattern (output of the DNN) characteristics of the proposed antenna system, visualized through a surface plot representing the effective radiated power (ERP) in dB. Critical system directions are marked by distinct indicators: the pointing direction is depicted with a dot, the target location (the desired signal source) with a square, and the interferer/BS directions with an asterisk and a cross.

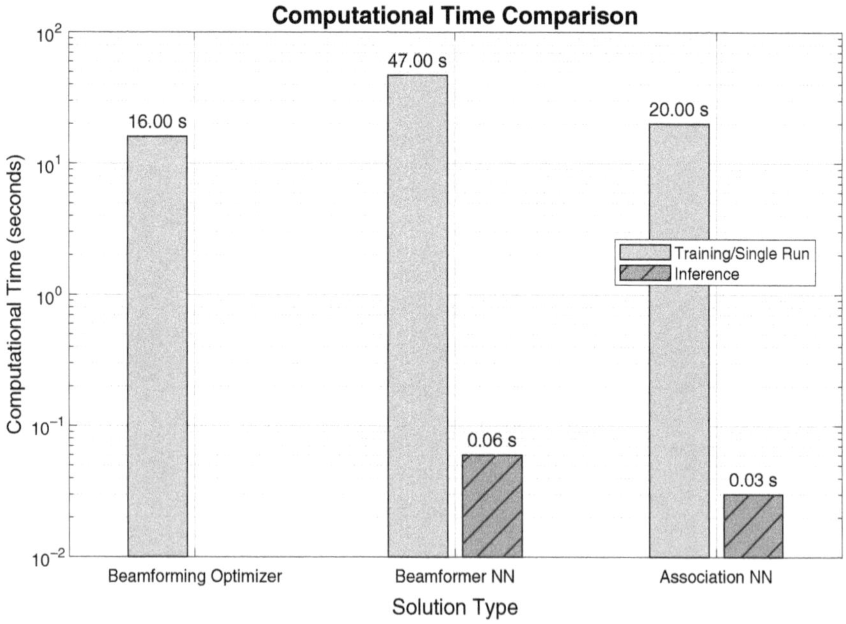

Figure 5.11 Comparison of computational times for different solutions. The plot shows both training/single run times and inference times (where applicable) in logarithmic scale.

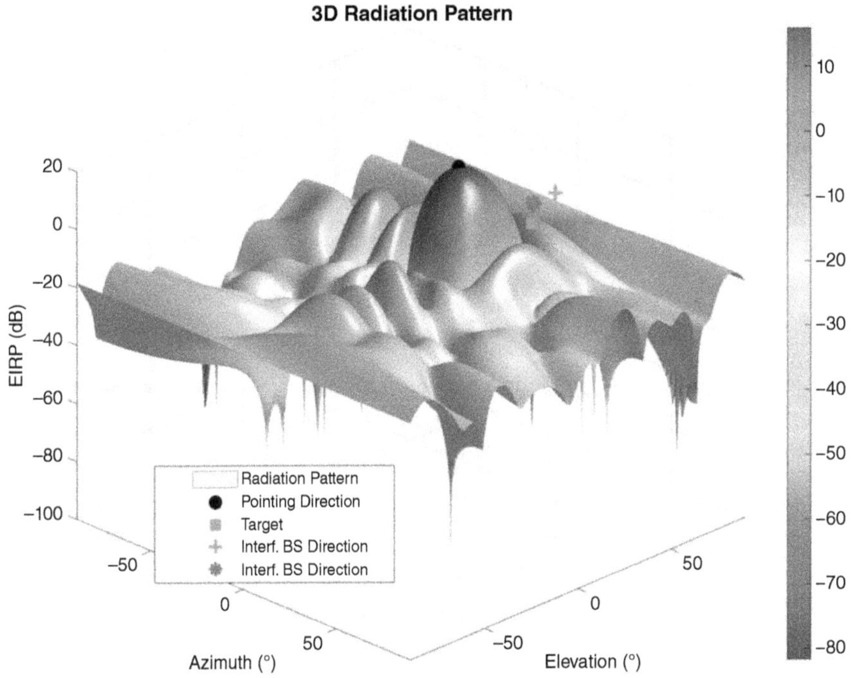

Figure 5.12 Example of 3D radiation pattern of the proposed antenna system. Critical system directions are marked by distinct indicators: (i) the circle stands for the pointing direction, i.e., the primary beam direction; (ii) the square indicates the target location, i.e., the desired signal source; (iii) an asterisk and a cross indicate potential interference sources.

The pattern demonstrates sophisticated spatial filtering capabilities through: (a) main beam formation: the highest peak indicating maximum gain, (b) null placement: regions with low EIRP values, showing interference suppression, and (c) sidelobe control: regions with intermediate EIRP values, showing controlled secondary radiation.

The surface visualization effectively captures the proposed DNN framework's ability to optimize signal formation while maintaining interference rejection across the spatial domain. More details on the ability of the DNN framework to select the most suitable GBS for association can be found in [24].

5.5 Hybrid Model- and Data-driven Learning

We now present a way in which RL can be used to enhance classical multi-agent navigation methods by combining the advantages of both solutions.

In Section 5.3, we have discussed exploiting model knowledge for a simplified multi-agent navigation problem, where information seeking is the sole objective. In practice, multi-agent navigation often includes multiple objectives, where information seeking is combined with other objectives such as goal approaching, sensing, and communication. These objectives may require different, sometimes even contradictory navigation strategies. Many of these objectives can be abstracted as reaching a goal $\boldsymbol{p}_{\mathrm{B}}$. We then use a goal-approaching objective to include any high-level mission objectives:

$$f_{\mathrm{g}}(\boldsymbol{u}_k) = \sum_{i\in\mathcal{N}} \| \boldsymbol{p}_{i,k+1}(\boldsymbol{u}_k) - \boldsymbol{p}_{\mathrm{B}} \|, \tag{5.28}$$

where $\boldsymbol{p}_{\mathrm{B}}$ is either the predefined mission destination or a waypoint generated from sensing or exploration functions. In addition, FI seeking can also be expressed as constraints with a higher priority:

$$h_{\mathrm{p},i}(\boldsymbol{u}_k) = \varepsilon - \mathrm{tr}\Big[\boldsymbol{\Gamma}_i\ \mathrm{CRB}[\boldsymbol{p}_{k+1}(\boldsymbol{u}_k)]\Big] \geq 0, \ \ \forall i \in \mathcal{N}, \tag{5.29}$$

where $\boldsymbol{\Gamma}_i = \mathrm{diag}\{\boldsymbol{0}_{1:2i-2},\ 1,\ 1,\ \boldsymbol{0}_{2i+1:2N}\}$ and ε is the maximally tolerable position mean square error (MSE) for each agent.

The FI seeking constraints can be rewritten with a compact vector form as $\boldsymbol{h}_{\mathrm{p}}(\boldsymbol{u}_k) \geq \boldsymbol{0}$. The overall navigation problem can be formulated as:

$$\min_{\boldsymbol{u}_k\in\mathcal{U}} \Big\{ \underbrace{f_{\mathrm{p}}(\boldsymbol{u}_k)}_{\text{(information seeking)}}, \underbrace{f_{\mathrm{g}}(\boldsymbol{u}_k)}_{\text{(goal approaching)}} \Big\}, \tag{5.30a}$$

$$\text{s.t. information seeking:} \quad \boldsymbol{h}_{\mathrm{p}}(\boldsymbol{u}_k) \geq \boldsymbol{0}, \tag{5.30b}$$

$$\text{collision avoidance:} \quad \boldsymbol{h}_{\mathrm{c}}(\boldsymbol{u}_k) \geq \boldsymbol{0}. \tag{5.30c}$$

We generate control commands $\boldsymbol{u}_{\mathrm{p},k}$ and $\boldsymbol{u}_{\mathrm{g},k}$ with the negative gradients of the cost functions of information seeking $f_{\mathrm{p}}(\boldsymbol{u}_k)$ and goal approaching $f_{\mathrm{g}}(\boldsymbol{u}_k)$, respectively, and linearly combine them to obtain an unconstrained control command $\boldsymbol{u}_k^*$, i.e.,

$$\boldsymbol{u}_k^* = -\mu \boldsymbol{W}_k \frac{\nabla_{\boldsymbol{u}_k} f_{\mathrm{p}}(\boldsymbol{u}_k)}{\|\nabla_{\boldsymbol{u}_k} f_{\mathrm{p}}(\boldsymbol{u}_k)\|} - \mu(\boldsymbol{I} - \boldsymbol{W}_k)\frac{\nabla_{\boldsymbol{u}_k} f_{\mathrm{g}}(\boldsymbol{u}_k)}{\|\nabla_{\boldsymbol{u}_k} f_{\mathrm{g}}(\boldsymbol{u}_k)\|}, \tag{5.31}$$

where μ is the step size and $\boldsymbol{W}_k = \mathrm{diag}\{\dots, w_{i,k}, \dots : \forall i \in \mathcal{N}\}$ the trade-off between information seeking and goal approaching.

In PSGD-based multi-agent navigation, control commands can be generated efficiently for large-scale agent networks. However, it is a greedy approach with step-by-step optimization. It cannot explicitly involve multi-step or long-term objectives like "all agents reaching the goal." More importantly, the control parameters like the weight on every state dimension in information seeking $\boldsymbol{\Lambda}_k$ in (5.9) and the trade-off between information seeking and goal approaching for every agent $\boldsymbol{W}_k$ have to be chosen manually at the beginning and are often fixed during the mission.

We consider the case study of Figure 5.13, where a swarm of rovers needs to navigate with high positioning accuracy, from their landing site on the left to an exploration site on the right. In each region, three anchors illustrated with square markers, such as a lander and static rovers and probes, are placed with limited signal coverage areas marked with dashed circles. Agents, illustrated with round markers, in the middle area outside the coverage of anchors cannot conduct agent-to-anchor measurements. Therefore, this area is referred to as the blind region. The mission is considered successful if all agents reach the region on the right while maintaining their positioning variance below ε. To that end, the swarm must autonomously adapt its formation into a rigid framework like a dynamic bridge sufficiently connected to anchors like in Figure 5.13a and let all its agents pass through the blind region. In Figure 5.13b, an unsuccessful mission snapshot is shown where the left agent can no longer reach the goal since there are not sufficient agents to support its navigation in the blind region. This snapshot illustrates the limitation of the PSGD algorithm with fixed control parameters. The drawbacks of the PSGD algorithm can be overcome by data-driven approaches, where

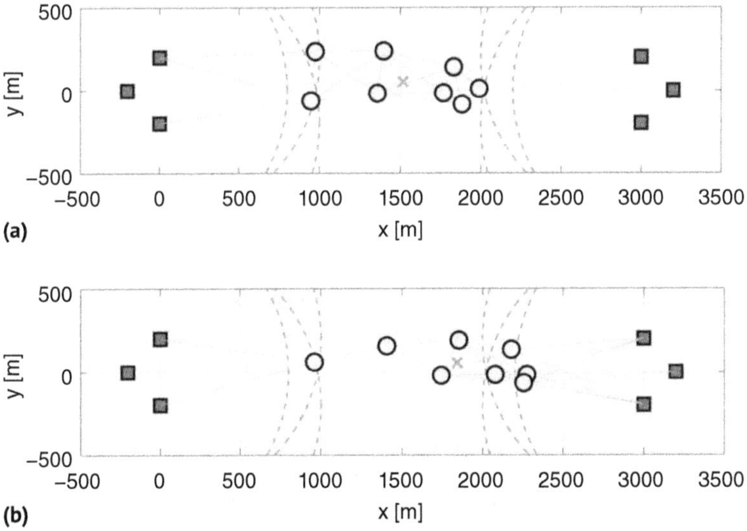

Figure 5.13 Resulting formations from PSGD-based multi-agent navigation. (a) Bridging the gap, (b) failure case.

ultimate mission objectives like goal approaching can be explicitly set as a long-term reward. By allowing the control parameters to dynamically adjust in response to changing agent situations, i.e., $\boldsymbol{\Lambda}_k = \boldsymbol{\Lambda}(\boldsymbol{p}_k)$ and $\boldsymbol{W}_k = \boldsymbol{W}(\boldsymbol{p}_k)$, we can add additional degrees of freedom, which make the multi-agent system more adaptive to handle complex navigation tasks. The optimal control parameters as functions of the current state can be learned, for example, with RL techniques.

5.5.1 Combining PSGD and Q-learning

One of the most successful applications of Q-learning is goal-directed tasks such as path planning and maze solving, which closely resemble our goal-approaching mission objective. However, in many implementations, the available actions are limited to movements in the cardinal directions (e.g., {N, E, S, W}), a choice that does not leverage any knowledge of the underlying model and thus reflects a model-free approach.

We propose a combination of PSGD and Q-learning approaches to benefit from the efficiency of model-based approaches and the long-term reward-seeking ability of data-driven approaches.

5.5.1.1 State

It is well known that Q-learning is impractical for high-dimensional systems like our multi-agent navigation system. Therefore, we need to choose a low-dimensional training state space that captures essential situational information of an agent. We select the agent-to-destination distance $d_{iB,k} = \|\boldsymbol{p}_{i,k} - \boldsymbol{p}_B\|$ and the mean distance to destination $d_{SB} = \sum_{i \in \mathcal{N}} \|\boldsymbol{p}_{i,k} - \boldsymbol{p}_B\|$ as the training state, i.e.,

$$\boldsymbol{s}^*_{i,k} = \text{vec}\{d_{iB,k}, d_{SB,k}\}. \tag{5.32}$$

These two dimensions describe the situation of the agent under investigation with respect to the multi-agent network and the exploration environment. The agent distance can be estimated using a positioning algorithm. The mean distance can be obtained in a decentralized fashion through a consensus algorithm. If needed, higher-order statistical moments of the multi-agent formation can be added to the state space, which capture more of the formation's collective characteristics.

5.5.1.2 Action

Each agent learns to select appropriate control parameters, namely the agent's information seeking weight $\lambda_{i,k}$ and the information seeking-to-goal approaching trade-off ratio $w_{i,k}$, according to its situation. The action space contains the discretized control parameters, i.e., $\boldsymbol{a}_{i,k} = \text{vec}\{\lambda_{i,k}, w_{i,k}\}$.

5.5.1.3 Q-table Update

We utilize the advantage of Q-learning and directly set the reward according to the ultimate mission objective, for example, all agents reaching region B, with a continuous exponential function:

$$r(\boldsymbol{s}_{i,k}^*, \boldsymbol{a}_{i,k}) = a \exp\left\{ -\frac{\max\left\{(d_{iB,k+1})^2 : \forall i \in \mathcal{N}\right\}}{d_{\mathrm{R}}^2} \right\}, \tag{5.33}$$

where a is a large scaling factor. The Q-table $Q(\boldsymbol{s}_{i,k}^*, \boldsymbol{a}_{i,k})$ contains discretized state $\boldsymbol{s}_{i,k}^*$ and actions $\boldsymbol{a}_{i,k}$, which is updated with the following procedures. First, if the state has changed, i.e., $\boldsymbol{s}_{i,k}^* \neq \boldsymbol{s}_{i,k-1}^*$, generate new actions $\boldsymbol{a}_{i,k}$ with ϵ-greedy algorithm either randomly or from the Q-table. Second, apply PSGD-based multi-agent control with the chosen $\lambda_{i,k}$ and $w_{i,k}$ and get the new state $\boldsymbol{s}_{i,k+1}^*$. Third, if the state has changed, i.e., $\boldsymbol{s}_{i,k+1}^* \neq \boldsymbol{s}_{i,k}^*$, update the Q-table with the Bellman equation. In the investigated scenario, we focus on a homogeneous multi-agent system, where the agents' explicit roles are identical, even though they may have different implicit roles like frontiers or bridges based on the current situation. Hence, agents are also interchangeable from a learning perspective. Therefore, the agents exchange their Q-tables to reach consensus across the network.

We verify the proposed RL-enhanced multi-agent navigation by simulating the swarm exploration mission. The two areas are 3,000 m apart. A 30×30 grid is used to discretize the training state. Signals from the anchors can be detected and exploited for ranging by an agent up to 1,000 m. Hence, the blind region spans over 1,000 m between the two regions. The agents can effectively perform ranging with each other up to a distance of 400 m. Eight agents depart from the left side and need to reach the vicinity of the destination on the right $\boldsymbol{p}_B = [3{,}000{,}0]^T$ with $d_{iB,K} < d_{\mathrm{R}}, \forall i \in \mathcal{N}$ within $K = 2{,}000$ steps, where $d_{\mathrm{R}} = 100$ m. The position CRB of every agent is constrained to $\varepsilon = 100\,\mathrm{m}^2$. The agents are trained for $E = 10{,}000$ episodes, with a learning rate, at the ith episode, $\alpha_i = i^{-1/2}$ and an exploration factor ϵ_i which linearly decreases from 1 to 0.1 in the first 2,000 episodes and remains at 0.1 until the end of training for ϵ-greedy action selection. The discount factor γ is set to 0.9. The entities in Q-table are initialized at the first time instant of the first episode uniformly within the interval $(0, 10^{-2}]$. The scaling factor of the instantaneous reward is $a = 10^4$. The agent's weight for FI seeking $\lambda_{i,k}$ can be selected from the discrete set $\{0.01, 1\}$, whereas the information seeking-to-goal approaching weights $w_{i,k}$ from $\{0.3, 0.7\}$.

In Figure 5.14, we selected two crucial formation snapshots that resulted from the RL-enhanced information seeking. Anchors are illustrated as square markers, and agents are illustrated with black and white round markers. Agent markers' colors show the trade-off factor between information seeking and goal

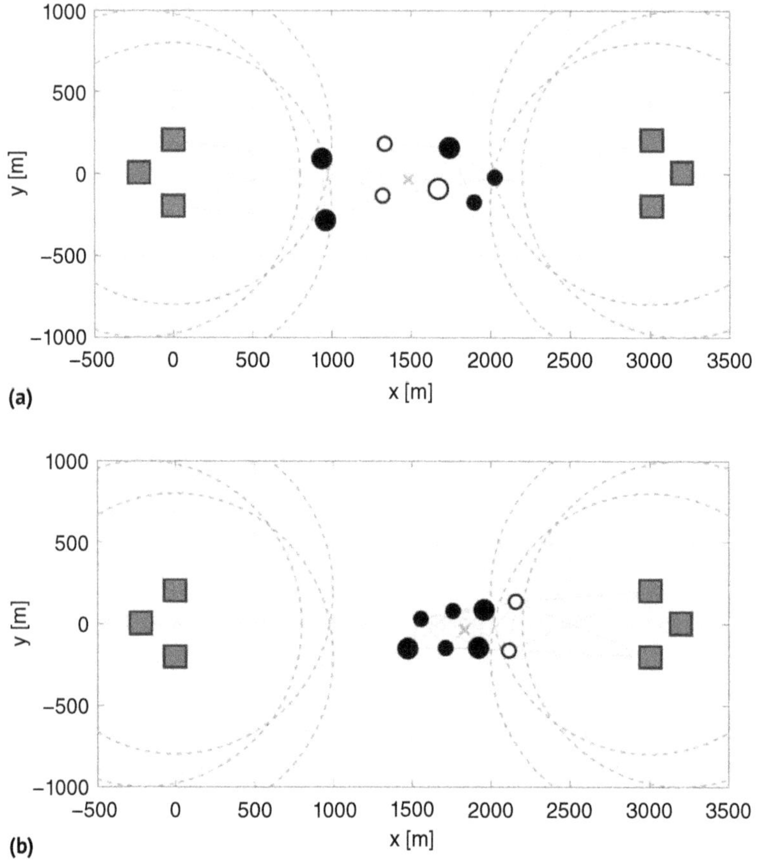

Figure 5.14 Resulting formations from RL-enhanced multi-agent navigation. (a) Snapshot at step 450, (b) snapshot at step 550.

approaching. A white marker indicates $w_{i,k} = 0.7$ whereas a black marker indicates $w_{i,k} = 0.3$. The markers' sizes show the weight of the agent in information seeking. A small marker indicates $\lambda_{i,k} = 0.01$, whereas a large marker indicates $\lambda_{i,k} = 1$. At time instant 450, the agents behind are the "pillars" for keeping the bridge rigid, yet ready to move forward. The agents in front just reach the right region. At time instant 550, the front agents change their implicit roles from goal approaching to information-seeking with a main objective of supporting the agents behind to cross the blind region. The success rate is defined as the percentage of episodes where all agents reach the right side within a 100-episode sliding window. This result verifies the effectiveness of our proposed RL-enhanced multi-agent information-seeking scheme. For a real-world mission, we may conduct training in a simulated environment first to learn an initial Q-table to speed up online training during the operation.

5.6 Conclusions

In this chapter, we examine how ML, particularly data-driven approaches, can significantly enhance the capabilities of AA networks for detection, localization, mapping, and navigation tasks. These functions are critical in radar-based systems operating in dynamic, uncertain, and time-sensitive environments.

While traditional model-based methods offer analytical rigor and interpretability, they rely heavily on accurate prior knowledge of the environment and system dynamics. This limits their applicability in scenarios where such models are unavailable, inaccurate, or too complex to derive in real time. In contrast, data-driven learning approaches can extract patterns directly from observations, enabling agents to adapt to changing conditions and learn from experience without requiring an explicit model of the system.

Through theoretical analysis and four case studies, we showed that data-driven methods, such as RL, transfer learning, and federated learning, can support more adaptive and robust decision-making. These techniques allow agents to handle uncertainty, share knowledge across different tasks and environments, and operate with limited prior assumptions, making them especially valuable in radar-based AA networks.

ML also offers powerful tools for enabling decentralized coordination and scalable behaviors through mechanisms such as swarm intelligence and predictive optimization. Moreover, techniques like real-time performance monitoring and distributed learning enable agents to self-assess and adjust their strategies collaboratively, contributing to more resilient and autonomous operations.

As the field progresses, the integration of data-driven ML with model-based control and signal processing will be key to unlocking the full potential of AA networks. This hybrid paradigm promises to deliver intelligent systems that are not only more flexible and efficient, but also capable of operating reliably in the complex and uncertain environments typical of modern radar applications.

Acknowledgments

This work was partially supported by the European Union under the Italian National Recovery and Resilience Plan (NRRP) of NextGenerationEU, partnership on "Telecommunications of the Future" (PE00000001 – program "RESTART"), under the NRRP of NextGenerationEU (Mission 4 – Component 2 – Investment 1.1) Prin 2022 (No. 104, 2/2/2022, CUP J53C24002790006), and under the European Research Council (ERC) Grant no. 101116257 (project CUE-GO: Contextual Radio Cues for Enhancing Decision Making in Networks of Autonomous Agents).

Appendix A—Detection and Tracking

Target detection is of primary importance in radar applications but can be particularly difficult, especially when dealing with miniature or microscale objects. Various approaches for AA detection can be imagined, based either on static systems or patrolling AAs. These methods can be categorized according to the type of sensors employed for the measurements, as outlined below:

RF-based Detection: This method utilizes low-cost, low-complexity RF sensors, such as software-defined radios (SDRs), which can be mounted on AAs to "sniff" the communication signals between the target AA and its controller. The collected RF signals are typically processed by classifiers, often employing ML techniques, to distinguish AA-specific RF signals. A major limitation of this method is the need for prior training to identify different AAs. Additionally, the effectiveness of these techniques depends on the transmit power and receiver sensitivity, and they cannot detect AAs that do not emit any signals, as they are not remotely controlled.

Radar-based Detection: Low-cost, lightweight frequency-modulated continuous-wave (FMCW) radars can be used to detect targets in the surrounding environment. Unlike vision-based methods, these radars perform well in NLOS conditions. However, radar detection can be challenging, particularly for micro-drones, as the RCS of the target may hinder accurate detection [60].

Sound-based Detection: This method uses one or more microphones to capture the characteristic noise produced by AAs during their mobility. A significant drawback of this technique is its ineffectiveness in urban or noisy environments, and it may not work well if the microphones are installed on patrolling AAs, which could generate similar noise.

Vision-based Detection: This approach employs cameras as sensors and can be extremely effective. The drawback is that it requires LOS and good weather conditions to operate properly and therefore faces difficulties when AAs are flying between buildings or hiding behind obstacles.

We now consider the tracking problem that consists of estimating the target's state (e.g., its position and velocity) from the collected measurements. We focus on possible statistical techniques that do not rely on geometric considerations or look-up tables (such as fingerprinting) but instead estimate the target's state over time based on the history of measurements.

Bayesian filtering methods, such as Kalman filters and particle filters (PFs) have proven to be practical tools for addressing the tracking problem. These techniques are particularly useful because they can handle heterogeneous measurements, model uncertainties statistically, and account for target mobility.

In the Bayesian framework, the goal is to estimate the full joint posterior probability of the target's state at time k, denoted $\mathbf{s}_k$, given the measurements $\mathbf{z}_{1:k}$.

If the target state and the measurements form a Markov sequence, a probabilistic state-space model can be defined using the following statistical models:

- **Prior Information**: This represents the statistical description of the state at time 0, such as after the detection process.
- **Measurement Model**: This model describes how the state relates to the measurements, expressed by the likelihood distribution $p(\boldsymbol{z}_k|\boldsymbol{s}_k)$.
- **Mobility Model**: This model describes the evolution of the state over time, given by $p(\boldsymbol{s}_k|\boldsymbol{s}_{k-1})$.

With this state-space model, Bayesian filtering is a recursive approach that enables estimating the marginal posterior distribution of the target's state based on the measurements. The process involves three steps [49]:

1. **Initialization**: The marginal at time step 0 is set to the prior.
2. **Prediction Step**: Using the mobility model, the predictive distribution of the state based on past measurements can be expressed as:

$$b_{k-1}(\boldsymbol{s}_k) = p(\boldsymbol{s}_k|\boldsymbol{z}_{1:k-1}), \tag{A.1}$$

3. **Update Step**: When a new measurement becomes available, the marginal posterior of the state:

$$b_k(\boldsymbol{s}_k) = p(\boldsymbol{s}_k|\boldsymbol{z}_{1:k}), \tag{A.2}$$

is computed using Bayes' rule.

Finally, a point estimate of the state can be derived as:

$$\hat{\boldsymbol{s}}_{k|k} = \mathbb{E}[\boldsymbol{s}_k|\boldsymbol{z}_{1:k}] = \int \boldsymbol{s}_k\, b_k(\boldsymbol{s}_k)\, d\boldsymbol{s}_k. \tag{A.3}$$

Typically, the observation functions are nonlinear (e.g., Euclidean distance), and the observation noise follows a Gaussian distribution. Two practical solutions to the filtering equations are the EKF and the unscented Kalman filter (UKF), which provide straightforward methods for solving the filtering problem. When an EKF is applied to each agent, the belief of the state can be approximated as:

$$b_{i,k}(\boldsymbol{s}_k) = p(\boldsymbol{s}_k|\boldsymbol{z}_{i,1:k}) \sim \mathcal{N}(\boldsymbol{\mu}_{i,k|k}, \boldsymbol{\Sigma}_{i,k|k}), \tag{A.4}$$

where $\boldsymbol{\mu}_{i,k|k}$ and $\boldsymbol{\Sigma}_{i,k|k}$ represent the posterior mean vector and covariance matrix of the target state [49], and $\boldsymbol{z}_{i,k}$ are the measurements obtained by the ith agent from its own radar and neighboring agents. In this case, the point estimate becomes:

$$\hat{\boldsymbol{s}}_{i,k|k} = \boldsymbol{\mu}_{i,k|k}. \tag{A.5}$$

Since in some networks the processing is distributed, like in DRNs, the measurements $\boldsymbol{z}_{i,k}$ may be delayed due to the number of hops between agents, resulting in outdated information about the target.

Appendix B—Mapping

We now provide insights about a possible mapping approach. For the sake of simplicity, let us assume that a single AA explores an unknown environment and can collect ranging-related measurements for different angular (steering) directions. Then, we assume that $m_k \in \mathbb{B}^{N_{\text{cell}}}$ is the actual map at time k, represented as a vector of N_{cell} cells that discretize the map. The environment is assumed to remain stationary, implying that $m_k = m$ for all k. The map is thus described as $\boldsymbol{m} = \left[m_1, \dots, m_j, \dots, m_{N_{\text{cell}}}\right]^{\text{T}}$, where each $m_j \in \mathbb{B}$ represents the occupancy state of the corresponding cell, which is 1 (or 0) if occupied (or empty).

A potential method for estimating an environmental map is to apply a BOG algorithm [53, 61]. This approach processes each cell individually and computes the log-odds of occupancy for the jth cell based on the given observations. The log-odds at time k can be expressed as:

$$v_k(m_j) = \log\left(\frac{b_k(m_j)}{1 - b_k(m_j)}\right), \tag{B.1}$$

where $b_k(m_j)$ represents the probability of the occupancy state of the jth cell, evaluated by the agent at time k. The algorithm can be summarized as follows [62].

1. **Initialization:** The map is initialized with $v_0(m_j) = \log\left(\dfrac{b_0(m_j)}{1 - b_0(m_j)}\right)$, where $b_0(m_j) = 0.5$, $\forall\, j = 1, 2, \dots, N_{\text{cell}}$, which indicates total uncertainty about the cell status.
2. **Measurement and Scan Vector Generation:** For each agent position, a new range-angle matrix is built based upon the collected measurements over each angular (steering) direction. Each row of the matrix is then compared to a threshold $0 < \gamma \leq 1$. The first distance value that exceeds γ is recorded in a vector $\boldsymbol{r}_k$. Simultaneously, all steering angles are collected in the vector $\boldsymbol{\phi} = [\phi_1, \phi_2, \dots, \phi_{N_{\text{steer}}}]$. The final scan vector at time k is given by $\boldsymbol{s}_k = \left[\boldsymbol{r}_k^{\text{T}}, \boldsymbol{\phi}^{\text{T}}\right]$.
3. **Log-odds Update:** Using $\boldsymbol{s}_k$, and by assuming the independence of each cell from all others (including neighboring cells), the beliefs are updated according to a standard occupancy grid method [61], resulting in:

$$v_k(m_j) = \log\left(\frac{p(\boldsymbol{s}_k | m_j = 1)}{p(\boldsymbol{s}_k | m_j = 0)}\right) + v_{k-1}(m_j), \quad \forall j = 1, \dots, N_{\text{cell}}$$

where $p(\boldsymbol{s}_k | m_j = 1)$ and $p(\boldsymbol{s}_k | m_j = 0)$ denote the likelihoods of observing the scan $\boldsymbol{s}_k$ given that the jth cell is occupied or free, respectively.

References

1 N. M. K. Dousai and S. Lončarić. Detecting humans in search and rescue operations based on ensemble learning. *IEEE Access*, 10:26481–26492, 2022.

2 M. Petrlík, T. Báča, D. Heřt, M. Vrba, T. Krajník, and M. Saska. A robust UAV system for operations in a constrained environment. *IEEE Robotics and Automation Letters*, 5(2):2169–2176, 2020.

3 P. Chhikara, R. Tekchandani, N. Kumar, M. Guizani, and M. M. Hassan. Federated learning and autonomous UAVs for hazardous zone detection and AQI prediction in IoT environment. *IEEE Internet of Things Journal*, 8(20):15456–15467, 2021.

4 Y. Liu, J. Nie, X. Li, S. H. Ahmed, W. Y. B. Lim, and C. Miao. Federated learning in the sky: Aerial-ground air quality sensing framework with UAV swarms. *IEEE Internet of Things Journal*, 8(12):9827–9837, 2020.

5 R. Ma, R. Wang, G. Liu, W. Meng, and X. Liu. UAV-aided cooperative data collection scheme for ocean monitoring networks. *IEEE Internet of Things Journal*, 8(17):13222–13236, 2021.

6 E. Staudinger, S. Zhang, R. Pöhlmann, and A. Dammann. The role of time in a robotic swarm: A joint view on communications, localization, and sensing. *IEEE Communications Magazine*, 59(2):98–104, 2021.

7 S. Zhang, R. Pöhlmann, T. Wiedemann, A. Dammann, H. Wymeersch, and P. A. Hoeher. Self-aware swarm navigation in autonomous exploration missions. *Proceedings of the IEEE*, 108(7):1168–1195, 2020.

8 A. Guerra, D. Dardari, and P. M. Djurić. Dynamic radar networks of UAVs: A tutorial overview and tracking performance comparison with terrestrial radar networks. *IEEE Vehicular Technology Magazine*, 15(2):113–120, 2020.

9 M. M. U. Chowdhury, F. Erden, and I. Guvenc. RSS-based Q-learning for indoor UAV navigation. In *Proceedings Military Communications Conference (MILCOM)*, pages 121–126, 2019.

10 A. Guerra, F. Guidi, D. Dardari, and P. M. Djurić. Networks of UAVs of low complexity for time-critical localization. *IEEE Aerospace and Electronic Systems Magazine*, 37(10):22–38, 2022.

11 X. Liu, Y. Liu, Y. Chen, and L. Hanzo. Trajectory design and power control for multi-UAV assisted wireless networks: A machine learning approach. *IEEE Transactions on Vehicular Technology*, 68(8):7957–7969, 2019.

12 Y. Zeng and R. Zhang. Energy-efficient UAV communication with trajectory optimization. *IEEE Transactions on Wireless Communications*, 16(6):3747–3760, 2017.

13 G. Fontanesi, A. Zhu, and H. Ahmadi. Deep reinforcement learning for dynamic band switch in cellular-connected UAV. In *2021 IEEE 94th Vehicular Technology Conference (VTC2021-Fall)*, pages 1–5. IEEE, 2021.

14 G. Fontanesi, A. Zhu, M. Arvaneh, and H. Ahmadi. A transfer learning approach for UAV path design with connectivity outage constraint. *IEEE Internet of Things Journal*, 10(6):4998–5012, 2022.

15 C. Sun, S. B. Chetty, G. Fontanesi, J. Zhang, A. Mohajerzadeh, D. Grace, and H. Ahmadi. Energy consumption reduction for UAV trajectory training: A transfer learning approach. *arXiv preprint arXiv:2501.11243*, 2025.

16 M. M. U. Chowdhury, S. J. Maeng, E. Bulut, and I. Güvenç. 3-D trajectory optimization in UAV-assisted cellular networks considering antenna radiation pattern and backhaul constraint. *IEEE Transactions on Aerospace and Electronic Systems*, 56(5):3735–3750, 2020.

17 J. Cui, Y. Liu, and A. Nallanathan. Multi-agent reinforcement learning-based resource allocation for UAV networks. *IEEE Transactions on Wireless Communications*, 19(2):729–743, 2020.

18 J. Hu, H. Zhang, and L. Song. Reinforcement learning for decentralized trajectory design in cellular UAV networks with sense-and-send protocol. *IEEE Internet of Things Journal*, 6(4):6177–6189, 2018.

19 X. Liu, Y. Liu, and Y. Chen. Reinforcement learning in multiple-UAV networks: Deployment and movement design. *IEEE Transactions on Vehicular Technology*, 68(8):8036–8049, 2019.

20 K. Arulkumaran, M. P. Deisenroth, M. Brundage, and A. A. Bharath. Deep reinforcement learning: A brief survey. *IEEE Signal Processing Magazine*, 34(6):26–38, 2017.

21 R. M. Annasamy and K. Sycara. Towards better interpretability in deep Q-networks. In *Proceedings of the AAAI Conference on Artificial Intelligence*, volume 33, pages 4561–4569, 2019.

22 A. Guerra, D. Dardari, and P. M. Djurić. Dynamic radar network of UAVs: A joint navigation and tracking approach. *IEEE Access*, 8:116454–116469, 2020.

23 A. Guerra, F. Guidi, D. Dardari, and P. M. Djurić. Multi-agent Q-learning in UAV networks for target detection and indoor mapping. In *Proceedings of the International Balkan Conference on Communications and Networking (BalkanCom)*, pages 80–84, 2021.

24 G. Fontanesi, A. Guerra, F. Guidi, J. A. Vásquez-Peralvo, N. Shlezinger, A. Zanella, E. Lagunas, S. Chatzinotas, D. Dardari, and P. M. Djurić. A deep-NN beamforming approach for dual function radar-communication THz UAV. *IEEE Transactions on Vehicular Technology*, 74(1):746–760, 2024.

25 S. M. Kay. *Fundamentals of Statistical Signal Processing, Vol. I: Estimation Theory*. Prentice Hall Upper Saddle River, NJ, USA, 1998.

26 Y. Zeng, R. Zhang, and T. J. Lim. Wireless communications with unmanned aerial vehicles: Opportunities and challenges. *IEEE Communications Magazine*, 54(5):36–42, 2016.

27 H. Yao, R. Qin, and X. Chen. Unmanned aerial vehicle for remote sensing applications—A review. *Remote Sensing*, 11(12):1443, 2019.

28 A. Rahmati, S. Hosseinalipour, Y. Yapici, X. He, I. Guvenc, H. Dai, and A. Bhuyan. Dynamic interference management for UAV-assisted wireless networks. *IEEE Transactions on Wireless Communications*, 21(4):2637–2653, 2021.

29 Y. Liu, Y. Wang, J. Wang, and Y. Shen. Distributed 3D relative localization of UAVs. *IEEE Transactions on Vehicular Technology*, 69(10):11756–11770, 2020.

30 E. Testi, E. Favarelli, and A. Giorgetti. Reinforcement learning for connected autonomous vehicle localization via UAVs. In *Proceedings IEEE International Workshop on Metrology for Agriculture and Forestry (MetroAgriFor)*, pages 13–17, 2020.

31 P. Popovski, J. J. Nielsen, C. Stefanovic, E. de Carvalho, E. Strom, K. F. Trillingsgaard, A.-S. Bana, D. M. Kim, R. Kotaba, J. Park, et al. Wireless access for ultra-reliable low-latency communication: Principles and building blocks. *IEEE Network*, 32(2):16–23, 2018.

32 Z. Sun, Z. Yu, B. Guo, B. Yang, Y. Zhang, and D. W. K. Ng. Integrated sensing and communication for effective multi-agent cooperation systems. *IEEE Communications Magazine*, 62(9):68–73, 2024.

33 S. Martínez and F. Bullo. Optimal sensor placement and motion coordination for target tracking. *Automatica*, 42(4):661–668, 2006.

34 S. Ragi and E. K. P. Chong. UAV path planning in a dynamic environment via partially observable Markov decision process. *IEEE Transactions on Aerospace and Electronic Systems*, 49(4):2397–2412, 2013.

35 Z. M. Kassas and T. E. Humphreys. Receding horizon trajectory optimization in opportunistic navigation environments. *IEEE Transactions on Aerospace and Electronic Systems*, 51(2):866–877, 2015.

36 K. Dogancay. UAV path planning for passive emitter localization. *IEEE Transactions on Aerospace and Electronic Systems*, 48(2):1150–1166, 2012.

37 C. Wang, J. Wang, Y. Shen, and X. Zhang. Autonomous navigation of UAVs in large-scale complex environments: A deep reinforcement learning approach. *IEEE Transactions on Vehicular Technology*, 68(3):2124–2136, 2019.

38 Y. Cai and Y. Shen. An integrated localization and control framework for multi-agent formation. *IEEE Transactions on Signal Processing*, 67(7):1941–1956, 2019.

39 R. Opromolla, G. Inchingolo, and G. Fasano. Airborne visual detection and tracking of cooperative UAVs exploiting deep learning. *Sensors*, 19(19):4332, 2019.

40 D. Ucinski. *Optimal Measurement Methods for Distributed Parameter System Identification*. CRC press, Cambridge, UK, 2004.

41 F. Meyer, H. Wymeersch, M. Fröhle, and F. Hlawatsch. Distributed estimation with information-seeking control in agent networks. *IEEE Journal on Selected Areas in Communications*, 33(11):2439–2456, 2015.

42 F. Meyer, P. Braca, P. Willett, and F. Hlawatsch. A scalable algorithm for tracking an unknown number of targets using multiple sensors. *IEEE Transactions on Signal Processing*, 65(13):3478–3493, 2017.

43 F. Meyer, O. Hlinka, H. Wymeersch, E. Riegler, and F. Hlawatsch. Distributed localization and tracking of mobile networks including noncooperative objects. *IEEE Transactions on Signal and Information Processing Over Networks*, 2(1):57–71, 2015.

44 S. Tang and V. Kumar. Autonomous flight. *Annual Review of Control, Robotics, and Autonomous Systems*, 1:29–52, 2018.

45 G. Gu, P. R. Chandler, C. J. Schumacher, A. Sparks, and M. Pachter. Optimal cooperative sensing using a team of UAVs. *IEEE Transactions on Aerospace and Electronic Systems*, 42(4):1446–1458, 2006.

46 S. K. Mitter. Filtering and stochastic control: A historical perspective. *IEEE Control Systems Magazine*, 16(3):67–76, 1996.

47 R. T. Haftka and Z. Gürdal. *Elements of Structural Optimization*. Springer Science+Business Media B.V., 1984.

48 M. Z. Win, Y. Shen, and W. Dai. A theoretical foundation of network localization and navigation. *Proceedings of the IEEE*, 106(7):1136–1165, 2018.

49 D. Dardari, P. Closas, and P. M. Djurić. Indoor tracking: Theory, methods, and technologies. *IEEE Transactions on Vehicular Technology*, 64(4):1263–1278, 2015.

50 Y. Bar-Shalom, X. R. Li, and T. Kirubarajan. *Estimation with Applications to Tracking and Navigation: Theory Algorithms and Software*. John Wiley & Sons, 2004.

51 D. Lee, N. He, P. Kamalaruban, and V. Cevher. Optimization for reinforcement learning: From a single agent to cooperative agents. *IEEE Signal Processing Magazine*, 37(3):123–135, 2020.

52 R. S. Sutton and A. G. Barto. *Reinforcement Learning: An Introduction*. MIT Press, Cambridge, UK, 2nd edition, 2018.

53 A. Guerra, F. Guidi, D. Dardari, and P. M. Djurić. Reinforcement learning for joint detection and mapping using dynamic UAV networks. *IEEE Transactions on Aerospace and Electronic Systems*, 60(3):2586–2601, 2024.

54 C. J. C. H. Watkins and P. Dayan. Q-learning. *Machine Learning*, 8(3–4):279–292, 1992.

55 A. Guerra, F. Guidi, D. Dardari, and P. M. Djurić. Real-time learning for THZ radar mapping and UAV control. In *Proceedings of the IEEE International Conference on Autonomous Systems (ICAS)*, pages 1–5, 2021.

56 G. Fontanesi, F. Ortíz, E. Lagunas, L. M. Garcés-Socarrás, V. M. Baeza, M. Á. Vázquez, J. A. Vásquez-Peralvo, M. Minardi, H. N. Vu, P. J. Honnaiah, et al. Artificial intelligence for satellite communication: A survey. *IEEE Communications Surveys & Tutorials*, 28:1381–1435, 2025.

57 J. A. Vásquez-Peralvo, G. Fontanesi, A. Guerra, F. Guidi, N. Shlezinger, A. Zanella, E. Lagunas, S. Chatzinotas, D. Dardari, and P. M. Djurić. Beam pattern optimization for integrated sensing and communication in UAV applications. In *European Conference on Antennas and Propagation (EuCAP)*, 2020.

58 H. Hua, J. Xu, and T. X. Han. Optimal Transmit Beamforming for Integrated Sensing and Communication. *IEEE Transactions on Vehicular Technology*, 72(8):10588–10603, 2023.

59 P. J. Freire, S. Srivallapanondh, A. Napoli, J. E. Prilepsky, and S. K. Turitsyn. Computational complexity evaluation of neural network applications in signal processing. *arXiv preprint arXiv:2206.12191*, 2022.

60 D. Solomitckii, M. Gapeyenko, V. Semkin, S. Andreev, and Y. Koucheryavy. Technologies for efficient amateur drone detection in 5G millimeter-wave cellular infrastructure. *IEEE Communications Magazine*, 56(1):43–50, 2018.

61 S. Thrun. Probabilistic robotics. *Communications of the ACM*, 45(3):52–57, 2002.

62 A. Guerra, F. Guidi, J. Dall'Ara, and D. Dardari. Occupancy grid mapping for personal radar applications. In *Proceedings IEEE Statistical Signal Processing Workshop (SSP)*, pages 766–770, 2018.

6

Sparse Antenna Array Interpolation with Model-based Machine Learning and Deep Frequency Attention Networks

Yunqiao Hu,[1] Ruxin Zheng,[1] Shunqiao Sun,[1,] and Yimin D. Zhang[2]*

[1] *Department of Electrical and Computer Engineering, The University of Alabama, Tuscaloosa, AL, USA*
[2] *Department of Electrical and Computer Engineering, Temple University, Philadelphia, PA, USA*

*Corresponding author: ssun21@eng.ua.edu

6.1 Introduction

Array signal processing is a cornerstone of modern sensing and communication systems, enabling the detection and localization of sources using radio frequency, acoustic, or ultrasonic signals. A critical task in array processing is direction-of-arrival (DOA) estimation, which determines the direction from which signals impinge on an array of sensors. Accurate DOA estimation is vital in applications such as radar, sonar, wireless communications, and audio systems, where precise localization supports mission-critical objectives and enhances overall system performance [1–5]. Traditionally, uniform linear arrays (ULAs) consisting of equally spaced sensors at half-wavelength inter-element spacing are widely considered to satisfy the Nyquist criterion. For ULAs, the array aperture is approximately proportional to the number of sensors. As a result, for practical implementations, achieving higher spatial resolution with a large array aperture demands an extensive number of sensors, thereby significantly driving up hardware costs [5]. Additionally, ULAs with half wavelength spacing may be highly susceptible to mutual coupling effects, thus deteriorating the accuracy of DOA estimates [6].

To address these challenges, sparse linear arrays (SLAs) have emerged as a cost-effective alternative to dense ULAs by strategically reducing the number of active elements while preserving the effective aperture. Sparse array configurations, such as the minimum redundancy array (MRA) [7], nested array [8], coprime array [9–11], and maximum inter-element spacing constraint (MISC) arrays [12, 13], have been widely studied. In the SLA context, traditional DOA estimation algorithms like multiple signal classification (MUSIC) and weighted subspace fitting [14], as well as covariance-based compressive sensing (CS) methods [15], have been adapted to exploit difference coarrays, i.e., virtual arrays synthesized from the pairwise differences of sensor positions. While difference coarrays allow more targets to be resolved than physical elements, they rely on accurate covariance estimation, which generally requires a large number of snapshots.

Automotive radar is a key technology for advanced driver-assistance systems and autonomous driving, where high-resolution perception is critical for situational awareness, collision avoidance, and safe navigation. For automotive radar, two-dimensional (2D) antenna arrays are employed to enable accurate estimation of both azimuth and elevation angles. The azimuth angle information is essential for precisely localizing objects such as vehicles, pedestrians, and obstacles in the horizontal plane. Elevation angle information, on the other hand, provides insights into the height of objects, supporting critical functions, such as determining whether it is safe to drive over or under obstacles. For example, while it is safe for a vehicle to drive over a metal can or under a pedestrian bridge, automotive radars with limited elevation resolution may misclassify these objects as hazardous stationary barriers. Meeting these stringent requirements in automotive radar demands large apertures in both the azimuth and elevation dimensions. At the same time, automotive radar systems must remain cost-effective. Consequently, 2D sparse arrays have been explored for their versatility and ability to balance performance with hardware efficiency [16–18]. Compared to linear sparse arrays, 2D sparse arrays offer greater design flexibility because of the additional degrees of freedom in array geometry [19]. This expanded spatial diversity enables higher sparsity, making 2D sparse arrays particularly attractive for automotive radar applications [3] (Figure 6.1).

Figure 6.1 Automotive radars need to provide elevation resolution to enable drive-over and drive-under functions.

Automotive radars typically operate in highly dynamic environments, where rapid motion of both the radar platform and surrounding objects prevents the collection of a large number of snapshots required for accurate covariance matrix estimation and difference coarray processing. In many cases, only a single snapshot can be acquired within the coherent time interval [3, 20]. Here, a snapshot refers to the array response corresponding to a specific range-Doppler bin [2]. Given these constraints, difference coarray-based methods are generally considered impractical for automotive radar systems. On the other hand, a cost-effective technology for synthesizing a large array aperture is the sum coarray concept enabled by multiple-input multiple-output (MIMO) radar. By employing multiple radar transceivers in a coherent cascade, MIMO radar systems can synthesize M_tM_r sum coarray elements using only M_t transmit and M_r receive antennas [21]. For example, a MIMO radar equipped with 12 transmit and 16 receive antennas can generate as many as 192 virtual elements in the resulting sum coarray [22–24]. This principle has been widely adopted in commercial products such as the ZF and Continental ARS540 radar systems [25, 26].

In both sum and difference coarrays, certain sparse array configurations can provide consecutive coarrays, effectively synthesizing a virtual array that behaves like a ULA. A prominent example is the nested array, which has been effectively applied in both sum and difference coarray frameworks [27, 28]. In contrast, other coarrays, such as those derived from coprime arrays, exhibit nonconsecutive coarrays, resulting in gaps in the virtual array. In these cases, gap-filling techniques can be employed to reconstruct a consecutive virtual ULA, thereby improving DOA estimation performance and enabling algorithms like MUSIC with spatial smoothing, which require a consecutive coarray structure [29, 30]. To support such reconstruction, several methods have leveraged the Toeplitz and Hankel properties of the ULA covariance matrix, applying structured CS and matrix completion techniques to recover missing covariance elements [15, 31–34].

In automotive radar systems, targets are typically separated in the range and Doppler domains, which results in a relatively small number of targets that need to be resolved in the angular domain within each range-Doppler bin [2]. Leveraging this observation, this work focuses on achieving a large virtual array aperture for high angular resolution in both azimuth and elevation directions, using sum coarrays while operating with only a single snapshot. The construction of a sum coarray in an active sensing scheme does not require covariance estimation and can be implemented with a single snapshot, provided that separable waveforms are used at different transmit antennas. This contrasts with the difference coarray approach, which relies on multiple snapshots to accurately estimate the covariance matrix. To further enhance the DOA estimation accuracy and reduce sidelobe levels, structured matrix completion techniques are applied to recover the missing elements of the covariance matrix and enable gridless DOA estimation [15, 35, 36]. Furthermore, recent advances in multifrequency sparse array processing have exploited multiple frequency bands to increase degrees of freedom and fill in missing coarray lags more reliably, requiring fewer snapshots while maintaining high-fidelity covariance recovery [37].

The interpolation and reconstruction methods described above are model-based and rely heavily on structured techniques. As a result, they typically require strong assumptions regarding precise array geometry and accurate sensor calibration. While these methods can perform well under ideal conditions, their effectiveness may degrade—or even fail—when such conditions are not strictly satisfied. Moreover, in single-snapshot scenarios, the lack of statistical averaging limits the ability to suppress noise and interference through subspace-based processing. Together, these factors make single-snapshot sparse array interpolation a highly challenging yet essential problem for achieving high-resolution and reliable radar perception in autonomous vehicles.

In modern radar signal processing, there is a growing emphasis on merging classical algorithmic insights with the flexibility of deep learning, a paradigm often referred to as model-based learning. Applying such techniques to sparse array interpolation effectively overcomes the limitations of traditional model-based signal processing as described above. A central concept in this direction is deep unrolling [38], which has shown strong effectiveness for tasks like sparse antenna array interpolation. Instead of treating a neural network as an opaque black box, deep unrolling maps each iteration of a well-established iterative algorithm onto a corresponding neural network layer. These layers preserve the mathematical structure and domain knowledge of the original algorithm while introducing learnable parameters, such as step sizes or thresholds, which can be optimized from training data. This hybrid approach achieves a balance between interpretability and adaptability, allowing faster convergence and more robust performance in complex environments.

Beyond unrolling classical algorithms, recent innovations have introduced deep learning architectures inspired by transformers, offering a more data-driven and flexible framework. Unlike unrolled networks that closely follow the structure and procedures of a specific algorithm, transformer-based models learn complex mappings directly from sparse measurements to complete array reconstructions. A representative example of this paradigm is the frequency-domain attention network (FA-Net) [39], which processes radar signals in the frequency domain and applies attention mechanisms to focus selectively on the most informative spectral components. In so doing, FA-Net can effectively suppress noise and artifacts while reconstructing missing antenna data, even from a single snapshot, making it particularly valuable for rapidly changing environments encountered by automotive radars.

These two paradigms, namely, deep unrolling and transformer-based architectures, represent complementary solutions to the sparse array interpolation problem. Deep unrolling provides strong interpretability and leverages domain knowledge, making it particularly well-suited when theoretical guarantees and insight into the reconstruction process are critical. Transformer-based deep networks, in contrast, deliver powerful data-driven learning capabilities and greater flexibility, enabling them to capture complex, nonlinear relationships that may be difficult to model analytically, albeit sometimes at the expense of reduced interpretability.

In this chapter, we explore both these cutting-edge approaches to sparse antenna array interpolation. We first delve into the principles and practical implementation of deep unrolling, with a focus on unrolling the iterative hard thresholding (IHT) algorithm and its neural network realizations. We then introduce transformer-inspired architectures, exemplified by FA-Net, and examine how frequency-domain attention mechanisms can enhance sparse signal reconstruction and DOA estimation performance. By presenting this dual perspective, we aim to provide a comprehensive understanding of how modern signal processing and deep learning can be integrated to push the boundaries of sparse array radar systems, offering significant improvements in resolution, accuracy, and robustness for real-world sensing applications.

6.2 Deep Neural Networks

Deep learning has revolutionized how we model complex data by constructing hierarchical representations that progress from raw inputs to increasingly abstract features. This capability has driven breakthroughs across fields such as computer vision, speech, and natural language understanding, powered by architectures like convolutional neural networks (CNNs), recurrent neural networks (RNNs), and, more recently, transformers.

Despite their impressive power, however, many deep models function as black boxes and often depend on massive datasets. Deep unrolling (or unfolding) addresses this by combining classic iterative algorithms with neural networks. In this approach, an established algorithm, e.g., one designed to solve an optimization or signal processing problem, is "unrolled" so that each of its iterations is mapped to a corresponding network layer. Each layer mimics one algorithmic step but incorporates trainable parameters. The resulting architecture retains the domain knowledge and interpretability of the original algorithm while gaining improved performance through data-driven learning.

For instance, unrolling approaches such as learned iterative shrinkage thresholding algorithm (LISTA) [40] and its derivatives, including learned fast iterative shrinkage-thresholding algorithm (LFISTA), adaptive LISTA [41], and group LISTA (GLISTA) [42, 43], have achieved state-of-the-art performance in compressed sensing and image reconstruction. These architectures deliver faster convergence and require significantly fewer trainable parameters compared to traditional deep neural networks. By embedding the structure of classic optimization algorithms within learnable deep models, deep unrolling techniques produce solutions that are data-efficient, interpretable, and firmly rooted in optimization theory.

In parallel, transformers have redefined sequence modeling by replacing recurrence with self-attention [44]. Transformers process entire input sequences simultaneously, leveraging multi-head attention to model interactions between all token pairs [45]. This architecture enables greater parallelism, more efficient training, and superior handling of long-range dependencies compared to RNNs and long short-term memory (LSTM) networks [46]. Nowadays, transformers have expanded far beyond language translation to encompass computer vision (e.g., Vision Transformer [47]), audio, multimodal tasks, and even scientific computing.

6.3 SLA Design

In this section, we first explore the design of SLAs as one-dimensional (1D) array configurations. The emphasis is on leveraging MIMO radar techniques to synthesize a sparse virtual sum coarray, thereby reducing the number of physical antennas while still achieving a large virtual array aperture. Extensions to a 2D array configurations will be discussed in Section 6.4.

Figure 6.2 illustrates a practical configuration of an automotive MIMO radar system that cascades two MIMO transceivers, each with 3 transmit and 4 receive antennas. All transmit and receive antennas are synchronized using a shared clock to ensure coherent signal processing. In this example, $M_t = 6$ transmit antennas and $M_r = 8$ receive antennas in the cascaded MIMO radar are arranged along the azimuth direction, and the antennas are arranged along the axis at:

$$\begin{aligned} l_{\text{TX}} &= [1, 19, 37, 55, 79, 91]\lambda/2, \\ l_{\text{RX}} &= [12, 22, 25, 39, 58, 62, 70, 73]\lambda/2, \end{aligned} \tag{6.1}$$

where λ denotes the wavelength of the signal. Each transmit antenna emits a distinct random sparse step-frequency waveform (RSSFW) [3], allowing the receive antennas to separate the contribution from each transmit antenna using direction of departure modulation [48]. By synthesizing the sum coarray, this configuration

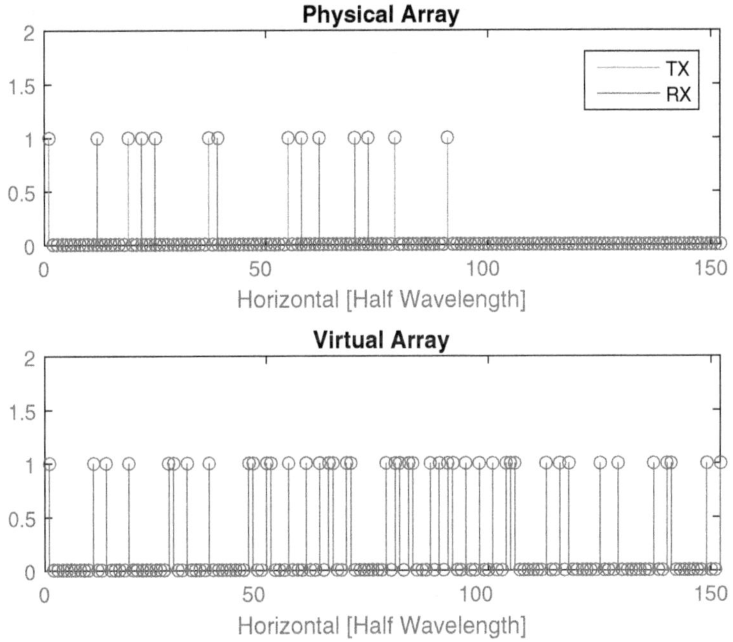

Figure 6.2 Example of an automotive MIMO radar cascaded with two 3×4 MIMO transceivers. The virtual array from the sum coarray contains 48 elements.

forms a virtual SLA consisting of 48 elements and achieves an aperture of 75λ, as depicted in the figure.

To model the array's response, consider a single snapshot of array data corresponding to K targets with DOAs $\theta_1, \theta_2, \dots, \theta_K$. In the absence of noise, the array response of the SLA is expressed as:

$$\boldsymbol{y}_s = \mathbf{A}_s\mathbf{s}, \tag{6.2}$$

where the array manifold matrix $\mathbf{A}_s = [\mathbf{a}_s(\theta_1), \mathbf{a}_s(\theta_2), \dots, \mathbf{a}_s(\theta_K)]$ consists of the steering vectors for all K targets with:

$$\mathbf{a}_s(\theta_k) = \left[1, e^{j\frac{2\pi}{\lambda}d_1 \sin(\theta_k)}, \dots, e^{j\frac{2\pi}{\lambda}(d_{M_tM_r-1})\sin(\theta_k)}\right]^{\mathrm{T}}, \tag{6.3}$$

where d_i denotes the position of the ith virtual sensor in the sum coarray for $i = 0, 1, \dots, M_tM_r - 1$, and $[\cdot]^{\mathrm{T}}$ denotes the transpose operation. Additionally, the vector of target responses is given by $\mathbf{s} = [s_1, \dots, s_K]^{\mathrm{T}}$.

Now, consider an M_o-element virtual ULA that spans the same aperture as the SLA, with inter-element spacing $d = \frac{\lambda}{2}$. The array response of this ULA configuration, in the absence of noise, is expressed as:

$$\boldsymbol{y}_o = \mathbf{A}_o \mathbf{s}, \tag{6.4}$$

where $\mathbf{A}_o = [\mathbf{a}_o(\theta_1), \mathbf{a}_o(\theta_2), \dots, \mathbf{a}_o(\theta_K)]$ is the array manifold matrix of the ULA with:

$$\mathbf{a}_o(\theta_k) = \left[1, e^{j\pi \sin(\theta_k)}, \dots, e^{j\pi(M_o-1)\sin(\theta_k)}\right]^{\mathrm{T}}. \tag{6.5}$$

Define $N_2 = \lfloor M_o/2 \rfloor$ and $N_1 = M_o - N_2$, where $\lfloor \cdot \rfloor$ indicates the floor function. We form an $N_1 \times N_2$ Hankel matrix $\boldsymbol{Y}$ from the $M_o \times 1$ array response vector $\boldsymbol{y}_o$ by dividing it into N_2 subarrays, each of length N_1. The (i, k)th element of the resulting Hankel matrix is given as:

$$\boldsymbol{Y}_{ik} = \boldsymbol{y}_{o,i+k-1}, \; i = 1, 2, \dots, N_1, \; k = 1, 2, \dots, N_2. \tag{6.6}$$

This Hankel matrix $\boldsymbol{Y}$ has a Vandermonde structure and can be factorized as [49]:

$$\boldsymbol{Y} = \mathbf{B}\boldsymbol{\Sigma}\mathbf{B}^{\mathrm{T}}, \tag{6.7}$$

where $\mathbf{B}$ is the subarray manifold matrix and $\boldsymbol{\Sigma}$ is a diagonal matrix containing the target responses. The rank of the Hankel matrix $\boldsymbol{Y}$ is K if $N_2 \geq K$.

For the SLA, a Hankel matrix $\boldsymbol{x}$ can be similarly constructed from the sparse measurement vector $\boldsymbol{y}_s$. Unlike the full ULA matrix $\boldsymbol{Y}$, however, the SLA matrix $\boldsymbol{x}$ contains missing entries, making it a subsampled version of $\boldsymbol{Y}$. Under suitable conditions, the recovery of missing elements in $\boldsymbol{x}$ can be formulated as the following nuclear norm minimization problem [50]:

$$\min \; \| \boldsymbol{x} \|_* \quad \text{s.t.} \;\; \mathcal{P}_\Omega(\boldsymbol{X}) = \mathcal{P}_\Omega(\boldsymbol{Y}), \tag{6.8}$$

where $\| \cdot \|_*$ denotes the nuclear norm, and $\mathcal{P}_\Omega(\boldsymbol{Y})$ is the sampling operator defined on the set of observed indices Ω, which is determined by the sparse array configuration.

In the presence of noise, the observed samples are given by $[\boldsymbol{X}]_{ij} = [\boldsymbol{Y}]_{ij} + [\boldsymbol{E}]_{ij}, (i, j) \in \Omega$, where $[\boldsymbol{E}]_{ij}$ represents the noise term. In this case, the matrix completion problem becomes:

$$\min \; \| \boldsymbol{X} \|_* \quad \text{s.t.} \quad \| \mathcal{P}_\Omega(\boldsymbol{X} - \boldsymbol{Y}) \|_{\mathrm{F}} \leq \delta, \tag{6.9}$$

where $\| \cdot \|_{\mathrm{F}}$ denotes the Frobenius norm and δ is a noise threshold.

Once an estimate of matrix $\boldsymbol{Y}$ is obtained as the solution of $\boldsymbol{X}$, the inherent Hankel property of true $\boldsymbol{Y}$ allows the full array response to be extracted by averaging along its anti-diagonal elements. Conventional array processing techniques can then be applied to perform DOA estimation based on the completed matrix $\boldsymbol{Y}$.

Nuclear norm minimization provides a convex relaxation of the rank minimization problem and is widely regarded for its strong theoretical guarantees in matrix completion. Beyond the Frobenius norm constraints used in (6.9), alternative formulations based on the ℓ_p-norm ($0 < p < 2$) have also been investigated due to their robustness against outliers. Examples include the ℓ_p-regression algorithm [51] and the ℓ_p-norm-based alternating projection algorithm [52]. In general, ℓ_1-norm has a better performance than ℓ_2-norm since it better addresses the minimization of large residuals.

6.4 Two-dimensional Sparse Array Design and Completion

To facilitate advanced driver-assistance functions, such as drive-over and drive-under capabilities, automotive radars must accurately determine the elevation angles of objects within the field of view (FOV). This requires high angular resolution in both the azimuth and elevation planes. Figure 6.3 illustrates an example configuration employing a MIMO radar setup comprising 12 transmit antennas and 16 receive antennas. This setup is realized by cascading four automotive MIMO radar transceiver modules, each equipped with 3 transmit antennas and 4 receive antennas.

To achieve a sufficiently large 2D array aperture with the highly limited number of transmit and receive antennas, the antennas are randomly distributed within a spatial region of $[0, 100]\,(\lambda/2) \times [0, 120]\,(\lambda/2)$ in azimuth and elevation directions. The physical 2D array configuration corresponds to a form factor of approximately 20 cm × 24 cm, assuming a carrier frequency of $f_c = 77$ GHz. It is important to recognize the tradeoff between angular resolution and radar form factor, as the physical size of the MIMO radar system must be constrained to ensure practical integration into a vehicle's bumper. The resulting 2D virtual sparse array obtained from the sum coarray contains 196 elements and has dimensions of $D_y \times D_x = 183(\lambda/2) \times 194(\lambda/2)$, effectively representing a random spatial sub-Nyquist sampling of a corresponding uniform rectangular array (URA) with the same 2D array aperture. In the fully sampled URA, all half-wavelength grid positions in both the horizontal and vertical directions are populated.

For this sum coarray configuration, the angular resolutions in the azimuth and elevation are, respectively, obtained as [53]:

$$\Delta\theta = 2\arcsin\left(\frac{1.4\lambda}{\pi D_x}\right) \approx 0.53^\circ, \tag{6.10}$$

$$\Delta\phi = 2\arcsin\left(\frac{1.4\lambda}{\pi D_y}\right) \approx 0.56^\circ. \tag{6.11}$$

The angular resolution of the described radar system is comparable to that of LiDAR products, such as Velodyne HDL-32E, which has a horizontal resolution

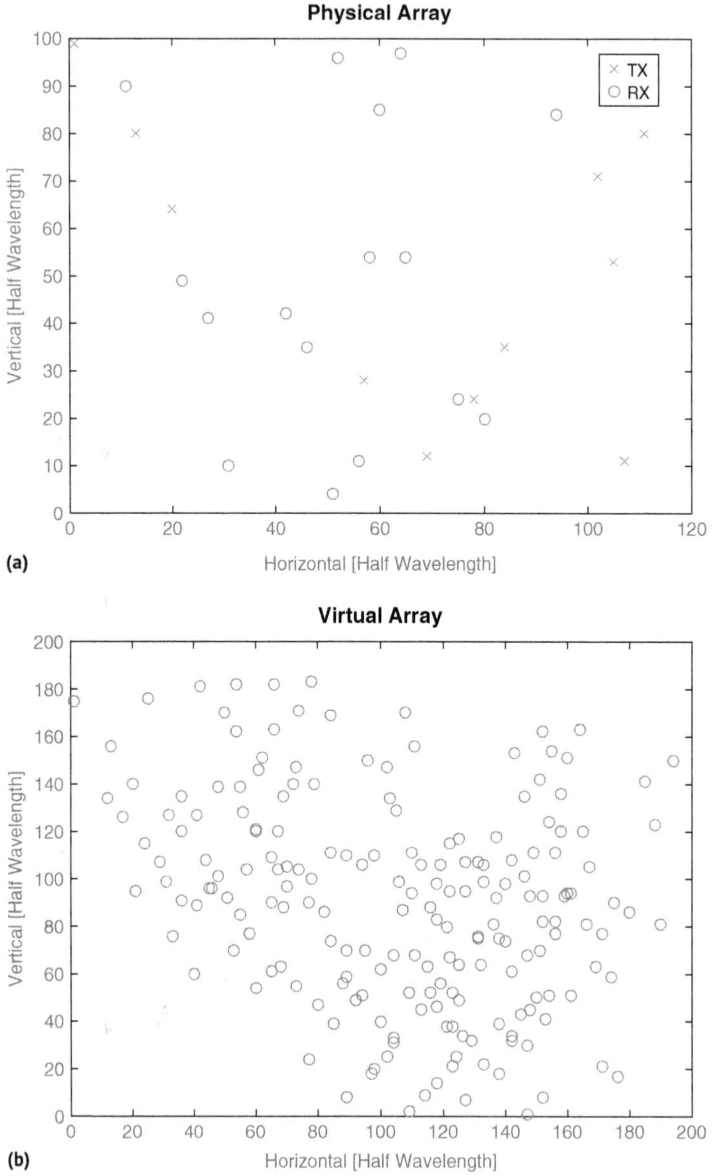

Figure 6.3 A MIMO radar configuration consisting of 12 transmit and 16 receive antennas, created by cascading four 3×4 automotive MIMO radar transceivers. The antennas are randomly arranged in a $[0, 100]\,(\lambda/2) \times [0, 120]\,(\lambda/2)$ region, generating a 2D MIMO virtual array with 196 elements. (a) Physical antenna placement; (b) Corresponding virtual array element positions.

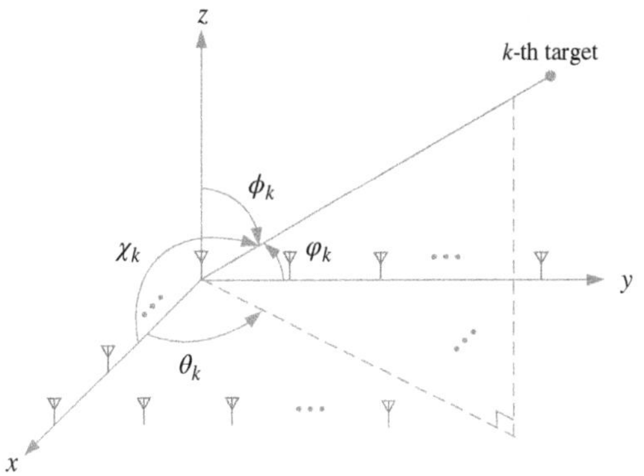

Figure 6.4 Geometry of a URA.

ranging between 0.1° and 0.4° depending on the rotation speed, and a vertical resolution of 1.33° [54]. Compared to LiDAR, radar systems can be manufactured at significantly lower cost while offering reliable sensing capabilities in all weather conditions.

Consider a general $M_1 \times M_2$ URA in the x–y plane with half-wavelength inter-element spacing, as depicted in Figure 6.4. Let the kth point target have an azimuth angle θ_k and an elevation angle ϕ_k, and denote χ_k as the angle between the kth target and the x axis, and φ_k as the angle between the kth target and the y axis. Then, these angles are related through $\cos(\chi_k) = \sin(\phi_k)\cos(\theta_k)$ and $\cos(\varphi_k) = \sin(\phi_k)\sin(\theta_k)$. Therefore, by measuring the angles χ_k and φ_k, the azimuth angle θ_k and elevation angle ϕ_k can be computed as:

$$\theta_k = \arctan\left(\frac{\cos(\varphi_k)}{\cos(\chi_k)}\right), \tag{6.12}$$

$$\phi_k = \arcsin\left(\sqrt{\cos^2(\chi_k) + \cos^2(\varphi_k)}\right). \tag{6.13}$$

For simplicity in signal modeling, we utilize the angles χ_k and φ_k in the URA setup.

The response of the (m_1, m_2)th element of the URA in the x-y plane to K targets, where each target has an angle χ_k relative to the x-axis and an angle φ_k relative to the y-axis for $k = 1, \ldots, K$, can be expressed as:

$$x_{m_1,m_2} = \sum_{k=1}^{K} s_k e^{j\pi[(m_1-1)\sin(\chi_k)+(m_2-1)\sin(\varphi_k)]}. \tag{6.14}$$

Given that $1 \leq m_1 \leq M_1$ and $1 \leq m_2 \leq M_2$, let $\boldsymbol{M} = \left[x_{m_1,m_2}\right]_{1\leq m_1\leq M_1, 1\leq m_2\leq M_2}$ represent the data matrix, where its entries correspond to the URA response as defined in (6.14). We form an $N_1 \times (M_1 - N_1 + 1)$ block Hankel matrix, given as:

$$\boldsymbol{Y}_E = \begin{bmatrix} \boldsymbol{Y}_0 & \boldsymbol{Y}_1 & \cdots & \boldsymbol{Y}_{M_1-N_1} \\ \boldsymbol{Y}_1 & \boldsymbol{Y}_2 & \cdots & \boldsymbol{Y}_{M_1-N_1+1} \\ \vdots & \vdots & \ddots & \vdots \\ \boldsymbol{Y}_{N_1-1} & \boldsymbol{Y}_{N_1} & \cdots & \boldsymbol{Y}_{M_1-1} \end{bmatrix}, \tag{6.15}$$

where

$$\boldsymbol{Y}_m = \begin{bmatrix} x_{m,0} & x_{m,1} & \cdots & x_{m,M_2-L} \\ x_{m,1} & x_{m,2} & \cdots & x_{m,M_2-L+1} \\ \vdots & \vdots & \ddots & \vdots \\ x_{m,L-1} & x_{m,L} & \cdots & x_{m,M_2-1} \end{bmatrix}, \tag{6.16}$$

is an $L \times (M_2 - L + 1)$ Hankel matrix. It can be verified that the rank of matrix $\boldsymbol{Y}_E$ is K if $N_1 \geq K$ and $L \geq K$ [55].

To design a 2D sparse array, the positions of the transmit and receive antennas are strategically chosen to achieve effective spatial subsampling of the underlying URA. The array response of the URA can be recovered from the sparse 2D array by completing the block Hankel matrix $\boldsymbol{Y}_E$, which is constructed from the measurements of the responses of the sparse arrays. The completion problem of the block Hankel matrix is formulated as:

$$\min \| \boldsymbol{X}_E \|_* \quad \text{s.t.} \quad \mathcal{P}_\Omega(\boldsymbol{X}_E) = \mathcal{P}_\Omega(\boldsymbol{M}), \tag{6.17}$$

where Ω represents the observation set corresponding to the locations of the 2D sparse virtual array elements, and $\boldsymbol{X}_E$ is the block Hankel matrix constructed from matrix $\boldsymbol{X}$ based on Equations (6.15) and (6.16). In the presence of noise, we define matrix $\boldsymbol{M}^o = \left[x^o_{m_1,m_2}\right]_{1\leq m_1\leq M_1, 1\leq m_2\leq M_2}$ with $x^o_{m_1,m_2} = x_{m_1,m_2} + n_{m_1,m_2}$ in lieu of $\boldsymbol{M}$, where $x^o_{m_1,m_2} = x_{m_1,m_2} + n_{m_1,m_2}$ denotes the observed signal and $\boldsymbol{E} = \left[n_{m_1,m_2}\right]_{0\leq m_1\leq M_1, 0\leq m_2\leq M_2}$ represents the noise term. We assume that the noise is bounded, i.e., $|\mathcal{P}_\Omega(\boldsymbol{E})|_\text{F} \leq \delta$. In the presence of noise, the block Hankel matrix completion problem becomes:

$$\min \| \boldsymbol{X}_E \|_* \quad \text{s.t.} \quad \| \mathcal{P}_\Omega\left(\boldsymbol{X}_E - \boldsymbol{M}^o\right) \|_\text{F} \leq \delta. \tag{6.18}$$

The optimization problem outlined above is convex and can be tackled using, e.g., the CVX toolbox [56]. In this work, we use the singular value thresholding (SVT) algorithm [57] to carry out matrix completion. The computational complexity for updating the low-rank matrix in each iteration is proportional to $m = |\Omega|$, the size of the observation set Ω.

6.5 IHT Algorithm for Hankel Matrix Completion

As discussed in Section 6.3, the Hankel matrix associated with an SLA configuration can be considered as a subsampled version of Hankel matrix $\boldsymbol{Y}$ corresponding to the full ULA. In this subsampled matrix, entries on or parallel to the anti-diagonals correspond to the SLA response $\boldsymbol{Y}_s$, while other entries are zeros. Given the low-rank structure of $\boldsymbol{Y}$, the missing entries of $\boldsymbol{X} = \mathcal{H}(\boldsymbol{y}_s)$ can be reconstructed by finding the minimum-rank Hankel matrix that matches the known entries [58], and the optimization problem is constructed in the same manner as shown in (6.9). It is important to note that the rank minimization problem in (6.8) is typically NP-hard [58]. To address this challenge, an IHT algorithm is introduced in [59] to enable practical recovery of low-rank Hankel matrices. By further incorporating tangent space projections, the convergence rate of the IHT algorithm can be accelerated, resulting in the fast iterative hard thresholding (FIHT) algorithm. The primary steps of the IHT algorithm at the ith iteration are outlined below.

The following equation describes the gradient descent update for the current estimate $\mathbf{x}_i$ using a fixed step size β on a Riemannian manifold [58]:

$$\boldsymbol{X}_i = \mathcal{H}(\boldsymbol{x}_i + \beta(\boldsymbol{x}_s - \boldsymbol{x}_i)), \tag{6.19}$$

$$\boldsymbol{x}_{i+1} = \mathcal{H}^{\dagger}(\mathcal{T}_r(\boldsymbol{X}_i)), \tag{6.20}$$

where $\mathcal{H}^{\dagger}(\cdot)$ is the inverse operator of $\mathcal{H}(\cdot)$ which transforms an $N_1 \times N_2$ Hankel matrix into an $M \times 1$ vector,

$$\mathcal{T}_r(\boldsymbol{X}_i) = \sum_{k=1}^{r} \sigma_k \mathbf{u}_k \mathbf{v}_k^{\mathrm{H}}, \qquad \sigma_1 \geq \sigma_2 \geq \cdots \geq \sigma_r, \tag{6.21}$$

represents the truncated singular value decomposition (t-SVD), which projects $\boldsymbol{X}_i$ onto the fixed-rank manifold to obtain a low-rank approximation of $\boldsymbol{X}_i$, with r representing the rank of matrix $\boldsymbol{X}_i$, and $[\cdot]^{\mathrm{H}}$ indicates the Hermitian transpose. The IHT algorithm operates iteratively and is known for its rapid convergence [59].

After recovering the complete array response, the target DOAs can be estimated using high-resolution DOA estimation algorithms designed for single-snapshot scenarios, such as single-snapshot MUSIC [60]. Single-snapshot MUSIC extends the classic MUSIC algorithm to cases where only one measurement (snapshot) is available. Originally developed for line-spectral estimation, this method constructs a Hankel data matrix from the single-snapshot vector, facilitating a Vandermonde decomposition akin to that used in multi-snapshot MUSIC. By applying SVD, the algorithm separates the signal and noise subspaces and identifies the target DOAs by locating the minima of the noise-subspace projection denominator, equivalently the maxima of the MUSIC pseudospectrum. Single-snapshot MUSIC method can be viewed as a spatial smoothing technique specifically designed for ULA. Once the full array response $\mathbf{y}$ is recovered, the Hankel matrix $\mathcal{H}(\mathbf{y}) \in \mathbb{C}^{(L+1)\times(M-L)}$ can be expressed as:

$$\mathcal{H}(\mathbf{y}) = \begin{bmatrix} y_1 & y_2 & \cdots & y_{M-L} \\ y_2 & y_3 & \cdots & y_{M-L+1} \\ \vdots & \vdots & \ddots & \vdots \\ y_{L+1} & y_{L+2} & \cdots & y_M \end{bmatrix}, \tag{6.22}$$

where $1 \leq L < M$. Usually, L is chosen such that Hankel matrix is a square or close to a square matrix. Performing SVD of $\mathcal{H}(\mathbf{y})$ as:

$$\mathcal{H}(\mathbf{y}) = \underbrace{[\mathbf{U}_\mathrm{s}, \mathbf{U}_\mathrm{n}]}_{\mathbf{U}} \underbrace{\mathrm{diag}(\sigma_1, \sigma_2, \dots, \sigma_r, \sigma_{r+1}, \dots)}_{\boldsymbol{\Sigma}} \underbrace{[\mathbf{V}_s, \mathbf{V}_n]^\mathrm{H}}_{\mathbf{V}^\mathrm{H}}, \tag{6.23}$$

where the columns of the square matrix $\mathbf{U} \in \mathbb{C}^{(L+1)\times(L+1)}$ with $\mathbf{U}_\mathrm{s} \in \mathbb{C}^{(L+1)\times r}$ and $\mathbf{U}_\mathrm{n} \in \mathbb{C}^{(L+1)\times(L+1-r)}$ are the signal and noise spaces, respectively, the diagonal matrix $\boldsymbol{\Sigma} \in \mathbb{C}^{(L+1)\times(M-L)}$ contains the singular with $\sigma_1 \geq \sigma_2 \geq \sigma_3 \geq \dots$, the columns of $\mathbf{V} \in \mathbb{C}^{(M-L)\times(M-L)}$ with $\mathbf{V}_s \in \mathbb{C}^{(M-L)\times r}$ and $\mathbf{V}_n \in \mathbb{C}^{(M-L)\times(M-L-r)}$ are the right singular vectors, and r is the number of potential target angles.

To estimate the DOA, the steering vectors $\mathbf{a}^{L+1}(\theta_k)$ that are orthogonal to the noise subspace must be identified, i.e., $\left|\mathbf{U}_\mathrm{n}^\mathrm{H}\mathbf{a}^{L+1}(\theta_k)\right| = \mathbf{0}$, where $\mathbf{0} = [0, \dots, 0]^\mathrm{T} \in \mathbb{R}^{(L+1-r)\times 1}$. The DOA can be identified with θ_k corresponding to the r largest local maxima of the pseudospectrum:

$$\mathbf{P}(\theta_k) = \frac{\|\mathbf{a}^{L+1}(\theta_k)\|_2}{\left\|\mathbf{U}_\mathrm{n}^\mathrm{H}\mathbf{a}^{L+1}(\theta_k)\right\|_2}, \tag{6.24}$$

where $\|\cdot\|_2$ is the ℓ_2-norm, and $\mathbf{a}^{L+1}(\theta_k) \in \mathbb{C}^{(L+1)\times 1}$ is the array steering vector with respect to the angle θ_k.

6.6 Model-based Deep Learning for Hankel Matrix Completion

Sparse array interpolation addresses the limitations originating from undersampling by reconstructing a virtual or augmented array response that emulates a denser array using only the measurements captured by the physically sparse sensors. The core idea is to infer the missing array elements or their covariances, thereby synthesizing a data structure compatible with classical high-resolution, gridless DOA algorithms such as MUSIC and ESPRIT. This process traditionally leverages the spatial correlation and redundancy among array elements, often exploiting mathematical constructs such as the difference coarray. By reconstructing a virtual full array, interpolation enhances angular resolution, enables identifying more sources than available sensors, and extends the applicability of advanced DOA estimation techniques to systems constrained by hardware resources.

Machine learning, and more recently deep learning, has emerged as transformative techniques for sparse array interpolation in the context of DOA estimation. The fundamental advantage lies in the exceptional ability of data-driven models, particularly neural networks, to learn complex nonlinear mappings between sparse array measurements and their corresponding full-array or virtual representations. This capability enables deep learning-based interpolation methods to address many challenges of traditional techniques, such as the requirement for multiple snapshots, extensive manual parameter tuning, and the limited adaptability to different operating conditions.

Deep neural networks, such as CNNs, are widely employed to process structured inputs, such as sample covariance matrices obtained from sparse sensor arrays. These deep neural network models are typically trained in a supervised manner using large datasets, where sparse input features are paired with ground-truth dense outputs and are optimized in terms of objective measures like the Cramér–Rao bound to enhance DOA estimation accuracy. An LSTM-based network was proposed in [61] to enlarge the effective array aperture. The network takes measurements from a small-aperture ULA as input and predicts the responses for an extrapolated virtual array, thereby generating a large-aperture virtual array to improve the angular resolution. A more recent study [62] leverages frequency-domain tokenization and attention mechanisms to selectively emphasize the most informative spectral features, thereby improving interpolation accuracy and noise resilience, particularly in single-snapshot scenarios.

Despite the advantages of IHT in terms of its performance and convergence, achieving its optimal performance remains challenging. This is primarily because the computation of the t-SVD $\mathcal{T}_r$ operations becomes resource-intensive, particularly when dealing with large Hankel matrix dimensions. In addition, the method requires careful tuning of parameters, such as the step size β and the matrix rank r.

To harness the strengths of the IHT algorithm while incorporating the flexibility of network-based approaches, iterative hard thresholding network (IHT-Net) translates the IHT update steps into a deep network structure composed of a fixed number of phases, with each phase corresponding to a single iteration of the traditional IHT method. In this framework, explicit mathematical operations like t-SVD are replaced by implicit computations embedded within the neural network's forward pass. Moreover, parameters that traditionally required manual design are now automatically optimized through backpropagation during training, as detailed in Sections 6.6.1 and 6.6.2.

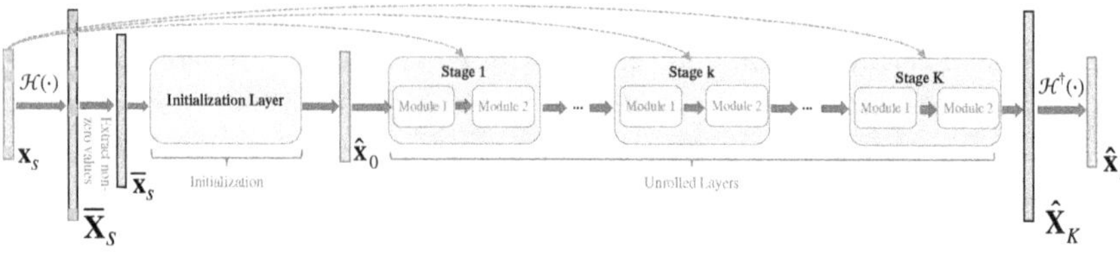

Figure 6.5 Illustration of an IHT-Net architecture.

A key advantage of this model-based design is that it maintains interpretability: each layer in IHT-Net is directly aligned with a known algorithmic step in the IHT procedure. The initialization layer mirrors the setup of the first estimate in IHT, while the subsequent unrolled layers emulate the iterative refinement process, ensuring that the network's operations remain faithful to the underlying optimization algorithm. As a result, IHT-Net does not behave as a purely black-box model, but rather as a structured architecture where the role of each component can be clearly traced back to its algorithmic counterpart. This alignment allows the network to inherit the theoretical insights of IHT while simultaneously gaining the flexibility and adaptability of data-driven learning.

As illustrated in Figure 6.5, the IHT-Net architecture comprises two main components: an initialization layer and a series of unrolled layers. The initialization layer generates the first estimate, analogous to the initialization step in the traditional IHT algorithm. This initial estimate is crucial, as it significantly affects both the convergence behavior and the quality of the final solution. The subsequent unrolled layers emulate the core iterative refinement process of IHT, progressively improving the estimate toward a more accurate solution.

6.6.1 Initialization Layer

In the proposed method, we substitute the t-SVD step in the IHT algorithm with shallow autoencoder layers, thereby eliminating the need for rank determination and SVD computations. Inspired by the impressive results of masked autoencoders [63], we implement an asymmetric structure in which the encoder processes only the observed values (without mask tokens) in the input long vector, while the decoder reconstructs the full signal from the latent representation using mask tokens. As depicted in Figure 6.5, the gray squares represent mask tokens, which are assigned zero values. Since $\boldsymbol{x}_s$ is complex-valued, we concatenate its real and imaginary parts along a single dimension to form a $2M \times 1$ vector. This concatenation doubles the length of the original vector, resulting in a long vector of size $2N_1N_2 \times 1$, denoted as $\bar{\boldsymbol{X}}_s$. The nonzero values are extracted from $\bar{\boldsymbol{X}}_s$, forming a

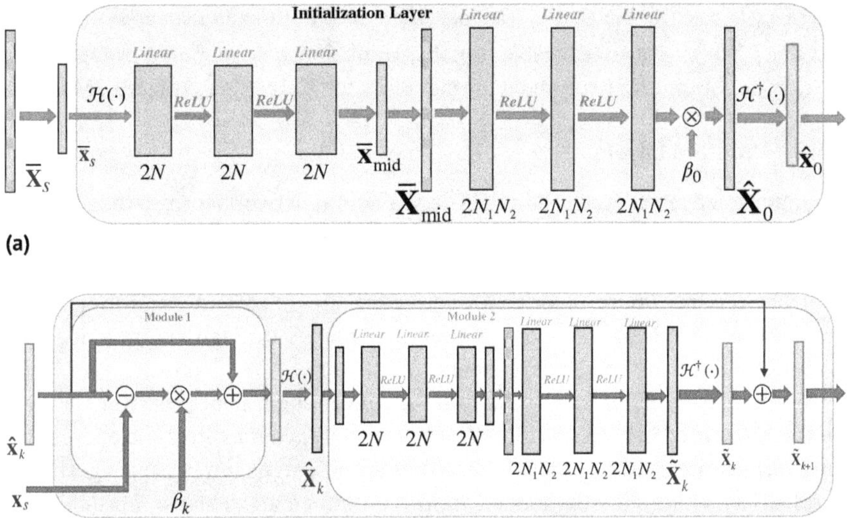

Figure 6.6 (a) Illustration of initialization layer of IHT-Net; (b) Illustration of the kth unrolled layer of IHT-Net.

new $2N \times 1$ vector $\bar{\boldsymbol{x}}_s$, with their positions represented as a list ϕ, which is shared across all layers, and N denotes the number of nonzero elements in $\boldsymbol{X}_s$.

Following the approach developed in [64], where multiple extra linear layers in deep neural networks act as implicit rank minimization, the encoder is designed with three linear layers (including bias) separated by ReLU activation functions. As shown in Figure 6.6a, these three linear layers share the same input and output dimensions, each set to $2N$, matching the length of the input vector. In our implementation, $2N$ is determined by the number of elements in Θ, as expressed in (6.8). Denoting the encoder as $\mathcal{F}_1^{(0)}(\cdot)$, its output is given by:

$$\bar{\boldsymbol{X}}_{\text{mid}} = \mathcal{F}_1^{(0)}(\bar{\boldsymbol{x}}_s). \tag{6.25}$$

Next, the output $\bar{\boldsymbol{x}}_{\text{mid}}$ is embedded into a $2N_1N_2 \times 1$ zero vector at the positions corresponding to the nonzero entries in the original long vector $\bar{\boldsymbol{X}}_s$, resulting in the vector $\bar{\boldsymbol{X}}_{\text{mid}}$ of size $2N_1N_2 \times 1$.

For the decoder, we follow the same structural design as the encoder but adjust the input and output dimensions to $2N_1N_2$. Denoting the decoder as $\mathcal{F}_2^{(0)}(\cdot)$, the final output of the initialization layer is expressed as:

$$\hat{\boldsymbol{X}}_0 = \eta_0 \mathcal{F}_2^{(0)}\left(\mathcal{F}_1^{(0)}(\bar{\boldsymbol{x}}_s)\right), \tag{6.26}$$

where η_0 is a learnable scalar. The parameters of both the encoder $\mathcal{F}_1^{(0)}(\cdot)$ and decoder $\mathcal{F}_2^{(0)}(\cdot)$ are also learnable. Finally, Hankel inverse mapping is applied to the decoder output $\hat{\boldsymbol{X}}_0$ to produce the signal vector $\hat{\boldsymbol{x}}_0 = \mathcal{H}^{\dagger}(\hat{\boldsymbol{X}}_0)$ of size $2M \times 1$.

6.6.2 Unrolled Layers

When unrolled, each layer of the IHT-based neural network corresponds exactly to one iteration of the classical algorithm, executing operations such as linear projections, thresholding, and residual computations. Unlike traditional implementations where algorithm parameters are fixed and hand-tuned, the unrolled network learns and adapts these parameters through learning, leading to more effective and robust performance.

Unrolled layers differ substantially from conventional neural network layers, such as those used in CNNs. While CNN layers apply generic filters and operations to extract features without explicit ties to any specific signal processing algorithm, unrolled layers are directly derived from the mathematical steps of established iterative methods. This close connection provides unrolled networks with strong interpretability, as each layer serves a clearly defined role in the reconstruction process. Additionally, unrolled networks typically require far fewer parameters than generic deep network models, making them more efficient and less demanding in terms of training data.

Indeed, interpretability is one of the most compelling benefits of unrolled layers. Because their architecture directly mirrors the logic of established algorithms, practitioners can understand not only the network's outputs but also the intermediate processes that produce them. This transparency is particularly valuable in sensitive applications like automotive radar, where insight into how decisions are made is critical. In addition, unrolled networks tend to be more parameter-efficient and data-efficient than purely data-driven models, leveraging domain knowledge to achieve high performance even with limited training data. In the context of sparse array interpolation, unrolled networks can accurately reconstruct missing antenna elements and significantly improve DOA estimation, enabling radar systems to achieve high resolution while using fewer physical antennas.

Nevertheless, employing unrolled networks comes with challenges. Determining the optimal depth, that is, the number of unrolled layers, is non-trivial; too few layers may limit the network's representation capability to capture necessary details, while too many can increase computational demands and the risk of overfitting. Training such networks can also present stability issues, as gradients may vanish or explode over many iterative steps. Moreover, while unrolling boosts interpretability, it can constrain the network's flexibility compared to fully data-driven approaches, which are free to discover entirely new representations unconstrained by prior algorithmic structures. Another limitation is that unrolling presumes the existence of a well-defined and effective iterative algorithm to model. For problems lacking such a foundation, designing an unrolled architecture may be less straightforward or even infeasible.

Despite these challenges, unrolled networks offer a powerful framework for radar signal processing, particularly in sparse antenna array interpolation. They elegantly combine the strengths of classical signal processing with modern machine learning, producing solutions that are both interpretable and highly effective. By learning optimal parameters within a well-understood algorithmic framework, unrolled networks enable radar systems to exceed traditional performance limits, achieving higher resolution and greater robustness in real-world scenarios while maintaining practical computational requirements. As a result, unrolled architectures represent a promising frontier in the development of next-generation radar sensing technologies.

Each unrolled stage consists of two components, respectively, referred to as the gradient descent module and the low-rank approximation module. These two modules are illustrated in Figure 6.6. The gradient descent module is analogous to equation (6.19) in the IHT algorithm. In the kth state ($k > 1$), given the input $\hat{\boldsymbol{x}}_k$ from the $(k-1)$th stage and the data vector $\boldsymbol{x}_s$, which is shared across all unrolled stages, the intermediate result at the kth stage is expressed as:

$$\hat{\boldsymbol{X}}_k = \mathcal{H}(\hat{\boldsymbol{x}}_k + \eta_k(\boldsymbol{x}_s - \hat{\boldsymbol{x}}_k)), \tag{6.27}$$

where the step size η_k is a learnable parameter.

The low-rank approximation module retains the same structure as the initialization layer but incorporates a skip connection between the layers. First, we extract $\hat{\boldsymbol{x}}_k$ from the output $\hat{\boldsymbol{X}}_k$ using the position list of nonzero values, ϕ. The output of this module is then obtained by passing it through the encoder-decoder layers, yielding:

$$\tilde{\boldsymbol{X}}_k = \mathcal{F}_2^{(k)}\left(\mathcal{F}_1^{(k)}(\hat{\boldsymbol{x}}_k)\right). \tag{6.28}$$

After performing the Hankel inverse operation, we obtain $\tilde{\boldsymbol{x}}_k = \mathcal{H}^{\dagger}(\tilde{\boldsymbol{X}}_k)$. It is noted that the truncated SVD step in the original IHT algorithm is replaced by an autoencoder–decoder architecture in our approach. By incorporating a skip connection between the input $\hat{\boldsymbol{x}}_k$ and the output $\tilde{\boldsymbol{x}}_k$, the final result of the kth unrolled stage is:

$$\hat{\boldsymbol{x}}_{k+1} = \tilde{\boldsymbol{x}}_k + \gamma_k(\tilde{\boldsymbol{x}}_k - \hat{\boldsymbol{x}}_k), \tag{6.29}$$

where γ_k is a learnable parameter that scales the residual term $(\tilde{\boldsymbol{x}}_k - \hat{\boldsymbol{x}}_k)$. The final estimate, $\hat{\boldsymbol{x}}_K$, is obtained after completing K stages of unrolled forward inference.

From a computational perspective, computing the truncated SVD of an $N_1 \times N_2$ Hankel matrix requires $\mathcal{O}(N_1 N_2 r)$ operations, which becomes prohibitive when both the matrix dimensions and the truncation rank r are large.

6.6.3 IHT-Net Training Specifics

To consider DOA estimation, we simulate P point-target signals within the same range-Doppler bin. The target angles are uniformly distributed within the FOV spanning $[-60°, 60°]$. The amplitudes of the targets follow a uniform distribution in the range $[0.5, 1]$, and their phases are uniformly distributed in $[0, 2\pi]$. Using the model in equation (6.2), we generate N_b noise-free training labels, denoted as $\{\boldsymbol{x}^q_{\text{label}}\}_{q=1}^{N_b}$, for a specific SLA configuration. We then create the inputs $\{\boldsymbol{x}^q_{\text{input}}\}_{q=1}^{N_b}$ for network training by adding varying levels of white Gaussian noise, with the signal-to-noise ratio (SNR) randomly selected from the range of [10 dB, 30 dB] for each training sample. In our experiments, we set $P = 2$ for both training and testing, with $N_b = 700{,}000$.

The learnable parameters in the kth phase of IHT-Net are given by $\{\eta_k, \gamma_k, \mathcal{F}_1^{(k)}, \mathcal{F}_2^{(k)}\}$ for $k = 1, \dots, K$. In addition, for $k = 0$, the learnable parameters are $\{\eta_0, \mathcal{F}_1^{(0)}, \mathcal{F}_2^{(0)}\}$.

Given the training data pairs $\{\boldsymbol{x}^q_{\text{label}}, \boldsymbol{x}^q_{\text{input}}\}_{q=1}^{N_b}$, where $\boldsymbol{x}^q_{\text{label}}$ and $\boldsymbol{x}^q_{\text{input}}$ for each q correspond to a block of data with N training sample pairs, IHT-Net takes $\boldsymbol{x}^q_{\text{input}}$ as the input and produces the output as the reconstructed estimate, denoted as $\hat{\boldsymbol{x}}^q_K$. The goal is to minimize the difference between $\hat{\boldsymbol{x}}^q_K$ and $\boldsymbol{x}^q_{\text{label}}$ while ensuring that the low-rank approximation condition is satisfied. Consequently, the loss function for IHT-Net is formulated as:

$$\text{Loss}_{\text{total}} = \text{Loss}_1 + \alpha \cdot \text{Loss}_2, \tag{6.30}$$

where α is a regularization parameter:

$$\text{Loss}_1 = \frac{1}{N_b N} \sum_{q=1}^{N_b} \left\| \hat{\boldsymbol{x}}^q_K - \boldsymbol{x}^q_{\text{label}} \right\|_2^2, \tag{6.31}$$

measures the discrepancy between the network's final output and the ground truth, and

$$\text{Loss}_2 = \frac{1}{N_b N} \sum_{q=1}^{N_b} \sum_{k=0}^{K} \left\| \mathcal{H}^\dagger\left(\mathcal{F}_1^{(k)}\left(\mathcal{F}_2^{(k)}\left(\hat{\boldsymbol{x}}^q_k\right)\right)\right) - \hat{\boldsymbol{x}}^q_k \right\|_2^2, \tag{6.32}$$

acts as an auxiliary, inner-layer penalty that encourages the encoder–decoder to learn a structured, low-rank approximation, effectively regularizing the latent representation toward a compact subspace.

For IHT-Net training, we set α to 0.01 and use the Adam optimization algorithm [65] with an initial learning rate of 10^{-4}. The learning rate is halved every 10 epochs during the training process. For the learnable parameters η_k in (6.27) and γ_k in (6.29), the initial values are set to 1 in practice, which provides stable performance across different runs and ensures convergence during training.

6.7 Transformer-based Sparse Array Interpolation

While traditional model-based methods are effective for single-snapshot sparse array interpolation, they require expert knowledge for hyperparameter tuning, lack task-specific adaptability, and suffer from high computational costs. To address these challenges, we propose a novel deep learning-based single-snapshot sparse array interpolation network that leverages a frequency-domain attention mechanism. In this approach, the sparse signal is transformed into the frequency domain, where the attention mechanism focuses on key spectral regions, enabling improved interpolation of missing elements. Frequency-domain attention is particularly advantageous because it can explicitly exploit the spectral sparsity and periodic structures inherent in many radar signals, allowing the network to better capture long-range dependencies and discriminate between true target signatures and noise. By working in the frequency domain, the model can more efficiently highlight informative frequency components while suppressing irrelevant or corrupted ones. By reducing computational costs and enhancing interpolation accuracy, the proposed method outperforms traditional approaches and is particularly well-suited for automotive radar applications.

Sparse arrays provide a practical and efficient solution for DOA estimation by strategically reducing the number of active antenna elements while maintaining the overall array aperture. In a sparse array, only a subset of the full array elements is utilized, resulting in measurements being available only at selected antenna positions. Mathematically, this selection can be represented by an $M \times N$ binary selection matrix $\mathbf{M}$ with $(M < N)$, which maps the responses of the full array to the corresponding sparse measurements. The received signal vector of the single-snapshot sparse array is expressed as $\boldsymbol{y}_s = \mathbf{M}\boldsymbol{y}$, where the observed sparse measurement vector $\boldsymbol{y}_s$ is obtained by selecting a subset of elements from the full measurement vector $\boldsymbol{y}$. The selection process inherently incorporates the noise term present in the sparse array. The sparsity level of the sparse array is defined as:

$$\text{Sparsity} = 1 - \frac{M}{N}, \tag{6.33}$$

which quantifies the fraction of removed array elements relative to the full array configuration.

6.8 FA-Net for Signal Reconstruction

In this section, we present the FA-Net, developed to reconstruct the original signal from sparse array measurements while simultaneously performing effective denoising. In addition, we describe the procedure for generating the training data and explain the methodology employed to train the network.

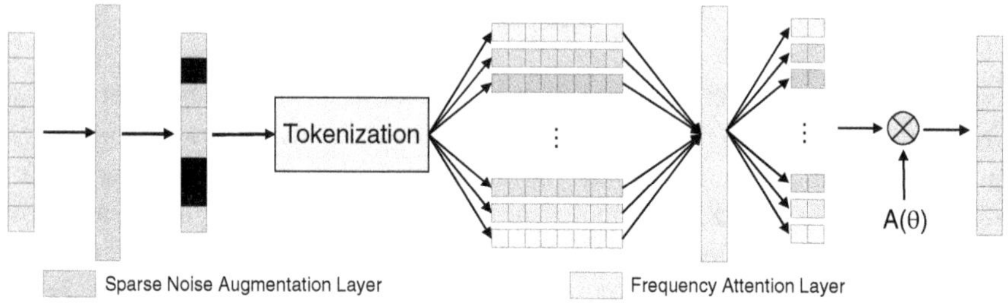

Figure 6.7 Network architecture integrating a sparse noise augmentation layer and a frequency-domain attention mechanism within a signal reconstruction framework.

6.8.1 The FA-Net Architecture

The FA-Net architecture, illustrated in Figure 6.7, consists of three core components: a sparse and noise augmentation layer, a frequency-domain tokenization module, and a frequency-domain attention mechanism. The process begins by augmenting the original data with additive noise and randomly masking elements to emulate sparse measurements. The data is then transformed into the frequency domain, where it is tokenized to enable structured and efficient processing. Finally, a frequency-domain attention mechanism is applied to assign importance weights to different frequency components, allowing the network to better distinguish true signal information from noise and thereby enhance reconstruction performance.

6.8.1.1 Sparse and Noise Augmentation Layer

Data augmentation is a well-established technique for improving model generalization and reducing overfitting [66]. In computer vision, common transformations such as flipping and rotation help models become more robust to variations. In the context of signal processing, the sparse noise augmentation layer serves a similar purpose by introducing structured sparsity and controlled noise, thereby improving signal reconstruction performance. This augmentation layer applies a random binary mask to the input signal, with the sparsity level controlled by a configurable maximum sparsity parameter. In addition, Gaussian noise is added to the signal, with its magnitude regulated by a user-defined SNR range. During training, the number of masked elements is randomly chosen for each batch, and the SNR value is sampled from the specified range to introduce variability.

By dynamically determining the number of active antennas during training, this augmentation layer exposes the network to diverse sparsity patterns and noise conditions, thus strengthening its generalization capability. In contrast, during evaluation, the sparsity pattern is fixed using a predefined thresholding algorithm to ensure consistent testing conditions. Notably, this sparse noise augmentation layer is applied only during training to improve the network's robustness to incomplete and noisy signals.

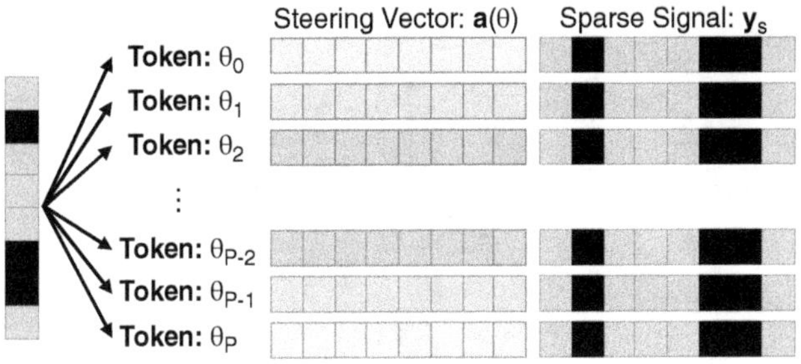

Figure 6.8 Frequency-domain tokenization.

6.8.1.2 Frequency-domain Tokenization

Random sparsity leads to missing signal elements, posing significant challenges for designing a deep learning network that can achieve robust signal reconstruction. To overcome this issue, the signal is transformed into the frequency domain using a tokenization process. As depicted in Figure 6.8, we define a frequency grid of size P, where each frequency bin is mapped to a unique token. For each token corresponding to θ_p, the steering vector $\mathbf{a}(\theta_p)$ is concatenated with the sparse signal. This representation enables the network to effectively capture both spatial and spectral characteristics of the underlying signal.

6.8.1.3 Attention Mechanism

In sparse array signal reconstruction, different frequency components contribute unequally to the underlying signal structure. Certain frequencies carry essential target information, whereas others are more susceptible to noise, distortions, or artifacts introduced by the sparse array geometry. The irregular spacing of the array elements creates nonuniform beampatterns, making some frequencies more vulnerable to sidelobes and missing data. To mitigate these challenges, an attention mechanism is incorporated to selectively emphasize informative frequency components while suppressing the influence of less reliable ones, thereby improving the overall reconstruction quality.

As shown in Figure 6.9, the attention mechanism captures frequency-domain dependencies by dynamically adjusting the contribution of each frequency token. Each frequency bin θ_p is represented as a token that integrates both spatial and spectral features. These tokens are then projected into query ($\mathbf{Q}$), key ($\mathbf{K}$), and value ($\mathbf{V}$) spaces, where the query and key representations determine the relevance between tokens, while the value encodes the corresponding feature information. The attention scores are computed using a scaled dot-product [47]:

$$\alpha = \text{Softmax}\left(\frac{\mathbf{QK}^{\mathrm{T}}}{\sqrt{d}}\right), \tag{6.34}$$

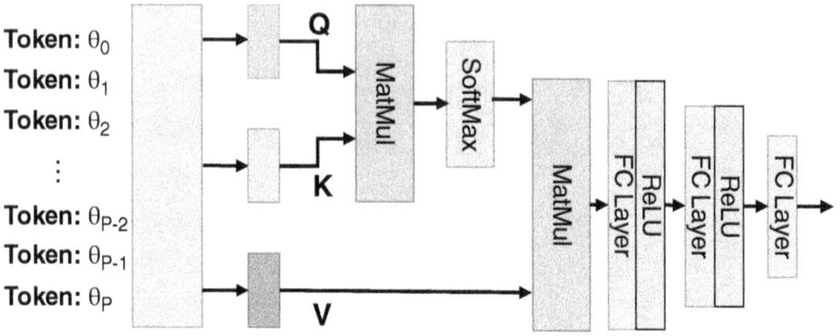

Figure 6.9 Attention layer.

where d is the key vector dimension. The final layer output is then obtained by weighting the value matrix $\mathbf{V}$ with these attention scores:

$$\text{Attention}(\mathbf{Q}, \mathbf{K}, \mathbf{V}) = \alpha \times \mathbf{V}. \tag{6.35}$$

The attention mechanism computes weighted combinations of the value representations, emphasizing significant frequency components while attenuating noise and irrelevant features. The resulting aggregated features are then refined through fully connected (FC) layers with ReLU activations to enhance feature extraction. For each θ_p, the network produces two outputs corresponding to the real and imaginary parts of the reconstructed complex signal. Finally, this output is multiplied by the array manifold matrix to recover the final signal.

This attention mechanism enables the network to adaptively prioritize the most informative frequency components, thereby improving its robustness to sparsity-induced distortions and enhancing the accuracy of signal reconstruction.

6.8.2 Dataset Generation and Training

We use a 20-element ULA with an inter-element spacing of half a wavelength to generate simulated single-snapshot beamforming vectors for up to two targets. The radar FOV is specified as $\boldsymbol{\phi}_{\text{FOV}} = [-30°, 30°]$ and is uniformly discretized in the frequency domain into 64 bins. The reflection coefficients s_k for each source are randomly generated as complex-valued variables, with the amplitudes drawn from a uniform distribution over the range $[0.5, 1]$ and the phases uniformly sampled from $[0, 2\pi]$. A total of 131,072 signals is simulated for training purposes.

The proposed FA-Net is trained for 500 epochs with a batch size of 512 using the Adam optimizer, a learning rate of 0.001, and a mean squared error (MSE) loss function. The training configuration includes a maximum sparsity of 0.4 and an

SNR ranging between 10 and 30 dB. The network model is trained end-to-end to minimize the MSE loss and enhance signal reconstruction accuracy. All experiments are implemented in Python 3.8 with PyTorch 1.10 and CUDA 11.1, running on four NVIDIA RTX A6000 GPUs.

6.9 Numerical Results

6.9.1 One-dimensional Sparse Array Completion

To obtain high azimuth angular resolution, several automotive radar transceivers are combined to form a large sparse array in the azimuth direction. Specifically, we examine the physical array configuration shown in Figure 6.2, where the transmit (Tx) and receive (Rx) positions are defined in Eq. (6.1). A virtual array consisting of 48 elements is created. The configuration of the transmit and receive antennas, along with the virtual array, is shown in Figure 6.2.

Two targets are positioned at the same range of $R = 100$ m with a velocity of $v = -10$ m/s. The azimuth angles of the targets are $\theta_1 = 0°$ and $\theta_2 = 20°$. Initially, the targets are separated in the range-Doppler domain using the RSSFW method [2, 3]. The complex peak values in the range-Doppler spectrum corresponding to each virtual sparse array are used to find the azimuth angles of the target signals.

The virtual SLA, depicted in Figure 6.2, serves as a deterministic sampler for a rank-2 Hankel matrix $\boldsymbol{Y} \in \mathbb{C}^{N \times N}$, where $N = 76$. This matrix is constructed from the array response of a ULA with 152 elements. The array response of the SLA is normalized with respect to its first element. The Hankel matrix $\boldsymbol{Y}$ is then completed using the SVT algorithm [57]. The completed Hankel matrix is denoted as $\boldsymbol{x}$. The full ULA response is reconstructed by averaging the anti-diagonal elements of $\boldsymbol{x}$. The completed array has an aperture size of 76λ. In this simulation, the matrix completion process provides a $10\log_{10}\left(\frac{152}{48}\right) \approx 5$ dB SNR improvement for array processing.

In Figure 6.10, the azimuth angle spectrum for the two targets is shown. The spectra for both targets are obtained by applying fast Fourier transform (FFT) to the original SLA (with gaps filled by zeros) and to the completed full array via matrix completion, respectively. Because of the large but sparse array aperture, the FFT of the SLA reveals two peaks corresponding to the correct azimuth directions but with high sidelobes, making it difficult to detect the targets in the azimuth direction. In contrast, the completed full array produces two distinct peaks corresponding to the accurate azimuth positions, with significantly reduced sidelobes.

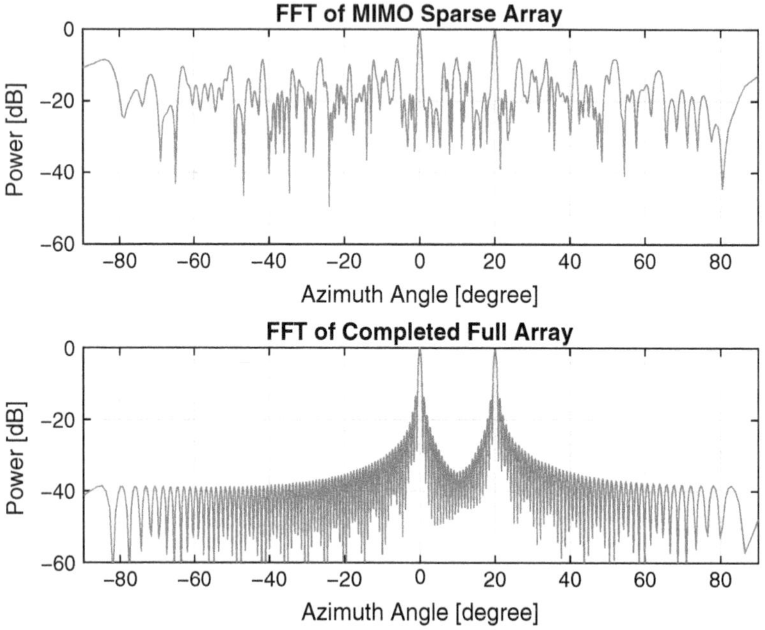

Figure 6.10 Spectral response of two targets located at azimuth angles $\theta_1 = 0°$ and $\theta_2 = 20°$ evaluated under both a MIMO sparse array and a fully populated array configuration.

6.9.2 Two-dimensional Sparse Array Completion

Figure 6.3a shows a MIMO system for joint high-resolution estimation of both azimuth and elevation angles, utilizing a setup of 4 automotive radar transceivers, each equipped with 3 transmit antennas and 4 receive antennas. The cascaded transceivers are equipped with 12 transmit antennas and 16 receive antennas, which are randomly positioned within an area of $[0, 100]\,(\lambda/2) \times [0, 120]\,(\lambda/2)$. This configuration forms a 2D MIMO virtual array consisting of 196 elements, as shown in Figure 6.3b. The physical dimensions of the cascaded radar setup are approximately 20×24 cm. In the diagram shown in Figure 6.3, the 2D sparse array has dimensions of $D_y \times D_x = 183(\lambda/2) \times 194(\lambda/2)$. To create a URA of equivalent size with half-wavelength inter-element spacing, 35,502 elements would be necessary. That is, the virtual sparse array utilizes only 0.54% of the elements that would be required by the fully populated URA.

Consider two targets that share the same range and Doppler bin. Their angles relative to the x and y directions are given by $(\chi_1, \varphi_1) = (-20°, 5°)$ and $(\chi_2, \varphi_2) = (20°, 10°)$, respectively. The snapshot from the sparse array is composed of the complex peak values from the range-Doppler spectrum for each individual array element. The array response is set with an input SNR of 20 dB, which is typical for automotive radar applications following the high processing gain achieved through fast-time and slow-time coherent processing. We then construct a block Hankel matrix $\boldsymbol{Y}_E$ of dimension $9{,}009 \times 8{,}928$ from a single snapshot of the 196-element 2D sparse array. Only 0.78% of the entries in the Hankel matrix are nonzero. Using this snapshot, the block Hankel matrix is completed with the SVT algorithm, leading to the reconstruction of the full URA. In this simulation, matrix completion results in an approximate $10\log_{10}(35{,}502/196) \approx 22.5$ dB processing gain for SNR enhancement.

Figures 6.11 and 6.12 display the azimuth-elevation spectra for the two targets using the 2D sparse array and the completed full URA, respectively. Both configurations produce two distinct peaks that correspond to the correct azimuth and elevation angles of the targets. However, the estimated spatial spectrum for the 2D sparse array exhibits significant sidelobes across the entire azimuth and elevation FOV. In contrast, the sidelobes are significantly reduced in the estimated spatial spectrum when the completed URA is used.

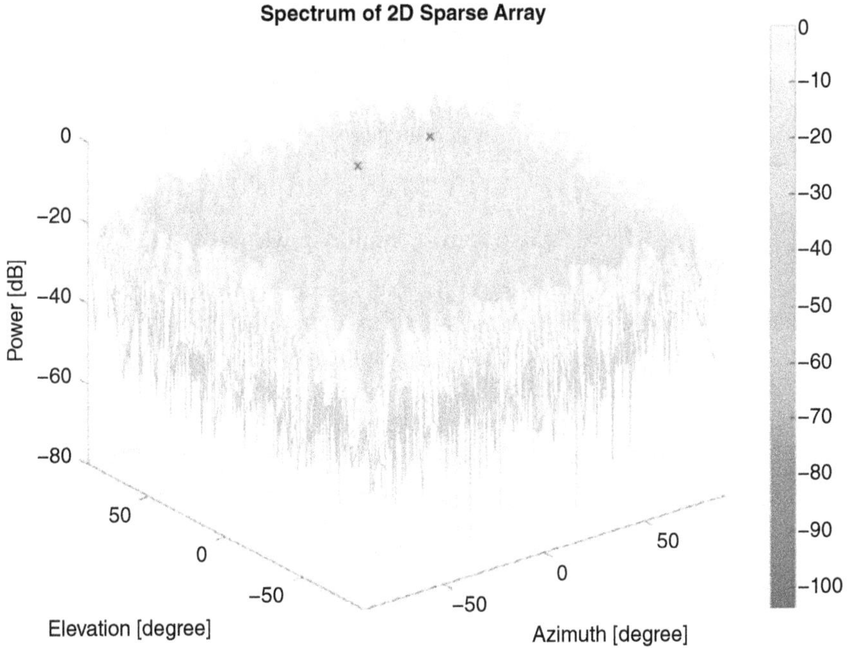

Figure 6.11 Spectral response of two targets located at azimuth and elevation angles $(\chi_1, \varphi_1) = (-20°, 5°)$, $(\chi_2, \varphi_2) = (20°, 10°)$, measured using a sparse array configuration. The target positions are marked by crosses on the spectrum. The plot demonstrates pronounced sidelobes, which arise due to the substantial gaps in the sparse array.

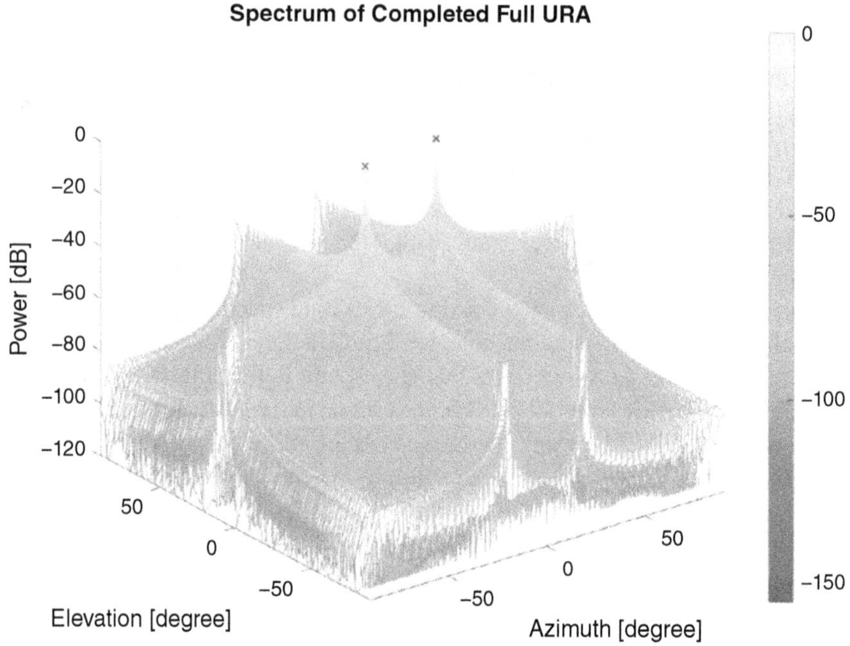

Figure 6.12 The spectrum of two targets with azimuth and elevation angles of $(\chi_1, \varphi_1) = (-20°, 5°)$, $(\chi_2, \varphi_2) = (20°, 10°)$ under the completed full URA. The targets' angles are marked with crosses.

6.9.3 One-dimensional Sparse Array Completion with IHT-Net

This section presents an evaluation of IHT-Net's performance through numerical simulations. A ULA with $N = 21$ elements is used, from which an SLA is derived by randomly selecting a subset of the ULA's antennas. We begin by testing with an 18-element SLA, using a training dataset generated as outlined in Section 6.6.3. The training process is carried out over 100 epochs. In Figure 6.13a, we observe a rapid decrease in training loss during the first 10 epochs, demonstrating that IHT-Net is capable of fast convergence. To assess the impact of the number of layers on IHT-Net's recovery performance, we train the model with various layer counts using the same dataset. We then generate 5,000 testing samples at a 20 dB input SNR for evaluation, with the testing loss computed as in equation (6.31). As shown in Figure 6.13b, the testing loss decreases as the number of unrolled phases increases, although the improvement plateaus after 8 phases. Therefore, we select 8 unrolled phases to achieve a good balance between reconstruction performance and computational efficiency.

Additionally, we compare IHT-Net with the FIHT algorithm [59] across different input SNR levels, using a testing set of 5,000 samples for each input SNR. In Figure 6.13c, IHT-Net consistently exhibits superior performance over FIHT,

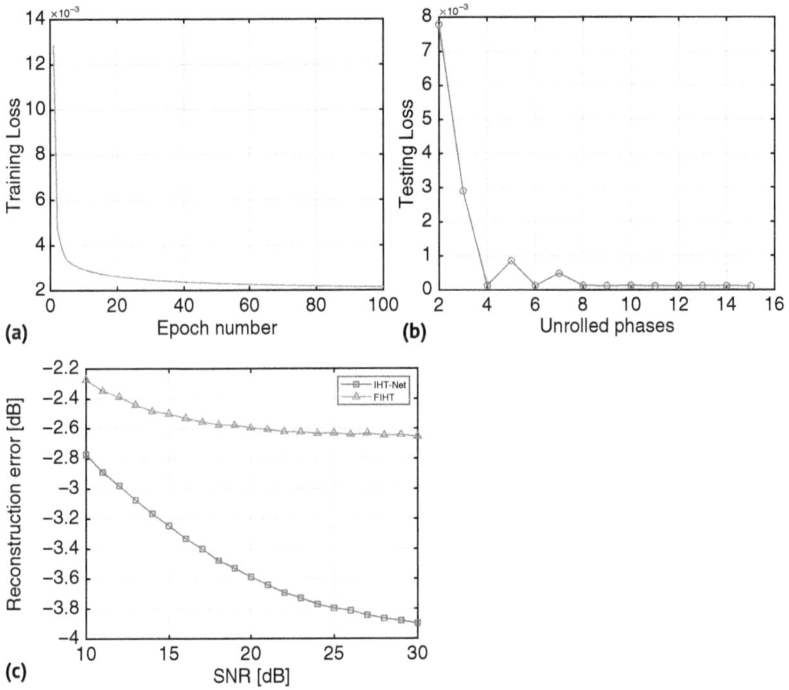

Figure 6.13 (a) IHT-Net training loss (defined in (6.31)) versus epoch for IHT-Net with 8 unrolled phases; (b) IHT-Net testing loss (defined in (6.31)) for varying numbers of unrolled phases of IHT-Net; (c) Comparison of signal reconstruction error between IHT-Net and FIHT [59] under different SNR levels.

especially at higher SNR levels, as indicated by the lower reconstruction loss. Additional experiments using a 10-element SLA are conducted to compare the recovered spatial spectra across different input SNR levels. Figure 6.14 shows the beampatterns of the response of the full array recovered by IHT-Net and FIHT, compared to the spectra of the full array response in both scenarios with and without noise. It is evident that IHT-Net exhibits denoising capabilities at lower SNR levels, such as 10 dB, suggesting that the components of IHT-Net function similarly to the t-SVD operation in FIHT. Moreover, both IHT-Net and FIHT perform well at higher SNR levels, such as 30 dB. Figure 6.14c, d shows that, with sparser SLAs, FIHT struggles to accurately recover the original signal, while IHT-Net consistently provides satisfactory recovered spectra, preserving the mainlobes and suppressing sidelobe levels, further confirming its superior recovery ability. Finally, we compare the MSE performance of the DOA estimates obtained from IHT-Net and FIHT reconstructions at different input SNR levels, where beamforming is used for DOA estimation. Using 5,000 testing samples per SNR level, the results

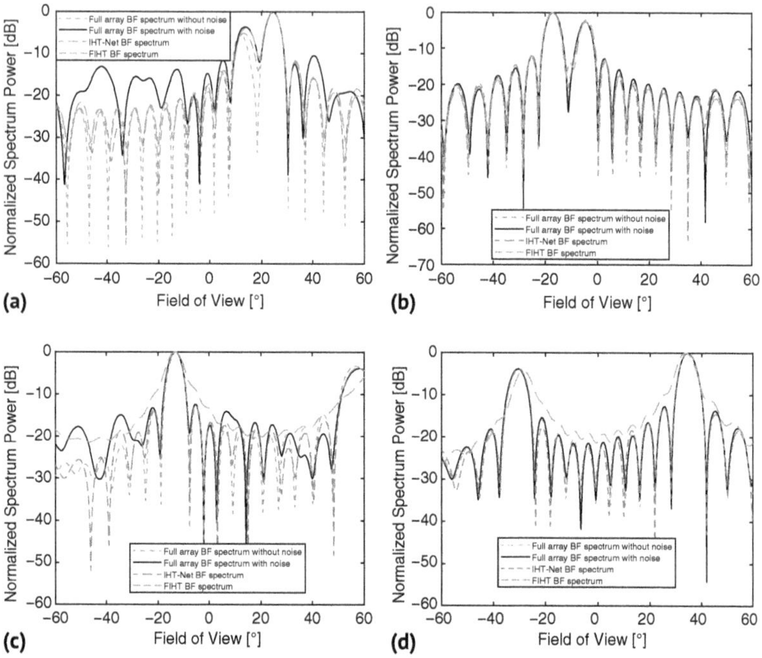

Figure 6.14 Examples of beamforming spectra at varying input SNR levels using different SLAs; (a) input SNR = 10 dB, 18-element SLA; (b) input SNR = 30 dB, 18-element SLA; (c) input SNR = 10 dB, 10-element SLA; (d) input SNR = 30 dB, 10-element SLA.

shown in Figure 6.15 illustrate that IHT-Net leads to more accurate DOA estimation compared to FIHT.

6.9.4 Performance Evaluation of Frequency Attention Network

We evaluate FA-Net's signal reconstruction using the MSE metric, comparing it to the IHT algorithm [59], with the ground-truth (GT) model order, referred to as IHT-GT. However, estimating the model order is a challenging task for single-snapshot sparse arrays and is beyond the scope of this chapter. A simple thresholding strategy is used in this chapter to identify the signal subspace, denoted as IHT-thresholding. Performance is assessed under 0.4 sparsity, with fixed missing element counts but randomly varying positions across trials. A Monte Carlo approach with 5,000 independent trials per SNR level ensures statistical reliability. Test signals contain two targets within the radar FOV. As shown in Figure 6.16, FA-Net consistently outperforms IHT-thresholding across all SNR levels. FA-Net performs better than IHT-GT at SNR levels above 20 dB.

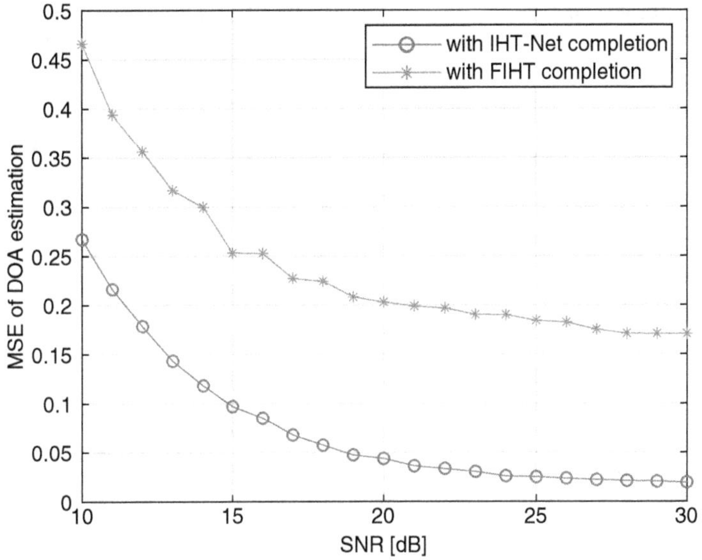

Figure 6.15 Comparison of DOA estimation errors using IHT-Net and FIHT [59] completion methods under different SNR levels.

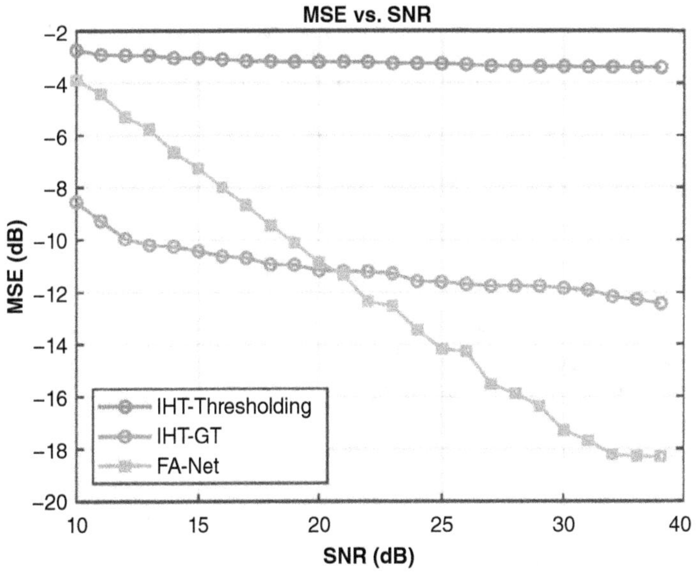

Figure 6.16 Signal reconstruction error versus SNR.

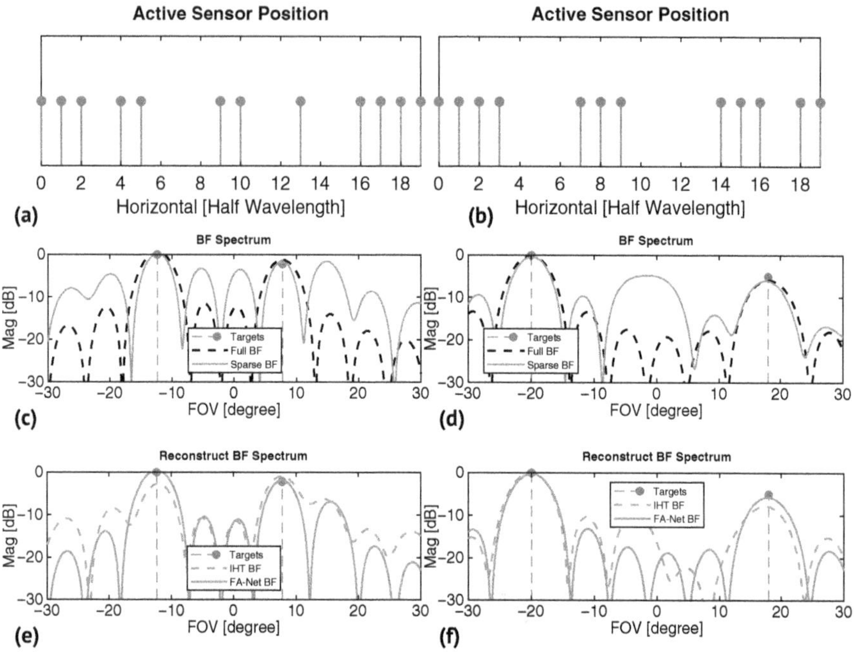

Figure 6.17 Signal reconstruction examples at 10 dB SNR ((a) (c) (e)) and 30 dB SNR ((b) (d) (f)). (a) and (b): Sparse array geometries. (c) and (d): Beamforming spectrum of the ground truth signal and sparse, noisy signal. (e) and (f): Reconstructed beamforming spectrum using the IHT algorithm and FA-Net.

Figure 6.17 demonstrates FA-Net's reconstruction performance across various SNR levels and sparse array conditions, utilizing beamforming to visualize spectral representations given the difficulty of direct signal visualization. Figure 6.17a, c and e shows results for 10 dB SNR, whereas Figure 6.17b, d and f shows those for 30 dB SNR, both with 0.4 sparsity (eight missing elements). Figure 6.17a, b displays the sparse array geometries. In Figure 6.17c, d, we compare the beamforming spectrum of the ground truth against the sparse, noisy signal, illustrating how sparsity and noise introduce increased sidelobes and spectral distortion. Figure 6.17e, f presents the beamforming spectrum of signals reconstructed by IHT and FA-Net, respectively. Examples using real-world radar data are shown in Figure 6.18. FA-Net clearly yields a more accurate beamforming spectrum, demonstrating its superior reconstruction capability and effective adaptation to random sparse array geometries over traditional algorithms.

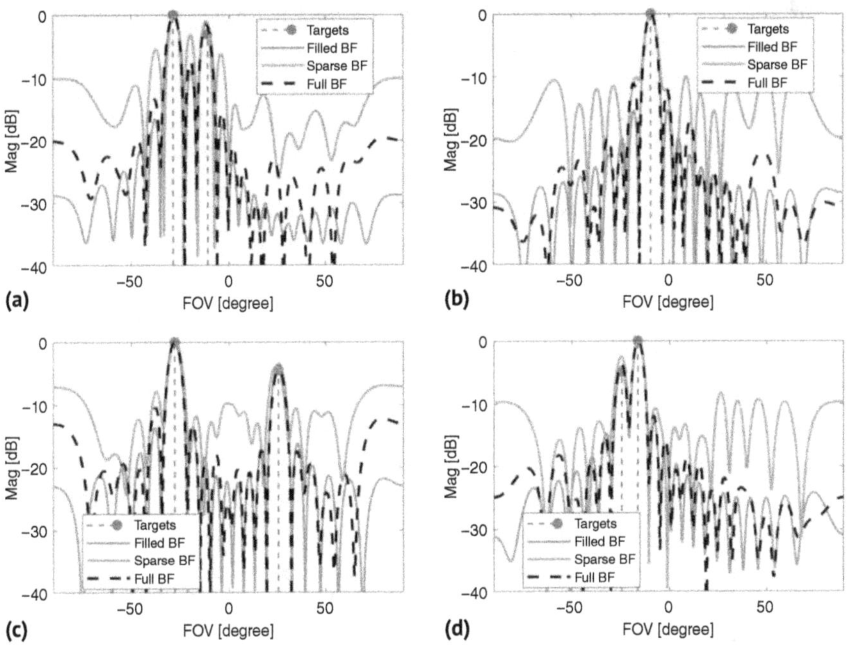

Figure 6.18 Reconstruction example based on real-world radar data. *Source*: [67] / with permission of IEEE.

6.10 Conclusion

This chapter has presented a comprehensive exploration of advanced methods for enhancing automotive radar systems through sparse array design and deep learning-based DOA estimation. We first examined the principles behind linear and 2D sparse array configurations, demonstrating how these architectures can maximize aperture while minimizing the number of physical antenna elements, thereby enabling efficient and precise DOA estimation across both azimuth and elevation.

To address the challenge of reconstructing missing array measurements, we introduced the IHT algorithm for Hankel matrix completion, establishing its effectiveness in recovering sparse array data. Building upon this, we developed IHT-Net, an unrolled network architecture that combines the mathematical rigor of IHT with the flexibility of deep learning. By embedding learnable parameters and nonlinear layers, IHT-Net achieves superior reconstruction and DOA estimation performance, as validated through extensive numerical evaluations.

Furthermore, we explored a complementary deep learning framework incorporating sparse and noise signal augmentation, frequency-domain tokenization, and attention mechanisms. This approach enhances the adaptability and robustness of automotive radar systems, achieving strong generalization across varying sparse geometries without the need for retraining. Simulation results confirmed the framework's ability to suppress sidelobe artifacts and improve signal fidelity, showcasing clear advantages over traditional techniques.

In summary, this chapter has highlighted the significant promise of integrating sparse array design with modern deep learning strategies to advance the capabilities of automotive radar. These innovations lay a solid foundation for achieving high-resolution, robust, and adaptable radar perception in future automotive applications.

Acknowledgments

The work of Y. Hu, R. Zheng, and S. Sun was supported in part by U.S. National Science Foundation (NSF) under Grants CCF-2153386, ECCS-2033433, and Alabama Transportation Institute (ATI). The work of Y. D. Zhang was supported in part by the NSF under Grant ECCS-2236023 and in part by the Air Force Office of Scientific Research under Grant FA9550-23-1-0255.

References

1 S. Sedighi, B. S. Mysore R, M. Soltanalian, and B. Ottersten. On the performance of one-bit DoA estimation via sparse linear arrays. *IEEE Transactions on Signal Processing*, 69:6165–6182, 2021.

2 S. Sun, A. P. Petropulu, and H. Vincent Poor. MIMO radar for advanced driver-assistance systems and autonomous driving: Advantages and challenges. *IEEE Signal Processing Magazine*, 37(4):98–117, 2020.

3 S. Sun and Y. D. Zhang. 4D automotive radar sensing for autonomous vehicles: A sparsity-oriented approach. *IEEE Journal of Selected Topics in Signal Processing*, 15(4):879–891, 2021.

4 J. Li and P. Stoica. MIMO radar with colocated antennas. *IEEE Signal Processing Magazine*, 24(5):106–114, 2007.

5 S. M. Patole, M. Torlak, D. Wang, and M. Ali. Automotive radars: A review of signal processing techniques. *IEEE Signal Processing Magazine*, 34(2):22–35, 2017.

6 B. Liao, Z. G. Zhang, and S.-C. Chan. DOA estimation and tracking of ULAs with mutual coupling. *IEEE Transactions on Aerospace and Electronic Systems*, 48(1):891–905, 2012.

7 A. Moffet. Minimum-redundancy linear arrays. *IEEE Transactions on Antennas and Propagation*, 16(2):172–175, 1968.
8 P. Pal and P. P. Vaidyanathan. Nested arrays: A novel approach to array processing with enhanced degrees of freedom. *IEEE Transactions on Signal Processing*, 58(8):4167–4181, 2010.
9 P. P. Vaidyanathan and P. Pal. Sparse sensing with co-prime samplers and arrays. *IEEE Transactions on Signal Processing*, 59(2):573–586, 2011.
10 S. Qin, Y. D. Zhang, and M. G. Amin. Generalized coprime array configurations for direction-of-arrival estimation. *IEEE Transactions on Signal Processing*, 63(6):1377–1390, 2015.
11 M. Wang and A. Nehorai. Coarrays, MUSIC, and the Cramer-Rao bound. *IEEE Transactions on Signal Processing*, 65(4):933–946, 2017.
12 Z. Zheng, W. Q. Wang, Y. Kong, and Y. D. Zhang. MISC array: A new sparse array design achieving increased degrees of freedom and reduced mutual coupling effect. *IEEE Transactions on Signal Processing*, 67(7):1728–1741, 2019.
13 W. Shi, Y. Li, and R. C. de Lamare. Novel sparse array design based on the maximum inter-element spacing criterion. *IEEE Signal Processing Letters*, 29:1754–1758, 2022.
14 M. Viberg, B. Ottersten, and T. Kailath. Detection and estimation in sensor arrays using weighted subspace fitting. *IEEE Transactions on Signal Processing*, 39(11):2436–2449, 1991.
15 C. Zhou, Y. Gu, X. Fan, Z. Shi, G. Mao, and Y. D. Zhang. Direction-of-arrival estimation for coprime array via virtual array interpolation. *IEEE Transactions on Signal Processing*, 66(22):5956–5971, 2018.
16 C. Schmid, R. Feger, C. Wagner, and A. Stelzer. Design of a linear non-uniform antenna array for a 77-GHz MIMO FMCW radar. In *Proceedings IEEE MTT-S International Microwave Workshop on Wireless Sensing, Local Positioning, and RFID*, Cavtat, Croatia, 2009.
17 J. Searcy and S. Alland. MIMO antenna with elevation detection. U.S. Patent 9 541 639 B2, 2017.
18 R. Rajamaki, S. P. Chepuri, and V. Koivunen. Hybrid beamforming for active sensing using sparse arrays. *IEEE Transactions on Signal Processing*, 68:6402–6417, 2020.
19 I. Aboumahmoud, A. Muqaibel, M. Alhassoun, and S. Alawsh. A review of sparse sensor arrays for two-dimensional direction-of-arrival estimation. *IEEE Access*, 9:92999–93017, 2021.
20 P. Hacker and B. Yang. Single snapshot DOA estimation. *Advances in Radio Science*, 8:251–256, 2010.
21 J. Li and P. Stoica. MIMO radar with colocated antennas. *IEEE Signal Processing Magazine*, 24(5):106–114, 2007.
22 I. Bilik, S. Villeval, D. Brodeski, H. Ringel, O. Longman, P. Goswami, C. Y. B. Kumar, S. Rao, P. Swami, A. Jain, et al. Automotive multi-mode cascaded radar

data processing embedded system. In *Proceedings IEEE Radar Conference*, Oklahoma City, OK, 2018.

23 V. Giannini, M. Goldenberg, A. Eshraghi, J. Maligeorgos, L. Lim, R. Lobo, D. Welland, C. K. Chow, A. Dornbusch, T. Dupuis, et al. A 192-virtual-receiver 77/79 GHz GMSK code-domain MIMO radar system-on-chip. In *Proceedings IEEE International Solid-State Circuits Conference (ISSCC)*, San Francisco, CA, 2019.

24 Texas Instruments Inc. Imaging radar using cascaded mmWave sensor reference design, 2020. https://www.ti.com/lit/ug/tiduen5a/tiduen5a.pdf.

25 ZF Friedrichshafen AG, 2021. https://www.zf.com/products/en/cars/products_58368.html.

26 AUMOVIO, 2021. https://www.continental-automotive.com.

27 P. Pal and P. P. Vaidyanathan. Nested arrays: A novel approach to array processing with enhanced degrees of freedom. *IEEE Transactions on Signal Processing*, 58(8):4167–4181, 2010.

28 J. Shi, F. Wen, and T. Liu. Nested MIMO radar: Coarrays, tensor modeling and angle estimation. *IEEE Transactions on Aerospace and Electronic Systems*, 57(1):573–585, 2021.

29 P. Pal and P. P. Vaidyanathan. Coprime sampling and the MUSIC algorithm. In *Proceedings Digital Signal Processing and Signal Processing Education Meeting (DSP/SPE)*. Sedona, AZ, 2011.

30 C.-L. Liu and P. P. Vaidyanathan. Remarks on the spatial smoothing step in coarray MUSIC. *IEEE Signal Processing Letters*, 22(9):1438–1442, 2015.

31 H. Qiao and P. Pal. Unified analysis of co-array interpolation for direction-of-arrival estimation. In *Proceedings IEEE International Conference on Acoustics, Speech, and Signal Processing (ICASSP)*, New Orleans, LA, 2017.

32 D. Romero, D. D. Ariananda, Z. Tian, and G. Leus. Compressive covariance sensing: Structure-based compressive sensing beyond sparsity. *IEEE Signal Processing Magazine*, 33(1):78–93, 2016.

33 S. Qin, Y. D. Zhang, M. G. Amin, and A. M. Zoubir. Generalized coprime sampling of Toeplitz matrices for spectrum estimation. *IEEE Signal Processing Magazine*, 65(1):81–94, 2017.

34 H. Qiao and P. Pal. Gridless line spectrum estimation and low-rank Toeplitz matrix compression using structured samplers: A regularization-free approach. *IEEE Transactions on Signal Processing*, 65(9):2221–2236, 2017.

35 T. Spreng, S. Yuan, V. Valenta, H. Schumacher, U. Siart, and V. Ziegler. Wideband 120 GHz to 140 GHz MIMO radar: System design and imaging results. In *Proceedings European Microwave Conference (EuMC)*, Paris, France, 2015.

36 C. Zhou, Y. Gu, Z. Shi, and Y. D. Zhang. Off-grid direction-of-arrival estimation using coprime array interpolation. *IEEE Signal Processing Letters*, 25(11):1710–1714, 2018.

37 S. Zhang, A. Ahmed, Y. D. Zhang, and S. Sun. DOA estimation exploiting interpolated multi-frequency sparse array. In *Proceedings IEEE Workshop on Sensor Array and Multichannel Processing (SAM),* Hangzhou, China, 2020.

38 V. Monga, Y. Li, and Y. C. Eldar. Algorithm unrolling: Interpretable, efficient deep learning for signal and image processing. *IEEE Signal Processing Magazine*, 38(2):18–44, 2021.

39 R. Zheng, S. Sun, and H. Liu. Deep frequency attention networks for single snapshot sparse array interpolation. In *Proceedings European Radar Conference (EuRAD)*, Utrecht, The Netherlands, 2025.

40 K. Gregor and Y. LeCun. Learning fast approximations of sparse coding. In *Proceedings of the 27th International Conference on International Conference on Machine Learning*, ICML'10, pages 399–406. Omnipress, Madison, WI, USA, 2010.

41 J. Liu, X. Chen, Z. Wang, and W. Yin. ALISTA: Analytic weights are as good as learned weights in LISTA. In *International Conference on Learning Representations*, 2019.

42 X. Chen, J. Liu, Z. Wang, and W. Yin. Theoretical linear convergence of unfolded ISTA and its practical weights and thresholds. *arXiv preprint arXiv:1808.10038*, 2018.

43 T. Chen, X. Chen, W. Chen, H. Heaton, J. Liu, Z. Wang, and W. Yin. Learning to optimize: A primer and a benchmark. *Journal of Machine Learning Research*, 23(189):1–59, 2022.

44 S. Islam, H. Elmekki, A. Elsebai, J. Bentahar, N. Drawel, G. Rjoub, and W. Pedrycz. A comprehensive survey on applications of transformers for deep learning tasks. *Expert Systems with Applications*, 241:122666, 2024.

45 A. Vaswani, N. Shazeer, N. Parmar, J. Uszkoreit, L. Jones, A. N. Gomez, Ł. Kaiser, and I. Polosukhin. Attention is all you need. In I. Guyon, U. Von Luxburg, S. Bengio, H. Wallach, R. Fergus, S. Vishwanathan, and R. Garnett, editors, *Advances in Neural Information Processing Systems*, volume 30. Curran Associates, Inc., 2017.

46 S. Hochreiter and J. Schmidhuber. Long short-term memory. *Neural Computation*, 9(8):1735–1780, 1997.

47 A. Dosovitskiy, L. Beyer, A. Kolesnikov, D. Weissenborn, X. Zhai, T. Unterthiner, M. Dehghani, M. Minderer, G. Heigold, S. Gelly, et al. An image is worth 16×16 words: Transformers for image recognition at scale. In *International Conference on Learning Representations*, 2021.

48 F. Xu, S. A. Vorobyov, and F. Yang. Transmit beamspace DDMA based automotive MIMO radar. *IEEE Transactions on Vehicular Technology*, 71(2):1669–1684, 2022.

49 J. Ying, J. F. Cai, D. Guo, G. Tang, Z. Chen, and X. Qu. Vandermonde factorization of Hankel matrix for complex exponential signal recovery-application in fast NMR spectroscopy. *IEEE Transactions on Signal Processing*, 66(21):5520–5533, 2018.

50 E. J. Candès and B. Recht. Exact matrix completion via convex optimization. *Foundations of Computational Mathematics*, 9(6):717–772, 2009.

51 W. J. Zeng and H. C. So. Outlier-robust matrix completion via ℓ_p -minimization. *IEEE Transactions on Signal Processing*, 66(5):1125–1140, 2018.

52 X. Jiang, Z. Zhong, X. Liu, and H. C. So. Robust matrix completion via alternating projection. *IEEE Signal Processing Letters*, 24(5):579–583, 2017.

53 M. A. Richards. *Fundamentals of Radar Signal Processing*. McGraw-Hill, New York, NY, 2nd edition, 2014.

54 Velodyne LiDAR, Inc. HDL-32E high definition real-time 3D LiDAR sensor. Datasheet, 2021. https://velodynelidar.com/products/hdl-32e/.

55 Y. Hua. Estimating two-dimensional frequencies by matrix enhancement and matrix pencil. *IEEE Transactions on Signal Processing*, 40(9):2267–2280, 1992.

56 M. Grant and S. Boyd. CVX: Matlab software for disciplined convex programming, version 2.1, CVX Research, Inc., 2014. http://cvxr.com/cvx.

57 J. F. Cai, E. J. Candès, and Z. Shen. A singular value thresholding algorithm for matrix completion. *SIAM Journal on Optimization*, 20(2):1956–1982, 2010.

58 B. Vandereycken. Low-rank matrix completion by riemannian optimization. *SIAM Journal on Optimization*, 23(2):1214–1236, 2013.

59 J. F. Cai, T. Wang, and K. Wei. Fast and provable algorithms for spectrally sparse signal reconstruction via low-rank Hankel matrix completion. *Applied and Computational Harmonic Analysis*, 46(1):94–121, 2019.

60 W. Liao and A. Fannjiang. MUSIC for single-snapshot spectral estimation: Stability and super-resolution. *Applied and Computational Harmonic Analysis*, 40(1):33–67, 2016.

61 I. Roldan, F. Fioranelli, and A. Yarovoy. Self-supervised learning for enhancing angular resolution in automotive MIMO radars. *IEEE Transactions on Vehicular Technology*, 72(9):11505–11514, 2023.

62 R. Zheng, S. Sun, and H. Liu. Deep frequency attention networks for single snapshot sparse array interpolation. In *2025 22nd European Radar Conference (EuRAD)*, 2025.

63 K. He, X. Chen, S. Xie, Y. Li, P. Dollár, and R. Girshick. Masked autoencoders are scalable vision learners. In *Proceedings of the IEEE/CVF Conference on Computer Vision and Pattern Recognition*, pages 16000–16009, 2022.

64 L. Jing, J. Zbontar, and Y. LeCun. Implicit rank-minimizing autoencoder. *Advances in Neural Information Processing Systems*, 33:14736–14746, 2020.

65 D. P. Kingma and J. Ba. Adam: A method for stochastic optimization. *arXiv preprint arXiv:1412.6980*, 2014.

66 C. Shorten and T. M. Khoshgoftaar. A survey on image data augmentation for deep learning. *Journal of Big Data*, 6(60):1–48, 2019.

67 R. Zheng, S. Sun, H Liu, H. Chen, and J. Li. Interpretable and efficient beamforming-based deep learning for single-snapshot DOA estimation. *IEEE Sensors Journal*, 24(14):22096–22105, 2024.

7

Tensorized Neural Networks in Array Signal Processing

Hang Zheng,[1] Zhiguo Shi,[1,] Sergiy A. Vorobyov,[2] and Yujie Gu[3]*

[1] *College of Information Science and Electronic Engineering, Zhejiang University, Hangzhou, Zhejiang, China*
[2] *Department of Information and Communications Engineering, Aalto University, Espoo, Finland*
[3] *Advanced Safety & User Experience, Aptiv, Agoura Hills, CA, USA*

*Corresponding author: shizg@zju.edu.cn

7.1 Introduction

Direction-of-arrival (DOA) estimation, as a fundamental aspect of array signal processing, is essential for direction finding and localization in applications, such as radar, sonar, acoustics, and wireless communications [1–8]. Over the past decades, model-based DOA estimation methods have been intensively studied. Among popular model-based DOA estimation methods are the multiple signal classification (MUSIC) [9], estimation of signal parameters via rotational invariance techniques (ESPRIT) [10], sparsity-based methods [11, 12], and tensor decomposition-based methods [13–18]. As electromagnetic propagation environments become increasingly complex, model-based methods are susceptible to degraded performance under nonideal signal conditions such as low signal-to-noise ratio (SNR) and limited number of snapshots. To address model deviations, sparse Bayesian learning and correlation matrix reconstruction techniques have been incorporated [19, 20]. Nevertheless, these methods entail high computational complexity and pose implementation challenges due to constraints in system resources.

To improve the robustness against challenging conditions while maintaining low computational complexity, machine learning techniques, such as radial basis function [21] and support vector regression [22], are leveraged for data-driven DOA estimation. Among these techniques, deep learning-based DOA estimation has emerged as a prominent research focus [23–27]. With carefully designed training strategies, deep neural network (DNN)-based DOA estimation demonstrates resilience to the challenges of array imperfection [28], source coherency [29], low SNR [30], mutual coupling [31], etc. However, for two-dimensional (2D) DOA estimation, commonly used DNNs typically follow a matrix processing framework, wherein vectorized inputs undergo multiplication with several parameter matrices to produce the output. The resulting long vector inputs require a proportionally large number of trainable parameters in multiple neural layers. As the dimension of signals expands, the existing DNN architectures for DOA estimation suffer from a significant computational burden, slow training speed, and potential system overload. Therefore, it becomes crucial to design resource-efficient neural network architectures tailored for multidimensional array signal processing.

Tensor decompositions, such as canonical polyadic (CP) decomposition (CPD) and Tucker decomposition, project tensor statistics into several low-order factors, thereby serving as powerful tools for efficiently extracting structural data characteristics [32]. The advantages offered by tensor decomposition have inspired the idea of tensorizing neural layers [33–36], where the input and hidden states are tensorized to capture intrinsic multiway data features. More importantly, the algebraic relationship between tensor decompositions and feature propagation over neural layers can be established for network compression, leading to high training efficiency.

The paradigms of tensorizing neural layers can generally be divided into two types. As shown in Figure 7.1a, the first paradigm is based on formulating fully connected (FC) parameters or convolutional kernel parameters as tensors, and directly decomposing them to generate lower-order factors. For example, tensor train decomposition (TTD) [37] and block-term decomposition (BTD) [38] are, respectively, applied to decompose kernels in convolutional neural layers and recurrent neural layers. As shown in Figure 7.1b, the second paradigm conceptualizes the neural layer as tensor decomposition, mapping hidden state and neural layer parameters to decomposed core tensor and factors. For example, in [39], the neural layers are factorized using CPD to accelerate the character-classification task, while in [40], the neural layers are formulated based on Tucker decomposition to achieve efficient gesture recognition.

Nevertheless, the aforementioned methods are developed for visual data analysis and are not suited for processing array signal statistics. In particular, the neural layer design must match the unique covariance tensor structure and address the specific issues of backpropagating DOA-associated loss function and tuning tensor ranks. To exploit the resource efficiency of tensorizing neural layers for DOA estimation, specific tensorized neural network architectures have been investigated with the deployments of either uniform rectangular array (URA) or sparse planar array (SPA) [41–43].

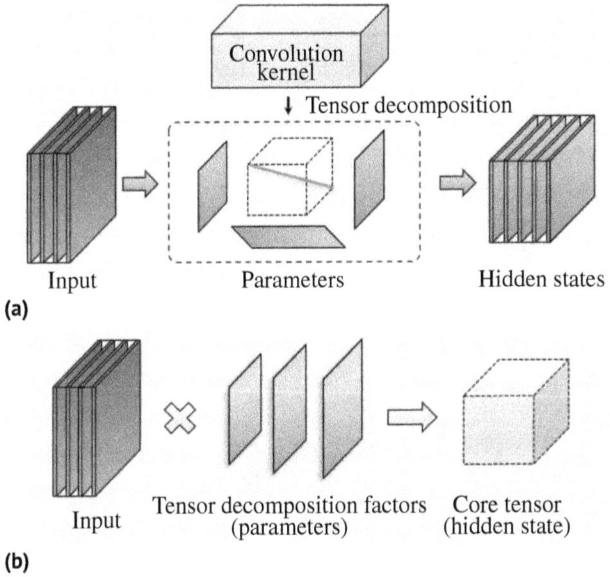

Figure 7.1 Paradigms of tensorizing neural layers. (a) decomposing neural layer parameters. (b) reformulating a neural layer as tensor decomposition.

In this chapter, we introduce two tensorized neural network architectures specifically designed for array signal processing applications. First, a tensorized DNN architecture for 2D DOA estimation with a URA is presented in Section 7.2. In this framework, the transformation from the covariance tensor to the hidden-state tensor is formulated as an inverse Tucker decomposition, enabling a compact representation of network parameters and significantly enhancing both training efficiency and computational scalability. Furthermore, a tensorized convolutional neural network (CNN) architecture for DOA estimation with an SPA is presented in Section 7.3. In the proposed tensorized CNN, convolution kernels are compressed via the CPD into a set of low-order factor tensors, substantially reducing model complexity. The coarray tensor statistics derived from the SPA signals are efficiently processed through sub-convolution operations. Finally, Section 7.4 summarizes the main contributions of this chapter and briefly discusses several promising directions for future research, including further model compression strategies and extensions to more complex array configurations and various array applications.

The notations listed in Table 7.1 are used throughout this chapter.

Table 7.1 List of notations.

Symbol	Description
a, $\boldsymbol{a}$, $\boldsymbol{A}$, $\mathcal{A}$	Scalar, vector, matrix, tensor
$\mathcal{A}_{(i_1,i_2,\ldots,i_I)}$	The $(i_1, i_2, \ldots, i_I)$th element in $\mathcal{A}$
$\jmath$	Imaginary unit
$\mathbb{C}$	Set of complex numbers
$\mathbb{R}$	Set of real numbers
$(\cdot)^{\mathrm{T}}$	Transpose
$(\cdot)^{\mathrm{H}}$	Hermitian transpose
$(\cdot)^{*}$	Conjugation
$(\cdot)^{\dagger}$	Pseudoinverse
$\times_i$	Mode-i tensor-matrix/vector product
$\langle \cdot \times_i \cdot \rangle$	Tensor contraction along the ith dimension
$[\,\cdot\,]_{\sqcup_i}$	Concatenation along the ith dimension
$\mathbb{E}\{\cdot\}$	Statistical expectation
$\circ$	Outer product
$\otimes$	Kronecker product
$\odot$	Khatri–Rao product
$\circledast$	Hadamard product
$f'(\cdot)$	Derivative of a function $f(\cdot)$
$\Re\{\cdot\}$, $\Im\{\cdot\}$	Real part and imaginary part of a complex number
$\mathrm{vec}(\cdot)$	Vectorization operator
$\|\cdot\|_1$	Taxicab norm
$\|\cdot\|_2$	Euclidean norm
$\lvert \cdot \rvert$	Absolute value of a scalar
$\boldsymbol{I}$, $\mathcal{I}$	Identity matrix and identity tensor
$\boldsymbol{0}$	Zero vector or matrix

7.2 Tensorized DNN for DOA Estimation with URA

In this section, we design a tensorized DNN for 2D DOA estimation with URA [42, 43]. The proposed network processes the signal covariance tensor as input and propagates it through multiple tensorized FC layers, generating tensorial signal features as hidden state tensors. The transformation from the covariance tensor to the hidden-state tensor is formulated as an inverse Tucker decomposition. Under this formulation, the Tucker core tensor, after nonlinear activation, serves as the hidden-state representation, while the inverse Tucker factor matrices encapsulate the compressed network parameters. To enable efficient optimization of these parameters, we develop tensorized feedforward and backpropagation procedures based on a compact tensorized FC layer design. Furthermore, the Tucker rank sequences are carefully fine-tuned via Bayesian optimization to minimize performance degradation caused by compression.

We demonstrate that the proposed tensorized DNN significantly reduces the number of parameters in the FC layers, thereby accelerating the training process and lowering overall system costs. Simulation results further confirm that the proposed tensorized neural network outperforms conventional matrix-based neural networks as well as model-based methods in terms of training speed and computational efficiency, while maintaining competitive estimation accuracy.

The roadmap of the proposed tensorized DNN is outlined in the following.

- We establish a connection between the tensor decomposition of the covariance tensor and compact FC layer design by mapping trainable parameters to tensor decomposition factors. This formulation enables efficient propagation of tensorial signal features and facilitates 2D DOA estimation within a tensor model–based DNN architecture.
- We develop tensorized feedforward and backpropagation procedures tailored to the DOA loss function to effectively train the compressed parameters. Furthermore, we fine-tune the tensor rank sequences to identify an optimal network topology for the tensorized DNN.
- We analyze the parameter compression ratio and computational complexity of the proposed tensorized DNN. The results demonstrate a substantial reduction in the number of trainable parameters, leading to improved network efficiency and reduced system cost.
- We demonstrate the adaptability and generalization capability of the trained tensorized DNN across different scenarios without retraining. In particular, we evaluate its performance under nonideal signal, propagation, and antenna array conditions to highlight its robustness.

7.2.1 Tensor Signal Model

To prevent spatial aliasing in 2D array signal processing, a URA with equally spaced sensors is typically adopted as the standard setup, as shown in Figure 7.2. In particular, an $M \times N$ URA with sensor positions set as:

$$\mathbb{S} = \{(x_{\mathbb{S}}, y_{\mathbb{S}}) | x_{\mathbb{S}} = [0, M-1]d, y_{\mathbb{S}} = [0, N-1]d\}, \tag{7.1}$$

is considered, where M and N, respectively, denote the number of sensors along the x-axis and the y-axis, and d is the inter-element spacing and is equal to half the signal wavelength. Let us also assume that there are K uncorrelated far-field narrowband signals impinging on the URA $\mathbb{S}$ from the spatial directions $\{(\theta_k, \phi_k), k = 1, 2, \dots, K\}$, where $\theta_k \in [0°, 180°]$ and $\phi_k \in [0°, 90°]$, respectively, represent the azimuth and elevation angles of the kth source.

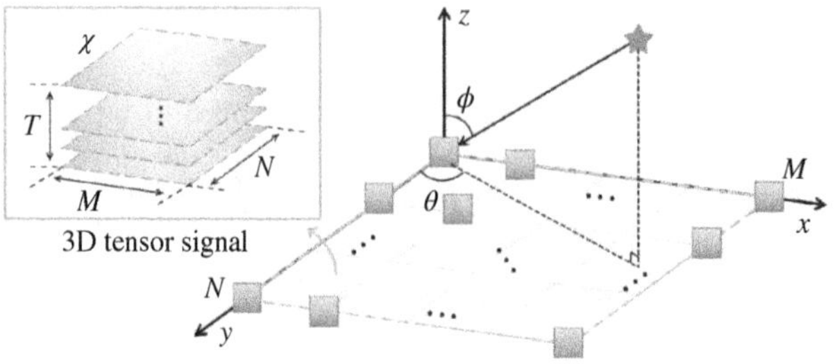

Figure 7.2 The deployed URA $\mathbb{S}$ and its received tensor signal.

To preserve the structure of signals received at the URA, these signals can be modeled as a three-dimensional (3D) tensor:

$$\begin{aligned}\mathcal{X} &\triangleq [\boldsymbol{X}(1), \boldsymbol{X}(2), \dots, \boldsymbol{X}(T)]_{\sqcup_3} \\ &= \sum_{k=1}^{K} \boldsymbol{a}(\mu_k) \circ \mathbf{b}(\nu_k) \circ \boldsymbol{s}_k + \mathcal{N} \in \mathbb{C}^{M\times N\times T},\end{aligned} \tag{7.2}$$

where

$$\mathbf{X}(t) = \sum_{k=1}^{K} [\boldsymbol{a}(\mu_k) \circ \mathbf{b}(\nu_k)] s_k(t) + \mathbf{N}(t) \in \mathbb{C}^{M\times N}, \tag{7.3}$$

is the signal matrix of the tth snapshot, $\forall t = 1, 2, \dots, T$. Here:

$$\boldsymbol{a}(\mu_k) = [1, e^{-\jmath\pi\mu_k}, \dots, e^{-\jmath\pi(M-1)\mu_k}]^{\mathrm{T}} \in \mathbb{C}^{M}, \tag{7.4}$$

and

$$\mathbf{b}(\nu_k) = [1, e^{-\jmath\pi\nu_k}, \dots, e^{-\jmath\pi(N-1)\nu_k}]^{\mathrm{T}} \in \mathbb{C}^{N}, \tag{7.5}$$

respectively, denote the steering vectors of $\mathbb{S}$ along the x-axis and the y-axis with:

$$\mu_k = \sin\phi_k \cos\theta_k, \tag{7.6}$$

and

$$\nu_k = \sin\phi_k \sin\theta_k, \tag{7.7}$$

$s_k(t) \in \mathbb{C}$ is the signal of the kth source at the tth snapshot, $\mathbf{N}(t) \sim \mathcal{CN}(\mathbf{0}, \sigma_n^2\mathcal{I})$ is an independent and identically distributed (i.i.d.) additive white Gaussian noise matrix with the noise power σ_n^2, $\boldsymbol{s}_k \triangleq [s_k(1), s_k(2), \dots, s_k(T)]^{\mathrm{T}} \in \mathbb{C}^{T}$ is the signal waveform of the kth source, and $\mathcal{N} \triangleq [\mathbf{N}(1), \mathbf{N}(2), \dots, \mathbf{N}(T)]_{\sqcup_3}$ is the additive white Gaussian noise tensor.

To obtain the second-order statistics of the tensor signal $\mathcal{X}$, a four-dimensional (4D) covariance tensor can then be derived as:

$$\begin{aligned}\mathcal{R} &\triangleq \mathbb{E}\{\boldsymbol{X}(t) \circ \boldsymbol{X}^*(t)\} \in \mathbb{C}^{M\times N\times M\times N} \\ &= \sum_{k=1}^{K} \sigma_k^2 \boldsymbol{a}(\mu_k) \circ \boldsymbol{b}(\nu_k) \circ \boldsymbol{a}^*(\mu_k) \circ \boldsymbol{b}^*(\nu_k) + \sigma_n^2 \mathcal{I},\end{aligned} \tag{7.8}$$

where $\sigma_k^2 \triangleq \mathbb{E}\{|s_k(t)|^2\}$ is the signal power of the kth source. The covariance tensor $\mathcal{R}$, embedded with the multidimensional signal characteristics, can then be fed into a neural network to retrieve the angular parameters.

Considering that a neural network can handle real-valued data with higher implementation efficiency, the real and imaginary parts of $\mathcal{R}$ are concatenated in an additional dimension to construct a five-dimensional (5D) real-valued tensor:

$$\mathcal{Y} = [\Re\{\mathcal{R}\}, \Im\{\mathcal{R}\}]_{\sqcup_5} \in \mathbb{R}^{M\times N\times M\times N\times 2}. \tag{7.9}$$

In practice, due to availability of finite snapshots, the ideal covariance tensor $\mathcal{R}$ is estimated by averaging the contraction between the 3D tensor signal $\mathcal{X}$ and its conjugation along the temporal dimension, i.e.,

$$\hat{\mathcal{R}} = \frac{1}{T}\langle \mathcal{X} \times_3 \mathcal{X}^* \rangle. \tag{7.10}$$

7.2.2 Principle of Tensorized FC Layer Design

In conventional neural networks based on matrix signal modeling and matrix products for signal feature propagation (i.e., matrix-based DNNs) [23–25], the URA signal is represented by a vector:

$$\boldsymbol{x}(t) = \sum_{k=1}^{K} [\boldsymbol{a}(\mu_k) \otimes \boldsymbol{b}(\nu_k)] s_k(t) + \boldsymbol{n}(t) \in \mathbb{C}^{MN}, \tag{7.11}$$

and the corresponding covariance matrix $\boldsymbol{R} \triangleq \mathbb{E}\{\boldsymbol{x}(t)\boldsymbol{x}^{\mathrm{H}}(t)\} \in \mathbb{C}^{MN\times MN}$ is first calculated (typically approximated by the sample covariance matrix) and further vectorized into a long vector input:

$$\boldsymbol{y} = \left[[\Re\{\mathrm{vec}(\boldsymbol{R})\}]^{\mathrm{T}}, [\Im\{\mathrm{vec}(\boldsymbol{R})\}]^{\mathrm{T}}\right]^{\mathrm{T}} \in \mathbb{R}^{2M^2N^2}. \tag{7.12}$$

Then, in a typical FC layer, the input $\boldsymbol{y}$ is multiplied by a parameter matrix $\boldsymbol{G}_0$ to yield a hidden state vector $\boldsymbol{h}_1 = f_1(\boldsymbol{G}_0\boldsymbol{y})$, where $f_1(\cdot)$ is a nonlinear activation function. The first hidden state vector $\boldsymbol{h}_1$ is propagated through multiple $(L-1)$ neural layers to sequentially generate $L-1$ hidden state vectors $\{\boldsymbol{h}_2, \boldsymbol{h}_3, \dots, \boldsymbol{h}_L\}$ using parameter matrices $\{\boldsymbol{G}_1, \boldsymbol{G}_2, \dots, \boldsymbol{G}_{L-1}\}$, i.e.,

$$\boldsymbol{h}_l = f_l(\boldsymbol{G}_{l-1}\boldsymbol{h}_{l-1}), \quad \forall l = 2, 3, \dots, L, \tag{7.13}$$

where L is the number of neural layers and $f_l(\cdot)$ is the nonlinear activation function in the lth layer. After that, the Lth hidden state vector $\boldsymbol{h}_L$ is multiplied by a parameter matrix $\boldsymbol{W}$ to estimate the DOAs.

In the matrix-based DNN, the numbers of parameters in the matrices $\{\boldsymbol{G}_0, \boldsymbol{G}_1, \dots, \boldsymbol{G}_{L-1}\}$ and $\boldsymbol{W}$ are proportional to the lengths of the vectors $\boldsymbol{y}$ and $\boldsymbol{h}_l$. The training of the network requires a high cost in terms of system memory and computation power, resulting in low training efficiency. To circumvent these drawbacks, tensorization in FC layers can be designed to efficiently extract tensorial hidden states from the covariance tensor input $\mathcal{Y}$ (7.9). The connection between tensor decomposition and the propagation of signal features across FC layers enables to dramatically reduce the number of trainable parameters into tensor decomposition factors, thereby facilitating an expedited training process.

Thus, to preserve the 5D tensorial structure of signal features, the covariance tensor $\mathcal{Y}$ is mapped to a 5D hidden state tensor $\mathcal{H}_1 \in \mathbb{R}^{I_{1,1}\times I_{1,2}\times I_{1,3}\times I_{1,4}\times I_{1,5}}$. Here, $\{I_{1,1}, I_{1,2}, I_{1,3}, I_{1,4}, I_{1,5}\}$ represents a sequence of Tucker ranks, which are, respectively, smaller than the sizes of the decomposed tensor, $\{M, N, M, N, 2\}$. In tensor decomposition models, such as Tucker decomposition and CPD, the tensor data is represented as a core tensor multiplied by multiple low-order factor matrices, enabling efficient extraction of essential multidimensional features. Leveraging this framework, we employ tensor decomposition to establish a computationally efficient transformation from $\mathcal{Y}$ to $\mathcal{H}_1$. In particular, the Tucker decomposition of $\mathcal{Y}$ can be represented as:

$$\mathcal{Y} = \mathcal{G} \times_1 \boldsymbol{V}_{0,1} \times_2 \boldsymbol{V}_{0,2} \times_3 \boldsymbol{V}_{0,3} \times_4 \boldsymbol{V}_{0,4} \times_5 \boldsymbol{V}_{0,5}, \tag{7.14}$$

where $\mathcal{G} \in \mathbb{R}^{I_{1,1}\times I_{1,2}\times I_{1,3}\times I_{1,4}\times I_{1,5}}$ is a 5D core tensor embedded with the extracted signal features, and $\boldsymbol{V}_{0,1} \in \mathbb{R}^{M\times I_{1,1}}$, $\boldsymbol{V}_{0,2} \in \mathbb{R}^{N\times I_{1,2}}$, $\boldsymbol{V}_{0,3} \in \mathbb{R}^{M\times I_{1,3}}$, $\boldsymbol{V}_{0,4} \in \mathbb{R}^{N\times I_{1,4}}$, $\boldsymbol{V}_{0,5} \in \mathbb{R}^{2\times I_{1,5}}$ are Tucker factor matrices serving as projection bases corresponding to the five dimensions.

Let the hidden state tensor $\mathcal{H}_1$ be the nonlinearly transformed core tensor, i.e.,

$$\mathcal{H}_1 \triangleq f_1(\mathcal{G}), \tag{7.15}$$

the transformation from $\mathcal{Y}$ to $\mathcal{H}_1$ can then be formulated as the inverse operation of (7.14), i.e., an inverse Tucker decomposition

$$\mathcal{H}_1 = f_1(\mathcal{Y} \times_1 \boldsymbol{V}^{\dagger}_{0,1} \times_2 \boldsymbol{V}^{\dagger}_{0,2} \times_3 \boldsymbol{V}^{\dagger}_{0,3} \times_4 \boldsymbol{V}^{\dagger}_{0,4} \times_5 \boldsymbol{V}^{\dagger}_{0,5}), \tag{7.16}$$

as depicted in Figure 7.3. Here, the inverse Tucker factor matrices $\boldsymbol{V}^{\dagger}_{0,1} \in \mathbb{R}^{M\times I_{1,1}}$, $\boldsymbol{V}^{\dagger}_{0,2} \in \mathbb{R}^{N\times I_{1,2}}$, $\boldsymbol{V}^{\dagger}_{0,3} \in \mathbb{R}^{M\times I_{1,3}}$, $\boldsymbol{V}^{\dagger}_{0,4} \in \mathbb{R}^{N\times I_{1,4}}$, and $\boldsymbol{V}^{\dagger}_{0,5} \in \mathbb{R}^{2\times I_{1,5}}$ contain the compressed parameters of the first FC layer.

In comparison with the matrix-based DNN which computes a matrix-vector product between the long input vector and a large-size parameter matrix to

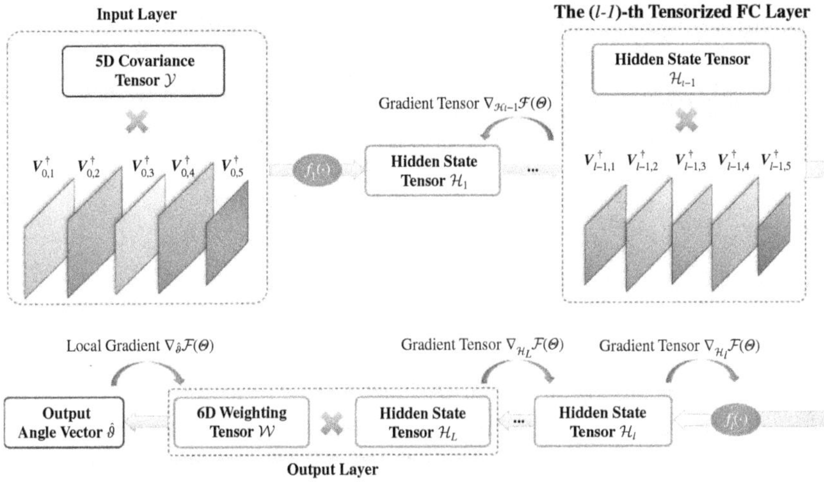

Figure 7.3 The architecture of a tensorized DNN for DOA estimation. Input: 5D covariance tensor; Trainable parameters: five inverse Tucker factors; Output: estimated DOAs.

yield the hidden state vector, the tensorized FC layer executes a multilinear tensor-matrix product between the covariance tensor $\mathcal{Y}$ and five inverse Tucker factor matrices $\{\boldsymbol{V}^{\dagger}_{0,h}, h = 1, 2, \dots, 5\}$ with significantly fewer multiplications. For example, for an 8×8 URA, the vectorized covariance input is $\boldsymbol{y} \in \mathbb{R}^{8,192}$, and the corresponding covariance tensor input is $\mathcal{Y} \in \mathbb{R}^{8\times8\times8\times8\times2}$. In the matrix-based FC layer, the multiplication of $\boldsymbol{y}$ with the parameter matrix $\boldsymbol{G}_0 \in \mathbb{R}^{8,192\times512}$ yields $\boldsymbol{h}_1 \in \mathbb{R}^{512}$. The number of parameters in $\boldsymbol{G}_0$ is 4, 194, 304, and the number of multiplications required for computing $\boldsymbol{h}_1$ is also 4, 194, 304. In contrast, the tensorized FC layer computes the tensor-matrix product between $\mathcal{Y}$ and five inverse Tucker factor matrices $\{\boldsymbol{V}^{\dagger}_{0,1}, \boldsymbol{V}^{\dagger}_{0,2}, \boldsymbol{V}^{\dagger}_{0,3}, \boldsymbol{V}^{\dagger}_{0,4} \in \mathbb{R}^{8\times4}, \boldsymbol{V}^{\dagger}_{0,5} \in \mathbb{R}^{2\times2}\}$ to yield the hidden state tensor $\mathcal{H}_1 \in \mathbb{R}^{4\times4\times4\times4\times2}$. The number of parameters in these matrices is $132 \ll 4,194,304$, and the number of required multiplications is $550{,}144 \ll 4,194,304$. This compact representation of parameters contributes to a faster training process and a computationally efficient propagation of signal features through the network.

7.2.3 Tensorized DNN Architecture for DOA Estimation

We describe the full architecture of the proposed tensorized DNN for DOA estimation in this section based on the principle of tensorized FC layer factorization for parameter compression. As illustrated in Figure 7.3, the tensorized DNN architecture includes a tensorized feedforward procedure to retrieve signal features, a backpropagation procedure to train the compressed parameters, and a rank selection approach to optimize the topology of the tensorized DNN. The resultant network proves to be resource-efficient and effective for 2D DOA estimation.

7.2.3.1 Tensorized Feedforward for DOA Estimation

To achieve 2D DOA estimation, the first hidden state tensor $\mathcal{H}_1$ is propagated through multiple tensorized FC layers to obtain the DOA estimates. Similar to the expression for $\mathcal{H}_1$ in (7.16), the lth hidden state tensor $\mathcal{H}_l \in \mathbb{R}^{I_{l,1}\times I_{l,2}\times I_{l,3}\times I_{l,4}\times I_{l,5}}$ can be obtained from the inverse Tucker decomposition of the $(l-1)$th hidden state tensor:

$$\mathcal{H}_l = f_l(\mathcal{H}_{l-1}\times_1 \boldsymbol{V}^{\dagger}_{l-1,1}\times_2 \boldsymbol{V}^{\dagger}_{l-1,2}\times_3 \boldsymbol{V}^{\dagger}_{l-1,3}\times_4 \boldsymbol{V}^{\dagger}_{l-1,4}\times_5 \boldsymbol{V}^{\dagger}_{l-1,5}), \forall l = 2, 3, \dots, L, \tag{7.17}$$

where the inverse Tucker factor matrices $\boldsymbol{V}^{\dagger}_{l-1,1} \in \mathbb{R}^{I_{l-1,1}\times I_{l,1}}$, $\boldsymbol{V}^{\dagger}_{l-1,2} \in \mathbb{R}^{I_{l-1,2}\times I_{l,2}}$, $\boldsymbol{V}^{\dagger}_{l-1,3} \in \mathbb{R}^{I_{l-1,3}\times I_{l,3}}$, $\boldsymbol{V}^{\dagger}_{l-1,4} \in \mathbb{R}^{I_{l-1,4}\times I_{l,4}}$, $\boldsymbol{V}^{\dagger}_{l-1,5} \in \mathbb{R}^{I_{l-1,5}\times I_{l,5}}$ contain the compressed parameters of the lth tensorized FC layer.

Note that, for the task of 2D DOA estimation, it is difficult to train a DNN designed as a classifier because the number of classification categories is extremely large. To be specific, the number of classification categories will be up to 180^2 for the required estimation accuracy of, for example, 1° at both azimuth and elevation. To cope with this issue, the tensorized DNN can be designed as a regression solver. To this end, the contraction between the Lth hidden state tensor $\mathcal{H}_L$ and a six-dimensional (6D) output layer weighting tensor $\mathcal{W} \in \mathbb{R}^{I_{L,1}\times I_{L,2}\times I_{L,3}\times I_{L,4}\times I_{L,5}\times 2K}$ is computed to output the estimate of the angle vector:

$$\boldsymbol{\vartheta} \triangleq [\theta_1, \theta_2, \dots, \theta_K, \phi_1, \phi_2, \dots, \phi_K]^{\mathrm{T}} \in \mathbb{R}^{2K}. \tag{7.18}$$

The HardTanh activation function is adopted in the output layer to constrain the estimated angle vector $\hat{\boldsymbol{\vartheta}} \triangleq [\hat{\theta}_1, \hat{\theta}_2, \dots, \hat{\theta}_K, \hat{\phi}_1, \hat{\phi}_2, \dots, \hat{\phi}_K]^{\mathrm{T}}$ as:

$$\hat{\boldsymbol{\vartheta}} = \mathrm{HardTanh}(\langle \mathcal{H}_L \times_{1,2,3,4,5} \mathcal{W}\rangle) = \begin{cases} 0, & \text{if } \langle \mathcal{H}_L \times_{1,2,3,4,5} \mathcal{W}\rangle < 0, \\ \pi, & \text{if } \langle \mathcal{H}_L \times_{1,2,3,4,5} \mathcal{W}\rangle \geq \pi, \\ \langle \mathcal{H}_L \times_{1,2,3,4,5} \mathcal{W}\rangle, & \text{otherwise}, \end{cases} \tag{7.19}$$

where $\hat{\theta}_k$ and $\hat{\phi}_k$, respectively, denote the estimated azimuth and elevation of the kth source.

The loss function of a regression network can be defined as either the mean absolute error (MAE) or the mean square error (MSE) between $\hat{\boldsymbol{\vartheta}}$ and $\boldsymbol{\vartheta}$. Nevertheless, the MSE function is sensitive to outliers in the training data, while the MAE function tends to have large derivatives as the loss decreases and becomes non-differentiable when the loss approaches 0. To overcome these drawbacks, a more robust and differentiable Huber loss function that combines the properties of the MAE and MSE is adopted. The Huber loss function with respect to (w.r.t.) the set of trainable FC layer parameters

$$\mathbf{\Theta} = \{\boldsymbol{V}^{\dagger}_{0,h}, \boldsymbol{V}^{\dagger}_{1,h}, \dots, \boldsymbol{V}^{\dagger}_{L-1,h}, \mathcal{W}, h = 1, 2, \dots, 5\} \tag{7.20}$$

is defined as:

$$\mathcal{F}(\mathbf{\Theta}) = \begin{cases} \frac{1}{4K}\|\boldsymbol{\vartheta} - \hat{\boldsymbol{\vartheta}}\|_2, & \text{if } \|\boldsymbol{\vartheta} - \hat{\boldsymbol{\vartheta}}\|_1/K \leq \gamma, \\ \frac{\gamma}{2K}(\|\boldsymbol{\vartheta} - \hat{\boldsymbol{\vartheta}}\|_1 - \frac{1}{2}\gamma), & \text{otherwise,} \end{cases} \tag{7.21}$$

where $\gamma > 0$ is a threshold value at which the computation is switched between the MAE and the MSE.

In practice, P input covariance tensors and their corresponding labels $\{(\mathcal{Y}_p, \boldsymbol{\vartheta}_p), p = 1, 2, \dots, P\}$ are generated to train the tensorized DNN. The gradients of their corresponding Huber losses $\{\mathcal{F}_p(\mathbf{\Theta}),\ p = 1, 2, \dots, P\}$ are computed to iteratively update the parameters during tensorized backpropagation. For the sake of simplicity, the subscript p will be omitted subsequently. Moreover, to demonstrate the backpropagation steps, the update of the parameter set $\mathbf{\Theta}$ for one input sample is referred to as one iteration. During the training process, all samples are fed into the tensorized DNN to iteratively update parameters, constituting one epoch of training. Then, multiple epochs are conducted to complete the network training.

7.2.3.2 Tensorized Backpropagation for Network Training

As described above, the Huber loss for each sample is backpropagated to update the inverse Tucker factor matrices and the output layer weighting tensor. This tensorized backpropagation involves computing the gradients of the Huber loss function $\mathcal{F}(\mathbf{\Theta})$ from the output layer backward through the hidden layers, and deriving the derivatives of $\mathcal{F}(\mathbf{\Theta})$ w.r.t. the trainable parameters.

Specifically, the local gradient of the Huber loss w.r.t. the estimated azimuth of the kth source can be calculated as:

$$\nabla_{\hat{\theta}_k}\mathcal{F}(\mathbf{\Theta}) \triangleq \frac{\partial \mathcal{F}(\mathbf{\Theta})}{\partial \hat{\theta}_k} = \frac{\partial \mathcal{F}(\mathbf{\Theta})}{\partial \hat{\boldsymbol{\vartheta}}}\frac{\partial \hat{\boldsymbol{\vartheta}}}{\partial \hat{\theta}_k} = \frac{\partial \mathcal{F}(\mathbf{\Theta})}{\partial \hat{\boldsymbol{\vartheta}}}\boldsymbol{e}_k, \tag{7.22}$$

where $\boldsymbol{e}_k \in \mathbb{R}^{2K}$ is an all-zero vector except the kth element being 1. Then, the derivative of $\mathcal{F}(\mathbf{\Theta})$ w.r.t. the output layer weighting tensor $\mathcal{W}$ corresponding to $\hat{\theta}_k$ can be calculated as:

$$\frac{\partial \mathcal{F}(\mathbf{\Theta})}{\partial \mathcal{W}_{(\cdot,\cdot,\cdot,\cdot,\cdot,k)}} = \frac{\partial \mathcal{F}(\mathbf{\Theta})}{\partial \hat{\theta}_k}\frac{\partial \hat{\theta}_k}{\partial \mathcal{W}_{(\cdot,\cdot,\cdot,\cdot,\cdot,k)}} = \nabla_{\hat{\theta}_k}\mathcal{F}(\mathbf{\Theta})\mathcal{H}_L. \tag{7.23}$$

Similarly, the local gradient of $\mathcal{F}(\mathbf{\Theta})$ w.r.t. the estimated elevation of the kth source can be calculated as:

$$\nabla_{\hat{\phi}_k}\mathcal{F}(\mathbf{\Theta}) \triangleq \frac{\partial \mathcal{F}(\mathbf{\Theta})}{\partial \hat{\phi}_k} = \frac{\partial \mathcal{F}(\mathbf{\Theta})}{\partial \hat{\boldsymbol{\vartheta}}}\frac{\partial \hat{\boldsymbol{\vartheta}}}{\partial \hat{\phi}_k} = \frac{\partial \mathcal{F}(\mathbf{\Theta})}{\partial \hat{\boldsymbol{\vartheta}}}\boldsymbol{e}_{k+K}, \tag{7.24}$$

and the derivative of $\mathcal{F}(\boldsymbol{\Theta})$ w.r.t. $\mathcal{W}$ corresponding to $\hat{\phi}_k$ can be calculated as:

$$\frac{\partial \mathcal{F}(\boldsymbol{\Theta})}{\partial \mathcal{W}_{(\cdot,\cdot,\cdot,\cdot,\cdot,K+k)}} = \nabla_{\hat{\phi}_k} \mathcal{F}(\boldsymbol{\Theta}) \mathcal{H}_L. \tag{7.25}$$

As such, at the $(p+1)$th iteration of network training (when updating parameters with the pth sample), the weighting tensor $\mathcal{W}$ can be updated as:

$$\mathcal{W}(p+1) = \mathcal{W}(p) - \eta \frac{\partial \mathcal{F}(\boldsymbol{\Theta})}{\partial \mathcal{W}}, \tag{7.26}$$

where $\eta > 0$ denotes the learning rate.

After updating the output layer weighting tensor $\mathcal{W}$, the local gradient of $\mathcal{F}(\boldsymbol{\Theta})$ w.r.t. the output angle vector $\hat{\boldsymbol{\vartheta}}$ defined as:

$$\begin{aligned} \nabla_{\hat{\boldsymbol{\vartheta}}} \mathcal{F}(\boldsymbol{\Theta}) \triangleq \big[& \nabla_{\hat{\theta}_1} \mathcal{F}(\boldsymbol{\Theta}), \nabla_{\hat{\theta}_2} \mathcal{F}(\boldsymbol{\Theta}), \dots, \nabla_{\hat{\theta}_K} \mathcal{F}(\boldsymbol{\Theta}), \\ & \nabla_{\hat{\phi}_1} \mathcal{F}(\boldsymbol{\Theta}), \nabla_{\hat{\phi}_2} \mathcal{F}(\boldsymbol{\Theta}), \dots, \nabla_{\hat{\phi}_K} \mathcal{F}(\boldsymbol{\Theta}) \big]^{\mathrm{T}} \in \mathbb{R}^{2K} \end{aligned} \tag{7.27}$$

is propagated back to the hidden layers. The local gradient of $\mathcal{F}(\boldsymbol{\Theta})$ w.r.t. the Lth hidden state tensor $\mathcal{H}_L$ is calculated as:

$$\nabla_{\mathcal{H}_L} \mathcal{F}(\boldsymbol{\Theta}) \triangleq \frac{\partial \mathcal{F}(\boldsymbol{\Theta})}{\partial \mathcal{H}_L} = \frac{\partial \mathcal{F}(\boldsymbol{\Theta})}{\partial \hat{\boldsymbol{\vartheta}}} \frac{\partial \hat{\boldsymbol{\vartheta}}}{\partial \mathcal{H}_L} = \mathcal{W} \times_6 \nabla_{\hat{\boldsymbol{\vartheta}}} \mathcal{F}(\boldsymbol{\Theta}). \tag{7.28}$$

The corresponding inverse Tucker factor matrices $\{\boldsymbol{V}^{\dagger}_{L-1,h}, h = 1, 2, \dots, 5\}$ at the $(p+1)$th iteration are then updated as:

$$\boldsymbol{V}^{\dagger}_{L-1,h}(p+1) = \boldsymbol{V}^{\dagger}_{L-1,h}(p) - \eta \frac{\partial \mathcal{F}(\boldsymbol{\Theta})}{\partial \boldsymbol{V}^{\dagger}_{L-1,h}}, \tag{7.29}$$

where the derivative of $\mathcal{F}(\boldsymbol{\Theta})$ w.r.t. $\boldsymbol{V}^{\dagger}_{L-1,h}$ is derived from $\nabla_{\mathcal{H}_L} \mathcal{F}(\boldsymbol{\Theta})$ as:

$$\begin{aligned} \frac{\partial \mathcal{F}(\boldsymbol{\Theta})}{\partial \boldsymbol{V}^{\dagger}_{L-1,h}} &= \frac{\partial \mathcal{F}(\boldsymbol{\Theta})}{\partial \mathcal{H}_L} \frac{\partial \mathcal{H}_L}{\partial \boldsymbol{V}^{\dagger}_{L-1,h}} \\ &= \langle \nabla_{\mathcal{H}_L} \mathcal{F}(\boldsymbol{\Theta}) \times_{1,2,3,4,5} f'_L(\mathcal{H}_{L-1} \times_1 \boldsymbol{V}^{\dagger}_{L-1,1} \times_2 \boldsymbol{V}^{\dagger}_{L-1,2} \cdots \times_5 \boldsymbol{V}^{\dagger}_{L-1,5}) \rangle \\ &\quad \cdot \mathcal{H}_{L-1} \times_1 \boldsymbol{V}^{\dagger}_{L-1,1} \cdots \times_{h-1} \boldsymbol{V}^{\dagger}_{L-1,h-1} \times_{h+1} \boldsymbol{V}^{\dagger}_{L-1,h+1} \cdots \times_5 \boldsymbol{V}^{\dagger}_{L-1,5}. \end{aligned} \tag{7.30}$$

The computation of the derivative mainly involves the multilinear tensor-matrix products as in the tensorized feedforward procedure, implying that the tensorized backpropagation is efficient as well.

After updating the inverse Tucker factor matrices $\boldsymbol{V}^{\dagger}_{L-1,h}$ for the Lth hidden layer, its corresponding local gradient tensor $\nabla_{\mathcal{H}_L} \mathcal{F}(\boldsymbol{\Theta})$ is propagated back to a local gradient tensor $\nabla_{\mathcal{H}_{L-1}} \mathcal{F}(\boldsymbol{\Theta}) \in \mathbb{R}^{I_{L-1,1} \times I_{L-1,2} \times I_{L-1,3} \times I_{L-1,4} \times I_{L-1,5}}$ in the

$(L-1)$th layer. More generally, the backpropagation from the local gradient tensor $\nabla_{\mathcal{H}_l}\mathcal{F}(\boldsymbol{\Theta})$ to $\nabla_{\mathcal{H}_{l-1}}\mathcal{F}(\boldsymbol{\Theta})$, $\forall l = L, L-1, \dots, 2$, can be found as:

$$\begin{aligned}\nabla_{\mathcal{H}_{l-1}}\mathcal{F}(\boldsymbol{\Theta}) &= f_l'(\mathcal{H}_{l-1} \times_1 \boldsymbol{V}^{\dagger}_{l-1,1} \times_2 \boldsymbol{V}^{\dagger}_{l-1,2} \cdots \times_5 \boldsymbol{V}^{\dagger}_{l-1,5}) \\ &\circledast \left[\nabla_{\mathcal{H}_l}\mathcal{F}(\boldsymbol{\Theta}) \times_1 (\boldsymbol{V}^{\dagger}_{l-1,1})^{\mathrm{T}} \times_2 (\boldsymbol{V}^{\dagger}_{l-1,2})^{\mathrm{T}} \cdots \times_5 (\boldsymbol{V}^{\dagger}_{l-1,5})^{\mathrm{T}}\right].\end{aligned} \tag{7.31}$$

Accordingly, the local gradient tensor $\nabla_{\mathcal{H}_{l-1}}\mathcal{F}(\boldsymbol{\Theta})$ is utilized to find the derivative of $\mathcal{F}(\boldsymbol{\Theta})$ w.r.t. $\boldsymbol{V}^{\dagger}_{l-2,h}$ as:

$$\begin{aligned}\frac{\partial \mathcal{F}(\boldsymbol{\Theta})}{\partial \boldsymbol{V}^{\dagger}_{l-2,h}} &= \langle(\mathcal{H}_{l-2} \times_1 \boldsymbol{V}^{\dagger}_{l-2,1} \cdots \times_{h-1} \boldsymbol{V}^{\dagger}_{l-2,h-1} \times_{h+1} \boldsymbol{V}^{\dagger}_{l-2,h+1} \cdots \times_5 \boldsymbol{V}^{\dagger}_{l-2,5}) \\ &\times_{1,\dots,h-1,h+1,\dots,5}\left[f_{l-1}'(\mathcal{H}_{l-2} \times_1 \boldsymbol{V}^{\dagger}_{l-2,1} \times_2 \boldsymbol{V}^{\dagger}_{l-2,2} \cdots \times_5 \boldsymbol{V}^{\dagger}_{l-2,5}) \right. \\ &\left. \circledast \nabla_{\mathcal{H}_{l-1}}\mathcal{F}(\boldsymbol{\Theta})\right]\rangle.\end{aligned} \tag{7.32}$$

This leads to the update rule for the inverse Tucker factor matrices $\{\boldsymbol{V}^{\dagger}_{l-2,h}, h = 1, 2, \dots, 5\}$ as:

$$\boldsymbol{V}^{\dagger}_{l-2,h}(p+1) = \boldsymbol{V}^{\dagger}_{l-2,h}(p) - \eta\frac{\partial \mathcal{F}(\boldsymbol{\Theta})}{\partial \boldsymbol{V}^{\dagger}_{l-2,h}}. \tag{7.33}$$

7.2.3.3 Selection of Tucker Rank Sequences

It is easy to see that the efficiency and performance of the tensorized DNN are simultaneously affected by the Tucker rank sequences $\chi_{\text{Tucker}} \triangleq \{I_{1,h}, I_{2,h}, \dots, I_{L,h}, h = 1, 2, \dots, 5\}$. In particular, a more compressed tensorized DNN has smaller Tucker ranks, albeit at the expense of compromised estimation performance to a greater extent. To ensure minimum performance degradation, automatic selection of the Tucker rank sequences is achieved via Bayesian optimization, which is a powerful tool for finding the optimum of objective functions that are expensive to evaluate [44].

A metric that reflects DOA estimation performance using the tensorized DNN must be defined to serve as the objective function w.r.t. the Tucker rank sequences to be optimized. It is denoted as $\Xi(\chi_{\text{Tucker}})$. Specifically, $\Xi(\chi_{\text{Tucker}})$ can be characterized by the averaged validation MSEs of DOA estimation based on the outcomes of cross-validation. In the process of cross-validation, the available P training samples are equally divided into several folds, say five folds. The first four folds are used as the training set for an initialized tensorized DNN with multiple epochs, whereas the last fold is used as the validation set to assess the DOA estimation performance of the trained network. The MSE between the output angle vector and its ground truth:

$$\epsilon_{\text{MSE}} = \| \boldsymbol{\vartheta} - \hat{\boldsymbol{\vartheta}} \|_2 / 2K, \tag{7.34}$$

is calculated as the validation metric. Then, by selecting the second to the fifth folds as the training set and the remaining one as the validation set, a newly initialized tensorized DNN is trained and validated to yield another MAE. After every fold has been used as the validation set, the corresponding MSEs $\epsilon_{\mathrm{MSE}_w}$, $w = 1, 2, \dots, 5$, are averaged as:

$$\Xi(\mathcal{X}_{\mathrm{Tucker}}) = \frac{1}{5} \sum_{w=1}^{5} \epsilon_{\mathrm{MSE}_w}, \tag{7.35}$$

to measure the overall performance of the network.

To obtain the best network performance, $\Xi(\mathcal{X}_{\mathrm{Tucker}})$ is then minimized to find the optimal $\mathcal{X}_{\mathrm{Tucker}}$, i.e.,

$$\mathcal{X}_{\mathrm{Tucker}}^{\mathrm{opt}} = \underset{\mathcal{X}_{\mathrm{Tucker}}}{\operatorname{argmin}}\ \Xi(\mathcal{X}_{\mathrm{Tucker}}). \tag{7.36}$$

Based on the principle of Bayesian optimization, (7.36) can be solved by presuming the prior distribution of $\Xi(\mathcal{X}_{\mathrm{Tucker}})$ and updating the posterior distribution of a new optimization result. Due to flexibility and tractability of typical Gaussian process, the prior knowledge about the distribution over $\mathcal{X}_{\mathrm{Tucker}}$ is characterized by:

$$\Xi(\mathcal{X}_{\mathrm{Tucker}}) \sim \mathcal{GP}(\varpi(\mathcal{X}_{\mathrm{Tucker}}), \Sigma(\mathcal{X}_{\mathrm{Tucker}}, \mathcal{X}_{\mathrm{Tucker}}^{\mathrm{H}})), \tag{7.37}$$

where $\varpi(\mathcal{X}_{\mathrm{Tucker}})$ and $\Sigma(\mathcal{X}_{\mathrm{Tucker}}, \mathcal{X}_{\mathrm{Tucker}}^{\mathrm{H}})$, respectively, denote the mean and covariance functions of $\mathcal{X}_{\mathrm{Tucker}}$. Based on the prior distribution of $\Xi(\mathcal{X}_{\mathrm{Tucker}})$ defined in (7.37), the predictive distribution (i.e., the posterior) over the newly sampled Tucker rank sequences $\bar{\mathcal{X}}_{\mathrm{Tucker}}$ is given by

$$\mathcal{P}(\Xi(\bar{\mathcal{X}}_{\mathrm{Tucker}}) | \mathcal{X}_{\mathrm{Tucker}}, \bar{\mathcal{X}}_{\mathrm{Tucker}}, \Xi(\mathcal{X}_{\mathrm{Tucker}})) = \mathcal{N}(\bar{\varpi}, \bar{\Sigma}), \tag{7.38}$$

where $\bar{\varpi}$ and $\bar{\Sigma}$ are, respectively, the predicted mean and variance of the predictive posterior distribution w.r.t. $\bar{\mathcal{X}}_{\mathrm{Tucker}}$.

Based on the posterior distribution, an acquisition function can be defined and maximized to determine the next search point of $\Xi(\mathcal{X}_{\mathrm{Tucker}})$ until finally obtaining the optimal $\mathcal{X}_{\mathrm{Tucker}}^{\mathrm{opt}}$. Note that some well-studied acquisition functions can be practically used [45].

7.2.4 Performance Analysis

7.2.4.1 Number of Trainable Parameters

As demonstrated in Section 7.2.2, the trainable parameters in a matrix-based DNN are the matrices $\left\{\boldsymbol{G}_{l-1} \in \mathbb{R}^{\prod_{h=1}^{5} I_{l,h} \times \prod_{h=1}^{5} I_{l-1,h}}\right\}_{l=2}^{L}$ and the output layer matrix $\boldsymbol{W} \in \mathbb{R}^{2K \times \prod_{h=1}^{5} I_{L,h}}$. The total number of parameters can be calculated as:

$$P_{\mathrm{MNN}} = 2M^2N^2 \prod_{h=1}^{5} I_{1,h} + \sum_{l=2}^{L} \prod_{h=1}^{5} I_{l-1,h} I_{l,h} + 2K \prod_{h=1}^{5} I_{L,h}. \tag{7.39}$$

In comparison, the parameters in a tensorized FC layers are represented in a compact format of the inverse Tucker factors. For the proposed tensorized DNN, the total number of parameters in the inverse Tucker factor matrices $\{\boldsymbol{V}_{0,h}^{\dagger}, \boldsymbol{V}_{1,h}^{\dagger}, \dots, \boldsymbol{V}_{L-1,h}^{\dagger}, h = 1, 2, \dots, 5\}$ and the output layer weighting tensor $\mathcal{W}$ can be calculated as:

$$P_{\text{TNN}} = \left(MI_{1,1} + NI_{1,2} + MI_{1,3} + NI_{1,4} + 2I_{1,5}\right) + \sum_{l=2}^{L}\sum_{h=1}^{5} I_{l-1,h}I_{l,h} + 2K\prod_{h=1}^{5} I_{L,h}. \tag{7.40}$$

Note that the product terms $2M^2N^2\prod_{h=1}^{5} I_{1,h}$ and $\sum_{l=2}^{L}\prod_{h=1}^{5} I_{l-1,h}I_{l,h}$ in (7.39) are, respectively, replaced by the summation terms $MI_{1,1} + NI_{1,2} + MI_{1,3} + NI_{1,4} + 2I_{1,5}$ and $\sum_{l=2}^{L}\sum_{h=1}^{5} I_{l-1,h}I_{l,h}$ in (7.40), which leads to $P_{\text{TNN}} \ll P_{\text{MNN}}$. Hence, the tensorized DNN enjoys an enhanced efficiency with reduced system cost compared to the matrix-based DNN.

7.2.4.2 Computational Complexity

After training the tensorized DNN, the DOAs of sources can be estimated by only executing the feedforward computation. Hence, the implementation of the tensorized DNN for DOA estimation includes the covariance tensor calculation and tensorized feedforward computation, and the computational complexity can be measured by:

$$C_{\text{TNN}} = \mathcal{O}\left((T + I_{1,1})M^2N^2\right). \tag{7.41}$$

Likewise, the implementation of the matrix-based DNN for DOA estimation [24] includes the covariance matrix calculation and the feedforward computation, and its computational complexity can be measured by:

$$C_{\text{MNN}} = \mathcal{O}\left(\left(T + \prod_{h=1}^{5} I_{1,h}\right)M^2N^2\right). \tag{7.42}$$

Meanwhile, the model-based MUSIC [9], ESPRIT [10] and CPD method [13] for 2D DOA estimation, respectively, have the computational complexities of $\mathcal{O}(M^3N^3 + MN(MN-K)I_{\text{SS}})$, $\mathcal{O}(M^3N^3K)$ and $\mathcal{O}(TM^2N^2 + KM^2N^2I_{\text{CPD}})$, where I_{SS} and I_{CPD} denote the number of grid points for spectral search and the number of iterations for solving the CPD optimization problem, respectively.

In comparison to the model-based methods, the tensorized DNN enjoys much lower computational complexity by replacing the time-consuming signal subspace processing and CPD optimization with the efficient feedforward computations. Moreover, the tensorized DNN is computationally more efficient than the matrix-based DNN, because the multilinear tensor-matrix product for signal feature propagation requires less multiplications than the product between vectorized signal statistics and parameter matrices.

7.2.5 Simulation Results

In the simulation example, a URA $\mathbb{S}$ with $M = 10$ and $N = 10$ is deployed. The tensorized DNN and the matrix-based DNN for DOA estimation are implemented with the PyTorch framework, and the Tucker rank selection of the tensorized DNN is executed with the Bayesian optimization tool [46]. The threshold value in the Huber loss function $\mathcal{F}(\boldsymbol{\Theta})$ (7.21) is set to γ=1. Unless otherwise specified, $K = 2$ sources are considered, the number of snapshots is fixed at $T = 200$, and the SNR is set to -5 dB. For each scenario, $I_{\text{MC}} = 3{,}000$ Monte Carlo trials are run.

For training the deployed DNNs, $P = 19,200$ covariance tensors are generated as training samples, which are divided into batches of equal size B for batch-wise processing. Each training sample is generated with source azimuth and elevation angles randomly drawn from the range $[0°, 90°]$, while the SNRs of source signals are randomly selected from the discrete set $\{-15$ dB, -10 dB, -5 dB, 0 dB$\}$. The number of snapshots is held constant at $T = 200$ for all samples. The parameter set $\{\boldsymbol{V}^{\dagger}_{0,h}, \boldsymbol{V}^{\dagger}_{1,h}, \boldsymbol{V}^{\dagger}_{2,h}, \mathcal{W}, h = 1, 2, \dots, 5\}$ is initialized according to the method presented in [47].

Following the Tucker rank selection methodology, Bayesian optimization is also employed to jointly tune additional hyperparameters. The search spaces of the learning rate η and the batch size B are, respectively, defined as $[10^{-5}, 10^{-1}]$ and $\{16, 32, 64, 128\}$. As detailed in section 7.2.3.3, the metric function w.r.t. (η, B), i.e., $\Xi(\eta, B)$, is minimized to simultaneously optimize the hyperparameters (η, B) and the Tucker rank sequences. This optimization yields the tuned hyperparameters $\eta = 8.317 \cdot 10^{-3}$ and $B = 32$ for the tensorized DNN, and $\eta = 0.273 \cdot 10^{-3}$ and $B = 32$ for the matrix-based DNN.

7.2.5.1 Comparison of Training Efficiency

A tensorized DNN with $L = 3$ hidden layers and predefined Tucker ranks is constructed, where each layer maintains Tucker-based tensor representations. Specifically, the network processes an input covariance tensor $\mathcal{Y} \in \mathbb{R}^{10\times10\times10\times10\times2}$ through the hidden state transformations as $\mathcal{H}_1 \in \mathbb{R}^{7\times7\times7\times7\times2}$, $\mathcal{H}_2 \in \mathbb{R}^{5\times5\times5\times5\times1}$ and $\mathcal{H}_3 \in \mathbb{R}^{3\times3\times3\times3\times1}$. The output layer employs a weighting tensor $\mathcal{W} \in \mathbb{R}^{3\times3\times3\times3\times1\times4}$ to produce the source DOA estimates. The nonlinear activation functions for the three hidden layers are chosen as tanh($\cdot$), tanh($\cdot$), and Leaky ReLU($\cdot$), respectively. For fair comparison, the counterpart matrix-based DNN also has three hidden layers. The size of the input vector $\boldsymbol{y}$ is $2 \cdot 10^4 = 20{,}000$, and the sizes of vectorized hidden states are $2 \cdot 7^4 = 4,802$, $5^4 = 625$, and $3^4 = 81$, respectively. Furthermore, the Tucker rank sequences of the tensorized FC layers are fine-tuned through Bayesian optimization as $(11, 15, 11, 15, 2)$, $(3, 9, 3, 9, 2)$, and $(3, 5, 3, 5, 1)$. The topology of the three DNN architectures can be compactly represented by the dimensions of hidden states in every layer as follows.

- The tensorized DNN with predefined ranks: (10, 10, 10, 10, 2)–(7, 7, 7, 7, 2)–(5, 5, 5, 5, 1)–(3, 3, 3, 3, 1)–4,
- The tensorized DNN with optimized ranks: (10, 10, 10, 10, 2)–(11, 15, 11, 15, 2)–(3, 9, 3, 9, 2)–(3, 5, 3, 5, 1)–4,
- The matrix-based DNN: 20,000–4,802–625–81–4.

To evaluate the training efficiency of the proposed network architecture, the training time and GPU memory consumption are compared between the tensorized DNN and the matrix-based DNN in Table 7.2. As can be observed by comparing (7.40) with (7.39), the tensorized DNN presents a remarkable parameter reduction from 99,092,199 parameters in the matrix-based DNN to just 811 parameters. The tensorized DNN achieves a compression ratio of trainable parameters by more than 122,000 times. This dramatic parameter reduction translates to significant improvements in both training speed and memory efficiency.

Specifically, the training time for the matrix-based DNN is 541.04 minutes, whereas the proposed tensorized DNN requires only 24.65 minutes. The GPU memory usage is 1,009 MiB for the tensorized DNN, reduced from 3,313 MiB for the matrix-based DNN. These efficiency improvements persist even in more challenging scenarios with 5 sources, where the tensorized DNN maintains low resource demands (34.92 minutes of training time and 1,009 MiB GPU memory usage). While the tensorized DNN with optimized Tucker rank sequences shows moderately higher resource requirements than the predefined-rank version, both configurations maintain substantial advantages over the matrix-based implementation. Overall, the proposed tensorized DNN architecture allows us to employ a more resource-efficient neural network or potentially much larger DNN for DOA estimation.

Table 7.2 Comparison of training time and memory usage.

Method	Topology	Number of parameters	Training time (minute)	GPU memory usage (MiB)
Matrix-based DNN (2 sources)	20,000–4,802–625–81–4	99,092,199	541.04	3, 313
Tensorized DNN with predefined ranks (2 sources)	(10, 10, 10, 10, 2)–(7, 7, 7, 7, 2)–(5, 5, 5, 5, 1)–(3, 3, 3, 3, 1)–4	811	24.65	1,009
Tensorized DNN with predefined ranks (5 sources)	(10, 10, 10, 10, 2)–(7, 7, 7, 7, 2)–(5, 5, 5, 5, 1)–(3, 3, 3, 3, 1)–10	1,299	34.92	1,009
Tensorized DNN with optimized ranks (2 sources)	(10, 10, 10, 10, 2)–(11, 15, 11, 15, 2)–(3, 9, 3, 9, 2)–(3, 5, 3, 5, 1)–4	1,874	107.45	1,103

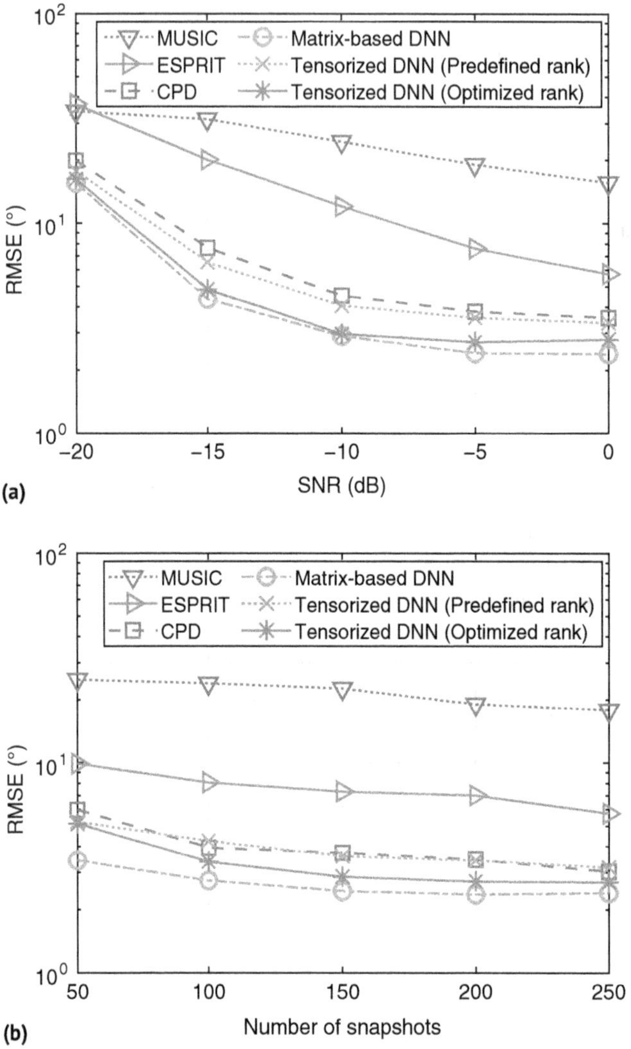

Figure 7.4 Comparison of estimation accuracy. (a) RMSE versus SNR. (b) RMSE versus number of snapshots.

7.2.5.2 Comparison of DOA Estimation Accuracy

In terms of DOA estimation accuracy, the proposed tensorized DNN is compared to model-based methods (i.e., MUSIC [9], ESPRIT [10], CPD method [13]) and the matrix-based DNN [24]. In Figure 7.4, the root-mean-square error (RMSE)

$$\text{RMSE} = \sqrt{\frac{1}{2KI_{\text{MC}}} \sum_{i_{\text{MC}}=1}^{I_{\text{MC}}} \| \hat{\boldsymbol{\vartheta}}_{i_{\text{MC}}} - \boldsymbol{\vartheta}_{i_{\text{MC}}} \|^2} \tag{7.43}$$

is shown for the evaluated methods versus SNR (Figure 7.4a) and number of snapshots (Figure 7.4b), where $\hat{\boldsymbol{\vartheta}}_{i_{\text{MC}}}$ and $\boldsymbol{\vartheta}_{i_{\text{MC}}}$ are, respectively, the estimated and true DOAs in the i_{MC}th trial. Both azimuth and elevation angles of sources are randomly generated within the interval $[0°, 90°]$ for each trial.

The MUSIC and ESPRIT methods perform subspace decomposition of the covariance matrix $\boldsymbol{R} \triangleq \mathbb{E}\{\boldsymbol{x}(t)\boldsymbol{x}^{\text{H}}(t)\}$, while the CPD method decomposes the covariance tensor $\mathcal{R} \triangleq \mathbb{E}\{\boldsymbol{X}(t) \circ \boldsymbol{X}^*(t)\}$ to retrieve the angular parameters μ_k and ν_k. The performance of these three methods depends on the reliability of signal statistical models. In contrast, DNNs do not require predefined statistical models tailored to specific noise conditions. Once extensively trained, DNNs exhibit superior adaptability and flexibility. As such, the deployed DNNs improve the estimation accuracy compared to the model-based DOA estimation methods.

The tensorized DNN with predefined Tucker ranks achieves comparable accuracy to the CPD method, while the one with optimized Tucker ranks demonstrates superior performance. Although the tensorized DNN shows slightly lower accuracy than the matrix-based DNN, this minor degradation is offset by a dramatic improvement in parameter efficiency. The tensorized DNN requires only 1,874 parameters compared to the matrix-based DNN's 99,092,199 parameters. This reduction exceeds four orders of magnitude. These results demonstrate that the proposed tensorized DNN architecture successfully prioritizes training efficiency while minimizing the inevitable accuracy trade-offs through tensor rank optimization.

7.2.5.3 Comparison of Angular Resolution

In Figure 7.5, the angular resolutions of all tested methods are compared. In this scenario, two sources impinge from spatially close directions (θ_1, ϕ_1) and (θ_2, ϕ_2). Both θ_1 and ϕ_1 are randomly generated within the interval $[0°, 90°]$, and the pair (θ_2, ϕ_2) is found to maintain a small angular spacing of $\Delta = \sqrt{|\theta_2 - \theta_1|^2 + |\phi_2 - \phi_1|^2}$ relative to (θ_1, ϕ_1). For a trial, a tested method is called to successfully estimate the DOAs of sources if $\sqrt{|\hat{\theta}_k - \theta_k|^2 + |\hat{\phi}_k - \phi_k|^2} < \Delta$ is satisfied. The probability of successful resolution (PSR) is calculated as the percentage of successful trials.

The results demonstrate that the tensorized DNN achieves superior angular resolution compared to conventional model-based methods. Notably, when optimized with Tucker rank reduction, the tensorized DNN outperforms the matrix-based DNN in resolution capability. This validates that the proposed architecture delivers enhanced angular resolution while simultaneously reducing computational resources.

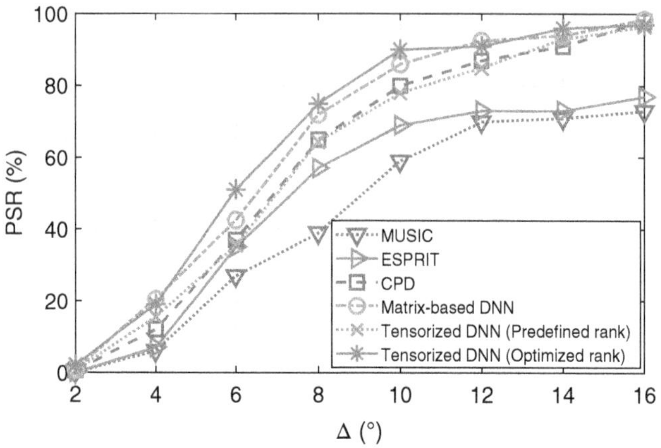

Figure 7.5 Comparison of angular resolution.

7.2.5.4 Comparison of Computational Time

To verify the computational efficiency of the proposed tensorized DNN, the computational time of all methods tested versus the number of sources and the number of snapshots is compared in Figure 7.6a, b, respectively. It is evident that the tensorized DNN is computationally faster than the model-based methods, since it replaces the complex subspace extraction, spectral search, and optimizations with efficient feedforward computation. Meanwhile, the tensorized DNN also owns increased computation speed compared to the matrix-based DNN (e.g., 0.0081 seconds for the matrix-based DNN and 0.0073 seconds for the tensorized DNN with $K = 1$, SNR $= -5$ dB, $T = 100$) thanks to the efficient computation of multilinear tensor-matrix product for signal feature propagation.

7.2.5.5 Comparison of Networks with Different Number of Layers

The effectiveness of the tensorized DNN is further assessed by increasing the number of FC layers. Following the definition of network topology in section 7.2.5.1, another three 5-layer DNNs with the topologies

- The tensorized DNN with predefined ranks: (10, 10, 10, 10, 2)–(8, 8, 8, 8, 2)–(6, 6, 6, 6, 2)–(5, 5, 5, 5, 1)–(4, 4, 4, 4, 1)–(3, 3, 3, 3, 1)–4
- The tensorized DNN with optimized ranks: (10, 10, 10, 10, 2)–(14, 10, 14, 10, 2)–(6, 11, 6, 11, 2)–(8, 2, 8, 2, 1)–(8, 8, 8, 8, 1)–(5, 3, 5, 3, 1)–4
- The matrix-based DNN: 20,000–8,192–2,592–625–256–81–4

are compared. The number of parameters in the 5-layer matrix-based DNN is 186,874,724, and it is reduced to 1,090 and 2,208, respectively, for the tensorized DNNs with predefined ranks and optimized ranks.

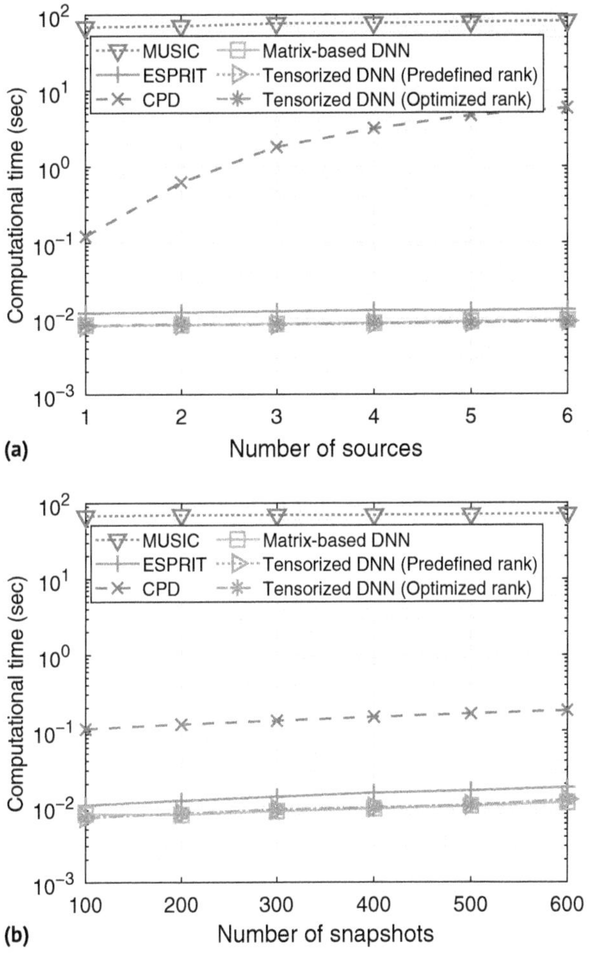

Figure 7.6 Comparison of computational complexity. (a) computational time versus number of sources. (b) computational time versus number of snapshots.

As shown in Figure 7.7, when the number of neural layers increases, the DOA estimation accuracy of the neural networks is enhanced. The tensorized DNN continues to exhibit performance comparable to that of the matrix-based DNN that demands substantial amount of resources. The 5-layer matrix-based DNN and the 5-layer tensorized DNN with optimized ranks are employed in the subsequent simulations.

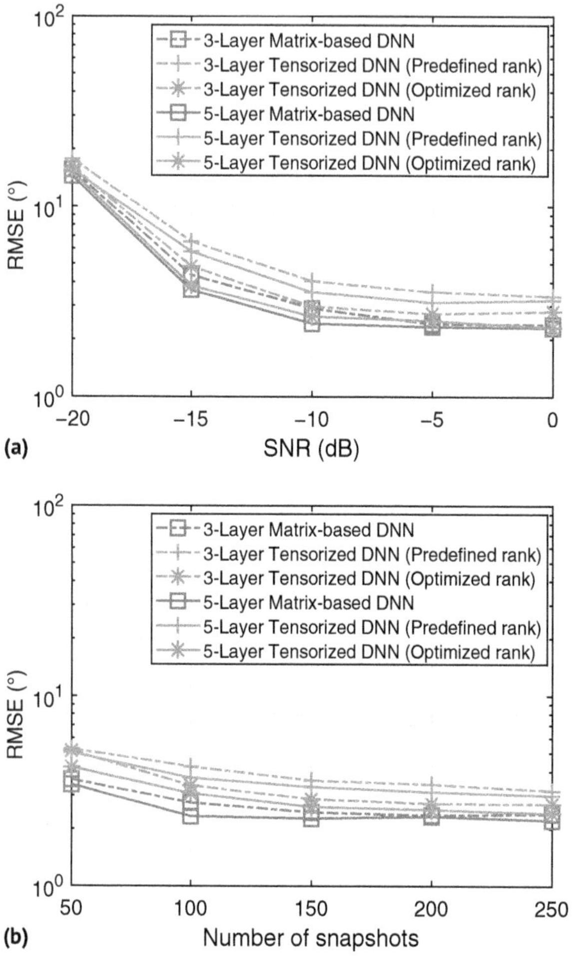

Figure 7.7 Comparison of neural networks with different number of layers. (a) RMSE versus SNR. (b) RMSE versus number of snapshots.

7.2.5.6 Adaptation to Different Number of Sources

For a regression network, the size of the output is typically fixed. To accommodate varying numbers of sources, we extend the output dimension to $2K_{\text{max}}$ and consistently select the top $2K$ outputs as the estimated DOAs. By reconstructing the training dataset to include samples with different numbers of sources (effectively implementing multitask learning), the trained neural networks gain the ability to adapt to varying number of sources.

In this simulation, K_{max} is set to 6, resulting in a fixed output size of 12. During training, the number of sources is randomly selected from $\{1, 2, 3\}$. Figure 7.8

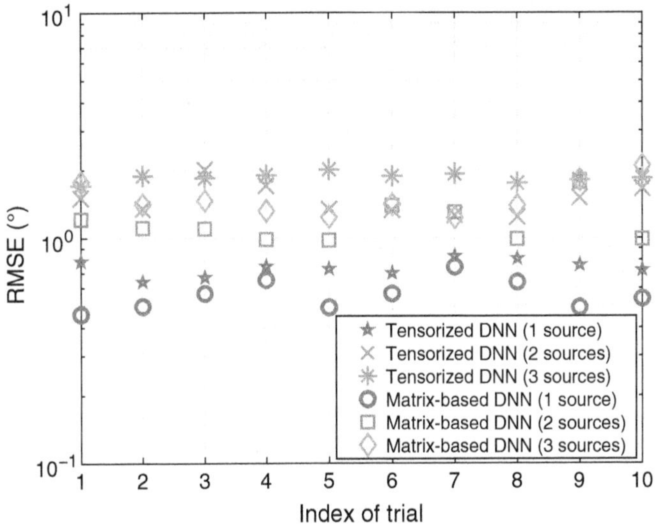

Figure 7.8 Adaptation to different number of sources.

evaluates the performance of the trained neural networks for $K = 1$ to $K = 3$, presenting results from 10 trials, each comprising 32 test samples. The SNR and number of snapshots are fixed at 0 dB and $T = 200$, respectively. The results demonstrate that the tensorized DNN successfully estimates varying numbers of sources while maintaining accuracy comparable to the matrix-based DNN. Furthermore, by expanding the training dataset and conducting extensive offline training, the networks can achieve even greater adaptability to larger numbers of sources.

7.2.5.7 Adaptation to Non-Gaussian Noise and Colored Signal

To validate the robustness of the tensorized DNN, we conduct a case study that simulates scenarios involving non-Gaussian noise and colored signals. The neural networks are trained using samples generated under ideal conditions and are subsequently evaluated under nonideal conditions to assess their generalization capability.

In certain applications, such as acoustics and communications, additive noise may deviate from the Gaussian assumption and instead exhibit impulsive characteristics, commonly referred to as non-Gaussian impulsive noise (IN). In this context, we assume that the noise vector $\boldsymbol{n}(t)$ or noise matrix $\boldsymbol{N}(t)$ follows an Alpha-stable distribution [48]. The parameters of the Alpha-stable distribution, namely, the characteristic exponent, index of symmetry, location parameter, and dispersion parameter, are, respectively, set to 1, 0.5, 0, and 1. Moreover, when the signal spectrum is biased toward one side, the resulting waveform is characterized as colored rather than white. To model this effect, we generate pink signals (PS) and evaluate the performance of the compared methods. As shown in Figure 7.9, under both non-Gaussian noise and colored-signal scenarios, the proposed tensorized DNN and the matrix-based DNN achieve comparable performance and consistently outperform the model-based ESPRIT and CPD methods.

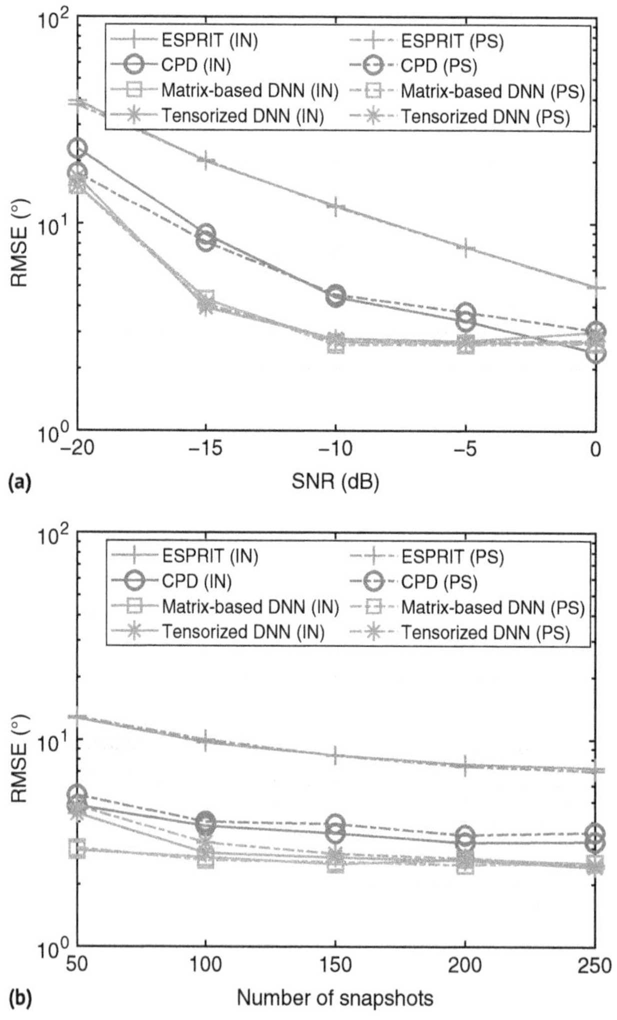

Figure 7.9 Comparison of performance with non-Gaussian noise and colored signal. (a) RMSE versus SNR. (b) RMSE versus number of snapshots.

7.2.5.8 Adaptation to Array imperfections

Furthermore, we evaluate the DOA estimation performance under array imperfections, including sensor position deviation (SPD) and sensor failure (SF). In the SPD scenario, the position of each sensor on the x-axis, denoted by $x_{\mathbb{S}}$, undergoes a random deviation within the range $\iota \in [0, 0.4d]$ for each trial. In the SF scenario, five sensors are randomly selected and assumed to be damaged in the deployed URA $\mathbb{S}$ for each trial. The results for both scenarios are presented in Figure 7.10.

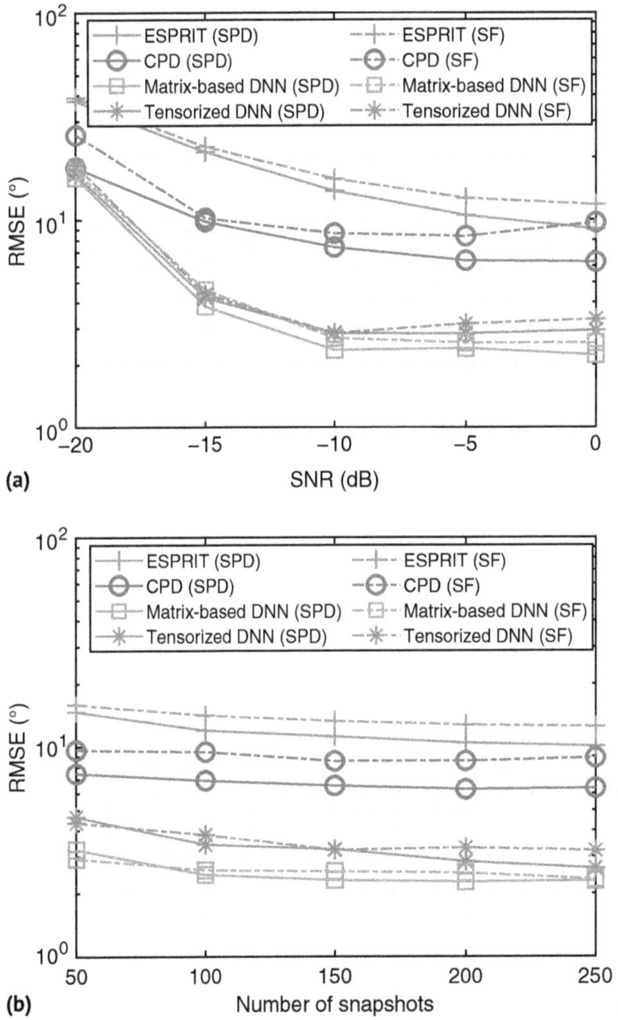

Figure 7.10 Comparison of performance with array imperfections. (a) RMSE versus SNR. (b) RMSE versus number of snapshots.

Since the ESPRIT and CPD methods rely on accurate signal covariance models, array imperfections introduce significant phase and gain mismatches, leading to noticeable performance degradation. In contrast, the trained neural networks exhibit greater adaptability and robustness to nonideal signal statistics induced by array imperfections. This robustness highlights a key advantage of employing neural networks for 2D DOA estimation.

7.3 Tensorized CNN for DOA Estimation with SPA

Compared to the traditional URA configuration, SPAs have garnered much attention due to their ability to achieve sub-Nyquist spatial sampling with nonuniformly spaced sensors while providing a larger aperture [4, 5, 20, 49–51]. To mitigate spatial aliasing in SPA DOA estimation, augmented coarray signal statistics are derived and utilized for Nyquist-matched coarray processing. Moreover, by integrating sub-Nyquist sampling with tensor modeling, multidimensional sub-Nyquist sampled signal structures have been preserved and exploited for enhanced 2D DOA estimation [52–56]. Based on the sub-Nyquist tensor signal statistics, an augmented coarray tensor is derived and exploited to achieve angle estimation for bistatic MIMO radar, demonstrating improved estimation accuracy, resolution, and source identifiability [56].

To improve the robustness of SPA DOA estimation, neural networks, including DNN and CNN, have been employed [57–59]. In particular, CNN can learn multi-way features from multidimensional data [60], making it applicable to high-accuracy angle classification [57, 61, 62]. To achieve gridless DOA estimation, a CNN-based Toeplitz covariance matrix reconstruction approach is reported in [26]. However, as demonstrated in Section 7.2.2, conventional matrix-based neural network architectures suffer from computational overload during training, limiting the applicability of deep learning in SPA DOA estimation.

To address the above problem, we introduce a tensorized CNN for DOA estimation with SPA [41] in the section. The proposed approach employs direct decomposition of convolution kernels to reduce the number of trainable parameters. In particular, based on the sub-Nyquist signal statistics, a coarray tensor corresponding to an augmented virtual array is derived. For efficient processing of this coarray tensor, convolution kernels are decomposed via CPD into low-rank CP factors. The feature map extraction is then formulated as a series of sub-convolutions between the coarray tensor and CP factors, ultimately yielding source DOA estimates. Consequently, the proposed tensorized CNN demonstrates significantly reduced training and computational requirements compared to conventional CNN architectures.

It is worth noting that the tensorized DNN introduced in Section 7.2 can also be adopted for DOA estimation using sub-Nyquist tensor statistics. The tensorized

CNN presented in this section serves as another illustrative example of integrating tensor models with standard neural network architectures to enhance resource efficiency. Moreover, the tensorized CNN and tensorized DNN can be jointly implemented, since conventional neural networks typically combine convolutional layers with FC layers in practice. This hybrid approach can potentially leverage the complementary strengths of both architectures for improved performance.

7.3.1 Sub-Nyquist Tensor Signal Model

The nested coprime planar array with compressed inter-element spacing (NCACIS), as a hybrid of nested and coprime arrays among various SPA configurations, is used here as an example array for its ability to produce a contiguous augmented virtual array [63]. As shown in Figure 7.11, the NCACIS consists of two URAs, i.e., a dense URA $\mathbb{U}_1$ with $M_1 \times N_1$ sensors and a sparse URA $\mathbb{U}_2$ with $M_2 \times N_2$ sensors. The inter-element spacings along the x-axis and the y-axis are both d for $\mathbb{U}_1$, and are $N_1 d$, $M_1 d$ for $\mathbb{U}_2$. Here, (M_1, M_2) and (N_1, N_2) are coprime integer pairs. The total number of sensors in the NCACIS is $M_1N_1 + M_2N_2 - 1$, since $\mathbb{U}_1$ and $\mathbb{U}_2$ share a sensor only at the origin $(0, 0, 0)$.

Similar to the previous section, we consider K uncorrelated far-field narrowband signals from the spatial directions $\{(\theta_k, \phi_k), k = 1, 2, \dots, K\}$. To maintain the structural properties of the sub-Nyquist sampled signals, the T snapshots are modeled in a CP tensorial format [13]. Specifically, the signals received at $\mathbb{U}_1$ and $\mathbb{U}_2$ are, respectively, represented as 3D sub-Nyquist tensors, i.e.,

$$
\begin{aligned}
\mathcal{X}_1 &= \sum_{k=1}^{K} \boldsymbol{a}_1(\mu_k) \circ \boldsymbol{b}_1(\nu_k) \circ \boldsymbol{s}_k + \mathcal{N}_1 \in \mathbb{C}^{M_1 \times N_1 \times T}, \\
\mathcal{X}_2 &= \sum_{k=1}^{K} \boldsymbol{a}_2(\mu_k) \circ \boldsymbol{b}_2(\nu_k) \circ \boldsymbol{s}_k + \mathcal{N}_2 \in \mathbb{C}^{M_2 \times N_2 \times T}.
\end{aligned}
\tag{7.44}
$$

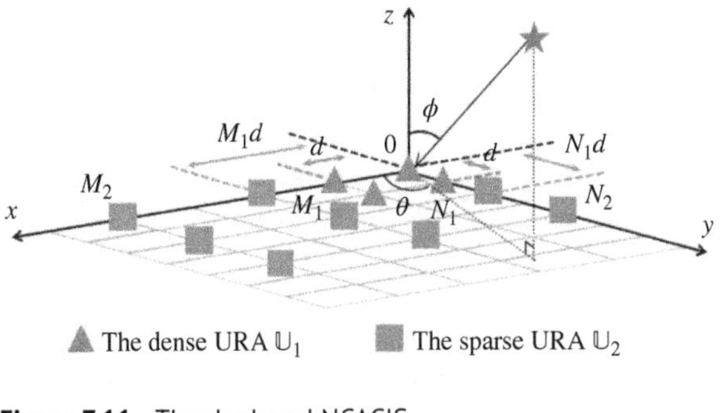

Figure 7.11 The deployed NCACIS.

Here,

$$\boldsymbol{a}_1(\mu_k) = [1, e^{-j\pi\mu_k}, \dots, e^{-j\pi(M_1-1)\mu_k}]^{\mathrm{T}} \in \mathbb{C}^{M_1}, \tag{7.45}$$

and

$$\boldsymbol{b}_1(\nu_k) = [1, e^{-j\pi\nu_k}, \dots, e^{-j\pi(N_1-1)\nu_k}]^{\mathrm{T}} \in \mathbb{C}^{N_1}, \tag{7.46}$$

are the steering vectors along the x-axis and the y-axis of $\mathbb{U}_1$,

$$\boldsymbol{a}_2(\mu_k) = [1, e^{-j\pi M_1\mu_k}, \dots, e^{-j\pi(M_2-1)M_1\mu_k}]^{\mathrm{T}} \in \mathbb{C}^{M_2}, \tag{7.47}$$

and

$$\boldsymbol{b}_2(\nu_k) = [1, e^{-j\pi N_1\nu_k}, \dots, e^{-j\pi(N_2-1)N_1\nu_k}]^{\mathrm{T}} \in \mathbb{C}^{N_2}, \tag{7.48}$$

are the steering vectors along the x-axis and the y-axis of $\mathbb{U}_2$, and $\mathcal{N}_1, \mathcal{N}_2$ are additive Gaussian white noise tensors.

To derive the second-order tensor signal statistics for the subsequent virtual array formulation, a 4D cross-correlation tensor between sub-Nyquist tensor signals is formulated as:

$$\begin{aligned}\mathcal{R} &\triangleq \mathbb{E}\{\boldsymbol{X}_1(t)\circ\boldsymbol{X}_2^*(t)\}\\ &= \sum_{k=1}^{K} \sigma_k^2 \boldsymbol{a}_1(\mu_k)\circ\boldsymbol{b}_1(\nu_k)\circ\boldsymbol{a}_2^*(\mu_k)\circ\boldsymbol{b}_2^*(\nu_k) + \tilde{\mathcal{N}} \in \mathbb{C}^{M_1\times N_1\times M_2\times N_2},\end{aligned} \tag{7.49}$$

where $\boldsymbol{X}_1(t)$ and $\boldsymbol{X}_2(t)$, $\forall t = 1, 2, \dots, T$, denote, respectively, the tth snapshot in the sub-Nyquist tensors $\mathcal{X}_1$ and $\mathcal{X}_2$, and $\tilde{\mathcal{N}}$ is the cross-correlation noise term with only the $(1, 1, 1, 1)$th element being σ_n^2 whereas all others being 0.

In practice, the cross-correlation tensor $\mathcal{R}$ is replaced by its sample version:

$$\hat{\mathcal{R}} = \frac{1}{T}\langle \mathcal{X}_1 \times_3 \mathcal{X}_2^* \rangle, \tag{7.50}$$

due to the limited number of snapshots.

7.3.2 Augmented Coarray Tensor Derivation

As demonstrated in [53, 54], the sub-Nyquist tensor signal statistics of the SPA can be transformed into an augmented coarray tensor corresponding to a virtual URA, leading to Nyquist-matched DOA estimation. Thus, to achieve spatial aliasing-free DOA estimation with the deployed NCACIS, the cross-correlation tensor $\mathcal{R}$ is partially unfolded, yielding a virtual URA $\mathbb{V}$ with $M_1M_2 \times N_1N_2$ virtual sensors and the inter-element spacing d.

Specifically, since pairs of steering vectors $\{\boldsymbol{a}_2^*(\mu_k), \boldsymbol{a}_1(\mu_k)\}$ and $\{\boldsymbol{b}_2^*(\nu_k), \boldsymbol{b}_1(\nu_k\}$, respectively, formulate difference coarrays along the x-axis and the y-axis, the corresponding dimension pairs $\{1, 3\}$, $\{2, 4\}$ of $\mathcal{R}$ are combined to obtain the coarray signal statistics:

$$\begin{aligned}\boldsymbol{R} &\triangleq \mathcal{R}_{\{1,3\},\{2,4\}} \\ &= \sum_{k=1}^{K} \sigma_k^2 [\boldsymbol{a}_2^*(\mu_k) \otimes \boldsymbol{a}_1(\mu_k)] \circ [\boldsymbol{b}_2^*(\nu_k) \otimes \boldsymbol{b}_1(\nu_k)] + \boldsymbol{N} \in \mathbb{C}^{M_1 M_2 \times N_1 N_2}.\end{aligned} \tag{7.51}$$

Here, $\boldsymbol{N}$ is the partially unfolded noise matrix. After rearranging elements in $\boldsymbol{R}$ to match the locations of virtual sensors in $\mathbb{V}$ [64], the equivalent second-order signal of $\mathbb{V}$ can be obtained as:

$$\boldsymbol{Y} = \sum_{k=1}^{K} \sigma_k^2 \boldsymbol{q}_x(k) \circ \boldsymbol{q}_y(k) + \bar{\boldsymbol{N}} \in \mathbb{C}^{M_1 M_2 \times N_1 N_2}, \tag{7.52}$$

where

$$\boldsymbol{q}_x(k) = [e^{-j\pi(-M_1 M_2 + M_1)\mu_k}, e^{-j\pi(-M_1 M_2 + M_1 + 1)\mu_k}, \dots, e^{-j\pi(M_1 - 1)\mu_k}]^{\mathrm{T}} \in \mathbb{C}^{M_1 M_2}, \tag{7.53}$$

and

$$\boldsymbol{q}_y(k) = [e^{-j\pi(-N_1 N_2 + N_1)\nu_k}, e^{-j\pi(-N_1 N_2 + N_1 + 1)\nu_k}, \dots, e^{-j\pi(N_1 - 1)\nu_k}]^{\mathrm{T}} \in \mathbb{C}^{N_1 N_2}, \tag{7.54}$$

represent the steering vectors along the x-axis and the y-axis of $\mathbb{V}$, respectively, and $\bar{\boldsymbol{N}}$ is the corresponding noise term.

Furthermore, the complex-valued coarray signal $\boldsymbol{Y}$ is transformed into a 3D coarray tensor by concatenating its real and imaginary parts along a new dimension, i.e.,

$$\mathcal{V} = [\Re\{\boldsymbol{Y}\}, \Im\{\boldsymbol{Y}\}]_{\sqcup_3} \in \mathbb{R}^{M_1 M_2 \times N_1 N_2 \times 2}. \tag{7.55}$$

Here, the real and imaginary components of the coarray signal serve as two channels of the coarray tensor $\mathcal{V}$, and it is treated as the input tensor for a CNN-based network architecture, which naturally takes multichannel real-valued tensors as inputs.

7.3.3 Tensorized CNN Architecture for DOA Estimation

In this section, we establish a resource-efficient tensorized CNN architecture for 2D DOA estimation with the NCACIS. Based on the low-rank structure of convolution kernels, these kernels can be decomposed via the CPD into a core tensor and several low-order CP factors, significantly reducing the network complexity. This decomposition facilitates efficient extraction of angular information from the Nyquist-matched coarray tensor statistics via lightweight sub-convolution operations. The resultant tensorized CNN architecture demonstrates lower computational burden during training and estimation accuracy comparable to conventional CNN architectures.

Note that, for the direct decomposition of convolution kernels, several tensor decomposition approaches beyond CPD can be adopted, such as Tucker decomposition and TTD. The CPD model requires tuning only a single rank parameter, whereas other tensor decomposition models typically require tuning multiple rank sequences. From the perspective of implementation simplicity, this section focuses on applying CPD to convolution kernels to demonstrate the fundamental mechanism.

7.3.3.1 Convolution Kernel Decomposition and Sub-convolutions for DOA Estimation

In conventional CNNs, a convolution kernel $\mathcal{C}_1 \in \mathbb{R}^{P_1 \times Q_1 \times H_0 \times H_1}$ is used to process the input coarray tensor $\mathcal{V}$ via a convolution operation, which produces the first feature map $\mathcal{M}_1 \in \mathbb{R}^{I_1 \times J_1 \times H_1}$ as:

$$\mathcal{M}_{1(i_1,j_1,h_1)} = \sum_{p_1=1}^{P_1} \sum_{q_1=1}^{Q_1} \sum_{h_0=1}^{2} \mathcal{C}_{1(p_1,q_1,h_0,h_1)} \mathcal{V}_{(i_1-1+p_1-\delta, j_1-1+q_1-\delta, h_0)}, \tag{7.56}$$

for $i_1 = 1, 2, \dots, I_1$, $j_1 = 1, 2, \dots, J_1$, $h_1 = 1, 2, \dots, H_1$. Here, (P_1, Q_1) denotes the size of the convolution kernel, (I_1, J_1) denotes the size of $\mathcal{M}_1$, $H_0 = 2$ stands for the number of the input channels (real/imaginary components), H_1 stands for the number of the output channels in the first convolutional layer, δ is the stride of zero-padding, and

$$\begin{aligned} I_1 &= M_1 M_2 - P_1 - 2\delta + 1, \\ J_1 &= N_1 N_2 - Q_1 - 2\delta + 1. \end{aligned} \tag{7.57}$$

Following the above convolution approach, subsequent layers employ a series of convolution kernels $\{\mathcal{C}_l \in \mathbb{R}^{P_l \times Q_l \times H_{l-1} \times H_l}\}_{l=2}^{L}$ to generate $L-1$ feature maps $\{\mathcal{M}_l \in \mathbb{R}^{I_l \times J_l \times H_l}\}_{l=2}^{L}$. However, similar to the FC layer as described in Section 7.2.2, these convolution kernels introduce an excessive number of trainable parameters, leading to significant memory consumption. Furthermore, the convolution operation in (7.56) incurs substantial computational complexity.

To reduce the number of convolution kernel parameters, we exploit the tensorial structures of the convolution kernels $\{\mathcal{C}_l, l = 1, 2, \dots, L\}$, which exhibit intrinsic low-rankness. To be specific, a general tensor decomposition approach approximates a given tensor as a core tensor that captures essential signal features and multiple low-rank factor matrices. Thus, as shown in Figure 7.12, applying the rank-R CPD to $\mathcal{C}_l$ yields:

$$\mathcal{C}_l \approx \mathcal{G}_l \times_1 \boldsymbol{V}_{l,1} \times_2 \boldsymbol{V}_{l,2} \times_3 \boldsymbol{V}_{l,3} \times_4 \boldsymbol{V}_{l,4}, \tag{7.58}$$

where $\mathcal{G}_l \in \mathbb{R}^{R \times R \times R \times R}$ denotes the diagonal core tensor, $\boldsymbol{V}_{l,1} \in \mathbb{R}^{P_l \times R}$, $\boldsymbol{V}_{l,2} \in \mathbb{R}^{Q_l \times R}$, $\boldsymbol{V}_{l,3} \in \mathbb{R}^{H_{l-1} \times R}$, $\boldsymbol{V}_{l,4} \in \mathbb{R}^{H_l \times R}$ are the CP factor matrices. Normally, the CP rank R is much smaller than P_l, Q_l, H_{l-1}, H_l. Note that, other tensor decomposition

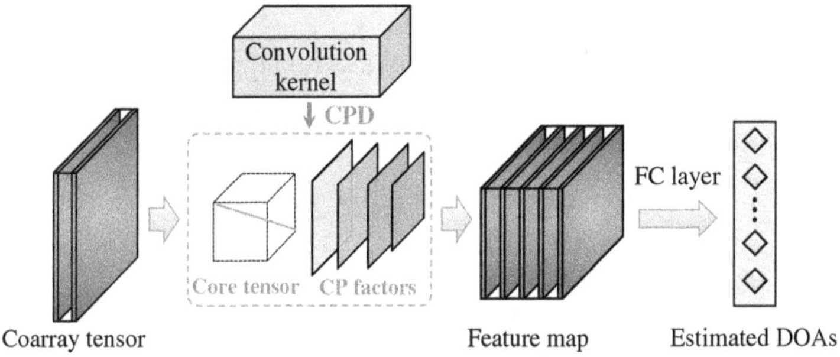

Figure 7.12 The architecture of a tensorized CNN for DOA estimation. Input: 3D coarray tensor; Trainable parameters: decomposed CP factors; Output: estimated DOAs.

approaches such as the Tucker decomposition and TTD can be implemented for convolution kernel decomposition as well.

By substituting the CPD representation of the convolution kernel (7.58) into (7.56), the convolution of $\mathcal{V}$ with $\mathcal{C}_1$ can be decomposed into the three sub-convolution operations, expressed as:

$$\mathcal{V}'_{(i,j,r_1)} = \sum_{h_0=1}^{2} \boldsymbol{V}_{1,3(h_0,r_1)} \mathcal{V}_{(i,j,h_0)}, \tag{7.59}$$

$$\mathcal{V}''_{(i_1,j_1,r_2)} = \sum_{p_1=1}^{P_1} \sum_{q_1=1}^{Q_1} \sum_{r_1=1}^{R} \left[\sum_{r=1}^{R} \sum_{r'=1}^{R} \mathcal{G}_{(r,r',r_1,r_2)} \boldsymbol{V}_{1,1(p_1,r)} \boldsymbol{V}_{1,2(q_1,r')} \right] \mathcal{V}'_{(i_1-1+p_1-\delta,j_1-1+q_1-\delta,r_1)}, \tag{7.60}$$

$$\mathcal{M}_{1(i_1,j_1,h_1)} = \sum_{r_2=1}^{R} \boldsymbol{V}_{1,4(h_1,r_2)} \mathcal{V}''_{(i_1,j_1,r_2)}, \tag{7.61}$$

where $i = 1, 2, \ldots, M_1M_2$, $j = 1, 2, \ldots, N_1N_2$, $r_1, r_2 = 1, 2, \ldots, R$. Similarly, the subsequent feature maps $\{\mathcal{M}_2, \ldots, \mathcal{M}_L\}$ can be efficiently extracted via these sub-convolutions. Note that all feature maps are further processed by nonlinear activation functions during feedforward propagation.

Following the feature map extraction via sub-convolutions, the final feature map $\mathcal{M}_L$ is fed into an FC layer to estimate the source DOAs $\hat{\boldsymbol{\vartheta}}$. Specifically, the vectorized feature map:

$$\boldsymbol{m}_L \triangleq \text{vec}(\mathcal{M}_L) \in \mathbb{R}^{I_L J_L H_L}, \tag{7.62}$$

is multiplied with a learnable weight matrix $\boldsymbol{W} \in \mathbb{R}^{I_L J_L H_L \times 2K}$ to yield $\hat{\boldsymbol{\vartheta}}$, i.e.,

$$\hat{\boldsymbol{\vartheta}} = \boldsymbol{W}^{\mathrm{T}} \boldsymbol{m}_L. \tag{7.63}$$

7.3.3.2 Resource-efficient Training of Tensorized CNN

To train the tensorized CNN, we employ the Huber loss function between the true source DOAs $\boldsymbol{\vartheta}$ and the estimated DOAs $\hat{\boldsymbol{\vartheta}}$ as defined in (7.21). Such Huber loss function is denoted by $\mathcal{F}(\boldsymbol{\Theta})$, where

$$\boldsymbol{\Theta} = \{\mathcal{G}_l, \boldsymbol{V}_{l,1}, \boldsymbol{V}_{l,2}, \boldsymbol{V}_{l,3}, \boldsymbol{V}_{l,4}, \boldsymbol{W}, \quad l = 1, 2, \dots, L\} \tag{7.64}$$

represents the complete set of trainable parameters. The gradients of $\mathcal{F}(\boldsymbol{\Theta})$ for each convolutional layer are derived and propagated backward to obtain the derivative of $\mathcal{F}(\boldsymbol{\Theta})$ w.r.t. $\boldsymbol{\Theta}$. As such, the compressed parameters in the tensorized CNN can be updated for each input coarray tensor via stochastic gradient descent. Note that, the detailed derivations for the derivatives of $\mathcal{F}(\boldsymbol{\Theta})$ w.r.t. $\boldsymbol{\Theta}$ and the corresponding parameter update steps follow the same formulations as presented in section 7.2.3. Therefore, we omit here their detailed derivation.

During batch processing, multiple coarray tensor samples are fed sequentially to iteratively refine the parameters. After sufficiently many training epochs, the optimized tensorized CNN enables efficient sub-Nyquist 2D DOA estimation by executing only the sub-convolution operations as defined in (7.59)–(7.61). As such, by leveraging the decomposed convolution kernel structure, this approach maintains high computational efficiency while preserving estimation accuracy.

7.3.3.3 Performance Analysis

To assess the effectiveness of network compression via convolution kernel decomposition, the total number of trainable parameters is evaluated. Specifically, the parameter count for the original convolution kernel $\mathcal{C}_l$ is given by:

$$B_l = P_l Q_l H_{l-1} H_l, \quad \forall l = 1, 2, \dots, L. \tag{7.65}$$

Under the proposed convolution kernel decomposition scheme, the parameter count for the core tensor $\mathcal{G}_l$ and CP factor matrices $\{\boldsymbol{V}_{l,1}, \boldsymbol{V}_{l,2}, \boldsymbol{V}_{l,3}, \boldsymbol{V}_{l,4}\}$ becomes:

$$\tilde{B}_l = R^4 + P_l R + Q_l R + H_{l-1} R + H_l R, \quad \forall l = 1, 2, \dots, L. \tag{7.66}$$

Given that the CP rank R satisfies $R < \min(P_l, Q_l, H_{l-1}, H_l)$, the proposed tensorized CNN achieves significant parameter reduction, with $\tilde{B}_l \ll B_l$ for all convolutional layers.

7.3.4 Simulation Results

In our simulation example, an NCACIS configuration with $M_1 = N_1 = 4$ and $M_2 = N_2 = 5$ is considered. Accordingly, an augmented virtual URA $\mathbb{V}$ with 20×20 virtual sensors is derived, and the corresponding 3D coarray tensor $\mathcal{V} \in \mathbb{R}^{20 \times 20 \times 2}$ is constructed. Similar to the simulation setting in section 7.2.5, $K = 2$ source signals are assumed. The training dataset consists of $P = 19{,}200$ samples, used to train neural networks for 60 epochs. For each training sample, azimuth and elevation angles are uniformly drawn from the interval $[10°, 65°]$, and the SNRs of two impinging signals are randomly selected from $\{-15, -10, -5, 0\}$ dB with a fixed number of snapshots $T = 150$. The learning rate and batch size for the proposed tensorized CNN are fine-tuned, respectively, as $0.572 \cdot 10^{-3}$ and 32 through Bayesian optimization [46]. Following the same procedure, these two hyperparameters for the conventional CNN are fine-tuned as $0.144 \cdot 10^{-3}$ and 32, respectively.

7.3.4.1 Comparison of Training Efficiency

A conventional CNN is deployed with $L = 3$ convolution kernels, whose sizes are $5 \times 5 \times 2 \times 16$, $5 \times 5 \times 16 \times 32$, and $3 \times 3 \times 32 \times 32$, respectively. The stride of zero-padding in the first convolutional layer is set to $\delta = 1$. The final feature map $\mathcal{M}_3 \in \mathbb{R}^{12\times12\times32}$ is multiplied with the output layer parameter matrix $\boldsymbol{W} \in \mathbb{R}^{4,608\times4}$. As the counterpart, a tensorized CNN is constructed, in which the CP ranks of the three convolutional layers are set to $\{3, 3, 2\}$. Consequently, the three convolution kernels are decomposed into core tensors and CP factors $\{\mathcal{G}_1 \in \mathbb{R}^{3\times3\times3}, \boldsymbol{V}_{1,1} \in \mathbb{R}^{5\times3}, \boldsymbol{V}_{1,2} \in \mathbb{R}^{5\times3}, \boldsymbol{V}_{1,3} \in \mathbb{R}^{2\times3}, \boldsymbol{V}_{1,4} \in \mathbb{R}^{16\times3}\}$, $\{\mathcal{G}_2 \in \mathbb{R}^{3\times3\times3}, \boldsymbol{V}_{2,1} \in \mathbb{R}^{5\times3}, \boldsymbol{V}_{2,2} \in \mathbb{R}^{5\times3}, \boldsymbol{V}_{2,3} \in \mathbb{R}^{16\times3}, \boldsymbol{V}_{2,4} \in \mathbb{R}^{32\times3}\}$, $\{\mathcal{G}_3 \in \mathbb{R}^{2\times2\times2}, \boldsymbol{V}_{3,1} \in \mathbb{R}^{3\times2}, \boldsymbol{V}_{3,2} \in \mathbb{R}^{3\times2}, \boldsymbol{V}_{3,3} \in \mathbb{R}^{32\times2}, \boldsymbol{V}_{3,4} \in \mathbb{R}^{32\times2}\}$. The nonlinear activation functions for the three convolutional layers are set as $\tanh(\cdot)$, $\mathrm{ReLU}(\cdot)$, and $\mathrm{ReLU}(\cdot)$, respectively.

As summarized in Table 7.3, the conventional CNN architecture requires 22, 816 parameters in its convolution kernels, while the proposed tensorized CNN architecture reduces this to merely 366 parameters, that is, a reduction exceeding 60 times. This substantial parameter compression yields significant improvements in computational efficiency. Specifically, the memory footprint for parameter storage decreases from 0.16 to 0.07 MB, while the computational complexity measured in floating point operations (FLOPs) drops dramatically from 4.114×10^6 to 1.8432×10^4. Here, FLOPs counts the number of addition, subtraction, multiplication, or division operations that involve floating-point numbers. As a result, the tensorized CNN architecture demonstrates superior hardware efficiency, reducing GPU utilization by more than 50%. These quantitative improvements collectively demonstrate the enhanced efficiency of the tensorized CNN architecture through effective convolution kernel decomposition.

Table 7.3 Comparison of network efficiency.

Method	Parameters in convolution kernels	Parameter memory	FLOPs	GPU utilization (%)
Conventional CNN	22,816	0.16 MB	4.114×10^6	24
Tensorized CNN	366	0.07 MB	1.843×10^4	11

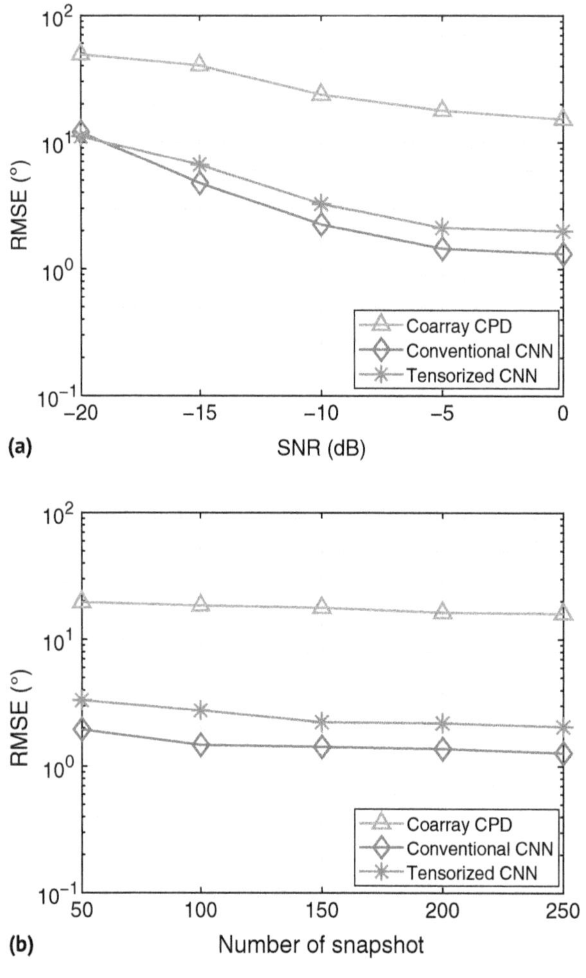

Figure 7.13 Comparison of estimation accuracy. (a) RMSE versus SNR. (b) RMSE versus number of snapshots.

7.3.4.2 Comparison of DOA Estimation Accuracy

In Figure 7.13, we compare the DOA estimation accuracy of the tensorized CNN against both the conventional CNN [57] and the coarray CPD-based method that directly decomposes the coarray tensor. Among the three approaches, the coarray CPD-based method exhibits the poorest performance. This degradation arises because the underlying tensor decomposition of the coarray signal model is highly sensitive to adverse conditions, particularly low SNR and limited number of snapshots, which significantly undermine the reliability of the decomposition.

In contrast, both CNN-based approaches exploit data-driven learning to extract robust tensorial features from large-scale training datasets, enabling them to achieve substantially higher estimation accuracy under challenging signal conditions. Notably, although the proposed tensorized CNN attains RMSE performance that is nearly identical to that of the conventional CNN as reflected by their closely overlapping error curves, it does so with dramatically reduced computational complexity and parameter amount. This outcome demonstrates that the tensorized CNN effectively preserves DOA estimation accuracy while delivering substantial efficiency gains through structured network compression, making it more suitable for practical and resource-constrained implementations.

7.4 Conclusion and Discussion

In summary, this chapter presents resource-efficient tensorized neural network architectures for DOA estimation. In the tensorized DNN, the transformation from the covariance tensor to the hidden state tensor is formulated as an inverse Tucker decomposition, significantly improving training and computational efficiency. In particular, for a 10×10 URA and 2 sources, the number of parameters in the tensorized FC layers is reduced by more than 122,000 times compared to the matrix-based DNN, cutting training time by 22 times and GPU memory usage by 3 times. Furthermore, in the tensorized CNN, convolution kernels are compressed via the CPD into low-order CP factors. The derived coarray tensor statistics are then efficiently processed using reformulated sub-convolution operations. In particular, for a SPA with 43 sensors and 2 sources, the convolution kernel parameters are compressed in the tensorized CNN by more than 50 times compared to the conventional CNN, thereby reducing system resource consumption by more than twice.

With dramatically reduced computational complexity, both tensorized neural network architectures achieve higher DOA estimation accuracy and better angular resolution, whether applied to URA or SPA, compared to conventional model-based methods. Compared to conventional matrix-based neural networks, both tensorized neural network architectures exhibit marginally lower estimation accuracy, as they prioritize computational and memory efficiency over absolute precision. However, this trade-off enables significant resource savings while maintaining practical usability. Crucially, the tensorized DNN and CNN remain highly robust under challenging signal conditions, including low SNRs and limited number of snapshots.

Despite a slight accuracy reduction, the efficiency and adaptability of the proposed tensorized neural network architectures make them viable alternatives for real-time DOA estimation in resource-constrained environments. Specifically, the drastic reductions in parameter counts and memory footprint make these architectures highly suitable for deployment on hardware-constrained platforms such as FPGAs and edge devices. The tensorized structures align naturally with parallel and low-rank computing primitives, which can be exploited in customized hardware accelerators to achieve real-time DOA estimation under stringent power and resource budgets.

The tensorized neural networks can be regarded as a plug-and-play solution, since the designed tensorized FC layer and tensorized convolutional layer can be seamlessly integrated into mainstream network architectures to reduce complexity. Thus, the tensorized neural networks summarized in this chapter are applicable to various array signal processing tasks besides DOA estimation. Beyond DOA estimation, these architectures have broad potential in various array signal processing tasks, including:

- Target localization and channel estimation in integrated sensing and communication (ISAC) systems [65–67]: The tensorized neural networks can efficiently and robustly estimate angle, Doppler, and distance parameters for ISAC.
- Data-driven beamforming [68–70]: The tensorized neural network architectures can be employed to generate beamforming vectors with higher efficiency.
- Deep learning-based array configuration optimization [71] and array diagnosis [72]: The tensorized neural networks present a promising alternative to conventional array optimization solutions due to their higher efficiency and lower implementation costs.

Acknowledgments

The work of H. Zheng and Z. Shi was supported in part by the National Natural Science Foundation of China under Grant U21A20456.

References

1 H. L. Van Trees. *Detection, Estimation, and Modulation Theory, Part IV: Optimum Array Processing*. Wiley, Hoboken, NJ, USA, 2002.
2 Y. Gu and Y. D. Zhang. *Information-theoretic Radar Signal Processing*. Wiley, Hoboken, NJ, USA, 2024.
3 M. Shaghaghi and S. A. Vorobyov. Subspace leakage analysis and improved DOA estimation with small sample size. *IEEE Transactions on Signal Processing*, 63(12):3251–3265, 2015.

4 C. Zhou, Y. Gu, X. Fan, Z. Shi, G. Mao, and Y. D. Zhang. Direction-of-arrival estimation for coprime array via virtual array interpolation. *IEEE Transactions on Signal Processing*, 66(22):5956–5971, 2018.

5 G. Qin, M. G. Amin, and Y. D. Zhang. DOA estimation exploiting sparse array motions. *IEEE Transactions on Signal Processing*, 67(11):3013–3027, 2019.

6 J. He, T. Shu, L. Li, and T.-K. Truong. Mixed near-field and far-field localization and array calibration with partly calibrated arrays. *IEEE Transactions on Signal Processing*, 70:2105–2118, 2022.

7 Z. Mao, S. Liu, Y. D. Zhang, L. Han, and Y. Huang. Joint DoA-range estimation using space-frequency virtual difference coarray. *IEEE Transactions on Signal Processing*, 70:2576–2592, 2022.

8 F. Wen, X. Li, S. Dang, A. N. D. B. Da Costa, and C. Yuen. Fast 2D-DOA estimation for polarized massive MIMO systems with irregularly spaced sensors. *IEEE Transactions on Communications*, 73(11):11 964–11 977, 2025.

9 R. O. Schmidt. Multiple emitter location and signal parameter estimation. *IEEE Transactions on Antennas and Propagation*, 34(3):276–280, 1986.

10 R. Roy and T. Kailath. ESPRIT-estimation of signal parameters via rotational invariance techniques. *IEEE Transactions on Acoustics, Speech, and Signal Processing*, 37(7):984–995, 1989.

11 D. Malioutov, M. Çetin, and A. S. Willsky. A sparse signal reconstruction perspective for source localization with sensor arrays. *IEEE Transactions on Signal Processing*, 53(8):3010–3022, 2005.

12 X. Wu and W.-P. Zhu. On efficient gridless methods for 2-D DOA estimation with uniform and sparse L-shaped arrays. *Signal Processing*, 191:108351, 2022.

13 N. D. Sidiropoulos, R. Bro, and G. B. Giannakis. Parallel factor analysis in sensor array processing. *IEEE Transactions on Signal Processing*, 48(8):2377–2388, 2000.

14 H. Zheng, C. Zhou, Z. Shi, and Y. Gu. Structured tensor reconstruction for coherent DOA estimation. *IEEE Signal Processing Letters*, 29:1634–1638, 2022.

15 F. Xu, M. W. Morency, and S. A. Vorobyov. DOA estimation for transmit beamspace MIMO radar via tensor decomposition with Vandermonde factor matrix. *IEEE Transactions on Signal Processing*, 70:2901–2917, 2022.

16 Y. Wang, L. Chao, S. Luo, X. Liu, and C. Tellambura. Smoothed sparse blind DOA estimation for MC-CDMA systems based on tensor decomposition. *IEEE Wireless Communications Letters*, 12(9):1548–1552, 2023.

17 T. Zhang, S. A. Vorobyov, and F. Xu. Transmit energy focusing for parameter estimation in slow-time transmit beamspace L-shaped MIMO Radar. *IEEE Transactions on Signal Processing*, 72:5228–5243, 2024.

18 F. Xu, H. Zheng, and S. A. Vorobyov. Tensor-based 2D DOA estimation for L-shaped nested array. *IEEE Transactions on Aerospace and Electronic Systems*, 60(1):604–618, 2024.

19 P. Gerstoft, S. Nannuru, C. F. Mecklenbräuker, and G. Leus. DOA estimation in heteroscedastic noise. *Signal Processing*, 161:63–73, 2019.

20 C. Zhou, Y. Gu, Z. Shi, and M. Haardt. Structured Nyquist correlation reconstruction for DOA estimation with sparse arrays. *IEEE Transactions on Signal Processing*, 71:1849–1862, 2023.

21 A. H. El Zooghby, C. G. Christodoulou, and M. Georgiopoulos. A neural network-based smart antenna for multiple source tracking. *IEEE Transactions on Antennas and Propagation*, 48(5):768–776, 2000.

22 M. Pastorino and A. Randazzo. A smart antenna system for direction of arrival estimation based on a support vector regression. *IEEE Transactions on Antennas and Propagation*, 53(7):2161–2168, 2005.

23 N. J. G. Fonseca, M. Coudyser, J. Laurinm, and J. Brault. On the design of a compact neural network-based DOA estimation system. *IEEE Transactions on Antennas and Propagation*, 58(2):357–366, 2010.

24 S. Zhang, Z. Yang, W. Shen, L. Zhang, J. Zhu, and Z. Zhao. Deep learning-based DOA estimation. *IEEE Transactions on Cognitive Communications and Networking*, 10(3):819–835, 2024.

25 P. Chen, Z. Chen, L. Liu, Y. Chen, and X. Wang. SDOA-Net: An efficient deep learning-based DOA estimation network for imperfect array. *IEEE Transactions on Instrumentation and Measurement*, 73:1–12, 2024.

26 X. Wu, X. Yang, X. Jia, and F. Tian. A gridless DOA estimation method based on convolutional neural network with Toeplitz prior. *IEEE Signal Processing Letters*, 29:1247–1251, 2022.

27 Y. Su, X. Wang, Y. Guo, and F. Gao. DOA estimation with deep learning: A limited training data framework. *IEEE Transactions on Communications*, 73(8):5993–6005, 2025.

28 Z.-M. Liu, C. Zhang, and P. S. Yu. Direction-of-arrival estimation based on deep neural networks with robustness to array imperfections. *IEEE Transactions on Antennas and Propagation*, 66(12):7315–7327, 2018.

29 D. T. Hoang and K. Lee. Deep learning-aided coherent direction-of-arrival estimation with the FTMR algorithm. *IEEE Transactions on Signal Processing*, 70:1118–1130, 2022.

30 G. K. Papageorgiou, M. Sellathurai, and Y. C. Eldar. Deep networks for direction-of-arrival estimation in low SNR. *IEEE Transactions on Signal Processing*, 69:3714–3729, 2021.

31 D. Meng, X. Li, and W. Wang. Deep learning-driven DOA estimation for distributed unmanned systems: Overcoming direction-dependent mutual coupling challenges. *IEEE Internet of Things Journal*, 12(21):44 002–44 011, 2025.

32 H. Chen, F. Ahmad, S. A. Vorobyov, and F. Porikli. Tensor decompositions in wireless communications and MIMO radar. *IEEE Journal of Selected Topics in Signal Processing*, 15(3):438–453, 2021.

33 I. Sutskever, J. Martens, and G. Hinton. Generating text with recurrent neural networks. In *Proceedings of International Conference on Machine Learning (ICML),* pages 1017–1024, Bellevue, WA, 2011.

34 R. Socher, D. Chen, C. D. Manning, and A. Ng. Reasoning with neural tensor networks for knowledge base completion. In *Advances in Neural Information Processing Systems (NIPS)*. Lake Tahoe, Nevad, 2013.

35 D. Wang, G. Zhao, H. Chen, Z. Liu, L. Deng, and G. Li. Nonlinear tensor train format for deep neural network compression. *Neural Networks*, 144:320–333, 2021.

36 W. Sun, S. Chen, L. Huang, H. C. So, and M. Xie. Deep convolutional neural network compression via coupled tensor decomposition. *IEEE Journal of Selected Topics in Signal Processing*, 15(3):603–616, 2021.

37 T. Garipov, D. Podoprikhin, A. Novikov, and D. Vetrov. Ultimate tensorization: Compressing convolutional and FC layers alike. *ArXiv Preprint arXiv:1611.03214*, 2016.

38 J. Ye, L. Wang, G. Li, D. Chen, S. Zhe, X. Chu, and Z. Xu. Learning compact recurrent neural networks with block-term tensor decomposition. In *Proceedings IEEE International Conferences Computer Vision Pattern Recognition (CVPR),* pages 9378–9387. Salt Lake City, UT, 2018.

39 V. Lebedev, Y. Ganin, M. Rakhuba, I. Oseledets, and V. Lempitsky. Speeding-up convolutional neural networks using fine-tuned CP-decomposition. In *Proceedings International Conference on Learning Representations (ICLR),* San Diego, CA, 2015.

40 J.-T. Chien and Y.-T. Bao. Tensor-factorized neural networks. *IEEE Transactions on Neural Networks and Learning Systems*, 29(5):1998–2011, 2018.

41 H. Zheng, C. Zhou, S. A. Vorobyov, Q. Wang, and Z. Shi. Decomposed CNN for sub-Nyquist tensor-based 2-D DOA estimation. *IEEE Signal Processing Letters*, 30:708–712, 2023.

42 H. Zheng, C. Zhou, S. A. Vorobyov, and Z. Shi. Tensorized neural layer decomposition for 2-D DOA estimation. In *Proceedings IEEE International Conferences Acoustics, Speech, and Signal Processing (ICASSP),* pages 1–5. Rhodes Island, Greece, 2023.

43 H. Zheng, Z. Shi, C. Zhou, S. A. Vorobyov, and Y. Gu. Deep tensor 2-D DOA estimation for URA. *IEEE Transactions on Signal Processing*, 72:4065–4080, 2024.

44 J. Snoek, H. Larochelle, and R. P. Adams. Practical Bayesian optimization of machine learning algorithms. In *Proceedings International Conferences Neural Information Processing Systems (NIPS),* Lake Tahoe, NV, 2012.

45 E. Brochu, V. M. Cora, and N. de Freitas. A tutorial on Bayesian optimization of expensive cost functions, with application to active user modeling and hierarchical reinforcement learning. *ArXiv Preprint arXiv:1012.2599*, 2010.

46 F. Nogueira. Bayesian optimization: Open source constrained global optimization tool for Python, 2014. https://github.com/fmfn/BayesianOptimization (accessed Dec 2022).

47 X. Glorot and Y. Bengio. Understanding the difficulty of training deep feedforward neural networks. In *Proceedings International Conference on Artificial Intelligence and Statistics (AISTATS)*, pages 249–256. Sardinia, Italy, 2010.

48 M. Shao and C. L. Nikias. Signal processing with fractional lower order moments: Stable processes and their applications. *Proceedings of the IEEE*, 81(7):986–1010, 1993.

49 P. P. Vaidyanathan and P. Pal. Sparse sensing with co-prime samplers and arrays. *IEEE Transactions on Signal Processing*, 59(2):573–586, 2011.

50 C. Zhou, Y. Gu, Z. Shi, and Y. D. Zhang. Off-grid direction-of-arrival estimation using coprime array interpolation. *IEEE Signal Processing Letters*, 25(11): 1710–1714, 2018.

51 X. Li, M. Jin, X.-T. Meng, B.-X. Cao, F.-G. Yan, M. S. Greco, and F. Gini. Sparse linear arrays for direction-of-arrival estimation: A tutorial overview. *IEEE Aerospace and Electronic Systems Magazine*, 41(4):32–64, 2025

52 H. Zheng, Z. Shi, C. Zhou, M. Haardt, and J. Chen. Coupled coarray tensor CPD for DOA estimation with coprime L-shaped array. *IEEE Signal Processing Letters*, 28:1545–1549, 2021.

53 H. Zheng, C. Zhou, Z. Shi, Y. Gu, and Y. D. Zhang. Coarray tensor direction-of-arrival estimation. *IEEE Transactions on Signal Processing*, 71:1128–1142, 2023.

54 H. Zheng, Z. Shi, C. Zhou, and A. L. F. de Almeida. Coarray tensor completion for DOA estimation. *IEEE Transactions on Aerospace and Electronic Systems*, 59(5):5472–5486, 2023.

55 Q. Xie, F. Wen, X. Xie, Z. Wang, and X. Pan. Coarray tensor train aided target localization for bistatic MIMO radar. *IEEE Signal Processing Letters*, 32:46–50, 2025.

56 S. Luo, Y. Wang, J. Li, C. Tellambura, and J. J. P. C. Rodrigues. Coarray tensor-based angle estimation for bistatic MIMO Radar with a dilated moving receive array. *IEEE Transactions on Aerospace and Electronic Systems*, 59(6):8995–9009, 2023.

57 R. Akter, V. S. Doan, T. Huynh-The, and D. S. Kim. RFDOA-Net: An efficient ConvNet for RF-based DOA estimation in UAV surveillance systems. *IEEE Transactions on Vehicular Technology*, 70(11):12 209–12 214, 2021.

58 C. Schuessler, M. Hoffmann, and M. Vossiek. Super-resolution radar imaging with sparse arrays using a deep neural network trained with enhanced virtual data. *IEEE Journal of Microwaves*, 3(3):980–993, 2023.

59 K. L. Chen and B. D. Rao. Subspace representation learning for sparse linear arrays to localize more sources than sensors: A deep learning methodology. *IEEE Transactions on Signal Processing*, 73:1293–1308, 2025.

60 Y. LeCun, B. Boser, J. S. Denker, D. Henderson, R. E. Howard, W. Hubbard, and L. D. Jackel. Backpropagation applied to handwritten zip code recognition. *Neural Computation*, 1(4):541–551, 1989.

61 S. Chakrabarty and E. A. P. Habets. Multi-speaker DOA estimation using deep convolutional networks trained with noise signals. *IEEE Journal of Selected Topics in Signal Processing*, 13(1):8–21, 2019.

62 T. N. T. Nguyen, W. S. Gan, R. Ranjan, and D. L. Jones. Robust source counting and DOA estimation using spatial pseudo-spectrum and convolutional neural network. *IEEE/ACM Transactions on Audio, Speech and Language Process*, 28:2626–2637, 2020.

63 S. Qin, Y. D. Zhang, and M. G. Amin. Generalized coprime array configurations for direction-of-arrival estimation. *IEEE Transactions on Signal Processing*, 63(6):1377–1390, 2015.

64 P. Pal and P. P. Vaidyanathan. Coprime sampling and the MUSIC algorithm. In *Proceedings IEEE Digital Signal Process. Signal Processing Education Meeting*, pages 289–294. Sedona, AZ, 2011.

65 L. Lin, H. Zheng, S. A. Vorobyov, C. Zhou, and Z. Shi. Sensing-aided communication channel estimation with tensor-based moving target localization. In *Proceedings IEEE International Conference on Acoustics, Speech, and Signal Processing (ICASSP)*, Seoul, Korea, pages 8606–8610, 2024.

66 L. Lin, J. Tong, H. Lin, H. Zheng, and Z. Shi. Joint target localization and channel estimation for ODDM-ISAC systems. *IEEE Signal Processing Letters*, 32:1525–1529, 2025.

67 L. Lin, C. Zhou, H. Zheng, Z. Shi, S. A. Vorobyov, and R. W. Heath. Sensing-aided precoding with high-dynamic moving scatterers. *IEEE Transactions on Signal Processing*, 73:3061–3078, 2025.

68 B. Luijten, R. Cohen, F. J. de Bruijn, H. A. W. Schmeitz, M. Mischi, and Y. C. Eldar. Adaptive ultrasound beamforming using deep learning. *IEEE Transactions on Medical Imaging*, 39(12):3967–3978, 2020.

69 J. M. J. Huttunen, D. Korpi, and M. Honkala. DeepTx: Deep learning beamforming with channel prediction. *IEEE Transactions on Wireless Communications*, 22(3):1855–1867, 2023.

70 G. Fontanesi, A. Guerra, F. Guidi, J. A. Vásquez-Peralvo, N. Shlezinger, A. Zanella, E. Lagunas, S. Chatzinotas, D. Dardari, and P. M. Djurić. A deep-NN beamforming approach for dual function Radar-communication THz UAV. *IEEE Transactions on Vehicular Technology*, 74(1):746–760, 2025.

71 K. Juretus, M. G. Amin, and S. A. Hamza. Deep learning of the sparse array configurations in optimum beamforming. *IEEE Transactions on Aerospace and Electronic Systems*, 61(3):5718–5730, 2025.

72 X. Yang, D. Yang, Y. Zhao, J. Pan, and Y. Chen. Synthesis of linear sparse array using DNN-based machine-learning method. *IEEE Transactions on Antennas and Propagation*, 71(8):6513–6522, 2023.

8

SAR Image Interpretation with Deep Neural Networks

Qian Guo, Feng Xu, and Haipeng Wang*

School of Information Science and Technology, Fudan University, Shanghai, China

*Corresponding author: fengxu@fudan.edu.cn

8.1 Introduction

Synthetic aperture radar (SAR) is an active microwave imaging radar system that operates independently of weather and lighting conditions, providing all-day and all-weather observation capabilities. SAR has been widely used in a variety of fields, including earth science, climate change, marine resources development, disaster monitoring, and target recognition [1], etc. It has demonstrated promising development potential in space-based earth observation. Since the 21st century, microwave-based remote sensing technologies have experienced rapid development, achieving significant breakthroughs in SAR with higher resolution, higher imaging dimension, and multipolarization, which play an important role in social development and economic construction.

Artificial intelligence (AI) techniques represented by deep neural networks (DNNs) have led a new generation of technological changes driven by big data, and have made disruptive progress in computer vision (CV) [2]. These methods have also been extended to intelligent interpretation with high-resolution SAR images [3], widely used in various tasks such as target classification, detection, terrain segmentation, change detection, three- (3D) and four-dimensional (4D) reconstruction, etc. [4]. However, unlike optical sensors, SAR operates in the microwave band and performs coherent imaging through side-looking imaging. The complexity of the imaging mechanism leads to the presence of speckle noise in SAR images, as well as geometric distortions such as range scale distortion, foreshortening, layover, and shadowing. The differences between SAR and optical data restrict the performance of DNNs in SAR image interpretation. To address the specificity of the radar imaging principle and the complexity of the physical mechanism of scattering, research on the information extraction theory and method applicable to the task of intelligent recognition of SAR images is essential for efficient analysis and utilization of SAR images.

Throughout the development process of SAR-automatic target recognition (SAR-ATR) techniques, from the clutter-statistics-based thresholding, template matching, and model-based feature extraction techniques to the automatic feature extraction techniques based on DNNs in recent years, the essence is to understand and extract the features of targets to achieve high-precision interpretation performance [1]. However, traditional methods based on hand-crafted features require the establishment of a unified statistical model, which limits their applicability. Meanwhile, data-driven SAR-ATR methods, primarily borrowed from CV, also face several critical bottlenecks, as follows:

1. **The inapplicability of directly applying CV models to SAR-ATR tasks.** Most of the DNN-based SAR-ATR algorithms are transferred from CV. However, due to the fundamental difference in the imaging principle of SAR and optical sensors, the two domains exhibit different characteristics. In optical sensors, targets typically appear as surfaces with continuous scattering and well-defined boundaries. In contrast, they manifest as discrete scattering points in SAR imagery. Additionally, due to the side-looking imaging mechanism, SAR images exhibit inherent geometric distortions. Directly transferring optical networks without considering SAR imaging mechanisms results in the loss of unique information about target scattering. Therefore, it is essential to incorporate SAR image scattering information and target semantic distributions into DNNs, as they can provide valuable insights for designing robust algorithms, enhancing the stability of models, and offering practical benefits.
2. **Data-driven DNNs are limited in addressing the fading effect in SAR images.** It is challenging for data-driven intelligent algorithms to handle the fading effect in SAR images due to the unique characteristics of radar sensors and the complex electromagnetic scattering mechanisms. Coherence is maintained throughout the imaging process, which results in significant variations in images of the same target due to phase-interference fading. The results of imaging can vary significantly with a slight change in observation parameters, such as the radar system, the observation mode, the platform's trajectory, the target position, etc. Especially for noncooperative high-value targets, scarce training data can easily result in an overfitting problem. These unique characteristics of SAR images severely limit the usefulness and effectiveness of historical data, further restricting the performance of data-driven deep models. Overall, the fading effect complicates the interpretation task with SAR images. Incorporating physical guidance into DNNs optimization is essential, as it clarifies the decision boundaries in complex environments, leading to more effective feature extraction and algorithm design.

3. **The lack of interpretability in DNNs hinders model optimization and practical implementation.** Current SAR-ATR algorithms rely on supervised training to optimize complex, highly parameterized models. Their 'black box' nature stems from intricate layers, nonlinear activations, and layer-by-layer transformation. This opacity significantly limits model interpretability, hindering operator comprehension of results and restricting the large-scale applications of AI-based SAR-ATR methods. However, SAR is underpinned by well-established theoretical foundations in electromagnetic wave propagation, geometric and radiometric modulation, and radar imaging algorithms. Consequently, improving interpretability has become a key priority, which involves: (1) investigating physical mechanisms of SAR to build interpretable DNNs; (2) designing inherently interpretable model architectures; and (3) analyzing the decision-making processes of models.

To address the problem of microwave radar imagery interpretation, we propose the development of cross-disciplinary microwave vision (MV) research, which further integrates electromagnetic physics and radar imaging mechanisms with human brain visual perception principles and CV technologies. The remainder of this chapter is organized as follows. In Section 8.2, we first present the perception model of MV. The forward and inverse problems of MV are introduced, and their possible solutions are further explored. Then, Section 8.3 presents a scattering information enhanced attention pyramid network (SIE-APN) for target detection in complex backgrounds and target-dense arrangement scenes. It integrates the scattering information modeled via Gaussian mixture model (GMM) with original images for effective enhancement to achieve precise detection performance. To address the few-shot learning (FSL) conditions in SAR-ATR, Section 8.4 further proposes the causal adversarial auto-encoder (CAE) based on disentangled factors of radar imaging and recognition process. It represents SAR imagery in a lower-dimensional manifold space, which significantly decreases the sample complexity for training. Besides, effective prior information is modeled based on the radar imaging process and merged with the DNN in a causally reasonable way. Both qualitative and quantitative experimental results and analysis are provided. Considering that the performance of DNN-based methods heavily relies on the quality of training data, Section 8.5 presents a recognition rate versus substitution rate curve (RSC) criterion for an objective and absolute assessment of AI-generated/electromagnetic-simulated datasets, which can be used as training data. Finally, Section 8.6 summarizes the main contributions of this work, and briefly discusses several promising directions for future research.

Notations: Throughout this chapter, upper(lower) bold face letters will be used for matrices (vectors); $\mathbb{C}^n/\mathbb{R}^n$ denotes spaces of n complex/real entries; The operators $(\cdot)^\star$, $\circ$, and $\otimes$ denote the optimal solution of any optimization problem, composition, and convolution, respectively; $\mathcal{N}(\boldsymbol{\mu}, \boldsymbol{\Sigma})$ denotes the Gaussian distribution with mean vector $\boldsymbol{\mu}$, and covariance matrix $\boldsymbol{\Sigma}$; and $\mathbb{E}[\boldsymbol{x}]$ denotes the expectation of $\boldsymbol{x}$.

8.2 The MV Cognition Model

To address the challenge of SAR image interpretation, the concept of MV is introduced [4]. It denotes the ability to intelligently perceive the microwave physical world in a manner analogous to human vision. From a visual cognition perspective, the primary task of MV is to extract semantic information about targets and scenes from two-dimensional (2D) radar images [4]. This section first presents the cognition model of MV, and then introduces the inverse problem defined in MV and its potential solving methods.

8.2.1 Statement of the Forward Problem

Visual cognition theory models human vision as an inverse problem of reasoning, using the Bayesian estimation theory to explain how visual reasoning resolves the ill-posed of inverse problems [5]. Therefore, we can establish MV as a machine reasoning inverse problem. First, a direct forward problem is established. The ultimate goal of intelligent model interpretation is to extract semantic knowledge from SAR images that aligns with human cognition, which differs from the physical and geometric information typically inverted in general perception inverse problems [6]. When observing a target, the inverse perception problem seeks to determine its 3D geometry. In contrast, semantic knowledge of target encompasses category, posture, and other semantic information of interest (IoI) to humans [7]. Semantic knowledge constitutes a subset of the geometric and physical information of target, whose space can be defined as a N_k-dimensional feature space with dimensions independent of each other, denoted as $\boldsymbol{k} \in \mathbb{C}^{N_k}$. The geometric and physical information of the target is represented as a N_x-dimensional real variable, denoted as $\boldsymbol{x} \in \mathbb{R}^{N_x}$. Based on the manifold distribution hypothesis [8], SAR images with high dimensions can be considered as concentrated on a low-dimensional manifold embedded in the high-dimensional space. The semantic knowledge represents the IoI corresponding to the intrinsic attributes of the target. Additionally, it is necessary to represent information that is not of interest, such as diversity denoted as $\boldsymbol{d} \in \mathbb{C}^{N_d}$ for generalization. Given $\boldsymbol{k}$ and $\boldsymbol{d}$, $\boldsymbol{x}$ can be uniquely determined, establishing a bijective function from the feature space to the target space:

$$\boldsymbol{x} = \mathrm{g}(\boldsymbol{k}, \boldsymbol{d}). \tag{8.1}$$

Here, the bijective function $\mathrm{g}(\cdot, \cdot)$ establishes a mapping from the feature to target space, referred to as the instancing process. This function characterizes the structure of the data manifold and the probability distribution of samples on it. It thereby embodies prior regularities, capturing fundamental properties such as the plausibility and veridicality of the samples. Determining the instancing function $g(\cdot, \cdot)$ is challenging due to its encapsulation of various physical laws and observational phenomena. But for a specific category of target, $g(\cdot, \cdot)$ may be constructed using prior knowledge and regularities, or it can be represented through generative models trained on extensive datasets. Further, a general perception model observed from target $\boldsymbol{x}$ to data $\boldsymbol{y}$ can be established as:

$$\boldsymbol{y} = f(\boldsymbol{x}, \theta, \delta), \tag{8.2}$$

where $\boldsymbol{y} \in \mathbb{R}^{N_y}$ represents the observation data, $\theta \in \mathbb{R}^{N_\theta}$ is the observation parameter, and $\delta \in \mathbb{R}^{N_y} \sim \mathcal{N}(\mathbf{0}, \boldsymbol{\Sigma})$ denotes the zero-mean Gaussian noise. Obviously, the perception function f is determined by the physical model of the perception system. Note that the perception function f must be injective and lossy, so that its inverse problem is ill-posed.

8.2.2 Inverse Problem and Its Solutions

The forward problem that integrates both the target generation and observation processes can be modeled as:

$$\boldsymbol{y} = f[g(\boldsymbol{k}, \boldsymbol{d}), \theta, \delta] = h(\boldsymbol{k}, \boldsymbol{d}, \theta, \delta). \tag{8.3}$$

Then, the inverse problem of perception can be derived as:

$$\boldsymbol{k} = h^{-1}(\boldsymbol{y}). \tag{8.4}$$

The function h is ill-posed. However, in the vast majority of cases, the observed data obtained under any observation conditions will have sufficient differences to distinguish between them for any two semantically distinct targets. Given a set of samples $[\boldsymbol{d}, \theta, \delta] \in \boldsymbol{\Theta}$, one can always find a mapping function m that satisfies:

$$\boldsymbol{k} = m(\boldsymbol{y}). \tag{8.5}$$

Actually, the DNN is a promising tool to fit the mapping function m, which can be defined with learnable parameters $\boldsymbol{\beta} \in \mathbb{R}^{N_\beta}$:

$$\boldsymbol{k} = m(\boldsymbol{y}|\boldsymbol{\beta}). \tag{8.6}$$

Hence, the learning process is:

$$\boldsymbol{\beta}^\star = \arg\min_{\boldsymbol{\beta}} L[\boldsymbol{k}, m(\boldsymbol{y}|\boldsymbol{\beta})], \tag{8.7}$$

where L represents the loss function. However, it is impossible to ensure that the $\boldsymbol{\Theta}$ covers all situations. The calculated mapping m based on $\boldsymbol{\Theta}$ is supposed to be generalized for another parameter set $\boldsymbol{\Theta}'$. That is, the following equations hold:

$$L'\left[\boldsymbol{k}, m(\boldsymbol{y}|\boldsymbol{\beta}^\star)\right] \approx L'\left[\boldsymbol{k}, m(\boldsymbol{y}'|\boldsymbol{\beta}^\star)\right], \tag{8.8}$$

$$\boldsymbol{y}' = h(\boldsymbol{k}, \boldsymbol{d}', \theta', \delta'), \forall[\boldsymbol{d}', \theta', \delta'] \in \boldsymbol{\Theta}', \tag{8.9}$$

where L' is the assessment criterion of learned mapping m. It does not require that the values of the $m(\mathbf{y}|\boldsymbol{\beta}^\star)$ and $m(\mathbf{y}'|\boldsymbol{\beta}^\star)$ be identical, but rather that their utility with respect to L' be consistent. Obviously, the learning objective of data-driven methods takes no account of prior distributions of targets. In contrast to CV, we can combine the electromagnetic physics and imaging mechanism of radar with data, developing a physics-informed theory with generalizability, interpretability, and reliability.

The relation between the forward and inverse problems in MV is presented in Figure 8.1. The forward problem is the process of obtaining observation data through instancing and perception based on the geometric-physical models of targets, scene diversity, and observational parameters. Conversely, the inverse problem aims to derive semantic knowledge abstracted from geometric-physical models based on observational data. Data-driven deep learning offers a pathway to solve the inverse problem when sufficient samples are available. However, due to the ill-posedness of inverse problems, the core challenge of their solution lies in constructing the physics-to-geometry mapping and achieving the fusion of models and data.

By referring to the Bayesian theory, the inverse problem in MV can be defined as a maximum a posterior estimation (MAP) shown below:

$$\begin{aligned}\mathbf{k}^\star &= \arg\max_{\mathbf{k}} P(\mathbf{k}|\mathbf{y}) \\ &= \arg\max_{\mathbf{k}} \frac{P(\mathbf{y}|\mathbf{k})P(\mathbf{k})}{P(\mathbf{y})} \\ &= \arg\max_{\mathbf{k}} P(\mathbf{y}|\mathbf{k})P(\mathbf{k}).\end{aligned} \tag{8.10}$$

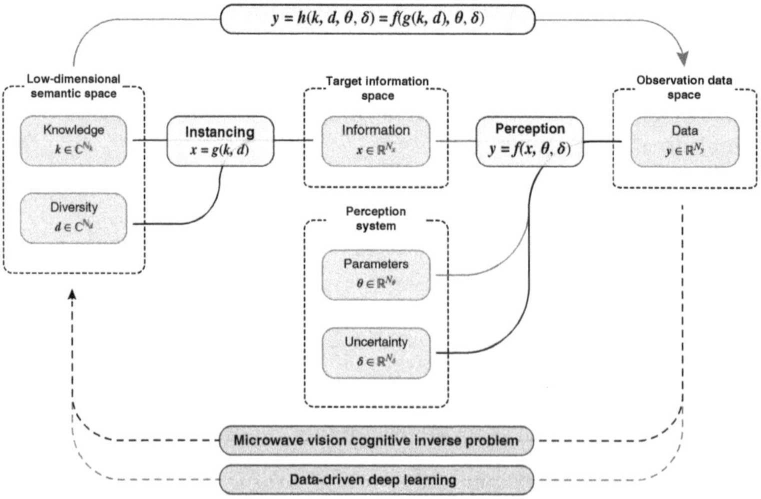

Figure 8.1 The perception model of MV.

It can be further expanded as:

$$\begin{aligned} \boldsymbol{k}^{\star} &= \arg\max_{\boldsymbol{k}} P(\boldsymbol{y}|\boldsymbol{x})P(\boldsymbol{x}|\boldsymbol{k})P(\boldsymbol{k}) \\ &= \arg\max_{\boldsymbol{k}} \mathbb{E}_{\theta,\delta}[P_f(\boldsymbol{y}|\boldsymbol{x},\theta,\delta)P(\theta)P(\delta)] \cdot \mathbb{E}_{d}[P_g(\boldsymbol{x}|\boldsymbol{k},\boldsymbol{d})] \cdot P(\boldsymbol{k}), \end{aligned} \tag{8.11}$$

where P_f and P_g, respectively, represent the observation and instancing processes, $P(\theta)$, $P(\delta)$, $P(\boldsymbol{d})$, and $P(\boldsymbol{k})$ are the prior distributions of datasets.

8.2.3 Possible Implementations of MV

The fundamental challenge in solving the MV cognitive model is to establish a mapping between the inverse and the forward problems, incorporating a well-defined, physics-based forward model into the inverse problem's solution strategy. The forward problem represents the electromagnetic physical mechanisms, while the inverse problem seeks the low-dimensional manifold geometry within a high-dimensional data space. Therefore, the core of physical intelligence is to establish a mapping between physics and geometry, achieving the integration of models and data. Suppose we have a representation function f and a generation function g, we can solve the above estimation problem through different strategies:

1. **Simulation Augmentation Method:** By efficiently sampling the forward problem and thereby augmenting more samples, we can train to obtain a more generalizable inverse mapping m.
2. **Physical Embedding Method:** We can design the structures or loss functions of networks using knowledge refined from the forward physical mechanisms, embedding physical laws into the DNNs to enhance their generalizability and interpretability.
3. **Interactive and Iterative Method:** A prediction model is first trained to estimate $\boldsymbol{k}$ from $\boldsymbol{y}$. Then, generating a new $\boldsymbol{y}'$ based on the estimated $\boldsymbol{k}$. Optimize the prediction model by comparing the original $\boldsymbol{y}$ and generated $\boldsymbol{y}'$. Iteratively optimizing this process to improve both models.

8.3 Scattering Information Enhanced Networks for Target Detection in SAR imagery

Influenced by the variability of SAR imaging results, identical targets often differ under different imaging parameters, which is unfavorable for neural networks that highly rely on training data. Enhancing the scattering information of targets is a feasible way to train a more generalizable inverse mapping for SAR-ATR tasks. We present two challenges faced by existing DNN-based strategies for SAR-ATR:

1. **Complex Scattering Characteristics of Target:** Targets in SAR images tend to appear as combinations of discrete scattering points, making it

difficult to achieve precise interpretation. SAR images are sensitive to observation parameters, resulting in substantial variations in the scattering characteristics of identical targets across different imaging conditions. This inherent variability poses a significant challenge to the development of robust data-driven models for SAR-ATR tasks [9].

2. **Infeasibility of Transferring from CV Models:** Contemporary state-of-the-art DNNs predominantly rely on optical imaging datasets for model pre-training. Nevertheless, SAR imagery exhibits fundamental disparities from optical imagery in physical characteristics, imaging mechanisms, and image morphology. This domain discrepancy leads to suboptimal feature representations when directly applying optical-based pretrained models to SAR image analysis, thereby constraining the models' generalization capacity.

8.3.1 Scattering Characteristics of Targets in SAR Images

The distinctive electromagnetic scattering information contained in SAR images is the basis for understanding and applying SAR images. As illustrated in Figure 8.2, these commonly used characteristics encompass geometric, gray statistics, texture, polarization, and noise characteristics. Traditional SAR-ATR algorithms are primarily based on a specific type or a combination of multiple features.

Taking aircraft as an example, the structures (excluding rotor structures and special types of aircraft) remain roughly similar and mainly consist of five components: nose, fuselage, wings, tail, and engine. Figure 8.3 shows the comparison of the imaging results of the components of an aircraft in the optical and Gaofen-3 SAR image (the pixel resolution is 1 m × 1 m). The main scattering mechanisms of each component are indicated.

As shown in Figure 8.3, due to the extraordinary detection performance on metal materials, targets in SAR images contain rich scattering information that primarily encompasses geometric, gray statistics, and texture characteristics.

1. **Geometric Characteristics:** The geometrical structure of aircraft serves as a distinctive feature in SAR images, enabling effective detection and recognition. For instance, aircraft typically manifest as discrete strong scattering points (SSPs) with distinct point-like scattering characteristics. Preliminary classification, localization, and identification can be achieved by detecting and analyzing such points. Additionally, the significant contrast in backscattering coefficients between aircraft targets (or their subcomponents) and the surrounding background induces abrupt grayscale transitions at target boundaries. These transitions generate linear edge features with pronounced edge information, further supporting target discrimination in complex SAR scenes.
2. **Gray Statistics Characteristics:** The grayscale intensity in SAR imagery is governed by the target's radar cross-section, which is influenced by its

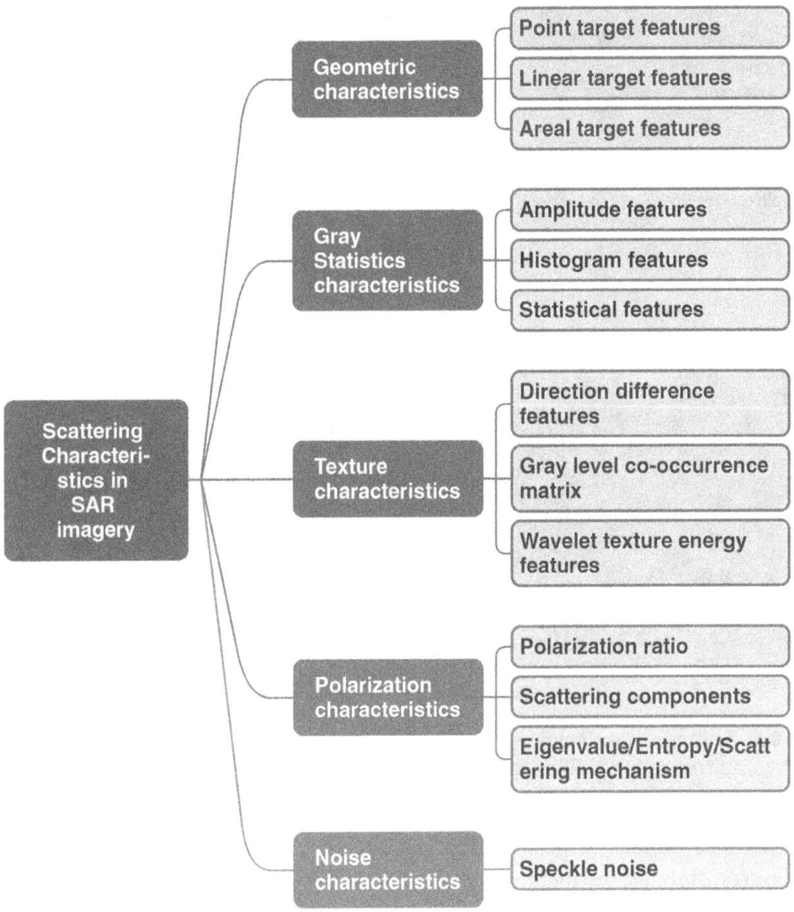

Figure 8.2 Scattering characteristics of SAR images.

structural and material properties under consistent imaging conditions. Aircraft targets exhibit significant structural and material disparities compared to their surrounding backgrounds, leading to pronounced differences in both grayscale amplitude and statistical distributions. Leveraging this contrast, aircraft extraction can be accomplished through two key strategies: modeling background clutter distributions, and statistically characterizing the grayscale distribution of targets.

3. **Texture Characteristics:** Texture refers to the spatial variation of grayscale intensity within an image region, determined by relative gray-level transitions rather than absolute intensity. In SAR images, texture patterns are influenced by radar system parameters (e.g., wavelength, resolution, incidence angle)

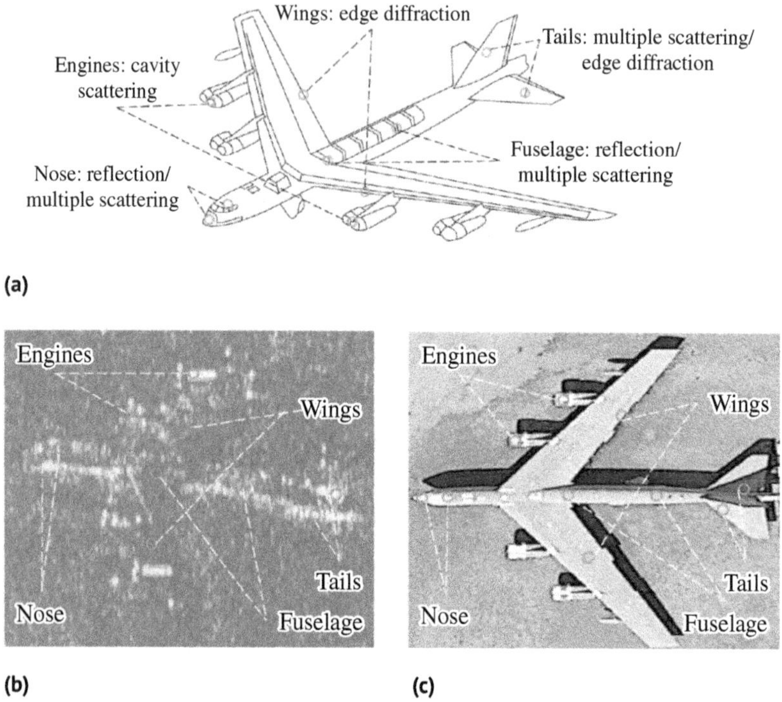

Figure 8.3 Scattering mechanism of each aircraft components. (a) 3D structure of aircraft, (b) corresponding SAR image, (c) corresponding optical image.

as well as target composition and the spatial arrangement of background features. Notably, distinctions among different types of aircraft primarily arise from texture variations rather than differences in grayscale intensity. Consequently, the accurate definition and calculation of local texture features are critical for suppressing false alarms and enabling precise classification of aircrafts.

8.3.2 Scattering Information Enhanced Attention Pyramid Network

The variability of targets in SAR images often results in significant differences under varying imaging conditions. Although DNN-based methods achieve satisfactory performance when there are abundant training samples [10, 11] or the target distribution is relatively uniform [12, 13], they remain fundamentally data-driven algorithms that struggle in complex environments or under few-shot conditions [14]. The key challenge lies in integrating DNNs with target scattering characteristics to improve the representational capability of DNNs for SAR images.

To validate this hypothesis, a detection network based on the SIE-APN in high-resolution SAR images is established [15], whose key contributions can be summarized as follows:

1. Integrating the grayscale and geometric features of airports in SAR images, an adaptive identification operator-based airport detection algorithm is proposed [16]. It can remove the irrelevant areas in advance, and has been demonstrated to be effective in the reduction of false alarms, and the improvement of efficiency.
2. Targets in SAR images are with plenty of scattering information, which implicitly represents the distinguished features of targets. Therefore, a scattering enhancement strategy is designed to address the discreteness and variability of targets that composed of the scattering information extraction and scattering region discrimination stages.
3. To circumvent the redundancy of deep convolution network and improve the detection performance of multiscale targets, an attention pyramid network based on the feature pyramid network (FPN) structure and an attention module (AM) is established. With the aim of mitigating the influence of unbalanced category distribution, the focal loss is implemented [17].

From an information augmentation perspective, the proposed SIE-APN is introduced. The structure of SIE-APN is presented in Figure 8.4 and consists of three stages:

1. Airport Detection Stage: An airport detection algorithm based on the adaptive identification operators is proposed, which can accurately and rapidly segment the airport area from a large scene of SAR imagery.
2. Scattering Information Enhancement Stage: SSPs are with satisfied significance and stability. Most scattering mechanisms of components can be essentially considered as a combination of SSPs. Specifically, the Harris–Laplace detector and the density-based spatial clustering of applications with noise (DBSCAN) algorithm are incorporated to eliminate noise. Subsequently, the GMM is utilized to model the extracted regions. Then, a discrimination process based on the Kullback–Leibler divergence (KLD) between distributions is implemented to extract and enhance candidate regions that may contain target.
3. Attention Pyramid Network Detection Stage: It is composed of the deep characteristic extraction and prediction. Firstly, an APN integrating multiscale feature extraction and attention mechanism through FPN and AM. The FPN architecture fuses high-resolution shallow features with high-semantic deep features, thereby enhancing detection accuracy through hierarchical context

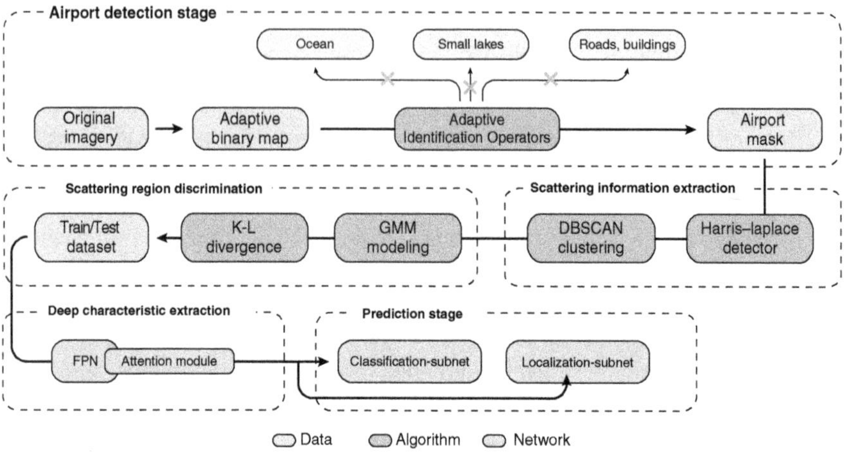

Figure 8.4 Framework of the proposed SIE-APN.

aggregation. Concurrently, the AM suppresses feature redundancy via cross-dimensional spatial-channel attention mapping. Besides, the focal loss is used as a replacement for cross-entropy loss, which ensures computational efficiency while optimizing decision boundaries between hard instances.

As localized maximum in scattering regions exhibiting strong gradient magnitudes, SSPs demonstrate consistent performance across varying observation conditions and geometric configurations. These properties establish SSPs as essential descriptors for the representation of scattering mechanisms. Stable extraction of SSPs can be achieved without prior knowledge. The Harris–Laplace detector is adopted for SSP extraction due to its inherent rotation and scale invariance. Firstly, the corner detection using the Harris operator across multiple scales is performed. Subsequently, characteristic scales are automatically determined by computing the Laplacian of Gaussian (LoG) at identified corner points. The multiscale second-moment matrix is calculated below:

$$\boldsymbol{M}(\boldsymbol{x}) = \sigma_D^2 G(\sigma_K) \otimes \begin{bmatrix} I_x^2(\boldsymbol{x}, \sigma_D) & I_x I_y(\boldsymbol{x}, \sigma_D) \\ I_x I_y(\boldsymbol{x}, \sigma_D) & I_y^2(\boldsymbol{x}, \sigma_D) \end{bmatrix}, \tag{8.12}$$

where $\boldsymbol{x}$ denotes the pixel position in the image, σ_K is the scale of Gaussian convolution kernel $G(\sigma_K)$, σ_D represents the local scale, $I_x(\boldsymbol{x}, \sigma_D)$ and $I_y(\boldsymbol{x}, \sigma_D)$, respectively, represent the first-order partial derivatives of the image at scale σ_D. Further, motivated by the circular/elliptical distribution of SSPs, the GMM is further integrated to enhance potential target regions prior. The GMM establishes clustering prototypes through multivariate Gaussian distributions, where spatial coordinates define cluster centroids, and the pixel intensities quantify probabilities, which has been demonstrated to have superior physical interpretability for SAR targets compared to deterministic approaches. The representation of GMM is given as:

$$p(\boldsymbol{x}) = \sum_{i=1}^{I} \omega_i \cdot p(\boldsymbol{x}|\boldsymbol{\mu}_i, \boldsymbol{\Sigma}_i), \tag{8.13}$$

where I represents the number of mixture models, $\boldsymbol{\mu}_i$ and $\boldsymbol{\Sigma}_i$ are, respectively, the mean vector and covariance matrix, and ω_i represents the mixture coefficient. The GMM is optimized through the expectation-maximization (EM) algorithm, which comprises two phases: the E step computes posterior probabilities of sample assignments to Gaussian components based on current parameter estimates; M step updates component parameters $\boldsymbol{\mu}_i$, $\boldsymbol{\Sigma}_i$, and ω_i through iterative updates. The Akaike's information criterion (AIC) is integrated to address high sensitivity to the initialization of component counts. The AIC is minus twice a penalized log-likelihood value, maximized using the maximum likelihood estimation (MLE) of the parameter vector $\boldsymbol{z}$. The likelihood function is denoted as:

$$\boldsymbol{A}\{f(\cdot;\boldsymbol{z})\} = -2\log L(\hat{\boldsymbol{z}}) + 2\dim(\boldsymbol{z}), \tag{8.14}$$

where $\dim(\boldsymbol{z})$ represents the length of parameter vector $\boldsymbol{z}$. Further, the KLD is adapted to discriminate the similarity among candidates and templates. Consequently, the Γ processing is empirically performed on the selected scattering regions for intensity enhancement of two channels, as defined in (8.15):

$$\Gamma(\boldsymbol{I}_{\text{in}}) = \begin{cases} \text{cat}(3, \boldsymbol{I}_{\text{in}}, \boldsymbol{I}_{\text{en}}, \boldsymbol{I}_{\text{en}}), D_{\text{KL}} \le T \\ \text{cat}(3, \boldsymbol{I}_{\text{in}}, \boldsymbol{I}_{\text{in}}, \boldsymbol{I}_{\text{in}}), D_{\text{KL}} > T, \end{cases} \tag{8.15}$$

where $\boldsymbol{I}_{\text{in}}$ is the input image corresponding to the selected region, $\boldsymbol{I}_{\text{en}}$ is the enhanced image of $\boldsymbol{I}_{\text{in}}$ by gamma distribution-based constant false alarm rate (CFAR), and T denotes the threshold. The two channels-enhanced processing is set experimentally. Through the SIE stage, effective information is enhanced in the image domain to construct multichannel image.

Then, the APN is trained with these processed samples to predict the accurate positions of target, whose structure is given in Figure 8.5. It is composed of the basic FPN structure incorporating AM. Specifically, FPN includes bottom-up, up-bottom, and lateral connection pathways to combine the high-resolution information of shallow layers with the high-semantic information of deep layers, thereby achieving precise detection performance for multiscale targets. Additionally, the AM is designed based on the CBAM module, which significantly improves feature discriminability with minimal computational cost. This module directs the network's focus toward critical aircraft attributes while filtering irrelevant features. We integrate the AM into each feature fusion layer of FPN, enabling scale-specific attention mapping to dynamically enhance salient features across various sizes.

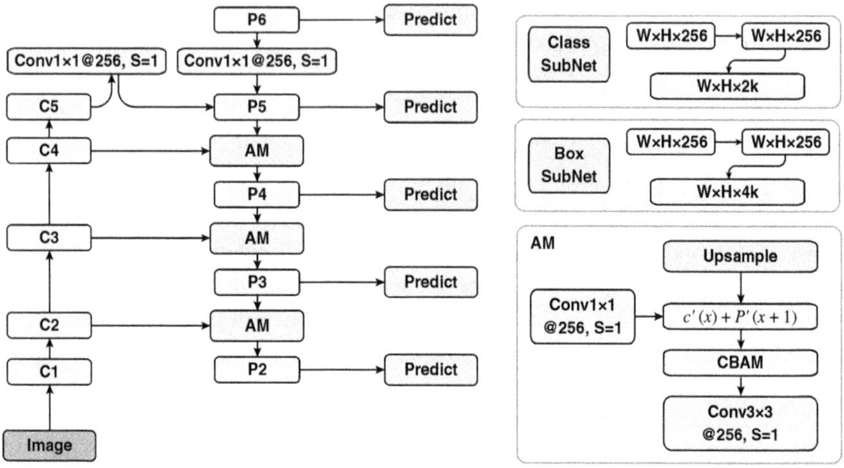

Figure 8.5 Structure of APN.

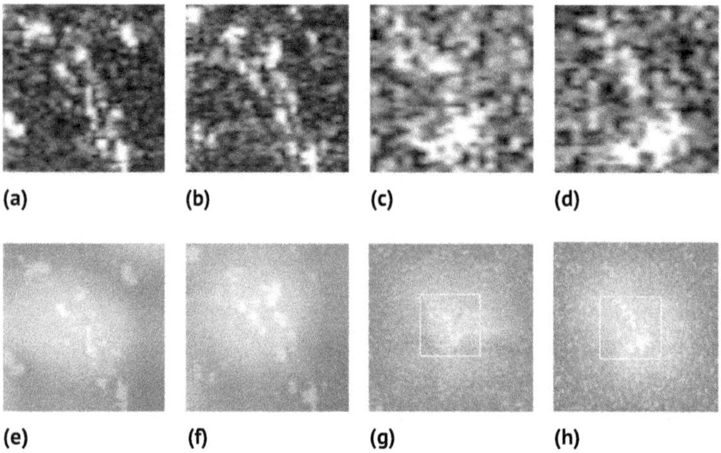

Figure 8.6 Visualizations of features for target with multi-scales. (a), (b) Samples of large aircraft. (c), (d) Samples of small aircraft. (e)–(h) Corresponding extracted features.

Feature visualizations indicate that large aircraft demands emphasis on component spatial relationships, whereas small aircraft requires comprehensive perception (see Figure 8.6).

8.3.3 Experimental Results and Analysis

Experimental datasets using Gaofen-3 and TerraSAR-X SAR images are constructed, named GF3-ADD and T-ADD, respectively, for evaluation. Ground-truth annotations were manually generated by cross-referencing the SAR imagery with co-registered optical imagery, ensuring high annotation accuracy for analysis.

In order to compare the performance of different algorithms, 8 representative scene slices are, respectively, selected based on the established datasets [18–23]: using the GF3-ADD to validate the performance of the model for target detection in complex backgrounds and using the T-ADD to validate the performance for densely arrayed target situations. The visualization of the detection effect of different algorithms is given in Figure 8.7 (the circles, pentagrams, and rectangles denote the false alarms, missed, and correctly detected targets, respectively). The experimental results based on the T-ADD data are given in Figure 8.8. It is demonstrated that the proposed SIE-APN achieves superior performance in complex and target-densely arrayed situations.

Evaluations of GF3-ADD and T-ADD datasets demonstrate the robustness of the proposed method, achieving 83.25% AP in complex environments with densely arranged targets. Compared with other representative algorithms, the SIE-APN comprises two key stages: (1) a Harris–Laplace detector extracts SSPs from the scene. DBSCAN clustering then generates candidate scattering regions. These regions are statistically modeled using GMMs, with KLD quantifying the similarity between pre-trained reference and test models. Scattering regions containing aircraft are enhanced by replacing two image channels with Gamma-CFAR-processed data. This process suppresses background clutter while emphasizing discriminative target features; (2) a multiscale architecture combining the FPN with adaptive AM addresses target discreteness and scale variations.

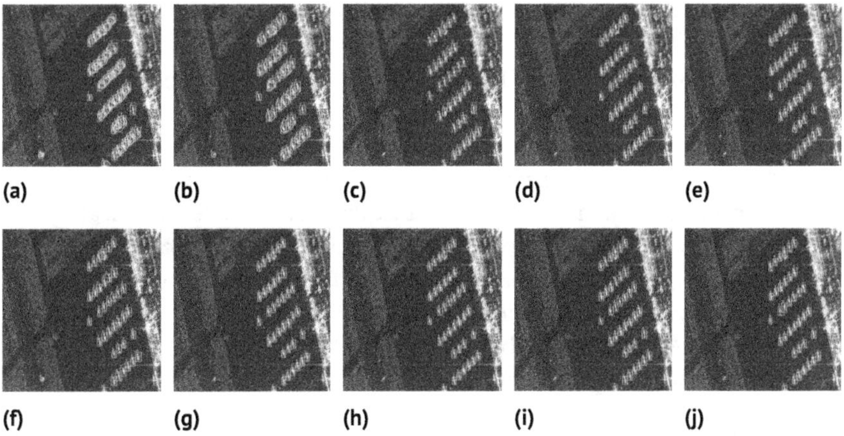

Figure 8.7 Aircraft detection results in several test samples of GF3-ADD. (a) SSD, (b) YOLO, (c) Faster R-CNN, (d) FPN, (e)FPN+SE, (f) APN-CBAM, (g) APN-FL, (h) APN, (i) SIE-FPN, (j) SIE-APN

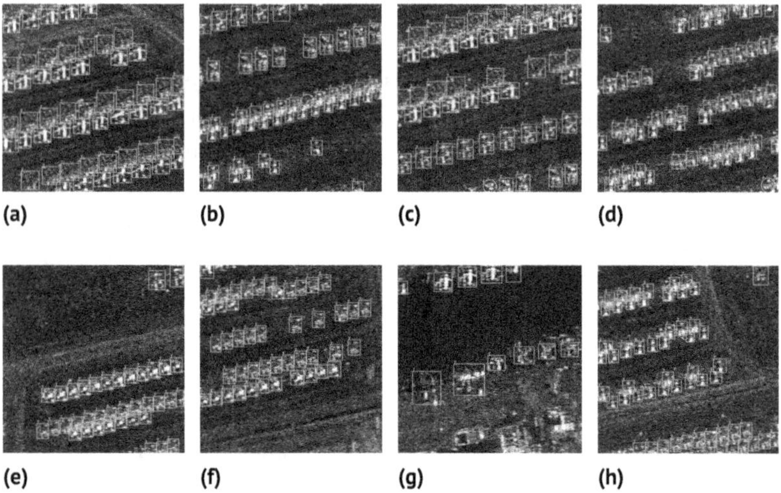

Figure 8.8 SIE-APN detection performance in T-ADD. (a) slice 1, (b) slice 2, (c) slice 3, (d) slice 4, (e) slice 5, (f) slice 6, (g) slice 7, (h) slice 8.

8.4 Physics-informed Models for SAR Image Representation

The SIE-APN provides an explicit way to combine scattering information of SAR images with the DNN for more robust detection performance with a large number of training samples. However, the decision-making mechanism of DNNs remains a black box, with an intricate hierarchical structure, highly nonlinear activation processing, and the extraction of features through layer-by-layer abstraction. Moreover, it is often difficult to obtain sufficient labeled data for model training in practical applications: (1) the long imaging period and high cost preclude the acquisition of sufficient SAR samples for model training in practical applications; (2) SAR images are susceptible to observation parameters, which is detrimental to data-driven DNNs. In such scenarios, another viable strategy involves integrating physical priors into both network architecture design and model optimization. To reduce the model's dependency on training samples while enhancing its generalization and representation capabilities. Establishing a physics-informed model for SAR image representation is precisely the motivation behind CAE. We summarize that the challenges encountered by existing SAR-ATR methods are predominantly manifested in the following two aspects:

1. **Insufficient Interpretability:** Although DNNs offer an automated end-to-end learning framework that significantly enhances the performance of the target interpretation task in SAR images, their internal decision logic remains opaque. The electromagnetic properties contained in SAR images, which

differ significantly from those in optical models, are not fully utilized. Lack of transparency seriously affects the reliability and applications.

2. **Poor Generalization Ability:** Due to the scarcity of raw data, the difficulty of sample annotation, and the sensitivity of sensor parameters, the number of available SAR training samples is much less than that of optical images in practical applications. Generalization of DNNs trained with sparse samples remains a challenging problem.

8.4.1 Causal-inspired Representation Model for SAR Image

8.4.1.1 Manifold Distribution Hypothesis of SAR Data

The distribution hypothesis of neural data can be attributed to two fundamental principles: (1) manifold distribution principle: high-dimensional data within a class inherently resides on a low-dimensional manifold, and (2) clustering distribution principle: distinct sub-classes exhibit separable probability distributions on this manifold with sufficient inter-class distance for effective classification [8, 24]. DNNs learn these manifolds through hierarchical nonlinear transformations. In the SAR imaging procedure, targets of identical category generate variable observations with varying wavelengths, depression angles, azimuth angles, and background clutter. Despite observational condition variations, the geometrically constrained continuity of these images implies that SAR targets occupy low-dimensional manifolds embedded in high-dimensional image spaces. Furthermore, the statistical appearance frequency of sub-classes samples suggests nonuniform probability distributions across the manifold structure.

Samples from the moving and stationary target acquisition and recognition (MSTAR) dataset are utilized to validate the manifold hypothesis [25]. Each 128×128 image is represented as a vector in $\mathbb{R}^{16384}$ space. A pretrained gold-standard network (trained on extensive data to encapsulate rich prior knowledge) serves as the feature extractor. Subsequently, the t-distributed stochastic neighbor embedding (t-SNE) is adapted to project the high-dimensional features onto a 2D manifold [26], revealing intrinsic data structures. As visualized in Figure 8.9a, the manifold of 10-class targets exhibits distinct color-coded clusters corresponding to categories. Figure 8.9b further demonstrates azimuth-dependent regularity within a single category: samples with adjacent angles maintain geometric proximity on the manifold, while angular variations induce progressive distribution shifts.

8.4.1.2 Causal Model for Disentangled Representation of SAR Image

8.4.1.2.1 Overview of the SAR Imaging Process

As illustrated in Figure 8.10, the composition of imaging radar typically encompasses a transmitter, a receiver, a converter, an antenna, and an assortment of additional components. The generation of high-power signals by the transmitter

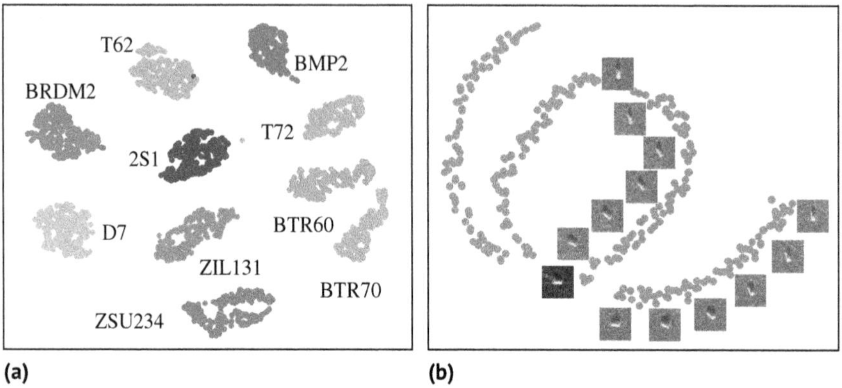

Figure 8.9 Visualization of the 2D manifold distribution of MSTAR. (a) embedded results of 10-class targets, (b) embedded results of azimuth attribute with a single category. *Source*: Adapted from MSTAR Overview / https://www.sdms.afrl.af.mil/index.php?collection=mstar / last accessed Feb. 13, 2026.

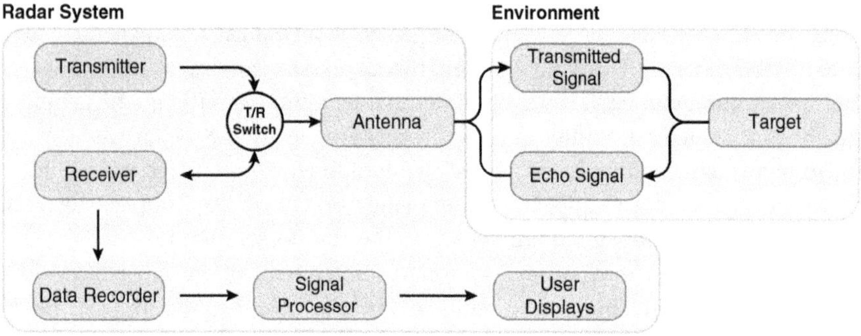

Figure 8.10 Composition of imaging radar.

initiates the process. These signals are then transmitted through the converter to the antenna. The antenna emits a narrow-beam short pulse, which, upon striking a ground target at a specific distance, generates an echo that is received by the antenna. This echo is subsequently processed by the converter and transmitted to the receiver. The receiver then stores the received signals in its memory. Subsequent to this, a series of imaging processors are employed to perform various imaging procedures. Given the complexity inherent in the realization of the hardware system, the high-frequency narrow pulse signal is typically obtained through post-processing methodologies.

Theoretical and experimental results demonstrate that the intensity of the SAR image is contingent on the magnitude of the back-scattering coefficient, denoted by σ. The factors influencing σ can be categorized into two distinct classes: radar system parameters, encompassing wavelength, angle of incidence, polarization mode, etc.; and inherent parameters, including target characteristics associated with the geometry and physical properties of the target and observation conditions. Specifically, SAR imaging attains high resolution through coherent processing of radar-transmitted waves with respect to the back-scattering signals. The entire signal returned from a ground unit is the coherent sum of the back-scattering signals in ground scattering. Given the random variation of the individual sub-scattered signals, the presence of both strong echoes and echo fading is possible when superimposed on the same resolution unit. This results in the manifestation of speckle noise in the SAR image. Furthermore, the sampling, quantization, compression, transmission, and other processing of SAR data invariably involves system noise. Consequently, the SAR image comprises two distinct components: IoI and noninterest information. The former encompasses the target intrinsic properties, while the latter includes background clutter, system noise, and speckle noise. Specifically, the IoI delineates the model's classification boundary, while the noninterest information is indispensable for the effective characterization of SAR images and representing the model's generalization performance.

8.4.1.2.2 Abstracted Causal Diagram of SAR Imaging

The efficacy of a neural network is contingent upon its capacity to disentangle and flatten the data manifold [27]. SAR images can also be considered concentrated on a low-dimensional manifold regulated by several intrinsic factors. Therefore, the causal graph model of SAR image generation, as analyzed through the SAR forward imaging process, can be represented by a network parameterized by $\boldsymbol{a} = \{a_1, a_2, ..., a_n\}$ and functions $\boldsymbol{b} = \{b_1, b_2, ..., b_n\}$, which includes both the external influencing factors of SAR imaging and the imaging mechanism. If $\boldsymbol{a}$ causes $\boldsymbol{c}$, then it can be expressed as $\boldsymbol{c} = u(\boldsymbol{a}, \boldsymbol{e})$ or $\boldsymbol{a} \to \boldsymbol{c}$. It indicates that the function u maps the variable $\boldsymbol{a}$ to the variable $\boldsymbol{b}$ based on some variables $\boldsymbol{e}$ that are independent of $\boldsymbol{a}$. These $\boldsymbol{e}$ are known as exogenous variables, as they are determined by conditions external to the system under study [28, 29]. The manipulation of exogenous variables is a crucial aspect of studying causal relationships. A causal graph of a basic SAR image generation process is illustrated in Figure 8.11. It encompasses the following key factors and causal relationships: radar system parameters, including frequency, power, observation angles, etc.; target characteristics, including category and azimuth angle, etc.; and background and system parameters. These factors and relationships can be further extended and refined to provide a more precise modeling of the actual SAR imaging process.

It is intuitive to assume that a latent space with less curvature will result in a smoother transition in representation than a hidden space with more curvature [30]. A disentangled representation model for SAR images to streamline the modeling process of the neural network and reduce the difficulty of modeling and optimization is proposed, as shown in Figure 8.12. It is primarily composed of two

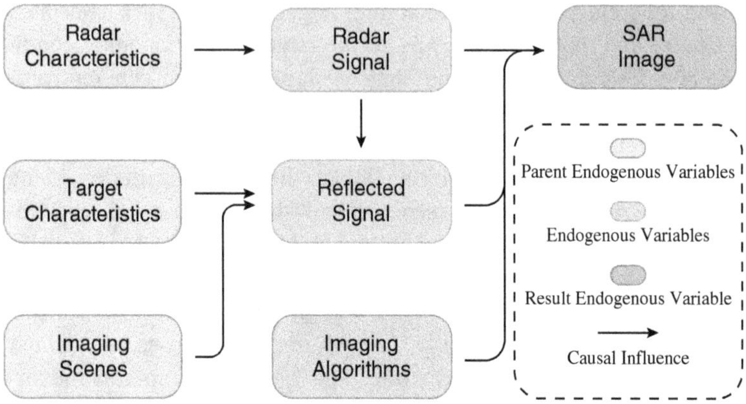

Figure 8.11 A causal graph for the generation process of SAR image.

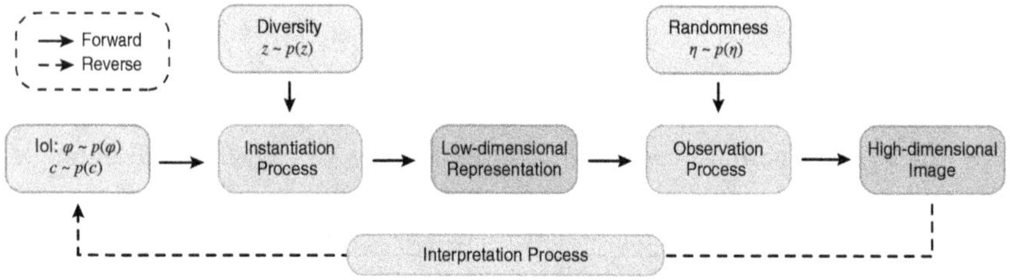

Figure 8.12 Causal modeling for disentangled factors.

semantic elements: IoI and noninterest information. The former focuses on the semantic features of the target, including azimuth and category information. The latter includes the diversity and randomness of the image for representation.

8.4.2 Interpretable Disentangled Adversarial Auto-encoder

8.4.2.1 Framework of the Proposed CAE

From a physical embedding perspective, a CAE is further proposed [31], guided by the disentangled feature representations illustrated in Figure 8.12. The framework (Figure 8.13) explicitly models three latent components: intrinsic (class-discriminative boundaries for classification), diversity (semantic-preserving image variations corresponding to scenes), and randomness (stochastic perturbations associated with coherent imaging system). This disentanglement enhances generalization by isolating invariant class characteristics from confounding factors. Inspired by the adversarial latent auto-encoder [32], we optimize latent space distributions under low-dimensional constraints to mitigate the dimensionality

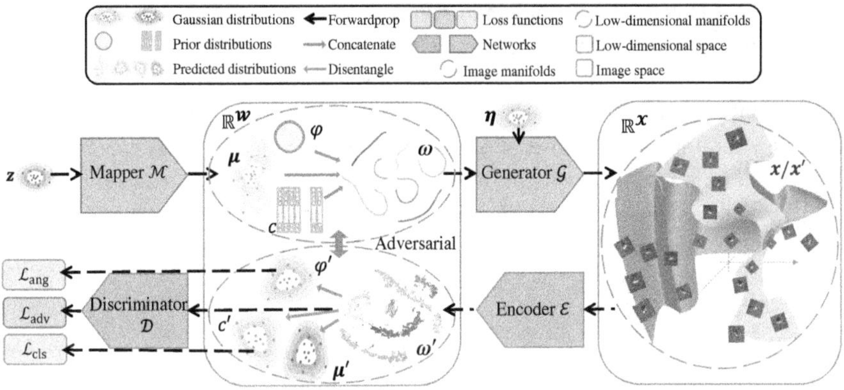

Figure 8.13 The architecture of CAE. The mapper $\mathcal{M}$ perform instancing process, and generator $\mathcal{G}$ is observation process. The encoder $\mathcal{E}$ represents the inverse perception process with the discriminator $\mathcal{D}$ as an auxiliary constraint. $\boldsymbol{z}$ and η are the diversity and randomness, respectively. ϕ and c jointly represent the IoI, and ω denotes the low-dimensional vectors.

curse in FSL scenarios. Additionally, SAR-specific prior distributions are integrated to regularize feature learning, leveraging domain knowledge for improved sample efficiency.

The proposed CAE leverages adversarial interactions among four core components, including mapper $\mathcal{M}$, generator $\mathcal{G}$, encoder $\mathcal{E}$, and discriminator $\mathcal{D}$, to achieve manifold learning and probabilistic distribution alignment. The framework operates in three stages: (1) $\mathcal{M}$ projects prior distributions to latent space, (2) $\mathcal{G}$ and $\mathcal{E}$ jointly establish bidirectional mappings between latent representations and image manifolds, and (3) $\mathcal{D}$ quantifies distributional discrepancies between the learned and the prior distributions of low-dimensional vectors. Disentangled prior injections follow a causally reasonable scheme (Figure 8.12):

$$\begin{aligned} &\text{azimuth:} && p_{\text{prior}}(\phi) = U(0, 2\pi) \\ &\text{category:} && p_{\text{prior}}(c) = \text{onehot}(c) \\ &\text{diversity:} && p_{\text{prior}}(\boldsymbol{z}) = \mathcal{N}(\boldsymbol{0}, \boldsymbol{I}) \\ &\text{randomness:} && p_{\text{prior}}(\eta) = \mathcal{N}(\boldsymbol{0}, \boldsymbol{I}). \end{aligned} \tag{8.16}$$

These priors are strategically deployed: $\boldsymbol{z}$ encodes target configuration variability independent of intrinsic features; ϕ and c modulate the pre-disentangled representation $\boldsymbol{\mu}$ transformed from $\boldsymbol{z}$; while η injects SAR-specific observation noise through the residual blocks of $\mathcal{G}$.

8.4.2.2 Loss Functions and Implementations

CAE optimizes high-dimensional image generation and interpretation problems as low-dimensional representation vector constraint problems. It realizes sample representation with a few of samples through the low-dimensional representation vectors. Therefore, the proposed CAE can be applied to tasks such as massive sample completion, target classification, and azimuth prediction. Ensuring that the distribution encoded by $\mathcal{E}$ is consistent with the prior distribution, essentially forcing the image distributions to be the same:

$$q_{\mathcal{E}}(\boldsymbol{\omega}') = q_{\text{prior}}(\boldsymbol{\omega}), \tag{8.17}$$

and

$$q_{\mathcal{G}}(\boldsymbol{x}') = p_{\text{data}}(\boldsymbol{x}). \tag{8.18}$$

In order to achieve Eq. (8.18), the adversarial learning loss $\mathcal{L}_{\text{adv}}$ and reconstruction loss $\mathcal{L}_{\text{rec}}$ are, respectively, designed as:

$$\begin{aligned}\mathcal{L}_{\text{adv}} &= \min_{\mathcal{M},\mathcal{G}} \max_{\mathcal{E},\mathcal{D}} (f_{\text{dis}}(\mathcal{D}\circ\mathcal{E}) + f_{\text{dis}}(-\mathcal{D}\circ\mathcal{E}\circ\mathcal{G}\circ\mathcal{M})) \\ &= \min_{\mathcal{M},\mathcal{G}} \max_{\mathcal{E},\mathcal{D}} (\mathbb{E}_{\boldsymbol{x}\sim p_{\text{data}}(\boldsymbol{x})}[f_{\text{dis}}(\mathcal{D}\circ\mathcal{E}(\boldsymbol{x}))|\boldsymbol{x}] \\ &\quad + \mathbb{E}_{\boldsymbol{z}\sim p_{\text{prior}}(\boldsymbol{z}),\phi\sim p_{\text{prior}}(\phi),c\sim p_{\text{prior}}(c)}[f_{\text{dis}}(-\mathcal{D}\circ\mathcal{E}\circ(\mathcal{G}\circ(\mathcal{M}(\boldsymbol{z}),\phi,c))|\boldsymbol{z},\phi,c]),\end{aligned} \tag{8.19}$$

and

$$\begin{aligned}\mathcal{L}_{\text{rec}} &= \min_{\mathcal{E},\mathcal{G},\mathcal{M}} \Delta(\mathcal{M}||\mathcal{E}\circ\mathcal{G}\circ\mathcal{M}) \\ &= \min_{\mathcal{E},\mathcal{G},\mathcal{M}} (\mathbb{E}_{\boldsymbol{z}\sim p_{\text{prior}}(\boldsymbol{z}),\phi\sim p_{\text{prior}}(\phi),c\sim p_{\text{prior}}(c)}[\Delta((\mathcal{M}(\boldsymbol{z}),\phi,c)|| \\ &\quad (\mathcal{E}\circ\mathcal{G}\circ(\mathcal{M}(\boldsymbol{z}),\phi,c)))|\boldsymbol{z},\phi,c]),\end{aligned} \tag{8.20}$$

where $f_{\text{dis}}(t) = t$ scores the output, and $\Delta(p||q)$ calculates the ℓ_2-norm discrepancy in implementations. A self-constrained optimization paradigm is employed in CAE. Furthermore, task-specific losses measuring the classification and azimuth angle prediction precision are jointly optimized through a unified learning framework:

$$\begin{aligned}\mathcal{L}_{\text{angd}} &= \min_{\mathcal{E}} (\mathbb{E}_{\boldsymbol{x}\sim p_{\text{data}}(\boldsymbol{x}),\phi_{\text{real}}\sim p_{\text{data}}(\phi)}[f_{\text{ang}}(\mathcal{E}(\boldsymbol{x})|\phi_{\text{real}})|\boldsymbol{x}] \\ &\quad + \mathbb{E}_{\phi\sim p_{\text{prior}}(\phi)}[f_{\text{ang}}(\mathcal{E}\circ\mathcal{G}(\phi)|\phi)] \\ &\quad - \mathbb{E}_{\boldsymbol{x}\sim p_{\text{data}}(\boldsymbol{x}),\phi_{\text{real}}\sim p_{\text{data}}(\phi),\phi\sim p_{\text{prior}}(\phi)}[f_{\text{ang}}(\mathcal{E}(\boldsymbol{x})|\phi\neq\phi_{\text{real}})|\boldsymbol{x}] \\ &\quad - \mathbb{E}_{\phi\sim p_{\text{prior}}(\phi),\phi_{\text{fake}}\sim p_{\text{prior}}(\phi)}[f_{\text{ang}}(\mathcal{E}\circ\mathcal{G}(\phi_{\text{fake}})|\phi\neq\phi_{\text{fake}})]),\end{aligned} \tag{8.21}$$

$$\mathcal{L}_{\text{angg}} = \min_{\mathcal{G}} \mathbb{E}_{\phi\sim p_{\text{prior}}(\phi)}[f_{\text{ang}}(\mathcal{E}\circ\mathcal{G}(\phi)|\phi)], \tag{8.22}$$

$$\begin{aligned}\mathcal{L}_{\text{clsd}} = \min_{\mathcal{E}}(&\mathbb{E}_{x\sim p_{\text{data}}(x),c_{\text{real}}\sim p_{\text{data}}(c)}[f_{\text{cls}}(\mathcal{E}(x)|c_{\text{real}})|x] \\ &+ \mathbb{E}_{c\sim p_{\text{prior}}(c)}[f_{\text{cls}}(\mathcal{E}\circ\mathcal{G}(c)|c)] \\ &- \mathbb{E}_{x\sim p_{\text{data}}(x),c_{\text{real}}\sim p_{\text{data}}(c),c\sim p_{\text{prior}}(c)}[f_{\text{cls}}(\mathcal{E}(x)|c\neq c_{\text{real}})|x] \\ &- \mathbb{E}_{c\sim p_{\text{prior}}(c),c_{\text{fake}}\sim p_{\text{prior}}(c)}[f_{\text{cls}}(\mathcal{E}\circ\mathcal{G}(c_{\text{fake}})|c\neq c_{\text{fake}})]),\end{aligned} \tag{8.23}$$

and

$$\mathcal{L}_{\text{clsg}} = \min_{\mathcal{G}} \mathbb{E}_{c\sim p_{\text{prior}}(c)}[f_{\text{cls}}(\mathcal{E}\circ\mathcal{G}(c)|c)], \tag{8.24}$$

where f_{ang} is ℓ_2-norm loss, f_{cls} represents cross-entropy loss, ϕ_{real} and c_{real}, respectively, represent the azimuth and category, and ϕ_{fake} and c_{fake} denote the ground-truth for fake images. Within each epoch, these loss functions are optimized in succession, with each one responsible for updating specific subnetworks.

A symmetrically conditional encoding (SCE) module is introduced to constrain the same positions of the low-dimensional representations of the samples to have consistent semantic representation properties. Based on a specified low-dimensional representation vector, the generator $\mathcal{G}$ gradually generates high-resolution 128×128 images from the 4×4 constant images, where angle and category information are introduced as conditional constraints for sample generation. Angle and category information are concatenated with the initially factorized low-dimensional representation. Because the encoder $\mathcal{E}$ with a symmetric structure is used to encode high-dimensional images, the resulting low-dimensional representation vector has the same semantic information in the corresponding positions as the pre-specified low-dimensional representation vector. Therefore, the target's angle and category information can be directly factorized from the vector output by encoder $\mathcal{E}$ based on the encoded positional information.

8.4.3 Evaluation for the Representation and Generalization Performances

Experiments are conducted on the MSTAR dataset. The training set and test set are randomly divided according to azimuth interval of the training set to construct various FSL conditions. As the azimuth interval gradually increases, the number of training samples gradually decreases. When the azimuth interval of the training set is 30°, there are 121 training samples with a depression angle of 17°, 2,625 test samples with a depression angle of 17°, and 2,426 test samples with a depression angle of 15°.

To validate the representation capability of the proposed method, we first train the CAE using 121 samples across 10 target classes. The trained model reconstructs and generates training/test samples, with real, reconstructed, and generated examples visualized in Figure 8.14. To demonstrate feature disentanglement, we synthesize two image groups sharing identical intrinsic attributes but varying diversity $\boldsymbol{z}$ and randomness $\boldsymbol{\eta}$. As shown in Figure 8.14, generated samples retain structural fidelity while exhibiting controlled variations in background texture. This controlled diversity confirms explicitly modeling $\boldsymbol{z}$ and $\boldsymbol{\eta}$ in enhancing sample quality.

The angular disentanglement capability is quantitatively evaluated through controlled interpolation experiments. Employing encoder $\mathcal{E}$, a fixed-category sample is initially generated. Progressive azimuth interpolation from the source to target angles is subsequently performed, with generator $\mathcal{G}$ synthesizing intermediate representations based on the interpolated latent vectors. Figure 8.15 illustrates representative cases from distinct categories of targets in MSTAR dataset, spanning azimuth ranges of [2°, 30°] and [67°, 96°].

To quantitatively evaluate label attribute disentanglement performance, target interpolation experiments using the proposed CAE are conducted. A fixed-angle sample of 2S1 is encoded into a latent representation via encoder $\mathcal{E}$. Subsequently, linear interpolation of category attributes generates target representations, which are decoded into synthesized images through generator $\mathcal{G}$. Figure 8.16 demonstrates two representative transformation sequences, where leftmost/rightmost

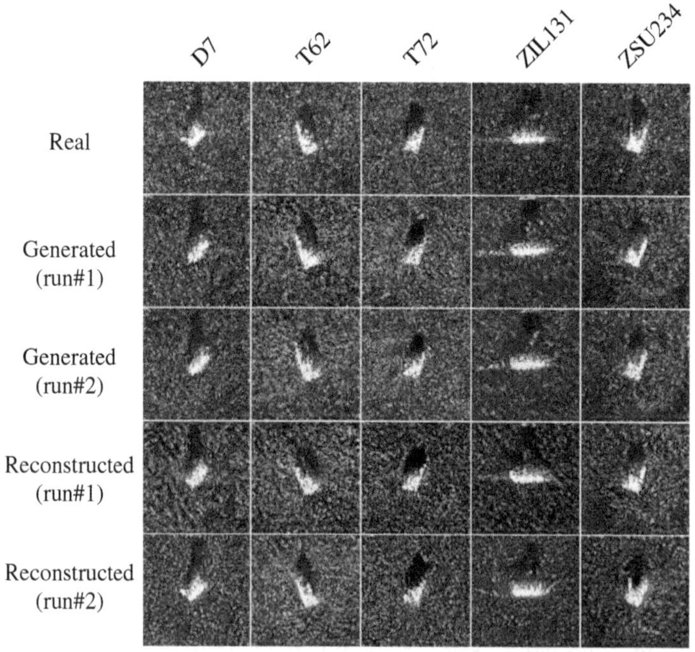

Figure 8.14 Illustrations of the real, generated, and reconstructed samples.

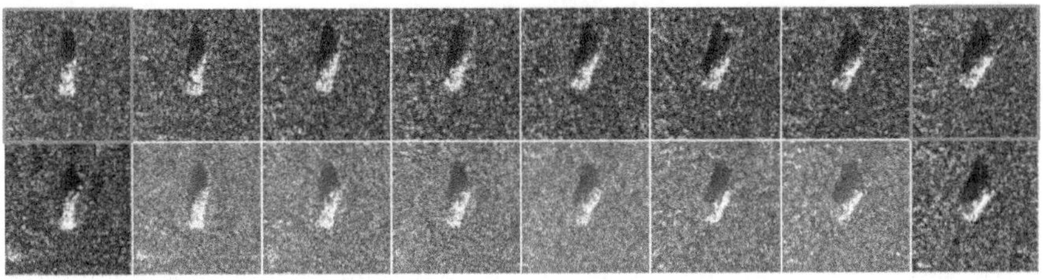

(a)

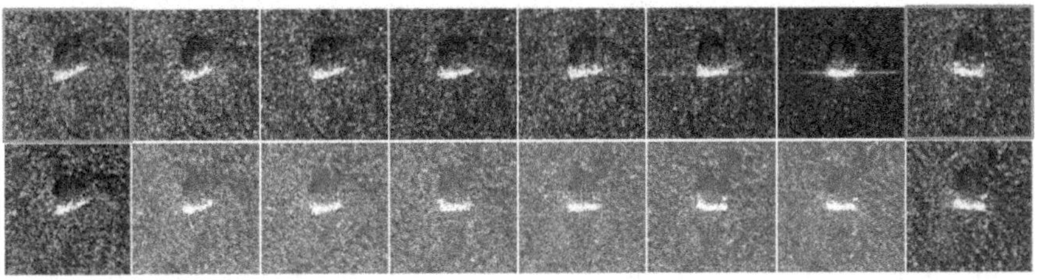

(b)

Figure 8.15 Illustrations of azimuth angle interpolation. The top and bottom rows are real and generated samples, with bolded boxes representing the training samples. (a) 2S1, (b) BRDM2.

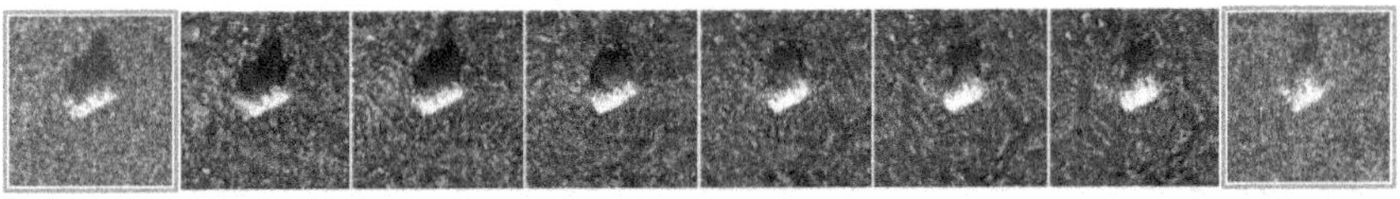

(a)

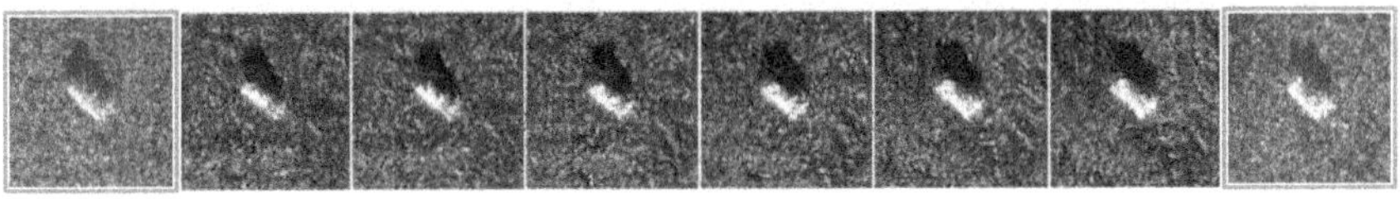

(b)

Figure 8.16 Progressively transformed samples with increasing $\gamma \in [0, 1]$, with bolded boxes representing the real samples. (a) transformed from 2S1 to D7 (azimuth angle = 60°), (b) transformed from 2S1 to T62 (azimuth angle = 315°).

positions display source/target samples. The proposed method generates photo-realistic imagery despite synthesizing objects absent from real-world domains. Notably, quantitative evaluations confirm the model's effectiveness in disentangling label-specific attributes while retaining intrinsic data characteristics, which is a critical capability for synthetic data generation systems.

To address the challenges of insufficient interpretability and poor generalization of current DNN-based methods, this section proposes a method for disentangling and representing SAR images using a CAE. It models SAR images as a low-dimensional manifold controlled by three semantic elements: intrinsic properties, diversity, and randomness. Unlike traditional adversarial generative models, CAE establishes an end-to-end framework that fuses image generation, classification, and azimuth prediction modules for optimization. It accomplishes the data manifold learning and the transformation of manifold probability distribution through a collaborative confrontation of four subnetworks. The disentangled feature distribution is used as prior information to constrain the network optimization process, reducing the model's demand for training samples. Specifically, an angle encoding method based on cyclic high-frequency information is used to improve the network's representation of rotational features; in addition, a symmetric network is used to ensure the semantic consistency. Experiments were carried out based on the MSTAR dataset. Qualitative analysis and quantitative comparison of the model's representation and generalization performance were carried out. The CAE establishes a causally interpretable model and demonstrates superior performance when the target azimuth is sparsely distributed. The proposed CAE not only can encode high-dimensional images into low-dimensional features to obtain the semantic attributes of the target, but also its superior sample generation capability can be used for high-quality sample completion.

8.5 An Assessment Criterion for Utility Evaluation of SAR Data

As mentioned above, synthetic datasets can effectively improve the poor generalization ability by introducing unseen attributes of measured samples. DNN-based models and physics-based simulators are promising techniques to generate massive SAR data when measured training data is scarce. In FSL scenarios, the quality of simulated data has a significant influence on the performance of the trained model. Therefore, we hope that there is an objective and absolute criterion for the quantitative measurement of the utility of simulated data.

Therefore, an evaluation criterion based on the RSC is designed to objectively assess simulated datasets during interactive and iterative processes [33]. It mainly includes two stages: construction of datasets based on the substitution rate (SR) and an evaluation stage based on the RSC. First, a training set of paired reference and simulated data is constructed based on a progressively increasing SR. Then, in the RSC-based evaluation phase, the constructed datasets are further utilized for the training of reference and comparison classifiers with/without simulated data.

Typical backbones are used as the evaluation networks. Furthermore, the metric is calculated using the k-fold cross-validation to reduce selection bias. The RSC score for a specific simulated dataset is defined as the difference between the integrals of the accuracies of the comparison and reference classifiers under different SRs.

8.5.1 Framework of the Proposed RSC

When evaluating simulated datasets in practical applications, on the one hand, the evaluation metrics are supposed to effectively describe visual similarity, and on the other hand, they are supposed to evaluate distribution consistency. Full reference-image quality assessment (FR-IQA) metrics require the information of reference images to be available during the evaluation process [34]. Nevertheless, two main limitations restrict the utilization of existing FR-IQA metrics: (1) the necessity for strict pairwise alignment between simulated and real data samples, posing implementation difficulties in real-world scenarios; (2) their inherent per-image assessment paradigm, which merely quantifies visual similarity at the pixel level while failing to capture distributional discrepancies at the dataset level.

Various statistical methods are used to model the distributions of SAR images. However, it's difficult to establish a general distribution model for SAR imagery with variable terrain [35]. Inspired by the average precision (AP) metric calculated based on the precision and recall curves at different confidence thresholds, this section introduces a data evaluation metric based on the RSC. It calculates the difference in recognition rates between the simulated data set and different real-world data sets under different replacement rates to obtain a quantitative evaluation metric for the simulated data set, which is characterized by authenticity, stability, and simplicity. The metric evaluation framework is presented in Figure 8.17.

It includes the substitution rate-based dataset allocation (SDA) and evaluation based on RSC criterion stages. In SDA phase, training data is first randomly divided into k equal-sized subsets for cross-validation. Meanwhile, the test dataset is updated accordingly. Then, based on a gradually increasing SR, the measured training set is replaced and combined with the simulated dataset to be evaluated to generate a training set corresponding to a uniformly varying $SR \in [0, 1]$. These divided datasets will be used to train the reference classifier and the comparison classifier. In the RSC-based evaluation phase, a pre-trained network structure for evaluation needs to be selected. Then, the reference and comparison classifiers corresponding to different SRs are trained based on the sub-training sets constructed in the SDA step without/with using the simulation data as an auxiliary training set, respectively. Subsequently, the difference between two RSCs is calculated as the RSC score.

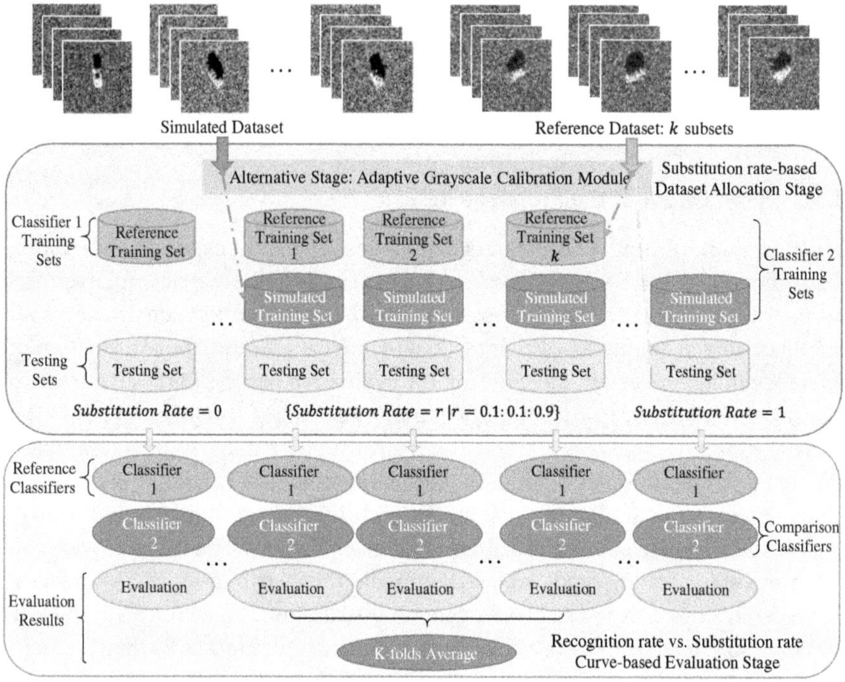

Figure 8.17 Evaluation flowchart of RSC.

8.5.2 Construction of Datasets Based on SR

Before evaluating a simulated dataset, the corresponding measured dataset should also be obtained. A complete simulated dataset can be generated by manual transformation methods, electromagnetic simulations, and deep models such as adversarial networks. The RSC-based evaluation index does not have limitations on sample amount, but the categories contained in the simulated and reference datasets should be consistent. It is evident that the increased number of reference samples directly correlates with enhanced representativeness, leading to more objective and effective evaluation results. The reference dataset is initially partitioned into k disjoint subsets, employing a k-fold cross-validation scheme, where one subset serves as the test set while the remaining $k-1$ constitute the training set [36]. Subsequently, the differential datasets are constructed. The SR denotes the ratio of amount of simulated data used to replace the reference data to it of the current training set. The dataset allocation for each experiment in k-fold cross-validation is shown in Figure 8.18. At ratio 0 (baseline configuration), both classifiers exclusively utilize the reference dataset, thus establishing pure

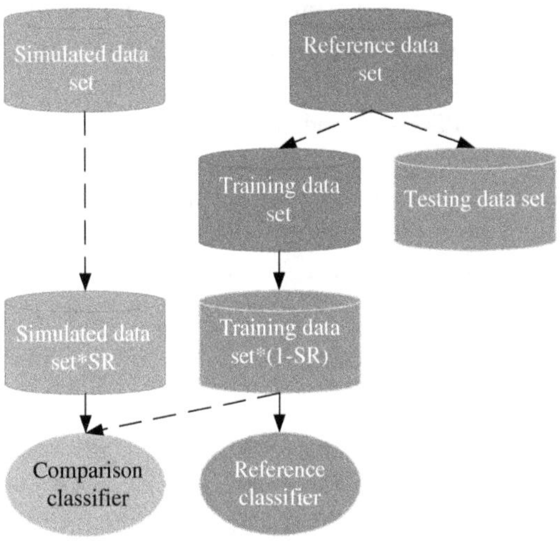

Figure 8.18 Substitution rate-based dataset allocation.

domain-specific training. Conversely, ratio 1 indicates complete replacement of the reference data with simulated data.

8.5.3 RSC-based Evaluation Stage

In this stage, a backbone structure is first selected for the feature extraction of samples. Then, the recognition rates of the same test set under different SRs are calculated, and the corresponding RSCs are plotted. Finally, the final RSC score of the simulated dataset can be obtained. The only distinguishing factor between the reference and comparison classifiers is the training data. Specifically, the reference classifier utilizes reference data exclusively for training. As the SR increases, the reference classifier experiences progressive scarcity of training data, resulting in a corresponding decline in test recognition rates. In contrast, the comparison classifier utilizes hybrid training data combining reference with simulated samples. When the simulated data accurately represents the target distribution, the comparison classifier demonstrates enhanced classification accuracy compared to its reference counterpart. Conversely, its performance degrades below baseline levels when there is a significant divergence between the simulation and measurement distributions. The RSC score Υ_S is calculated as:

$$\Upsilon_S = \frac{1}{k} \int_0^1 \left[\sum_{i=1}^{k} \theta_i(t) - \sum_{i=1}^{k} \gamma_i(t) \right] dt, \tag{8.25}$$

where k denotes the cross-folds, t represents the SR, $\theta_i(t)$ and $\gamma_i(t)$ are the recognition rates for test set of the comparison and reference classifiers with t, respectively. The theoretical range of RSC is $[1/m - 1, 1 - 1/m]$, where m is the number of categories. A nonpositive RSC score indicates non-supplementary gain of the simulated dataset to the reference data, disqualifying it as an auxiliary dataset. Conversely, a positive RSC score demonstrates its enhancement efficacy, where the magnitude quantitatively reflects the utility of simulated dataset in domain adaptation tasks. The calculation of the RSC score is illustrated in Figure 8.19.

8.5.4 Experimental Results and Analysis

In order to evaluate the rationality and effectiveness of RSC, experiments were carried out on MSTAR. The imaging mode is spotlight, with a resolution of 0.3×0.3 m and a pixel size of 128×128 pixel. Three categories of targets: BMP2, BTR70, and T72 with a depression angle of 15° are selected as the experimental dataset.

A multisource simulated dataset is collected to evaluate the performance of RSC. The experimental simulated datasets generated by electromagnetic simulation are provided by Southeast University. They are named S15, S17, and SX, which all include BMP2, BTR70, and T72. These datasets differ in terms of model settings, depression angles, and slice counts.

Specifically, the experimental configuration specifies depression angles of 15° (S15) and 17° (S17), both maintaining an azimuth sampling interval of 1°. Notably, the SX preserves full angular consistency with the benchmark dataset. In addition, a dataset generated based on a deep generation model named AG is also used [37]. The DA dataset is constructed using conventional data augmentation techniques

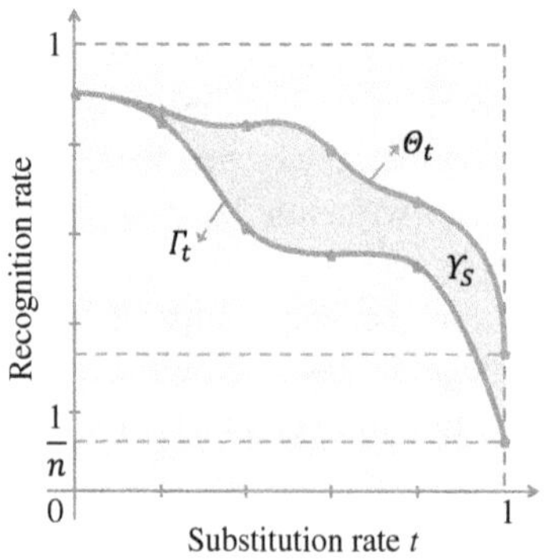

Figure 8.19 Illustration of the calculation of RSC score.

(e.g., random rotation, translation, and cropping). A Gaussian noise-based dataset (denoted as Noise) is specifically designed for ablation studies.

Experimental simulated datasets exhibit high visual fidelity; thereby, an objective visual similarity evaluation is hard to conclude from the appearance. The proposed RSC is a bounded quantitative metric that concurrently evaluates visual similarity and distribution consistency. The RSC metric demonstrates monotonically increasing scores with enhanced dataset realism and practical applicability when evaluated under consistent network configurations. Experimental results of different simulated datasets using AconvNet [38] are presented in Figure 8.20, where triangle markers with solid lines denote the recognition rate of the reference classifier, while circle markers with dotted lines curves indicate comparison results. The filled area represents RSC scores between the two classifiers. Statistical distributions of sub-experiments are presented using boxplot visualizations.

When the SR is 0, all comparison classifiers demonstrate equivalent accuracy to their reference classifiers. With progressive SR increases, the proportion of measured training samples reduces, leading to a gradual decline in reference classifier performance with minor fluctuations. When the SR is 1, the reference classifier attains its minimum recognition rate (empirically set to 0.333 for the 3-class classification task). Simulated datasets are subsequently employed as auxiliary inputs to maintain the amount of reference data during comparison model training. As illustrated in Figure 8.20, the performance of the comparison classifier is improved

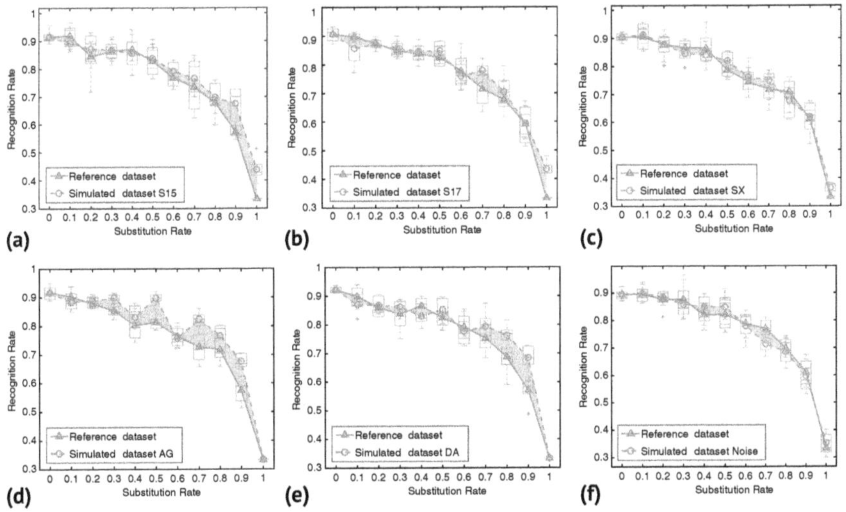

Figure 8.20 Evaluation results of different datasets with AConvNet. (a) evaluation result of S15, (b) evaluation result of S17, (c) evaluation result of SX, (d) evaluation result of AG, (e) evaluation result of DA, (f) evaluation result of noise.

Table 8.1 Comparison with representative IQA criteria.

Dataset	MSE	PSNR	SSIM [%]	MSSIM [%]	IFC [%]	VIF [%]	RSC [%]
S15	4903.636	11.235	3.063	16.993	9.576	0.917	2.278
S17	4927.304	11.214	2.849	15.775	8.927	0.856	1.236
SX	4995.255	11.153	2.896	16.108	8.872	0.848	0.386
AG	4308.877	11.854	5.708	25.338	15.907	1.520	5.470

by adding the corresponding simulated data as auxiliary training sets under different SRs. It is also remarkable that the effects of simulated datasets are not noticeable when the SR is small. The recognition gain caused by the simulated dataset increases in proportion to the SR becoming larger, which can be attributed to a decline in the number of reference training samples, accompanied by an increase in the proportion of simulation training samples. Comparisons among the RSC and representative FR-IQA criteria are presented in Table 8.1. Compared with traditional FR-IQA criteria, the RSC is an objective and absolute criterion that includes utility evaluation ability, flexibility, and extensibility.

8.6 Conclusion

DNN-based algorithms have accelerated the development and application of in various SAR-ATR tasks, which outperform traditional methods on standard SAR datasets. However, the DNNs are data-driven, whose performances hardly rely on the training data. To enhance the applicability of DNNs for SAR images and improve the generalization and interpretability of the models under FSL conditions, our work encompasses three primary aspects: algorithm design, model construction, and the formulation of evaluation metrics. The main contributions can be summarized as: (1) a robust representation learning of targets in SAR image guided by the scattering information named SIE-APN is proposed, which incorporates the scattering information of targets as prior knowledge for model optimization; (2) a causally inspired disentangled representation model of SAR images named CAE is designed, which integrates end-to-end forward sample generation and reverse image interpretation processes into a causally plausible framework; (3) a similarity criterion of SAR images named RSC is tailored to the utility assessment of SAR simulated datasets.

However, there are still many issues worth further exploration in the research on MV theory and applications inspired by physical intelligence, including (1) research on target interpretation methods that integrate SAR scattering mechanisms. Developing generative models utilizing AI-Generated Content (AIGC) technologies for microwave data. By learning parameterized generative models from massive datasets, we can construct prior distributions for inverse problems of MV, enabling high-precision generative model training and robust SAR data representation; (2) construction of physically interpretable representation models for SAR data. Develop systematic foundational theories for knowledge and data fusion, enabling integration across different scales and granularities; (3) interaction verification between intelligent agents and simulators. It is worthwhile to research the interaction mechanisms between intelligent agents and simulators, exploring human-in-the-loop continuous learning collaboration models for agent-expert interactions.

References

1 X. X. Zhu, D. Tuia, L. Mou, G. S. Xia, L. Zhang, F. Xu, and F. Fraundorfer. Deep learning in remote sensing: A comprehensive review and list of resources. *IEEE Geoscience and Remote Sensing Magazine*, 5(4):8–36, 2017.

2 C. Szegedy, W. Liu, Y. Jia, P. Sermanet, S. Reed, D. Anguelov, D. Erhan, V. Vanhoucke, and A. Rabinovich. Going deeper with convolutions. In *Proceedings of the IEEE Conference on Computer Vision and Pattern Recognition*, pages 1–9, 2015.

3 S. Chen, H. Wang, F. Xu, and Y. Q. Jin. Target classification using the deep convolutional networks for SAR images. *IEEE Transactions on Geoscience and Remote Sensing*, 54(8):4806–4817, 2016.

4 F. Xu and Y. Q. Jin. Microwave vision and intelligent perception of radar imagery. *Journal of Radars*, 13(2):285–306, 2024.

5 Z. Pizlo. Perception viewed as an inverse problem. *Vision Research*, 41(24):3145–3161, 2001.

6 C. Oliver, S. Quegan, C. Ding, J. Chen, and G. He. *Understanding Synthetic Aperture Radar Images*. Publishing House of Electronics Industry, 2009.

7 D. Marr. *Vision: A Computational Investigation into the Human Representation and Processing of Visual Information*. The MIT Press, Cambridge, UK, 2010.

8 N. Lei, D. An, Y. Guo, K. Su, S. Liu, Z. Luo, S. T. Yau, and X. Gu. A geometric understanding of deep learning. *Engineering*, 6(3):361–374, 2020.

9 Q. Guo, H. Wang, and F. Xu. Research progress on aircraft detection and recognition in SAR imagery. *Journal of Radars*, 9(R20020):497, 2020.

10 F. Dou, W. Diao, X. Sun, Y. Zhang, and K. Fu. Aircraft reconstruction in high-resolution SAR images using deep shape prior. *ISPRS International Journal of Geo-Information*, 6(11):330, 2017.

11 Q. An, Z. Pan, L. Liu, and H. You. DRBox-v2: An improved detector with rotatable boxes for target detection in sar images. *IEEE Transactions on Geoscience and Remote Sensing*, 57(11):8333–8349, 2019.

12 C. He, M. Tu, X. Liu, D. Xiong, and M. Liao. Mixture statistical distribution based multiple component model for target detection in high resolution sar imagery. *ISPRS International Journal of Geo-Information*, 6(11):336, 2017.

13 C. He, M. Tu, D. Xiong, F. Tu, and M. Liao. Adaptive component selection-based discriminative model for object detection in high-resolution SAR imagery. *ISPRS International Journal of Geo-Information*, 7(2):72, 2018.

14 K. Fu, F. Z. Dou, H. C. Li, W. H. Diao, X. Sun, and G. L. Xu. Aircraft recognition in sar images based on scattering structure feature and template matching. *IEEE Journal of Selected Topics in Applied Earth Observations and Remote Sensing*, 11(11):4206–4217, 2018.

15 Q. Guo, H. Wang, and F. Xu. Scattering enhanced attention pyramid network for aircraft detection in SAR images. *IEEE Transactions on Geoscience and Remote Sensing*, 59(9):7570–7587, 2020.

16 Q. Guo, H. Wang, L. Kang, Z. Li, and F. Xu. Aircraft target detection from spaceborne SAR image. In *IGARSS 2019 - 2019 IEEE International Geoscience and Remote Sensing Symposium*, pages 1168–1171, 2019.

17 T. Y. Lin, P. Goyal, R. Girshick, K. He, and P. Dollár. Focal loss for dense object detection. In *2017 IEEE International Conference on Computer Vision (ICCV)*, pages 2999–3007, 2017.

18 S. Ren, K. He, R. Girshick, and J. Sun. Faster R-CNN: Towards real-time object detection with region proposal networks. *IEEE Transactions on Pattern Analysis and Machine Intelligence*, 39(6):1137–1149, 2017.

19 J. Redmon, S. K. Divvala, R. B. Girshick, and A. Farhadi. You only look once: Unified, real-time object detection. *CoRR, abs/1506.02640*, 2015.

20 T. Y. Lin, P. Dollár, R. B. Girshick, K. He, B. Hariharan, and S. J. Belongie. Feature pyramid networks for object detection. *CoRR, abs/1612.03144*, 2016.

21 W. Liu, D. Anguelov, D. Erhan, C. Szegedy, S. Reed, C. Y. Fu, and A. C. Berg. SSD: Single shot multibox detector. In *European Conference on Computer Vision*, pages 21–37. Springer, 2016.

22 J. Hu, L. Shen, and G. Sun. Squeeze-and-excitation networks. In *Proceedings of the IEEE Conference on Computer Vision and Pattern Recognition*, pages 7132–7141, 2018.

23 S. Woo, J. Park, J. Y. Lee, and I. S. Kweon. CBAM: Convolutional block attention module. In *Proceedings of the European Conference on Computer Vision (ECCV)*, pages 3–19, 2018.

24 H. S. Seung and D. D. Lee. The manifold ways of perception. *Science*, 290(5500):2268–2269, 2000.

25 Sandia National Laboratory. The air force moving and stationary target recognition database, 2016. https://www.sdms.afrl.af.mil/datasets/mstar/

26 L. Van der Maaten and G. Hinton. Visualizing data using t-SNE. *Journal of Machine Learning Research*, 9(11):2579–2605, 2008.

27 P. P. Brahma, D. Wu, and Y. She. Why deep learning works: A manifold disentanglement perspective. *IEEE Transactions on Neural Networks and Learning Systems*, 27(10):1997–2008, 2015.

28 M. Kocaoglu, C. Snyder, A. G. Dimakis, and S. Vishwanath. CausalGAN: Learning causal implicit generative models with adversarial training. *arXiv preprint arXiv:1709.02023*, 2017.

29 M. Yang, F. Liu, Z. Chen, X. Shen, J. Hao, and J. Wang. CausalVAE: Disentangled representation learning via neural structural causal models. In *Proceedings of the IEEE/CVF Conference on Computer Vision and Pattern Recognition*, pages 9593–9602, 2021.

30 T. Karras, S. Laine, and T. Aila. A style-based generator architecture for generative adversarial networks. In *Proceedings of the IEEE/CVF Conference on Computer Vision and Pattern Recognition*, pages 4401–4410, 2019.

31 Q. Guo, H. Xu, and F. Xu. Causal adversarial autoencoder for disentangled SAR image representation and few-shot target recognition. *IEEE Transactions on Geoscience and Remote Sensing*, 61:1–14, 2023.

32 S. Pidhorskyi, D. A. Adjeroh, and G. Doretto. Adversarial latent autoencoders. In *Proceedings of the IEEE/CVF Conference on Computer Vision and Pattern Recognition*, pages 14104–14113, 2020.

33 Q. Guo, Y. Qian, H. Wang, W. Yu, F. Xu, T. J. Cui, and Y. Q. Jin. Recognition rate versus substitution rate curve: An objective utility assessment criterion of simulated training data. *IEEE Transactions on Geoscience and Remote Sensing*, 60:1–15, 2022.

34 Z. Wang, A. C. Bovik, H. R. Sheikh, and E. P. Simoncelli. Image quality assessment: From error visibility to structural similarity. *IEEE Transactions on Image Processing*, 13(4):600–612, 2004.

35 Y. Li and L. Liu. Review of statistical models of SAR image. *Computer Engineering and Applications*, 49(13):180–186+227, 2013.

36 S. Arlot and A. Celisse. A survey of cross-validation procedures for model selection. *Statistics Surveys*, 4:40–79, 2010.

37 Q. Song, F. Xu, X. X. Zhu, and Y. Q. Jin. Learning to generate sar images with adversarial autoencoder. *IEEE Transactions on Geoscience and Remote Sensing*, 60:1–15, 2022.

38 M. Amrani and F. Jiang. Deep feature extraction and combination for synthetic aperture radar target classification. *Journal of Applied Remote Sensing*, 11:042616, 2017.

9

Data-driven Kalman Filter for Target Tracking

Xiaoyong Ni,[1] Nir Shlezinger,[2,] Ruud J. G. van Sloun,[3] Tirza Routtenberg,[2] and Yonina C. Eldar[4]*

[1] *School of Engineering, EPFL Campus Biotech, Geneva, Switzerland*
[2] *School of Electronic and Computer Engineering, Ben-Gurion University of the Negev, Be'er Sheva, Israel*
[3] *Department of Electronic Engineering, Eindhoven University of Technology, Eindhoven, The Netherlands*
[4] *Faculty of Mathematics and Computer Science, Weizmann Institute of Science, Rehovot, Israel*

*Corresponding author: nirshl@bgu.ac.il

9.1 Introduction

Kalman filter (KF) is among the most widely used algorithms in signal processing and control theory, forming the backbone of numerous estimation and tracking systems [1]. Originally introduced by R. E. Kalman in the 1960s [2], KFs offer a computationally efficient and statistically optimal solution to the problem of estimating the hidden states of linear dynamical systems affected by Gaussian noise [3]. Their recursive nature and ability to provide real-time state estimation have made them highly useful for radar-based target tracking, navigation, and guidance [4], as well as many other applications in defense and aerospace industries [5].

While the KF and its variants [6] are highly effective, they are fundamentally statistical model-based. In particular, they rely on the availability of a well-calibrated statistical model in the form of a state-space (SS) model [7], which specifies both the system dynamics and the observation model. This reliance on explicit models introduces limitations in practical settings where model mismatch, nonlinear dynamics, non-Gaussian noise, and parameter uncertainties degrade the performance of classical Kalman-type filters. In real-world radar tracking scenarios, the system dynamics and noise covariances are often only imprecisely known, challenging the robustness and optimality of purely model-driven methods.

In recent years, deep learning has emerged as a powerful data-driven alternative for model-based solutions to many tasks, including those associated with time sequences [8], such as state estimation and target tracking. Deep learning approaches, such as recurrent neural networks (RNNs) [9] and transformers [10], can learn complex dynamics from data, enabling model-free solutions to state estimation. However, despite their flexibility, deep learning-based trackers have notable shortcomings. They typically require large datasets and lengthy training to generalize to unseen conditions; they rely on computationally demanding and highly overparameterized architectures; and they often lack the interpretability and reliability of classical methods.

Accordingly, while end-to-end deep neural networks (DNNs) can be trained to carry out tasks such as estimation in some settings, replacing classic model-based algorithms such as Kalman-type filters entirely with deep learning models may lead to significant performance bottlenecks. Model-based methods are generally reliable and robust, degrading gracefully when assumptions are mildly violated, but their performance can deteriorate sharply when the underlying assumptions—such as linearity or Gaussianity—are severely mismatched. In contrast, DNNs can learn complex inference models directly from data, effectively bypassing such assumptions, but they are typically less robust: small shifts in the data distribution at test time can result in dramatic and unpredictable drops in performance, as they tend to degrade ungracefully. This has motivated the rise of model-based deep learning approaches [11, 12], which aim to combining the domain knowledge of classical signal processing with the expressive power of DNNs. Among the various approaches to model-based deep learning, a framework that is particularly suitable for target tracking is based on DNN augmentation [13], converting Kalman-type algorithms into machine learning architectures by augmenting them with data-driven modules, thus enabling improved performance under model mismatch and limited-data regimes.

This chapter introduces and details a representative family of model-based deep learning algorithms inspired by Kalman-type filters, focusing on the KalmanNet framework [14]. KalmanNet and its variants [15–18] combine SS model-driven recursions with learnable components, providing robustness to model uncertainties while retaining the interpretability of the Kalman filtering paradigm. The KalmanNet paradigm has already been successfully applied in a variety of domains, ranging from brain–machine interface [19] to acoustic echo cancellation [20, 21], financial monitoring [22], robotic arm tracking [23], velocity tracking in autonomous racing cars [24], beam tracking in wireless systems [25], and drone-based monitoring systems [26].

The importance of Kalman filtering and its variants in the context of target tracking stems from its widespread usage in such applications. From radar and sonar systems to autonomous navigation and sensor fusion, Kalman-type algorithms have long served as the foundational tool for real-time state estimation under uncertainty. Their recursive nature, low computational overhead, and well-characterized statistical properties make them ideally suited for tracking moving objects in dynamic, partially observable environments. However, the increasing complexity of modern sensing systems—such as high-resolution radar, multimodal perception, and distributed sensor arrays—introduces challenges that strain the assumptions underlying classical filters. In these cases, artificial intelligence (AI)-based augmentation of Kalman filtering is not merely an enhancement but a necessity: it enables robust operation in the face of nonlinearities, unmodeled dynamics, and non-Gaussian noise, while preserving the interpretability and modular structure of the classical estimator. The KalmanNet framework thus offers a principled and practical approach to modernizing target tracking pipelines, extending the trusted Kalman filtering paradigm into the data-rich, model-imperfect era.

The goal of this chapter is to present a systematic introduction to KalmanNet and its extensions, with a particular emphasis on their relevance and potential in the field of radar-based target tracking. KalmanNet stands at the intersection of classical estimation theory and modern deep learning, offering a unified framework that enhances traditional Kalman-type filters through principled data-driven augmentation. This hybrid approach allows one to preserve the strengths of classical model-based tracking—such as interpretability, robustness, and recursive efficiency—while incorporating the flexibility and learning capacity of neural networks. As such, KalmanNet provides a compelling solution for real-world radar systems, where nonlinearity, partial observability, and model uncertainties often challenge conventional tracking pipelines.

To guide the reader through this emerging paradigm, the chapter is organized as follows. We begin with a concise review of the model-based Kalman filter and its well-known extensions in Section 9.2, establishing the theoretical foundation upon which the KalmanNet family is built. Section 9.3 then introduces the KalmanNet architecture, including several notable variants tailored for different tracking contexts. In Section 9.4, we provide a detailed assessment of KalmanNet's capabilities through both qualitative insights and quantitative evaluations. This section highlights not only measurable performance improvements—such as enhanced accuracy and robustness under model mismatch—but also conceptual benefits that stem from integrating structured prior knowledge into deep learning systems. Together, these contributions position KalmanNet as a powerful and flexible tool for next-generation AI-aided radar tracking.

Table 9.1 Notations.

Notation	Description
$\boldsymbol{x}$	Boldface lower-case letters denote vectors
$\boldsymbol{X}$	Boldface upper-case letters denote matrices
$\mathcal{X}$	Calligraphic letters denote sets
$[\boldsymbol{x}]_i$	ith entry of vector $\boldsymbol{x}$
$[\boldsymbol{X}]_{i,j}$	(i, j)th entry of matrix $\boldsymbol{X}$
$(\cdot)^\top$	Matrix or vector transpose
$\|\cdot\|_2$	Euclidean (ℓ_2) norm
$\mathbb{E}[\cdot]$	Statistical expectation operator
$\mathrm{Tr}(\cdot)$	Trace of a square matrix
$\mathrm{diag}(\boldsymbol{x})$	Diagonal matrix with $\boldsymbol{x}$ on its main diagonal
$\det(\cdot)$	Determinant of a matrix
$\odot$	Element-wise (Hadamard) product
$\nabla_{(\cdot)}$	Gradient operator with respect to the argument
$\boldsymbol{I}$	Identity matrix of appropriate dimension
$\mathcal{N}(\boldsymbol{\mu}, \boldsymbol{\Sigma})$	Multivariate Gaussian distribution with mean $\boldsymbol{\mu}$ and covariance $\boldsymbol{\Sigma}$
$\mathbb{R}$	Set of real numbers
$\mathbb{N}^+$	Set of positive integers

9.1.1 Notations

Throughout the chapter, we adopt the following notational conventions, summarized in Table 9.1.

9.2 Model-based Kalman Filter

9.2.1 State-space Model

In this section, we present the statistical modeling framework that lies at the core of the classical family of Kalman-type algorithms: the SS model [7]. We begin by formulating the general model and then specialize it into widely used cases relevant for target tracking applications.

9.2.1.1 Formulation

The SS model is a probabilistic model that characterizes the temporal evolution of a latent state $\boldsymbol{x}_t \in \mathbb{R}^n$ and its relationship to noisy observations $\boldsymbol{y}_t \in \mathbb{R}^m$. While SS models are frequently used to describe closed-loop control systems [27], we here focus on open-loop formulations relevant in tracking. The generic form of an SS model is defined by two conditional probability distributions:

$$x_t \sim p(x_t \mid x_{t-1}), \quad \text{(state transition)}, \tag{9.1a}$$

$$y_t \sim p(y_t \mid x_t), \quad \text{(measurement model)}, \tag{9.1b}$$

for $t = 1, 2, \ldots, T$, where x_0 is assumed to be drawn from a prior distribution $p(x_0)$.

A common specialization is the Gaussian noise model, which specializes (9.1) into:

$$x_t = f(x_{t-1}) + w_{t-1}, \qquad w_{t-1} \sim \mathcal{N}(\mathbf{0}, Q), \tag{9.2a}$$

$$y_t = h(x_t) + v_t, \qquad v_t \sim \mathcal{N}(\mathbf{0}, R), \tag{9.2b}$$

where $f(\cdot)$ and $h(\cdot)$ are potentially nonlinear functions modeling the system dynamics and observation model, respectively. $w_t \in \mathbb{R}^n$ and $v_t \in \mathbb{R}^m$ are state transition noise and observation noise, respectively. Here, $Q \in \mathbb{R}^{n \times n}$ and $R \in \mathbb{R}^{m \times m}$ are the corresponding covariance matrices. This formulation encompasses many practical applications where physical processes exhibit nonlinearities. While the formulation in (9.2) considers the functions $f(\cdot)$ and $h(\cdot)$ as well as the distribution of the noise signals to be time-invariant, it can generally extend to a time-varying setting.

The most classical and widely adopted form of an SS model assumes *linear Gaussian* dynamics and observations, yielding the following special case of (9.1):

$$x_t = F x_{t-1} + w_{t-1}, \qquad w_{t-1} \sim \mathcal{N}(\mathbf{0}, Q), \tag{9.3a}$$

$$y_t = H x_t + v_t, \qquad v_t \sim \mathcal{N}(\mathbf{0}, R), \tag{9.3b}$$

where $F \in \mathbb{R}^{n \times n}$ is the state transition matrix and $H \in \mathbb{R}^{m \times n}$ is the observation matrix.

Examples in Target Tracking

In radar-based target tracking, formulations based on SS models, and particularly linear Gaussian ones as in (9.3), are widely adopted [4]. Classical models such as constant velocity (CV) and constant acceleration (CA) are typical choices:

1. **CV Model:**

$$x_t = \begin{bmatrix} p_t \\ v_t \end{bmatrix}, \quad F = \begin{bmatrix} I & \Delta t \cdot I \\ \mathbf{0} & I \end{bmatrix}, \tag{9.4}$$

where $\boldsymbol{p}_t$ and $\boldsymbol{v}_t$ are position and velocity vectors, Δt is the sampling interval, and $\boldsymbol{I}$ is the identity matrix.

2. **CA Model:**

$$\boldsymbol{x}_t = \begin{bmatrix} \boldsymbol{p}_t \\ \boldsymbol{v}_t \\ \boldsymbol{a}_t \end{bmatrix}, \quad \boldsymbol{F} = \begin{bmatrix} \boldsymbol{I} & \Delta t \cdot \boldsymbol{I} & 0.5\Delta t^2 \cdot \boldsymbol{I} \\ \boldsymbol{0} & \boldsymbol{I} & \Delta t \cdot \boldsymbol{I} \\ \boldsymbol{0} & \boldsymbol{0} & \boldsymbol{I} \end{bmatrix}, \tag{9.5}$$

where $\boldsymbol{a}_t$ denotes acceleration.

These models are essential building blocks for classical tracking algorithms. While they are inherently approximations of the true dynamics encountered in target tracking, they lay the foundation for more sophisticated, data-driven approaches, as discussed in the sequel.

9.2.1.2 Tasks

The (open-loop) SS model formulation enables several key tasks. The main tasks that fall under the umbrella of *state estimation* include:

1. Filtering

Filtering refers to the sequential estimation of the state $\boldsymbol{x}_t$ at time t using all observations up to and including time t, i.e., $\boldsymbol{y}_{1:t}$. The goal is to compute an estimate $\hat{\boldsymbol{x}}_t$. When estimation is evaluated in terms of the mean squared error (MSE), the optimal filter is:

$$\hat{\boldsymbol{x}}_t = \mathbb{E}[\boldsymbol{x}_t \mid \boldsymbol{y}_{1:t}], \tag{9.6}$$

where the expectation is taken with respect to the filtering distribution $p(\boldsymbol{x}_t \mid \boldsymbol{y}_{1:t})$. In (9.6), $\hat{\boldsymbol{x}}_t$ provides the minimum MSE estimation of the current state, which is crucial for real-time target tracking applications.

2. Smoothing

Smoothing aims to estimate the state $\boldsymbol{x}_t$ at time t using the full set of observations $\boldsymbol{y}_{1:T}$, where $T > t$. The optimal smoothed estimate (in terms of the MSE) is given by:

$$\hat{\boldsymbol{x}}_t^{\text{smooth}} = \mathbb{E}[\boldsymbol{x}_t \mid \boldsymbol{y}_{1:T}]. \tag{9.7}$$

The state estimate in (9.7) generally leads to more accurate state estimates than filtering by leveraging future information. Smoothing is often applied in post-processing tasks where latency is less critical.

9.2.1.3 Other Tasks

Beyond filtering and smoothing, SS models support additional tasks such as:

- **System Identification:** Estimating the unknown system parameters (e.g., $\boldsymbol{F}, \boldsymbol{Q}, \boldsymbol{H}, \boldsymbol{R}$) from observed data.
- **Imputation**: Reconstructing missing observations or states based on the available data, which is critical in radar systems encountering intermittent measurement gaps.

These tasks are fundamental in both classical and modern machine learning-enhanced approaches to radar signal processing.

9.2.2 State Estimation Algorithms

We now turn to present the classical state estimation algorithms designed to solve the filtering and smoothing tasks for SS models. This section focuses on the celebrated KF and smoother, followed by their extensions to nonlinear systems (Table 9.2).

9.2.2.1 Kalman Filtering

The KF is the MSE optimal state estimator for linear Gaussian SS models as in (9.3) [7]. It provides a recursive solution for estimating the latent state $\boldsymbol{x}_t$ given the observations up to time t.

Specifically, the KF proceeds via a prediction-update cycle, which recursively updates the estimates of the states' first- and second-order moments [1]. These steps are stated as follows:

- **Prediction Step:** This step utilizes the previous state estimate to predict the first- and second-order moments of the state variable via:

$$\hat{\boldsymbol{x}}_{t|t-1} = \boldsymbol{F}\hat{\boldsymbol{x}}_{t-1|t-1}, \tag{9.8a}$$

$$\boldsymbol{P}_{t|t-1} = \boldsymbol{F}\boldsymbol{P}_{t-1|t-1}\boldsymbol{F}^{\top} + \boldsymbol{Q}, \tag{9.8b}$$

Table 9.2 Assumptions of Kalman filters and Kalman smoothers.

Algorithm	Assumption
KF/KS	Linear functions, Gaussian noise
EKF/EKS	Linear/nonlinear functions, Gaussian noise
UKF/UKS	Linear/nonlinear functions, Gaussian noise
CKF/CKS	Linear/nonlinear functions, Gaussian noise
PF/PS	Linear/nonlinear functions, Gaussian/non-Gaussian noise

predict the first- and second-order moments of the observation via:

$$\hat{\boldsymbol{y}}_{t|t-1} = \boldsymbol{H}\hat{\boldsymbol{x}}_{t|t-1}, \tag{9.9a}$$

$$\boldsymbol{S}_{t|t-1} = \boldsymbol{H}\boldsymbol{P}_{t|t-1}\boldsymbol{H}^\top + \boldsymbol{R}. \tag{9.9b}$$

where $\boldsymbol{S}_{t|t-1}$ is the covariance of the predicted observation $\hat{\boldsymbol{y}}_{t|t-1}$.

- **Update Step:** The second step uses the current observation $\boldsymbol{y}_t$, and particularly its innovation $\Delta\boldsymbol{y}_t \triangleq \boldsymbol{y}_t - \hat{\boldsymbol{y}}_{t|t-1}$, to update the state moments via:

$$\boldsymbol{K}_t = \boldsymbol{P}_{t|t-1}\boldsymbol{H}^\top\boldsymbol{S}_{t|t-1}^{-1}, \tag{9.10a}$$

$$\hat{\boldsymbol{x}}_{t|t} = \hat{\boldsymbol{x}}_{t|t-1} + \boldsymbol{K}_t\Delta\boldsymbol{y}_t, \tag{9.10b}$$

$$\boldsymbol{P}_{t|t} = (\boldsymbol{I} - \boldsymbol{K}_t\boldsymbol{H})\boldsymbol{P}_{t|t-1}. \tag{9.10c}$$

Here, $\hat{\boldsymbol{x}}_{t|t}$ is the filtered estimate of the state at time t, $\boldsymbol{P}_{t|t}$ is its covariance, and $\boldsymbol{K}_t$ is the Kalman gain (KG) matrix. The Kalman filter achieves optimality as in (9.6) under Gaussianity and linearity assumptions and is widely used in radar-based tracking.

9.2.2.2 Kalman Smoothing

While filtering estimates $\boldsymbol{x}_t$ using $\boldsymbol{y}_{1:t}$, smoothing refines the estimate by also leveraging future observations $\boldsymbol{y}_{t+1:T}$. The Rauch–Tung–Striebel (RTS) smoother [28] provides an efficient backward pass that builds on the forward KF recursion. It achieves optimality as in (9.7) under linear Gaussian assumptions.

Given filtered estimates $\{\hat{\boldsymbol{x}}_{t|t}, \boldsymbol{P}_{t|t}\}$ from the KF, the RTS smoother computes smoothed estimates $\hat{\boldsymbol{x}}_{t|T}$ and covariances $\boldsymbol{P}_{t|T}$ for $t = T-1, \dots, 1$ via:

- **Backward Smoothing Step:**

$$\boldsymbol{G}_t = \boldsymbol{P}_{t|t}\boldsymbol{F}^\top\boldsymbol{P}_{t+1|t}^{-1}, \tag{9.11a}$$

$$\hat{\boldsymbol{x}}_{t|T} = \hat{\boldsymbol{x}}_{t|t} + \boldsymbol{G}_t(\hat{\boldsymbol{x}}_{t+1|T} - \hat{\boldsymbol{x}}_{t+1|t}), \tag{9.11b}$$

$$\boldsymbol{P}_{t|T} = \boldsymbol{P}_{t|t} + \boldsymbol{G}_t(\boldsymbol{P}_{t+1|T} - \boldsymbol{P}_{t+1|t})\boldsymbol{G}_t^\top. \tag{9.11c}$$

The RTS smoother refines the state trajectory by combining forward (filtering) and backward (smoothing) information, yielding superior estimates compared to filtering alone.

9.2.2.3 Nonlinear Extensions

When the state transition $\boldsymbol{f}(\cdot)$ or the measurement function $\boldsymbol{h}(\cdot)$ is nonlinear, the KF and RTS smoother are no longer directly applicable. In such cases, even if the state transition noise is Gaussian, the posterior distribution of the latent state after nonlinear transformation becomes non-Gaussian and generally intractable to compute exactly. As a result, optimal filtering and smoothing are no longer analytically feasible. To address this, several approximate methods have been developed for nonlinear Gaussian SS models, i.e., dynamics characterized via (9.2). Arguably, the most common approach is the Extended KF (EKF) [6].

9.2.2.3.1 *Extended Kalman Filter*

The extended Kalman filter (EKF) linearizes the nonlinear functions around the current estimate using first-order Taylor expansions:

$$\boldsymbol{x}_t = \boldsymbol{f}(\boldsymbol{x}_{t-1}) + \boldsymbol{w}_{t-1}, \quad \boldsymbol{w}_{t-1} \sim \mathcal{N}(\boldsymbol{0}, \boldsymbol{Q}), \tag{9.12}$$

$$\boldsymbol{y}_t = \boldsymbol{h}(\boldsymbol{x}_t) + \boldsymbol{v}_t, \quad \boldsymbol{v}_t \sim \mathcal{N}(\boldsymbol{0}, \boldsymbol{R}). \tag{9.13}$$

The EKF replaces the system matrices in (9.8)–(9.10) with Jacobians:

$$\boldsymbol{F}_{t-1} = \left.\frac{\partial \boldsymbol{f}}{\partial \boldsymbol{x}}\right|_{\hat{\boldsymbol{x}}_{t-1|t-1}}, \quad \boldsymbol{H}_t = \left.\frac{\partial \boldsymbol{h}}{\partial \boldsymbol{x}}\right|_{\hat{\boldsymbol{x}}_{t|t-1}}. \tag{9.14}$$

These Jacobians are used within the Kalman filter recursion to approximate the nonlinear model locally.

9.2.2.3.2 *Alternative Nonlinear Filters*

Beyond the EKF, several other techniques have been developed for nonlinear filtering. Representative examples include [7, Ch. 10]:

- **Unscented KF (UKF):** The UKF avoids linearization by propagating a set of sigma points through the nonlinear functions [29]. This captures posterior moments more accurately than EKF, especially in strongly nonlinear regimes.
- **Cubature KF (CKF):** The CKF uses cubature rules to approximate Gaussian-weighted integrals in nonlinear models [30]. It can be interpreted as a generalization of the UKF with improved numerical stability.
- **Particle filter (PF):** PFs represent the filtering distribution using a set of weighted particles [31]. While flexible and able to approximate arbitrary distributions, PFs often require a large number of particles, making them computationally expensive.

It is emphasized that among the above detailed algorithms, the EKF and UKF are geared toward SS models with Gaussian noise signals, while algorithms such as the CKF and PF can be utilized for generic non-Gaussian formulations as in (9.1). A similar strategy, when applied to smoothing, results in extended Kalman smoother (EKS), unscented Kalman smoother (UKS) [32], cubature Kalman smoother (CKS) and particle smoother (PS) [33]. Arguably the most common nonlinear smoother is the EKS.

While these algorithms address nonlinearity, they rely on approximations (e.g., linearization or sigma points) or sampling strategies (e.g., particles) that may degrade performance under model mismatch, high-dimensional states, or long time horizons. This has inspired a new wave of machine learning-based methods that aim to learn effective filtering rules directly from data.

9.2.2.4 From Classical to Data-driven Approaches: Motivation for KalmanNet

Classical filtering and smoothing algorithms, such as the KF, the EKF, and the RTS smoother, rely heavily on accurate knowledge of the underlying system and noise models. Specifically, these algorithms assume precise access to the functions governing the system dynamics and measurement models—typically denoted $\boldsymbol{f}(\cdot)$ and $\boldsymbol{h}(\cdot)$—as well as to the statistical properties of the process and observation noise, encoded via the covariance matrices $\boldsymbol{Q}$ and $\boldsymbol{R}$. Under the assumption of linearity and Gaussianity, this model-based formulation offers provable guarantees of optimality in terms of minimum mean squared error (MMSE). However, in practice, these assumptions are often violated. For example, the true dynamics of a maneuvering target may be nonlinear or time-varying; measurements may be corrupted by clutter, interference, or other sources of non-Gaussian noise; and parameters such as transition matrices or noise covariances may be uncertain, only partially known, or slowly drifting over time.

These practical challenges frequently arise in radar-based target tracking and other real-world signal processing applications. Consider the commonly used CV or CA models: while these serve as useful approximations, they often fail to capture abrupt maneuvers or environmental disturbances. Furthermore, traditional filters struggle in situations involving sensor calibration errors, missing measurements, or temporal correlation in the noise processes. Nonlinear extensions of the KF, such as the EKF or UKF, can partially mitigate these issues, but they are limited by their reliance on analytical approximations (e.g., Jacobian linearizations or sigma-point propagation), which may break down in highly nonlinear or data-limited settings. PFs provide a more general alternative by relying on sampling-based inference but tend to suffer from high computational cost and degeneracy in high-dimensional spaces.

In light of these limitations, data-driven techniques have emerged as a promising direction to enhance the flexibility and robustness of classical filters. Among these, *model-based deep learning* approaches [11]—such as KalmanNet—have gained traction for their ability to combine the structural rigor of classical estimation algorithms with the expressive capacity of DNNs. Rather than discarding the well-established Kalman recursion, KalmanNet strategically augments it with learnable components, most notably by replacing the analytically derived KG with a trainable function. This allows the filter to compensate for modeling errors and capture complex, nonlinear interactions that are difficult to describe analytically. At the same time, the recursive architecture of the filter is retained, ensuring computational efficiency, interpretability, and the ability to operate in real time.

By learning specific components from data—while preserving the overall filter structure—KalmanNet offers a principled compromise between model-based reliability and data-driven flexibility. This hybrid design allows it to generalize better under model mismatch and adapt to partially known or evolving environments, all while requiring significantly less training data and fewer parameters than fully end-to-end networks. As such, KalmanNet represents a robust and scalable framework for modern state estimation tasks in radar and beyond. In the following section, we delve into the KalmanNet architecture, exploring its formulation, training procedure, and several notable extensions designed for filtering, smoothing, adaptation, and structure-aware signal processing.

9.3 Hybrid Model-based/Data-driven Kalman Filtering

In this section, we introduce *KalmanNet*, a model-based deep learning approach that enhances classical Kalman-type algorithms with learnable components. KalmanNet exemplifies the broader principle of *model-based deep learning*, where domain-specific knowledge is embedded into neural network architectures to improve learning efficiency, generalization, and robustness. Unlike end-to-end black-box neural networks, KalmanNet retains the recursive structure and interpretability of the classical Kalman filter, while learning key components—such as the KG—from data. This fusion of structure and flexibility enables KalmanNet to perform well under practical conditions of model mismatch, nonlinearity, or unknown parameters that degrade the performance of classical filters.

The motivation for such hybrid architectures arises from the inherent challenges in modern target tracking systems. In many radar and sensing applications, the dynamics of the tracked target, noise statistics, or measurement models may not be fully known or may vary over time. Classical Kalman filters assume access to accurate models and statistical priors, and while they degrade gracefully under mild mismatch, they are vulnerable to strong deviations from their assumptions. On the other hand, deep learning models can adapt to unknown conditions by learning from data but typically require large labeled datasets, suffer from overfitting, and often lack transparency. KalmanNet is designed to address these limitations: by constraining the learning to specific components within the filtering pipeline, it provides a structured yet adaptive solution, suitable for both data-limited and nonstationary tracking environments.

This section begins with an overview of the model-based deep learning paradigm and how it applies to filtering in SS models. We then present the KalmanNet architecture, illustrating how it builds upon the EKF while replacing only the analytically challenging components with learnable neural networks. We describe its training methodology under supervised settings and highlight variants that extend KalmanNet to smoothing (RTSNet), adaptation (Adaptive KalmanNet), and structured signal domains (GSP-KalmanNet). Together, these approaches (Table 9.3) offer a comprehensive suite of hybrid solutions for robust and efficient target tracking in the presence of real-world complexities.

Table 9.3 Comparison of tasks for KalmanNet variants.

Algorithm	Task
KalmanNet [14]	Filtering
RTSNet [15]	Smoothing
Adaptive KalmanNet [16]	Adaptive filtering
GSP-KalmanNet [34]	Filtering for graph signals

9.3.1 Model-based Deep Learning

Generally speaking, deep learning methods excel in learning complex representations from data. However, they often lack interpretability and may require large datasets for successful training. In the context of target tracking, classical model-based algorithms such as the KF are derived from first principles and leverage a solid mathematical characterization of the task at hand. In doing so, they offer provable optimality when the model assumptions hold. However, in many radar signal processing applications, these assumptions are often violated due to nonlinearities, non-Gaussian noise, or mismatched system parameters. Nonlinearities arise naturally from factors such as target motion over curved trajectories, Doppler effects, or angle-dependent measurements in polar coordinates. The assumption of Gaussian noise often fails due to the presence of spurious detections, clutter, or interference, which introduce heavy-tailed or multimodal noise characteristics. Additionally, system parameters such as target dynamics, process noise covariances, or sensor calibration settings are often only partially known or subject to mismatch due to modeling inaccuracies or environmental variability. These challenges significantly degrade the performance of classical filters and motivate the development of more flexible, data-driven approaches.

Model-based deep learning encompasses methodologies aiming to combine the strengths of both worlds by embedding domain knowledge (e.g., SS model structure) within a deep learning framework [12]. Rather than fully replacing the model-based algorithm with a generic DNN, the idea is to *augment* classical algorithms by learning components that are difficult to specify analytically [35]. This approach transforms classical algorithms into trainable discriminative machine learning architectures that can be learned directly from data representing their desired task [36].

The model-based deep learning paradigm is particularly attractive for Kalman-type algorithms: while the recursive nature of the KF provides an efficient computational structure, its performance can be degraded by inaccurate knowledge of the system dynamics or noise statistics. Augmenting specific parts of the KF with neural networks allows leveraging data to compensate for model mismatches, yielding hybrid solutions that are both interpretable and robust. Throughout this section, we focus on converting Kalman-type algorithms into machine learning architectures, focusing here on supervised learning, i.e., when training is carried out based on a dataset comprised of multiple pairs of trajectories of both measurements and state values, i.e.,

$$\mathcal{D}_s = \{\boldsymbol{x}_{1:T}^{(i)}, \boldsymbol{y}_{1:T}^{(i)}\}_{i=1}^{N}. \tag{9.15}$$

9.3.2 Filtering: KalmanNet

KalmanNet [14] is a model-based neural network architecture designed to perform filtering in nonlinear SS models, using the structure of the EKF as a backbone.

9.3.2.1 Rationale

KalmanNet retains the recursive structure of the EKF but replaces the Jacobian-based computation of the KG with a learned module. This allows the algorithm to adaptively infer a gain matrix that accounts for unknown nonlinearities and model mismatches, while still benefiting from the sequential prediction-update cycle of the KF and the available characterization of the state transition and measurement functions.

9.3.2.2 Architecture

The KalmanNet architecture as shown in Figure 9.1 is composed of:

- A **predictor** block, which applies the known nonlinear transition model $\boldsymbol{f}(\cdot)$ to generate predicted states $\hat{\boldsymbol{x}}_{t|t-1}$.
- A **learned KG** module, which is a DNN with parameters θ that outputs the KG $\boldsymbol{K}_t(\theta)$ based on a set of instantaneous features. To properly estimate the KG, with (9.10) as an example for the linear Gaussian case, these instantaneous input features should capture the relevant statistics of the observations and the state process. They typically include
  - innovation $\Delta\boldsymbol{y}_t = \boldsymbol{y}_t - \hat{\boldsymbol{y}}_{t|t-1}$.
  - forward update difference $\Delta\hat{\boldsymbol{x}}_{t-1} = \hat{\boldsymbol{x}}_{t-1|t-1} - \hat{\boldsymbol{x}}_{t-1|t-2}$.

  The architecture of the DNN is based on conventional architectures suitable for time sequences [8], and can be comprised of an RNN [14], a pair of RNNs [37], and even transformers [17].
- An **update** block, which follows the KF update rule in (9.10), namely:

$$\hat{\boldsymbol{x}}_{t|t}(\boldsymbol{y}_{1:t};\theta) = \hat{\boldsymbol{x}}_{t|t-1} + \boldsymbol{K}_t(\theta)\big(\boldsymbol{y}_t - \boldsymbol{h}(\hat{\boldsymbol{x}}_{t|t-1})\big). \tag{9.16}$$

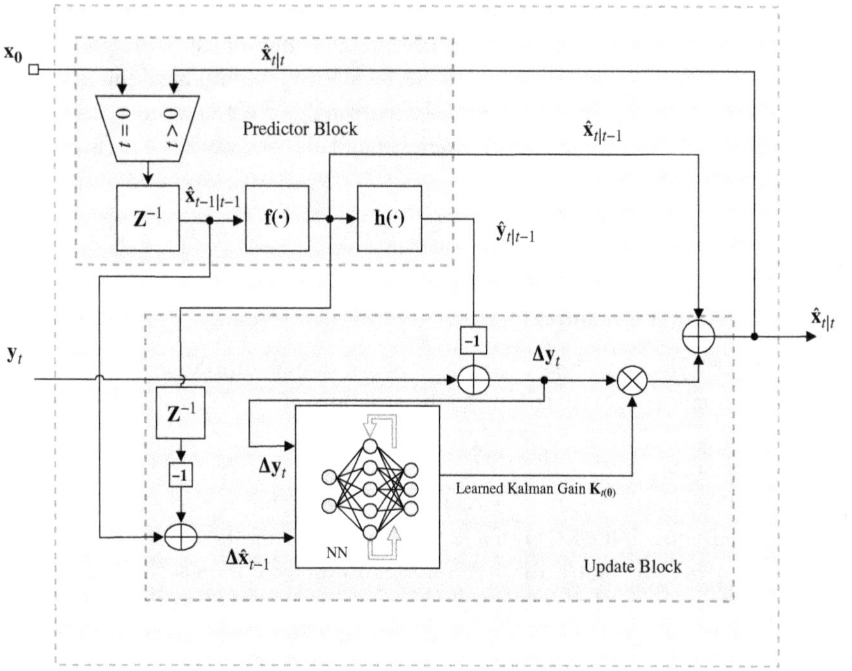

Figure 9.1 KalmanNet architecture: following the same computation flow as the classic KF, KalmanNet only replaces the KG with a carefully designed DNN.

9.3.2.3 Training Algorithm

The model is trained end-to-end using supervised learning. Given a dataset of the form (9.15), KalmanNet is trained to minimize a loss function such as the regularized MSE, i.e.,

$$\mathcal{L}_{\mathcal{D}_\mathrm{S}}(\boldsymbol{\Theta}) = \frac{1}{|\mathcal{D}_\mathrm{S}|}\sum_{i=1}^{|\mathcal{D}_\mathrm{S}|}\frac{1}{T}\sum_{t=1}^{T}\left\|\hat{\boldsymbol{x}}_{t|t}^{(i)}(\boldsymbol{y}_{1:t}^{(i)};\boldsymbol{\Theta}) - \boldsymbol{x}_t^{(i)}\right\|_2^2 + \lambda\|\boldsymbol{\Theta}\|_2. \tag{9.17}$$

Here, $\hat{\boldsymbol{x}}_{t|t}^{(i)}(\boldsymbol{y}_{1:t};\boldsymbol{\Theta})$ is the estimate obtained with DNN parameters $\boldsymbol{\Theta}$ via (9.16), while λ is a regularization coefficient. The regularization term $\lambda\|\boldsymbol{\Theta}\|_2$ plays a critical role in preventing overfitting by discouraging overly complex models with large parameter values, thereby promoting generalization to unseen data. The loss in (9.17) is used to guide the training of the internal DNN of KalmanNet to compute the KG not based on "ground-truth" KG matrices, but based on the performance of the overall state estimation algorithm. Given this formulation and the differentiability of the EKF [38], KalmanNet can be trained via conventional training methods based on, e.g., mini-batch stochastic gradient descent (SGD), in which backpropagation through time (BPTT) is used to update the weights of the neural network components.

9.3.3 Smoothing: RTSNet

RTSNet [15] extends KalmanNet by incorporating a smoothing capability based on the RTS smoother framework, providing enhanced performance for offline state estimation tasks.

9.3.3.1 Rationale

RTSNet enhances KalmanNet's filtering capabilities by implementing a backward pass that refines the estimates by leveraging future observations, which is analogous to the classical RTS smoother. RTSNet learns to approximate the smoothing gain via neural networks, replacing the analytical expressions used in the RTS smoother.

9.3.3.2 Architecture

The RTSNet architecture, as shown in Figure 9.2, mirrors the structure of the classical RTS smoother. Each forward–backward pass consists of a forward (KF-type) pass and a backward recursion. To augment the classical RTS smoother, RTSNet replaces the forward and backward KG computations with dedicated RNNs. As such, it builds upon KalmanNet with additional backward-processing modules:

- **Forward Pass:** Identical to KalmanNet's filtering pass, producing $\hat{\boldsymbol{x}}_{t|t}$ and $\hat{\boldsymbol{x}}_{t+1|t}$ using a KG DNN.
- **Learned Backward Gain:** A DNN, typically an RNN, that processes features extracted from the sequence backward in time to produce a learned smoothing gain $\boldsymbol{G}_t(\Theta)$.
- **Smoothed Update:** Refinement of the state estimate using:

$$\hat{\boldsymbol{x}}_{t|T}(\boldsymbol{y}_{1:T};\Theta) = \hat{\boldsymbol{x}}_{t|t}(\boldsymbol{y}_{1:t};\Theta) + \boldsymbol{G}_t(\Theta)\big(\hat{\boldsymbol{x}}_{t+1|T}(\boldsymbol{y}_{1:T};\Theta) - \hat{\boldsymbol{x}}_{t+1|t}\big). \tag{9.18}$$

9.3.3.3 Unfolded RTSNet Inference Algorithm

Since the conventional single forward–backward operation of the RTS algorithm can be shown to be optimal only for linear Gaussian SS models, RTSNet can be extended to learn to carry out multiple passes as a form of deep unfolding as shown in Figure 9.3. Notably, RTSNet performs $K \geq 1$ forward–backward passes (indexed by $k \in \{1, \dots, K\}$) using a deep unfolding strategy. For each pass, the outputs are increasingly refined state sequences $\left\{\hat{\boldsymbol{x}}^{(k)}_{1|T}(\boldsymbol{y}_{1:T};\Theta), \dots, \hat{\boldsymbol{x}}^{(k)}_{T|T}(\boldsymbol{y}_{1:T};\Theta)\right\}_{k=1}^{K}$.

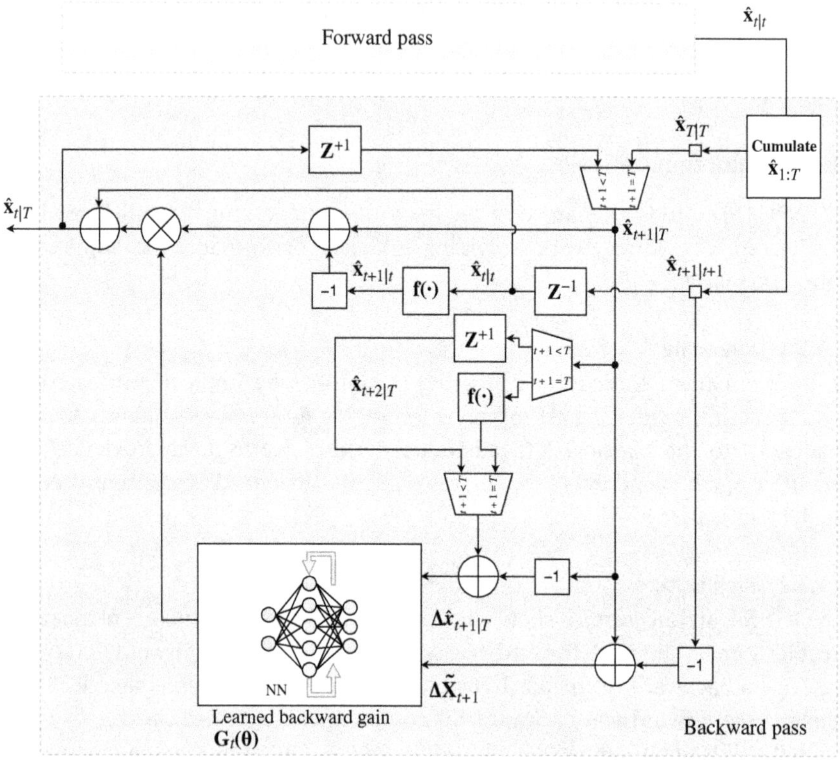

Figure 9.2 RTSNet architecture: composed of a forward and a backward pass, RTSNet follows the same computation flow as the classic RTS smoother, with the forward and backward KGs replaced by two carefully designed DNNs.

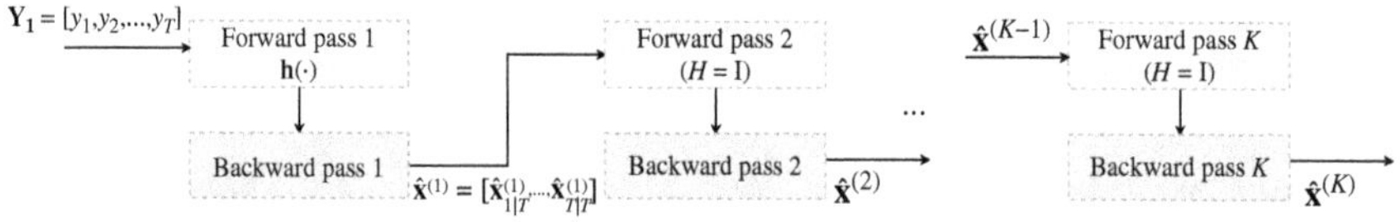

Figure 9.3 Unfolded RTSNet architecture: Unfolded RTSNet concatenates several forward and backward passes together. Note that the observation matrix $\boldsymbol{H}$ becomes an identity matrix starting from the 2nd pass. Taking the output sequence from previous pass as observation, the following passes of unfolded RTSNet aim to further refine the sequence.

9.3.3.4 Training Algorithm

RTSNet is trained end-to-end in a supervised manner by minimizing the MSE between $\hat{\boldsymbol{x}}_{t|T}^{(k)}$ and the ground truth state $\boldsymbol{x}_t$ across all k passes. Training can follow two paradigms:

- **Joint Training:** Aggregating the loss across all K passes:

$$\mathcal{L}_{\mathcal{D}_s}(\theta) = \frac{1}{|\mathcal{D}_s|}\sum_{i=1}^{|\mathcal{D}_s|}\frac{1}{T}\sum_{t=1}^{T}\sum_{k=1}^{K}\alpha_k\left\|\hat{\boldsymbol{x}}_{t|T}^{(k,i)}(\boldsymbol{y}_{1:T}^{(i)};\theta) - \boldsymbol{x}_t^{(i)}\right\|_2^2 + \lambda\|\theta\|_2. \quad (9.19)$$

  where $\{\alpha_k\}_{k=1}^{K}$ control the relative weighting of each pass, with a suitable setting being $\alpha_k = \log(1+k)$ [39].
- **Sequential Training:** Training each pass k sequentially, while freezing the previous passes.

In both cases, training is done via conventional training methods based on, e.g., mini-batch SGD. Note that the training approaches outlined above both leverage the interpretable operation of RTSNet, and particularly the ability to cast the output of each forward–backward pass as what should be a gradually improved estimate of the state trajectory.

9.3.4 Adaptive KalmanNet

Adaptive KalmanNet [16] extends the original KalmanNet by incorporating online adaptation mechanisms that enable robust filtering in environments with time-varying noise statistics and model mismatches. By leveraging a compact hypernetwork for context-dependent modulation, Adaptive KalmanNet adjusts its KG computation dynamically, eliminating the need for costly retraining when the noise conditions change.

9.3.4.1 Rationale

Standard KalmanNet is designed for filtering under fixed noise distributions and known system parameters, limiting its performance when these conditions vary. Adaptive KalmanNet addresses this limitation by introducing an adaptation mechanism that utilizes a scalar indicator, termed the *State of the World* (SoW), which captures the scaling ratio between the process and observation noise variances:

$$\mathrm{SoW}_t = \frac{n \cdot \mathrm{Tr}(\boldsymbol{Q}_t)}{m \cdot \mathrm{Tr}(\boldsymbol{R}_t)}. \quad (9.20)$$

A dedicated hypernetwork receives this SoW as input and outputs modulation parameters (gain and shift) that fine-tune the internal KG DNN. This approach retains the majority of the model-based structure of KalmanNet while providing the flexibility to adapt to a continuously changing environment without requiring retraining.

9.3.4.2 Architecture

The overall architecture of Adaptive KalmanNet, as shown in Figure 9.4, mirrors that of KalmanNet with the addition of a hypernetwork for adaptation:

- **Predictor:** Implements the known state transition model $\boldsymbol{f}(\cdot)$ to generate the predicted state $\hat{\boldsymbol{x}}_{t|t-1}$.
- **Learned KG Module:** A DNN that computes the KG. In Adaptive KalmanNet, this module is augmented with context-dependent modulation (CM). The hypernetwork with parameters ψ processes the SoW to produce modulation parameters:

$$g = d_{\psi}(\text{SoW}, 1), \quad s = d_{\psi}(\text{SoW}, 0), \tag{9.21}$$

  where g and s denote the multiplicative gain and additive shift, respectively. These parameters modulate the linear layer computations in the KG DNN as follows:

$$\sigma\Big((Wx + b) \odot g + s\Big), \tag{9.22}$$

- **Update:** Adopts the standard Kalman update rule:

$$\hat{\boldsymbol{x}}_{t|t}(\boldsymbol{y}_{1:t}; \theta, \psi) = \hat{\boldsymbol{x}}_{t|t-1} + \boldsymbol{K}_t(\theta, \psi)\Big(\boldsymbol{y}_t - \boldsymbol{h}(\hat{\boldsymbol{x}}_{t|t-1})\Big), \tag{9.23}$$

  where $\boldsymbol{K}_t(\theta, \psi)$ is the adapted KG.

9.3.4.3 Training Algorithm

Adaptive KalmanNet is trained in a supervised manner using the regularized MSE loss function:

$$\mathcal{L}_{\mathcal{D}_{\text{s}}}(\theta, \psi) = \frac{1}{|\mathcal{D}_{\text{s}}|} \sum_{i=1}^{|\mathcal{D}_{\text{s}}|} \frac{1}{T} \sum_{t=1}^{T} \| \hat{\boldsymbol{x}}_{t|t}^{(i)}(\boldsymbol{y}_{1:t}^{(i)}; \theta, \psi) - \boldsymbol{x}_t^{(i)} \|_2^2 + \lambda\|\theta\|_2 + \lambda\|\psi\|_2. \tag{9.24}$$

The training proceeds in two stages:

1. **Stage One—KalmanNet Training:** The CM layers are fixed to output unit gains and zero shifts, effectively disabling the adaptation. Only the original KalmanNet parameters θ are trained using data with stationary or similar noise distributions.
2. **Stage Two—Hypernetwork Training:** With θ fixed, the hypernetwork parameters ψ are trained using a dataset that includes diverse, time-varying noise conditions. This stage tunes the CM parameters so that the adapted KG $\boldsymbol{K}_t(\theta, \psi)$ can accurately reflect changes in the noise statistics.

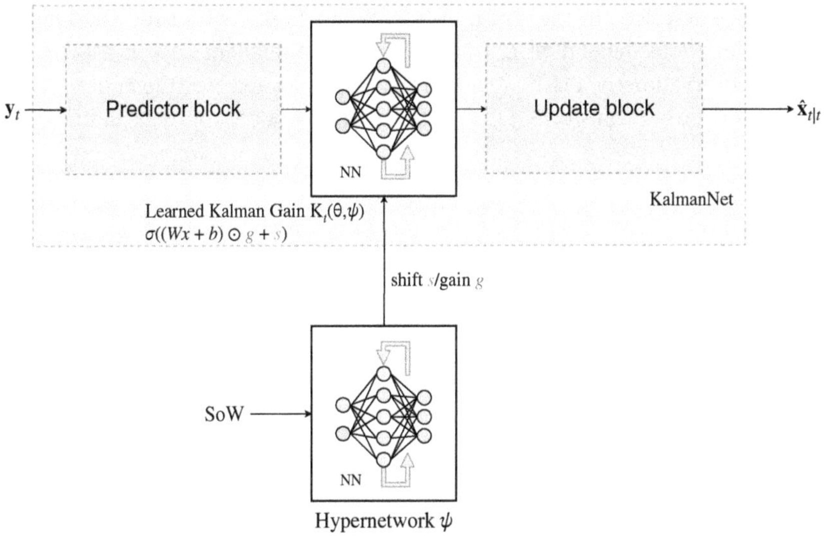

Figure 9.4 Adaptive KalmanNet architecture: adaptive KalmanNet uses a hypernetwork to fine-tune the weights of the KG DNN of KalmanNet.

This two-stage training strategy enables Adaptive KalmanNet to achieve robust state estimation while rapidly adapting to changes in the underlying noise environment.

9.3.5 GSP-KalmanNet: Kalman Filtering for Graph Signals

While Kalman-type filters are widely used for tracking time-varying signals, many real-world signals are inherently defined over graphs. Such *graph signals* arise in applications ranging from power system monitoring to social and transportation networks, where each node represents a spatial location or entity and the graph encodes structural dependencies. The irregular domain and high dimensionality of these signals, along with possible model mismatches, challenge conventional filtering methods. To address this, the graph signal processing *(GSP)-KalmanNet* architecture [34] extends the KalmanNet framework by incorporating tools from GSP. The central idea is to represent and process the KG in the graph frequency domain by modeling it as a graph filter, which naturally aligns with the graph topology and enables scalable inference.

9.3.5.1 Graph Frequency Domain Modeling

Let $\boldsymbol{L} \in \mathbb{R}^{N\times N}$ be the graph Laplacian associated with the underlying graph, with eigendecomposition $\boldsymbol{L} = \boldsymbol{V}\boldsymbol{\Lambda}\boldsymbol{V}^{\top}$. We consider a nonlinear SS model over graphs, which can be described by the model in (9.2), where both the transition and measurement functions depend on $\boldsymbol{L}$. In addition, both the latent state $\boldsymbol{x}_t \in \mathbb{R}^N$ and the noisy observations $\boldsymbol{y}_t \in \mathbb{R}^N$ are graph signals, measured over the nodes.

The graph Fourier transform (GFT) of the graph signals $\boldsymbol{x}_t$ and $\boldsymbol{y}_t$ is defined as:

$$\tilde{\boldsymbol{x}}_t = \boldsymbol{V}^\top \boldsymbol{x}_t. \tag{9.25}$$

The EKF for the SS model in (9.2) (assuming known noise distributions) can be implemented in the graph frequency domain. To this end, we transform (9.2) to the graph frequency domain by left-multiplying these equations by $\boldsymbol{V}^\top$. By using (9.25), we obtain the graph frequency SS model:

$$\tilde{\boldsymbol{x}}_t = \boldsymbol{V}^\top \boldsymbol{f}_t(\boldsymbol{L}, \boldsymbol{V}\tilde{\boldsymbol{x}}_{t-1}) + \boldsymbol{V}^\top \boldsymbol{e}_t, \tag{9.26a}$$

$$\tilde{\boldsymbol{y}}_t = \boldsymbol{V}^\top \boldsymbol{h}_t(\boldsymbol{L}, \boldsymbol{V}\tilde{\boldsymbol{x}}_t) + \boldsymbol{V}^\top \boldsymbol{v}_t. \tag{9.26b}$$

Using the conventional model assumptions, it can be verified that the state-evolution and measurement noises in the graph frequency domain, $\boldsymbol{V}^\top \boldsymbol{e}_t \in \mathbb{R}^N$ and $\boldsymbol{V}^\top \boldsymbol{v}_t \in \mathbb{R}^N$, respectively, are mutually independent at time t, and each one of them is also independent between different time samples, since they are obtained from the noises in (9.2) by a linear transformation. In addition, they still have zero mean and known covariance matrices that are now given by $\boldsymbol{V}^\top \boldsymbol{R}\boldsymbol{V}$ and $\boldsymbol{V}^\top \boldsymbol{Q}\boldsymbol{V}$, respectively.

9.3.5.2 Graph-filtered Kalman Gain

In GSP-KalmanNet, we use the EKF where we model the KG as a *frequency-domain graph filter*, i.e., a diagonal operator in the GFT domain:

$$\boldsymbol{K}_t = \boldsymbol{V}\ \mathrm{diag}(\boldsymbol{g}_t)\ \boldsymbol{V}^\top, \tag{9.27}$$

where $\boldsymbol{g}_t \in \mathbb{R}^N$ denotes the learned frequency response. This is a diagonal matrix that is only a function of the eigenvalues of the Laplacian matrix, $\boldsymbol{\Lambda}$. This structure ensures that the KG respects the graph topology and allows efficient implementation. The resulting algorithm is the GSP-EKF [40], which can be implemented as in Subsection 9.2.2.3 with the KG in (9.27).

9.3.5.3 Learned Gain Computation via RNN

To account for partial model knowledge and adapt to unknown or time-varying dynamics, we design GSP-KalmanNet to learn the KG from data in a similar manner to KalmanNet by combining the learned KG with the overall GSP-EKF flow. In particular, we use a trainable RNN to output the coefficient of a frequency-domain graph filter, enforcing a diagonal matrix manipulation in the graph frequency domain. This architecture is illustrated in Figure 9.5.

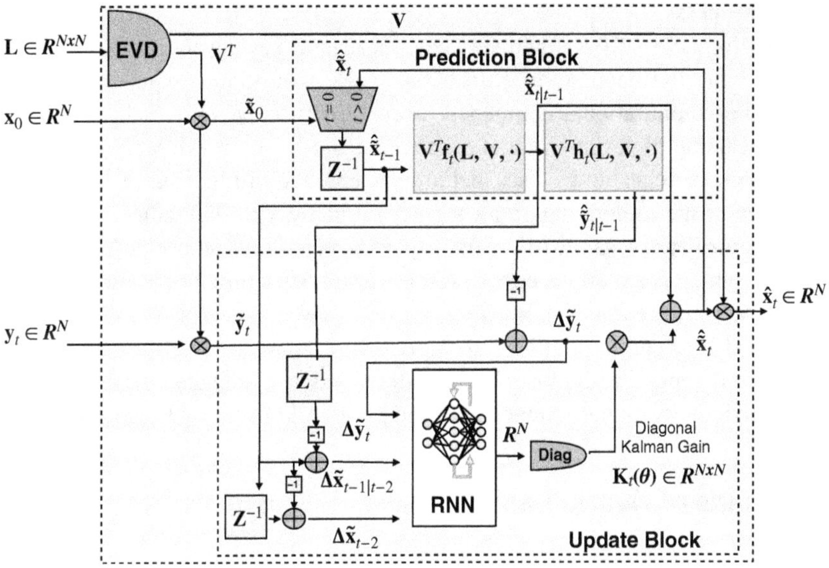

Figure 9.5 GSP-KalmanNet block diagram.

The graph filter $\boldsymbol{g}_t$ is learned via an RNN $\mathcal{R}_\theta$ parameterized by θ:

$$\boldsymbol{g}_t = \mathcal{R}_\theta(\Delta\tilde{\boldsymbol{y}}_t, \Delta\tilde{\boldsymbol{x}}_t), \tag{9.28}$$

where the features include innovation in the GFT domain:

$$\Delta\tilde{\boldsymbol{y}}_t = \boldsymbol{V}^\top(\boldsymbol{y}_t - \hat{\boldsymbol{y}}_{t|t-1}), \quad \Delta\tilde{\boldsymbol{x}}_t = \boldsymbol{V}^\top(\hat{\boldsymbol{x}}_{t-1|t-1} - \hat{\boldsymbol{x}}_{t-2|t-2}).$$

The state update follows:

$$\hat{\boldsymbol{x}}_{t|t} = \hat{\boldsymbol{x}}_{t|t-1} + \boldsymbol{K}_t(\boldsymbol{y}_t - \boldsymbol{h}(\boldsymbol{L}, \hat{\boldsymbol{x}}_{t|t-1})), \tag{9.29}$$

where $\boldsymbol{K}_t$ is a graph filter with the structure given in (9.27).

9.3.5.4 Training

Given a training dataset of state-observation sequences $\mathcal{D} = \left\{(\boldsymbol{x}_{1:T}^{(i)}, \boldsymbol{y}_{1:T}^{(i)})\right\}_{i=1}^{N}$, the parameters θ are learned by minimizing the MSE:

$$\mathcal{L}(\theta) = \frac{1}{NT}\sum_{i=1}^{N}\sum_{t=1}^{T}\left\|\boldsymbol{x}_t^{(i)} - \hat{\boldsymbol{x}}_{t|t}^{(i)}(\theta)\right\|^2 + \lambda\|\theta\|^2, \tag{9.30}$$

where $\lambda > 0$ is a regularization parameter to prevent overfitting.

9.3.5.5 Benefits

By leveraging GSP, GSP-KalmanNet achieves several key advantages. First, it is highly scalable: the diagonal structure of the KG $\boldsymbol{K}_t$ in the graph frequency domain reduces the computational complexity from $\mathcal{O}(N^3)$ to $\mathcal{O}(N)$, which is critical when processing high-dimensional graph signals. Second, it exhibits robustness to model mismatches and noise, including scenarios with partial or uncertain knowledge of the underlying graph topology. Finally, GSP-KalmanNet is inherently structure-aware, as the use of the Laplacian eigenbasis allows the algorithm to exploit spatial correlations encoded in the graph structure, improving estimation accuracy and generalization.

Empirical evaluations demonstrate that GSP-KalmanNet achieves improved accuracy and generalization over both model-based and fully data-driven benchmarks, particularly in high-dimensional and partially known graph signal models.

9.4 Gains of Hybrid Design

The KalmanNet framework exemplifies a broader class of hybrid architectures that integrate classical model-based estimation with modern data-driven learning [12]. While prior sections have focused on algorithmic formulation and measurable improvements in tracking accuracy, this section aims to highlight the broader advantages of such hybrid designs—beyond raw performance metrics. By embedding learning components within the principled structure of Kalman-type filters, KalmanNet not only enhances estimation robustness under model mismatch but also introduces several practical benefits that make it highly suitable for real-world deployment. These include improved interpretability, uncertainty quantification, adaptability to dynamic environments, and simple integration into existing modular pipelines.

In what follows, we provide a structured discussion of these qualitative gains. We begin by highlighting the simplicity and efficiency of KalmanNet's hybrid design. Then, we discuss how uncertainty quantification, a fundamental aspect of classical filtering, is naturally preserved and enhanced, after which we show how KalmanNet enables learning even in the absence of supervision. We next explore the ability of the hybrid architecture to bridge pre- and post-estimation modules through trainable interfaces, breaking the traditional separation principle. We conclude by illustrating how prior structural knowledge can be seamlessly incorporated into KalmanNet for efficient and scalable design.

9.4.1 Simplicity

A key advantage of the KalmanNet framework lies in its simplicity stemming from its hybrid design, which augments classical Kalman filter structures with deep learning, rather than replacing them. Since Kalman-type filters are already deeply integrated into numerous radar and control systems, this augmentation-based approach provides a seamless and natural pathway for introducing AI into existing pipelines. Leveraging the established recursive structure of Kalman filters not only simplifies the implementation but also preserves interpretability and robustness, avoiding the need to reinvent proven estimation mechanisms from scratch. This hybrid paradigm thereby offers an intuitive transition path for engineers and practitioners, requiring minimal changes to the overall architecture of deployed systems.

In addition, augmenting the Kalman filter with neural networks yields much simpler and more efficient deep learning models compared to purely end-to-end alternatives. Rather than training large black-box networks to learn full SS inference from raw measurements, the hybrid architecture delegates the core estimation tasks to the structured Kalman recursion, limiting the role of the neural network to learning specific, hard-to-model components—such as the KG. This results in reduced model complexity, allowing the learned module to be lightweight and more easily deployed on hardware-limited platforms. A representative example of this benefit is demonstrated in the autonomous racing scenario detailed in [24], where KalmanNet was successfully implemented on the limited microcontroller of a high-speed race car. In contrast, previous deep learning-based estimators with comparable performance were too complex for on-board execution. This highlights how hybridization with Kalman filters leads to practical, efficient, and scalable AI-aided tracking systems.

9.4.2 Uncertainty Extraction

In many radar-based target tracking applications, the ability to quantify the uncertainty in the estimated state is almost as important as tracking the state itself. Such information is not only essential for robust decision-making and sensor fusion but also critical in safety-critical systems such as navigation, collision avoidance, or sensor scheduling. Classical Kalman-type filters inherently provide a principled uncertainty estimate in the form of the error covariance matrix, which reflects the confidence in the estimated state.

In contrast, deep learning-based tracking methods, particularly end-to-end discriminative neural networks, typically produce point estimates without a natural or interpretable way to assess uncertainty. In fact, while deep learning models can be extended to capture uncertainty, doing so often requires careful design. Simple approaches such as training with a negative log-likelihood loss to regress both the mean and covariance, MC dropout [41], or ensembling can introduce uncertainty estimates with minimal added complexity. More sophisticated mechanisms [42], including Bayesian deep learning (which was combined with KalmanNet in [43]) and conformal prediction [44], offer additional tools. However, the core challenge lies in calibrating it: ensuring that the predicted uncertainty accurately reflects the true posterior, especially under distribution shifts or in regions outside the training regime. This issue remains a key obstacle for deploying deep learning models in real-world tracking scenarios, where reliable uncertainty estimates are often as critical as the predictions themselves.

9.4.2.1 Uncertainty Extraction in KalmanNet

The KalmanNet framework tackles this limitation by retaining an interpretable structure where one of the key learned features corresponds to the KG. This allows us to leverage the structure of the Kalman filter to recover an approximation of the error covariance matrix [45]. Specifically, in the classical EKF, the filtered error covariance $\boldsymbol{\Sigma}_{t|t}$ is related to the predicted error covariance $\boldsymbol{\Sigma}_{t|t-1}$ and the KG $\boldsymbol{K}_t$ via:

$$\boldsymbol{\Sigma}_{t|t} = (\boldsymbol{I} - \boldsymbol{K}_t\boldsymbol{H}_t)\boldsymbol{\Sigma}_{t|t-1}, \tag{9.31}$$

where $\boldsymbol{H}_t$ is the Jacobian of the measurement function at time t. When the KG is learned, as in KalmanNet, and $\boldsymbol{H}_t$ is known or can be estimated, this expression can be used to derive the posterior error covariance from the learned gain. Moreover, even when the prior covariance is not explicitly tracked, it can be recovered from the gain under mild conditions using the inverse relation:

$$\boldsymbol{\Sigma}_{t|t-1} = (\boldsymbol{I} - \boldsymbol{K}_t\boldsymbol{H}_t)^{-1}\boldsymbol{K}_t\boldsymbol{R}\boldsymbol{H}_t^{\top}(\boldsymbol{H}_t^{\top}\boldsymbol{H}_t)^{-1}, \tag{9.32}$$

where $\boldsymbol{R}$ is the observation noise covariance. Once the prior covariance is recovered, it can be used in Eq. (9.31) or the Joseph form to yield a robust error covariance estimate.

9.4.2.2 Uncertainty-aware Training

To ensure that the learned KG leads to a faithful estimation of the error covariance, the training procedure can be modified to explicitly encourage accurate uncertainty prediction. Two principled alternatives can be employed. The first is an *empirical averaging* loss, where the predicted diagonal entries of the error covariance are matched to the squared residuals of the state estimates. Specifically, defining the instantaneous prediction error as $\boldsymbol{e}_t = \boldsymbol{x}_t - \hat{\boldsymbol{x}}_t$ and the error covariance estimated from the KG via (9.31) as $\hat{\boldsymbol{\Sigma}}_{t|t}$, the training loss augments the standard ℓ_2 loss denoted $\mathcal{L}_{\ell_2}$, e.g., as in (9.17), with a covariance alignment term:

$$\mathcal{L}_{\text{Emp}} = (1-\beta)\mathcal{L}_{\ell_2} + \beta\sum_{j=1}^{m}\left|e_{t,j}^2 - \left[\hat{\boldsymbol{\Sigma}}_{t|t}\right]_{j,j}\right|, \tag{9.33}$$

where $\beta \in [0, 1]$ controls the emphasis on uncertainty estimation. This loss guides the network to match its predicted uncertainty with observed estimation errors.

The second approach assumes a Gaussian prior on the state and trains the network using the negative log-likelihood of a Gaussian distribution, whose mean and covariance are, respectively, given by the predicted state and its inferred error covariance. This results in the following loss:

$$\mathcal{L}_{\mathrm{GP}} = \boldsymbol{e}_t^\top \hat{\boldsymbol{\Sigma}}_{t|t}^{-1} \boldsymbol{e}_t + \log \det(\hat{\boldsymbol{\Sigma}}_{t|t}). \tag{9.34}$$

This formulation encourages both accurate tracking and calibrated uncertainty, as it penalizes deviations from the assumed predictive distribution. Empirical results demonstrate that incorporating such loss terms into the training of KalmanNet yields improved fidelity in error covariance extraction without compromising estimation performance.

9.4.3 Unsupervised Learning

A key advantage of the hybrid and interpretable nature of KalmanNet is that its internal operations can be directly associated with meaningful quantities in the Kalman filtering process. This structure not only facilitates uncertainty estimation but also enables effective unsupervised learning. In traditional deep learning-based state estimation methods, training typically requires a large corpus of labeled data comprising ground-truth states. However, in many real-world radar tracking scenarios, such ground-truth information is unavailable or expensive to obtain. KalmanNet mitigates this challenge by enabling learning based solely on observation sequences, without needing access to the true hidden states [46]. This is achieved by leveraging the fact that KalmanNet internally predicts the next observation using the structure of the SS model—specifically through the innovation process—which can be computed from data.

Unsupervised KalmanNet exploits this by defining a loss function based on the discrepancy between the actual observation $\boldsymbol{y}_t$ and its prediction $\hat{\boldsymbol{y}}_{t|t-1}$:

$$\hat{\boldsymbol{y}}_{t|t-1} = h(\hat{\boldsymbol{x}}_{t|t-1}), \tag{9.35}$$

where $\hat{\boldsymbol{x}}_{t|t-1}$ is the predicted state computed as:

$$\hat{\boldsymbol{x}}_{t|t-1} = f(\hat{\boldsymbol{x}}_{t-1}). \tag{9.36}$$

The innovation, $\Delta \boldsymbol{y}_t = \boldsymbol{y}_t - \hat{\boldsymbol{y}}_{t|t-1}$, quantifies how well the internal state estimation aligns with the observations, and can thus be used as a training signal. The unsupervised loss for the ith trajectory is defined as:

$$\tilde{\ell}_i(\Theta) = \frac{1}{T_i} \sum_{t=1}^{T_i} \left\| \boldsymbol{y}_t^{(i)} - \hat{\boldsymbol{y}}_{t|t-1}^{(i)}(\Theta) \right\|^2 + \gamma \cdot \|\Theta\|^2, \tag{9.37}$$

where Θ are the trainable parameters of KalmanNet (typically within the RNN computing the KG), and γ is a regularization coefficient. Training can then proceed using SGD and BPTT. This procedure allows KalmanNet to learn from purely observational data by minimizing the innovation energy, effectively tuning itself to perform accurate filtering without labeled supervision. We summarize the considered training objectives and supervision types of KalmanNet and its variants in Table 9.4.

The interpretable architecture of KalmanNet thus enables an adaptive, online training regime, which is complementary to the hypernetwork-based extension detailed earlier. Here, a model trained offline can be continually refined post-deployment as the environment or the underlying SS model evolves. Specifically, once every predefined amount of time steps, the unsupervised loss (9.37) is computed over the most recent observations, and the parameters Θ are updated accordingly. This semi-supervised paradigm allows KalmanNet to track not only the hidden state but also to adapt to temporal changes in system dynamics—such as changes in process or observation noise—making it highly suitable for long-term deployment in realistic radar tracking settings.

Table 9.4 Comparison of loss functions for KalmanNet variants.

Algorithm	**Loss function**	**Supervision**	**Notes**
KalmanNet [14]	ℓ_2 loss: $\|\hat{\boldsymbol{x}}_t - \boldsymbol{x}_t\|^2$	Supervised	Standard MSE loss; regularization may be added
RTSNet [15]	Aggregated ℓ_2 loss: $\sum_k \alpha_k \|\hat{\boldsymbol{x}}_t^{(k)} - \boldsymbol{x}_t\|^2$	Supervised	Weights α_k (e.g., $\log(1+k)$) to encourage better smoothing
Uncertainty-aware KalmanNet [43]	ℓ_2 + covariance penalty (see (9.33) and (9.34))	Supervised	Adds penalty to align predicted error covariance with residuals. Encourages accurate covariance estimation
Unsupervised KalmanNet [46]	Innovation loss: $\|\boldsymbol{y}_t - \hat{\boldsymbol{y}}_{t\|t-1}\|^2$	Unsupervised	Uses observations only; no need for ground-truth states
Adaptive KalmanNet [16]	ℓ_2 loss: $\|\hat{\boldsymbol{x}}_t - \boldsymbol{x}_t\|^2$	Supervised (2-stage)	Stage 1: train KalmanNet. Stage 2: train hypernetwork using SoW input

9.4.4 Break Separation

Another powerful advantage of using a trainable Kalman filter such as KalmanNet is its capacity to break the conventional separation between signal acquisition and tracking, while still adhering to the modular structure favored in many practical systems. In traditional radar and signal processing architectures, the system is typically designed following a separation principle, where the high-dimensional raw sensor signals are first passed through a dedicated preprocessing or feature extraction stage, and only then forwarded to a state estimator such as a Kalman filter, after which the tracked state is further processed for, e.g., action taking. This design simplifies modular development, but it also imposes a rigid interface that may not be optimal for the downstream task. The interpretability and modularity of KalmanNet, however, allow it to retain this structured separation, while its trainability enables the entire processing pipeline to be optimized end-to-end.

9.4.4.1 Preprocessing

In particular, this means that the preprocessing stage can be tuned specifically to produce features that are most useful for accurate state estimation, rather than for generic reconstruction or classification. This paradigm is exemplified by the recent development of Latent-KalmanNet [47], which demonstrates how KalmanNet can be extended to jointly learn with a trainable encoder that maps high-dimensional observations, such as images or range-Doppler maps, into low-dimensional latent features suitable for filtering. In Latent-KalmanNet, a DNN encoder processes the complex raw measurement $\boldsymbol{y}_t$ into a latent vector $\boldsymbol{z}_t$, which is then used by KalmanNet to infer the hidden state $\boldsymbol{x}_t$. Crucially, both the encoder and KalmanNet are trained together using a discriminative loss function based on the final state estimate. This encourages the encoder to generate latent features that facilitate accurate tracking, even when the original measurement model is unknown or highly nonlinear. This methodology proves particularly valuable in applications such as video-based object tracking or radar signal processing, where the raw observations are extremely high-dimensional and difficult to model analytically. By enabling joint learning across the acquisition and estimation stages, KalmanNet-based architectures like Latent-KalmanNet offer a data-driven means to optimize the full pipeline while preserving the structure, interpretability, and efficiency of model-based filtering.

9.4.4.2 Post-processing

KalmanNet's flexibility to preserve separation in system design while deviating from it in training is also particularly beneficial when the Kalman filter serves as a component within a broader system that includes downstream modules. These include, for example, when the estimated state is used for decision-making or control. In such settings, the classical separation principle of stochastic control posits that the state estimation (e.g., via Kalman filtering) should be performed independently of the downstream controller. However, KalmanNet enables one to break this separation during training by optimizing the estimator not solely for state estimation accuracy, but rather for performance in the downstream task. This ability to align the output of the Kalman filter with the needs of the task-specific objective leads to better end-to-end system performance.

A clear example of this idea is found in the recent development of LQGNet [48, 49], which integrates KalmanNet as the state estimator within a linear-quadratic-Gaussian (LQG) control framework. In classic LQG control, the optimal policy decomposes into a Kalman filter followed by a linear-quadratic regulator (LQR), relying on a fully known system model. However, in practical scenarios where the model is only partially known, applying the classical separation-based solution can lead to sub-optimal control performance. LQGNet addresses this by using a trainable KalmanNet to estimate the state and feeding it to a fixed LQR. While this design maintains the structural modularity of the classic separation-based controller, training is performed end-to-end with respect to the overall control cost, allowing the KalmanNet module to adapt its internal estimation behavior to better support the controller. In this way, LQGNet learns to output state estimates—or more generally, features—that may not minimize estimation error but are best suited for achieving effective control. This paradigm is particularly powerful in stochastic and partially known systems, enabling robust and efficient learning of feedback policies while retaining interpretability and modularity in deployment. Similar methodologies based on KalmanNet were proposed in, e.g., tracking of financial assets for reward maximization [22].

9.4.5 Structure Awareness

The standard KalmanNet architecture is designed to operate generically, without making any assumptions about the structural properties of the signals or their domain. This generality enables wide applicability, but it may also result in unnecessary complexity and inefficiency when the signals exhibit known structure. A powerful benefit of KalmanNet's interpretable and modular hybrid design is the ability to incorporate prior knowledge about the underlying system into the deep learning component. In scenarios where the signal lies on a known irregular domain—such as a network or a graph—one can exploit this structure to design more efficient, scalable, and robust architectures. In such cases, rather than relying on generic dense neural networks, one can tailor the KalmanNet architecture to process signals in structure-aware domains such as the graph frequency space.

This concept is exemplified in the GSP-KalmanNet algorithm [34], discussed in Section 9.3.5. GSP-KalmanNet leverages this by processing both the prediction and update steps in the graph frequency domain using tools from GSP, such as the GFT and frequency-domain graph filters. The key insight is to constrain the KG to a diagonalizable graph filter in the GFT basis, significantly reducing the number of learnable parameters and the computational burden, while still allowing the filter to be optimized from data. By augmenting a graph-structured version of the extended Kalman filter (GSP-EKF) with a neural network that learns the parameters of this graph filter, GSP-KalmanNet achieves efficient, interpretable, and scalable tracking. Empirical studies in [34] show that this structure-aware approach yields performance comparable or superior to conventional KalmanNet, especially in high-dimensional or partially modeled settings, while requiring fewer resources and exhibiting improved robustness. This demonstrates how the KalmanNet framework can flexibly incorporate domain knowledge into its architecture, yielding better generalization and performance through structure-aware learning.

9.5 Quantitative Analysis

We now turn to a rigorous quantitative evaluation of KalmanNet and its variants. This section provides empirical evidence demonstrating their enhanced robustness and accuracy, particularly under conditions of noise and model mismatch. By benchmarking KalmanNet and its variants against the classical model-based methods within a controlled setting, we aim to clarify the numerical advantages of hybrid approaches in concrete terms.

Throughout the empirical study and unless stated otherwise, in the experiments involving synthetic data, the SS model is generated using diagonal noise covariance matrices; i.e.,

$$\boldsymbol{Q} = \mathrm{q}^2 \cdot \boldsymbol{I}, \quad \boldsymbol{R} = \mathrm{r}^2 \cdot \boldsymbol{I}, \quad \nu \triangleq \frac{\mathrm{q}^2}{\mathrm{r}^2}. \tag{9.38}$$

By (9.38), setting ν to be 0 dB implies that both the state noise and the observation noise have the same variance.

9.5.1 KalmanNet

In this section, we first verify that in the linear Gaussian setting, KalmanNet achieves performance comparable to the theoretically optimal KF, and even surpasses it in the presence of model mismatch. We then evaluate KalmanNet on the more challenging nonlinear model, showing that it performs robustly and consistently outperforms all model-based counterparts, both with full model information and under model mismatch. Finally, in this challenging nonlinear scenario with model mismatch, we demonstrate KalmanNet's transfer learning ability, showing that it can be trained on short trajectories and successfully generalize to longer ones.

9.5.1.1 Linear SS Model

The state vector $\boldsymbol{x}_t = [p_t, v_t, a_t]^\top \in \mathbb{R}^3$ describes the one-dimensional motion of an object, where p_t, v_t, and a_t correspond to its position, velocity, and acceleration at time t. Noisy observations of the position are obtained at regular time steps of $\Delta\tau = 10^{-2}$. This setup results in a linear Gaussian SS model, where the observation model is given by $\boldsymbol{H} = [1, 0, 0]$, $\boldsymbol{R} = \mathrm{r}^2$. The state transition model is:

$$\boldsymbol{F}_{CA} = \begin{bmatrix} 1 & \Delta\tau & \frac{1}{2}\Delta\tau^2 \\ 0 & 1 & \Delta\tau \\ 0 & 0 & 1 \end{bmatrix}; \qquad \boldsymbol{Q}_{CA} = \mathrm{q}^2 \cdot \begin{bmatrix} \frac{1}{20}\Delta\tau^5 & \frac{1}{8}\Delta\tau^4 & \frac{1}{6}\Delta\tau^3 \\ \frac{1}{8}\Delta\tau^4 & \frac{1}{3}\Delta\tau^3 & \frac{1}{2}\Delta\tau^2 \\ \frac{1}{6}\Delta\tau^3 & \frac{1}{2}\Delta\tau^2 & \Delta\tau \end{bmatrix},$$

where $\boldsymbol{Q}_{CA}$ can be deduced [14] from the discretization of the stochastic differential equation $\frac{d\boldsymbol{x}_t}{dt} = \begin{bmatrix} 0 & 1 & 0 \\ 0 & 0 & 1 \\ 0 & 0 & 0 \end{bmatrix} \boldsymbol{x}_t + \begin{bmatrix} 0 \\ 0 \\ \boldsymbol{w}_t \end{bmatrix}$. The scalars q^2 and r^2 serve as tunable scaling factors that allow one to simulate various noise conditions commonly encountered in real-world target tracking scenarios.

9.5.1.2 Model Mismatch

While the synthetic dataset is generated using the CA model, we evaluate the algorithms under the assumption that they are plugged in with a CV model, thereby mimicking the motion modeling errors commonly encountered in real-world target tracking scenarios. The CV model is defined as follows:

$$\boldsymbol{F}_{\mathrm{CV}} = \begin{bmatrix} 1 & \Delta\tau \\ 0 & 1 \end{bmatrix}, \qquad \boldsymbol{Q}_{\mathrm{CV}} = \mathrm{q}^2 \cdot \begin{bmatrix} \frac{1}{3}\Delta\tau^3 & \frac{1}{2}\Delta\tau^2 \\ \frac{1}{2}\Delta\tau^2 & \Delta\tau \end{bmatrix}, \tag{9.39}$$

where $\boldsymbol{Q}_{\mathrm{CV}}$ can similarly be deduced [14] from the discretization of the stochastic differential equation $\frac{d\boldsymbol{x}_t}{dt} = \begin{bmatrix} 0 & 1 \\ 0 & 0 \end{bmatrix} \boldsymbol{x}_t + \begin{bmatrix} 0 \\ \boldsymbol{w}_t \end{bmatrix}$. The observation model is given by $\boldsymbol{H} = [1, 0]$, $\boldsymbol{R} = \mathrm{r}^2$.

9.5.1.3 Representative Results

The performance of the hybrid model KalmanNet and the classical model-based KF is evaluated under model mismatch and varying observation noise

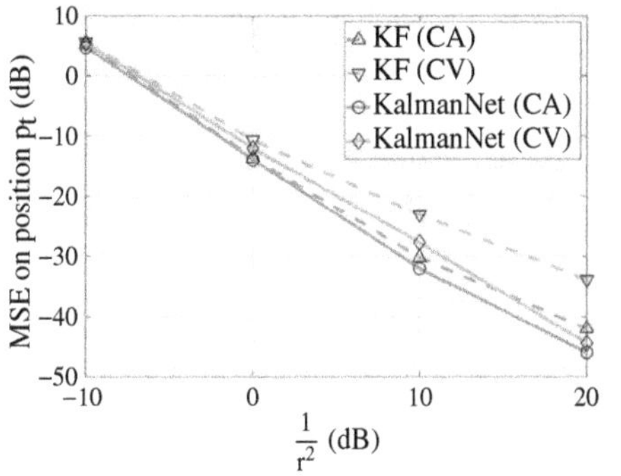

Figure 9.6 Linear SS model with or without model mismatch: solid lines represent KalmanNet while dashed lines are for KF. The dataset is generated using CA model. The (CA) notation in the legend stands for plugging in the true CA model, while the (CV) represents plugging in the mismatched CV model.

settings. The results, illustrated in the following figures, highlight the practical advantages of hybrid designs in scenarios that mimic real-world target tracking conditions—where system models are often imperfect and noise characteristics are uncertain.

Figure 9.6 compares the MSE on the position state for both KalmanNet and KF in two distinct scenarios: one where the true motion model $\boldsymbol{F}_{\text{CA}}$ is used, and another where a mismatched model $\boldsymbol{F}_{\text{CV}}$ is applied. In both cases, the systems are tested under a range of observation noise variances. The results show that KalmanNet consistently outperforms the classical Kalman filter when faced with model mismatch, maintaining lower MSE values across all tested conditions. This is particularly relevant in real-world radar-based tracking systems, where exact target dynamics are rarely known and filters must operate robustly under imperfect assumptions.

More interestingly, even in this linear Gaussian setting—where the Kalman filter using the correct CA model is theoretically optimal in minimizing the MSE for the full state vector $[p_t, v_t, a_t]$—KalmanNet can still achieve better performance on the position component p_t if it is trained using a loss function that targets only position. This highlights the potential of task-driven learning to prioritize specific estimation goals, a flexibility not available in purely model-based filters.

9.5.1.4 Nonlinear SS Model

The Lorenz attractor is a three-dimensional chaotic trajectory arising from the Lorenz system of continuous-time ordinary differential equations. This synthetic example illustrates the challenge of online tracking for highly nonlinear dynamics, as well as the practical difficulty introduced when converting a continuous-time signal into its discrete-time representation [50].

In particular, the noiseless state-evolution of the continuous-time process $\boldsymbol{x}_\tau$ with $\tau \in \mathbb{R}^+$ is given by:

$$\frac{\partial}{\partial \tau}\boldsymbol{x}_\tau = \boldsymbol{A}(\boldsymbol{x}_\tau) \cdot \boldsymbol{x}_\tau, \;\; \boldsymbol{A}(\boldsymbol{x}_\tau) = \begin{pmatrix} -10 & 10 & 0 \\ 28 & -1 & -\boldsymbol{x}_{1,\tau} \\ 0 & \boldsymbol{x}_{1,\tau} & -\dfrac{8}{3} \end{pmatrix}. \tag{9.40}$$

To obtain a discrete-time state-evolution model, we follow the procedure outlined in [51].We first sample it at intervals of $\Delta\tau$, assuming that $\boldsymbol{A}(\boldsymbol{x})_\tau$ can be kept constant in a small neighborhood of $\boldsymbol{x}_\tau$; i.e., $\boldsymbol{A}(\boldsymbol{x})_\tau \approx \boldsymbol{A}(\boldsymbol{x}_{\tau+\Delta\tau})$. Next, the continuous-time solution of the differential system (9.40), which holds in the vicinity of $\boldsymbol{x}_\tau$ over the short interval $\Delta\tau$, is given by:

$$\boldsymbol{x}_{\tau+\Delta\tau} = \exp\left(\boldsymbol{A}(\boldsymbol{x}_\tau) \cdot \Delta\tau\right) \cdot \boldsymbol{x}_\tau. \tag{9.41}$$

Finally, by expanding (9.41) into a Taylor series and truncating it to a finite sum with J terms, we obtain:

$$\boldsymbol{F}(\boldsymbol{x}_\tau) \triangleq \exp\left(\boldsymbol{A}(\boldsymbol{x}_\tau) \cdot \Delta\tau\right) \approx \boldsymbol{I} + \sum_{j=1}^{J} \frac{\left(\boldsymbol{A}(\boldsymbol{x}_\tau) \cdot \Delta\tau\right)^j}{j!}. \tag{9.42}$$

The resulting discrete-time evolution process is given by:

$$\boldsymbol{x}_{t+1} = \boldsymbol{f}(\boldsymbol{x}_t) = \boldsymbol{F}(\boldsymbol{x}_t) \cdot \boldsymbol{x}_t. \tag{9.43}$$

The discrete-time state-evolution model in (9.43), augmented with process noise, is employed to generate the simulated Lorenz attractor data. Unless noted otherwise, the data were produced using a Taylor series of order $J = 5$ and a sampling interval of $\Delta\tau = 0.02$.

9.5.1.5 Partial Information

We next evaluate KalmanNet and compare its performance with that of the model-based counterparts under conditions of partial model knowledge. In the Lorenz attractor setup, we consider two potential sources of model mismatch: first, a state transition model mismatch resulting from the low-order Taylor series approximation, and second, an observation model mismatch caused by matrix rotation.

State Transition Model Mismatch: In this study, both KalmanNet and the model-based algorithms are run using a coarse approximation of the evolution dynamics, obtained by evaluating (9.42) with $J = 2$, whereas the data are generated using a Taylor series of order $J = 5$. The results, shown in Figure 9.7 and summarized in Table 9.5, indicate that KalmanNet is able to partially compensate for this model mismatch, outperforming the model-based methods given the same limited model information.

Observation Model Mismatch: In this scenario, mismatch in the observation model is introduced by generating data using an identity matrix rotated by only $\theta = 1°$, corresponding to a sensor misalignment of approximately 0.55%. As shown in Figure 9.8 and summarized in Table 9.6, even this seemingly small rotation can significantly degrade the performance of the model-based filters. In contrast, KalmanNet is able to leverage the data to compensate for such mismatches, substantially outperforming its model-based counterparts, which are sensitive to model uncertainty. For this experiment, KalmanNet was trained on short trajectories of $T = 100$ time steps and tested on longer trajectories of $T = 1{,}000$ time steps, demonstrating the transferability of its learned model.

Table 9.5 MSE [dB]–Lorenz attractor with state transition model mismatch $J = 2$.

$1/r^2$[dB]		10	20	30	40
EKF	$\hat{\mu}$	**−20.37**	**−30.40**	**−40.39**	**−49.89**
$J = 5$	$\hat{\sigma}$	±0.25	±0.24	±0.24	±0.20
EKF	$\hat{\mu}$	−19.47	−23.63	−33.51	−41.15
$J = 2$	$\hat{\sigma}$	±0.25	±0.11	±0.18	±0.12
UKF	$\hat{\mu}$	−11.95	−20.45	−30.05	−39.98
$J = 2$	$\hat{\sigma}$	±0.87	±0.27	±0.09	±0.09
PF	$\hat{\mu}$	−17.95	−23.47	−30.11	−33.81
$J = 2$	$\hat{\sigma}$	±0.18	±0.09	±0.10	±0.13
KalmanNet	$\hat{\mu}$	**−19.71**	**−27.07**	**−35.41**	**−41.74**
$J = 2$	$\hat{\sigma}$	±0.29	±0.18	±0.20	±0.11

Bold numbers indicate (1) the performance upper bound (EKF with the true model), and (2) the best-performing algorithm among the methods using the mismatched model.

Table 9.6 MSE [dB]–Lorenz attractor with observation rotation.

$1/r^2$[dB]		0	10	20	30
EKF	$\hat{\mu}$	**−10.40**	**−20.41**	**−30.50**	**−40.45**
$\theta = 0°$	$\hat{\sigma}$	±0.35	±0.37	±0.34	±0.34
EKF	$\hat{\mu}$	**−9.80**	−16.50	−18.19	−18.57
$\theta = 1°$	$\hat{\sigma}$	±0.54	±6.51	±0.22	±0.21
UKF	$\hat{\mu}$	−2.08	−6.92	−7.89	−8.09
$\theta = 1°$	$\hat{\sigma}$	±1.73	±0.53	±0.59	±0.62
PF	$\hat{\mu}$	−8.48	−0.18	15.24	19.87
$\theta = 1°$	$\hat{\sigma}$	±3	±8.21	±3.50	±0.80
KalmanNet	$\hat{\mu}$	−9.63	**−18.17**	**−27.32**	**−34.04**
$\theta = 1°$	$\hat{\sigma}$	±0.53	±0.42	±0.67	±0.77

Bold numbers indicate (1) the performance upper bound (EKF with the true model), and (2) the best-performing algorithm among the methods using the mismatched model.

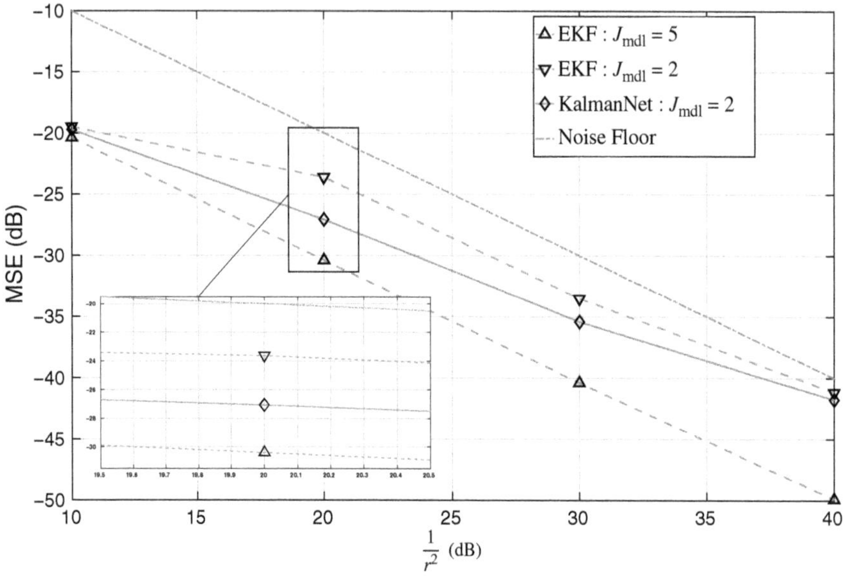

Figure 9.7 Lorenz attractor: state transition mismatch, identity $\boldsymbol{h}$, $T = 2{,}000$.

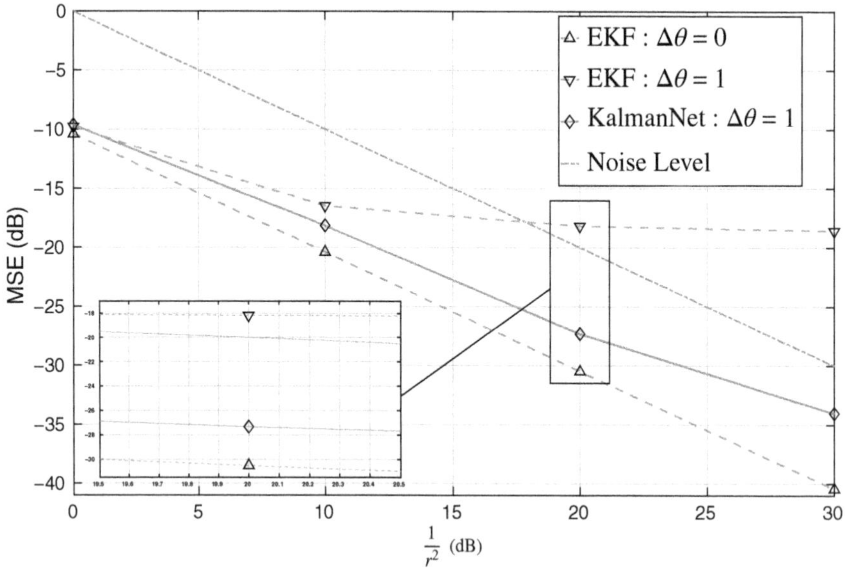

Figure 9.8 Lorenz attractor: observation mismatch, $\Delta\theta = 1°$, $T = 1{,}000$.

9.5.2 RTSNet

In this section, we evaluate RTSNet on both linear and nonlinear SS models, comparing its performance against model-based Kalman smoothers and neural network approaches. In the challenging nonlinear scenario with sampling mismatch, we show that the unfolded RTSNet can further enhance performance over the standard RTSNet.

9.5.2.1 Linear SS Model

We consider a 2×2 SS model where $\boldsymbol{F}$ takes a canonical form and $\boldsymbol{H}$ is set to be the identity matrix, namely,

$$\boldsymbol{F} = \begin{pmatrix} 1 & 1 \\ 0 & 1 \end{pmatrix}, \quad \boldsymbol{H} = \boldsymbol{I}. \tag{9.44}$$

We use multiple noise levels, in [dB] scale, of $\mathrm{r}^2 \in \{10, 0, -10, -20, -30\}$, and $\nu \in \{-20, -10, 0\}$. The results provided in Figure 9.9, show that RTSNet converges to the minimum MSE estimate produced by the RTS smoother in the first two moments. This indicates that RTSNet successfully learns to implement the RTS smoother when it is MSE optimal.

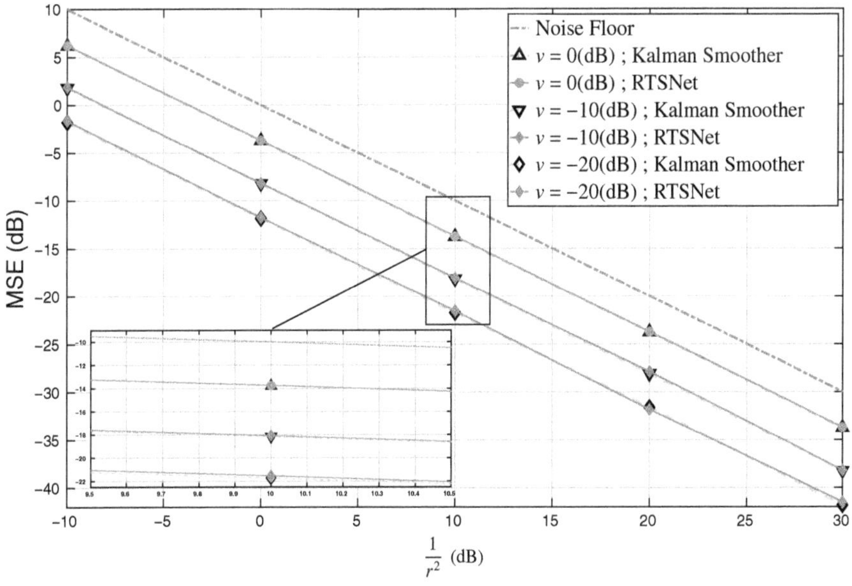

Figure 9.9 Linear SS model—full information.

9.5.2.2 Nonlinear SS Model

We consider the nonlinear Lorenz attractor model described in (9.41) and introduce a sampling mismatch scenario. This setup reflects a practical situation in which a physical process evolves in continuous-time, but the smoother only has access to noisy discrete-time observations, resulting in an inherent mismatch in the SS model. To generate the data, we first simulate the approximate continuous-time noise-free state evolution $\boldsymbol{F}(\boldsymbol{x}_\tau)$ with a high-resolution time step of $\Delta\tau = 10^{-5}$. The process is then subsampled by a factor of 1/2,000, producing a decimated sequence with $\Delta t = 0.02$. Finally, noisy observations of the true state are generated using an identity observation matrix $\boldsymbol{h}(\cdot) = \boldsymbol{I}$ and uncorrelated observation noise with $\mathrm{r}^2 = 0[\mathrm{dB}]$.

The MSE for sequences of length $T = 3{,}000$, reported in Table 9.7, shows that RTSNet effectively mitigates the mismatch caused by representing a continuous-time SS model in discrete-time, achieving significant performance gains over both model-based and data-driven methods. Here, RTSNet-1 denotes the standard RTSNet, while RTSNet-2 refers to the unfolded RTSNet with two folds.

9.5.3 Adaptive KalmanNet

Adaptive KalmanNet is designed to enable KalmanNet to adapt to time-varying noise statistics. In this section, we show that, even when trained under limited noise conditions, Adaptive KalmanNet generalizes effectively to previously unseen noise settings, for both Gaussian and non-Gaussian noise.

Gaussian Noise: We first test adaptive KalmanNet on a linear Gaussian SS model. The initial noise covariance matrices, $\boldsymbol{Q}_0$ and $\boldsymbol{R}_0$, are randomly chosen, requiring only that they be positive definite, without imposing diagonal constraints. The dataset includes various noise variance pairs $(\mathrm{q}_t^2, \mathrm{r}_t^2)$. In Figures 9.10 and 9.11, the dashed lines show the performance of the KF on these noise settings, serving as the optimal baseline for the linear Gaussian scenario. On the other hand, adaptive KalmanNet is trained only on four noise settings, marked as circles in Figure 9.10. The results indicate that adaptive KalmanNet not only matches the KF for the four noise settings seen during training but also generalizes to unseen configurations. This generalization covers SS models with the same

Table 9.7 MSE [dB]—Lorenz attractor with sampling mismatch.

Noise	EKF	PF	KalmanNet	EKS	PS	RNN	RTSNet-1	RTSNet-2
−0.024	−6.316	−5.333	−11.106	−10.075	−7.222	−2.342	−15.436	−16.803
± 0.049	± 0.135	± 0.136	± 0.224	± 0.191	± 0.202	± 0.092	± 0.329	± 0.301

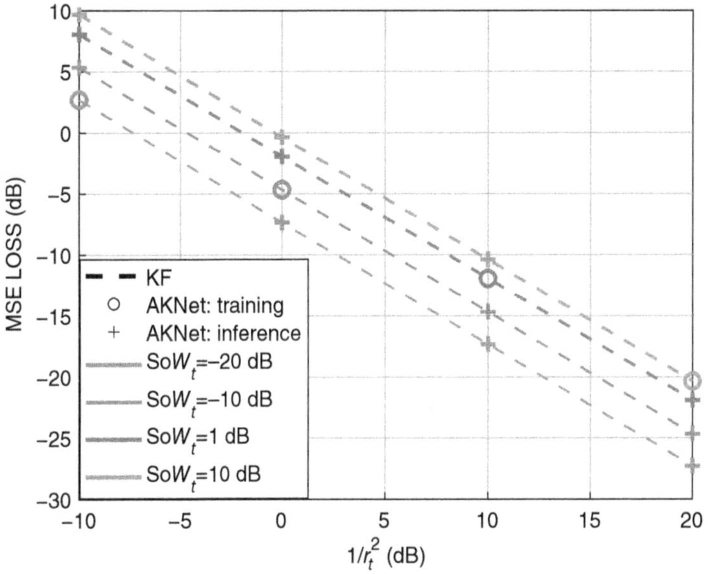

Figure 9.10 Linear Gaussian system: adaptive KalmanNet trained on four discrete value pairs of (q_t^2, r_t^2) (marked as circles) and tested on shifted noise preserving the same $\frac{q_t^2}{r_t^2}$ ratios.

SoW_t ratio but different scaling values q_t^2 and r_t^2, as well as models with SoW_t ratios not encountered during training. These findings demonstrate that adaptive KalmanNet requires only a small amount of training data across a limited set of noise settings to effectively handle a wide range of noise statistics.

Non-Gaussian Noise: Next, we consider exponentially distributed noise that is spatially uncorrelated, setting $\boldsymbol{Q}_0$ and $\boldsymbol{R}_0$ as scaled identity matrices. Again, adaptive KalmanNet is trained on only four distribution pairs and tested on SS models with different distributions, including those preserving the same SoWs as well as those encountered only during inference. As shown in Figures 9.12 and 9.13, adaptive KalmanNet maintains its generalization capabilities even in the non-Gaussian setting. Moreover, it significantly outperforms the classic KF, reflecting the advantage of KalmanNet in handling non-Gaussian noise.

9.5.4 GSP-KalmanNet

The high dimensionality and irregular structure of graph signals, along with potential model mismatches, challenge conventional filtering methods. To address this, GSP-KalmanNet [34] extends KalmanNet by incorporating GSP techniques. Here, we consider power system monitoring [52], focusing on the task of power

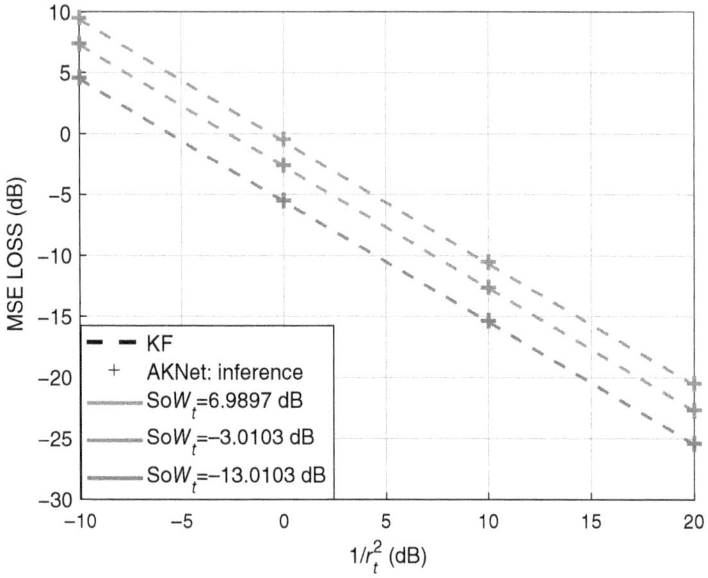

Figure 9.11 Linear Gaussian system: adaptive KalmanNet trained on four discrete value pairs of (q_t^2, r_t^2) (marked as circles) and tested on shifted noise with unseen $\frac{q_t^2}{r_t^2}$ ratios.

system state estimation. A power system can be modeled as an undirected weighted graph $\mathcal{G}(\mathcal{V}, \mathcal{E}, \boldsymbol{W})$, where nodes $\mathcal{V}$ correspond to buses (generators or loads) and edges $\mathcal{E}$ represent transmission lines. The weight matrix is typically derived from the system's admittance matrix, representing complex line weights. We use the Laplacian $\boldsymbol{L}$ computed from the susceptance matrix $\boldsymbol{B}$ [53, 54]. The graph for the IEEE 14-bus system is shown in Figure 9.14.

The state vector $\boldsymbol{x}_t$ comprises the voltage phases at the different buses, while $\boldsymbol{y}_t$ represents the active power injections. The widely used nonlinear AC power flow model, derived from Kirchhoff's and Ohm's laws, defines the relationship between voltage phasors and complex power injections. Under this model, the measurement function can be expressed as:

$$[\boldsymbol{h}(\boldsymbol{x}_t)]_i = \sum_{j=1}^{N}([\boldsymbol{G}]_{i,j} \cos([\boldsymbol{x}_t]_i - [\boldsymbol{x}_t]_j) + [\boldsymbol{B}]_{i,j} \sin([\boldsymbol{x}_t]_i - [\boldsymbol{x}_t]_j), \tag{9.45}$$

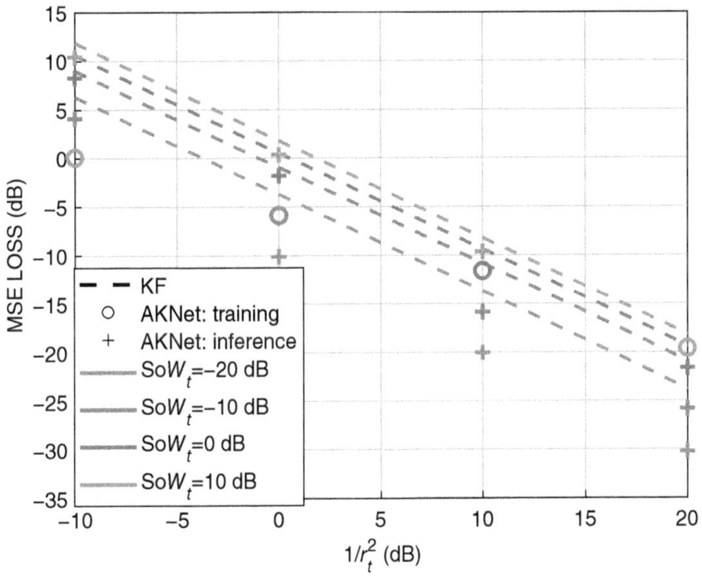

Figure 9.12 Linear non-Gaussian system: adaptive KalmanNet trained on four discrete value pairs of (q_t^2, r_t^2) (marked as circles) and tested on shifted noise preserving the same $\frac{q_t^2}{r_t^2}$ ratios.

$i = 1, \dots, N$, where $[\boldsymbol{G}]_{i,j}$ and $[\boldsymbol{B}]_{i,j}$ are the conductance and susceptance of the transmission line $(i, j) \in \mathcal{E}$ [55]. The elements of $\hat{\boldsymbol{H}}_t = \frac{\partial h}{\partial \boldsymbol{x}_t}$ is given by:

$$[\hat{\boldsymbol{H}}_t]_{i,j} = [\boldsymbol{G}]_{i,j} \sin([\boldsymbol{x}_t]_i - [\boldsymbol{x}_t]_j) - [\boldsymbol{B}]_{i,j} \cos([\boldsymbol{x}_t]_i - [\boldsymbol{x}_t]_j), \quad i \neq j, \tag{9.46}$$

and

$$[\hat{\boldsymbol{H}}_t]_{i,j} = \sum_{m=1}^{N} (-[\boldsymbol{G}]_{i,m} \sin([\boldsymbol{x}_t]_i - [\boldsymbol{x}_t]_m) + [\boldsymbol{B}]_{i,m} \cos([\boldsymbol{x}]_i - [\boldsymbol{x}]_m)), i = j, \tag{9.47}$$

In particular, since $[\boldsymbol{G}]_{i,j} = 0$ and $[\boldsymbol{B}]_{i,j} = 0$ for any $(i, j) \notin \mathcal{E}$, it can be seen that $[\hat{\boldsymbol{H}}_t]_{i,j} = 0$ for any $(i, j) \notin \mathcal{E}$. Thus, $\hat{\boldsymbol{H}}_t$ is a sparse matrix with the sparsity pattern of the graph. In the following, we examine two types of measurement noises: Gaussian and non-Gaussian.

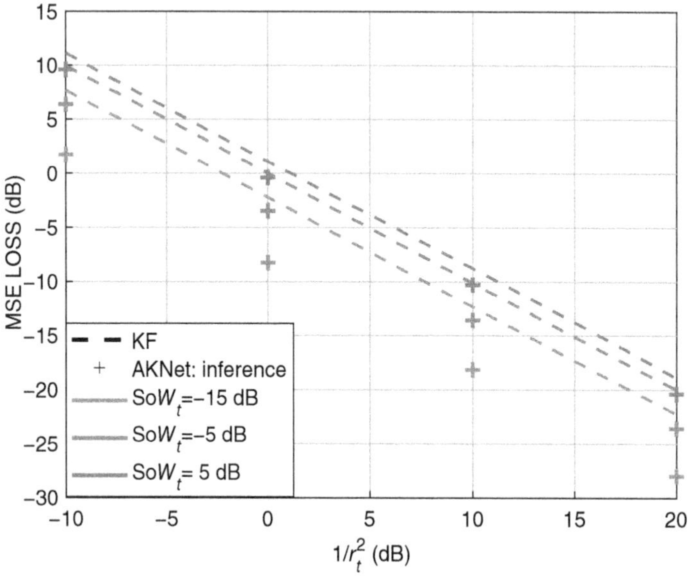

Figure 9.13 Linear non-Gaussian system: adaptive KalmanNet trained on four discrete value pairs of (q_t^2, r_t^2) (marked as circles) and tested on shifted noise with unseen $\frac{q_t^2}{r_t^2}$ ratios.

Gaussian Measurement Noise: In the first setup, the observations are corrupted by independent and identically distributed (i.i.d) Gaussian noise with different variance value, r^2. We use the state-evolution function

$$f_t(\boldsymbol{W}, \boldsymbol{x}_t) = 1 - (\boldsymbol{x}_t + \boldsymbol{W}\boldsymbol{x}_t), \tag{9.48}$$

which is influenced by the weighted graph with $N = 14$ nodes shown in Figure 9.14, where $\boldsymbol{x}_t + \boldsymbol{W}\boldsymbol{x}_t$ represents the weighted sum of first-order neighbors. We evaluate two scenarios: full information, where the assumed graph $\boldsymbol{W}$ matches the data-generating model, and partial information, where $\boldsymbol{W}$ is perturbed. The resulting MSEs are reported in Figure 9.15.

In the full information case (solid lines), learnable methods outperform classical ones by over 10dB, with GSP-KalmanNet exceeding both KalmanNet and graph neural network (GNN)+RNN. Moreover, while KalmanNet and GNN+RNN required multiple initializations, GSP-KalmanNet trained stably across all runs.

Under partial information (dashed lines), where 10 edges are randomly removed from $\boldsymbol{W}$, trainable methods maintained an advantage of about 8 dB over model-based baselines.

Exponential Measurement Noise: In the second setting, observations are corrupted by exponential noise with varying parameter λ_{exp}, reflecting temporal

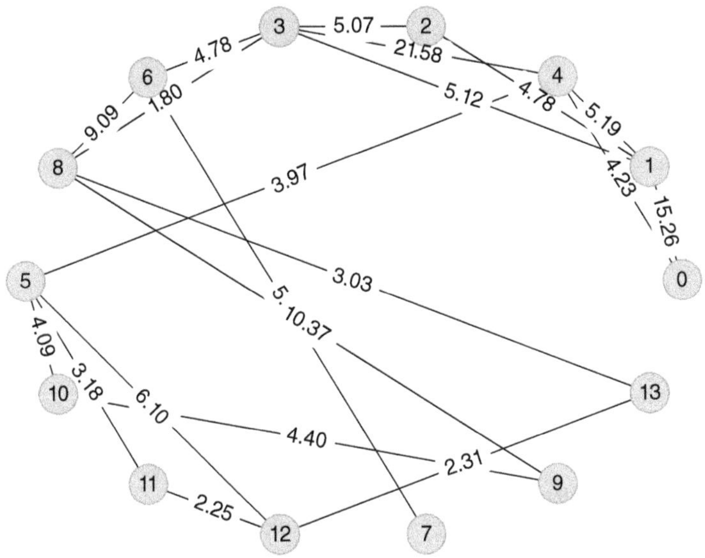

Figure 9.14 Power grid graph with $N = 14$ nodes according to the IEEE 14-bus system.

variations in $\boldsymbol{v}_t$. We use the same measurement function (9.45), governed by the weighted graph in Figure 9.14, while the state evolution is modeled by a simple linear mapping:

$$f_t(\boldsymbol{x}_t) = \boldsymbol{x}_t + 0.05. \tag{9.49}$$

We first consider the full information case, where the SS model parameters $(\boldsymbol{G}, \boldsymbol{B}, \boldsymbol{f}(\cdot), \boldsymbol{h}(\cdot))$ match those used for data generation. The model-based baselines were provided with Gaussian noise variances for $\tilde{\boldsymbol{e}}_t$ and $\tilde{\boldsymbol{v}}_t$, selected via grid search. As shown in Figure 9.16, these methods exhibit degraded performance, with large errors for nearly all λ_{exp} values. In contrast, learnable approaches achieve over 10 dB improvement, with GSP-KalmanNet outperforming both KalmanNet and the purely data-driven GNN+RNN. Notably, at very low noise levels (rightmost point), data-driven methods converge to the same mapping as the classical ones.

9.6 Conclusions

This chapter presented *KalmanNet*, a model-based deep learning framework designed to augment classical Kalman filtering with trainable neural network modules. The motivation behind this framework stems from the complementary strengths and weaknesses of traditional statistical estimators and modern deep

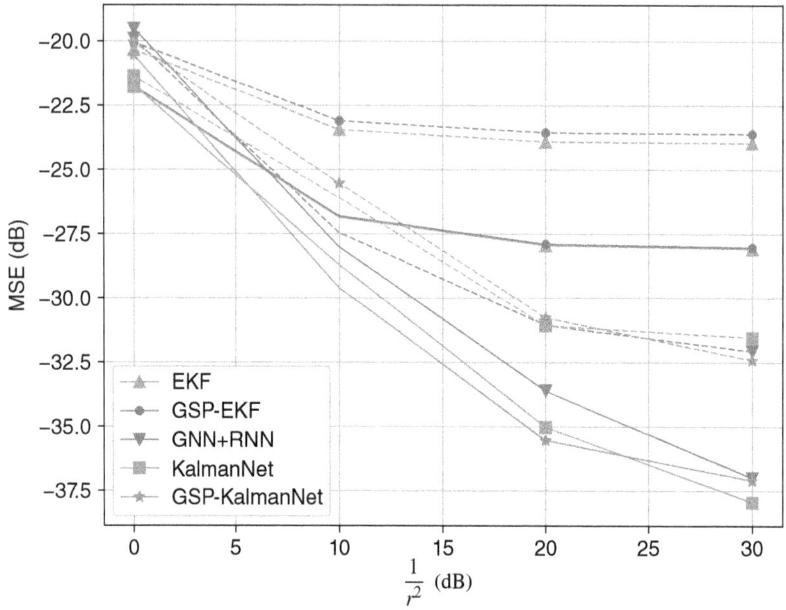

Figure 9.15 The measurement noise $\boldsymbol{v}_t$ is Gaussian. We report the benchmark MSE as a function of the Gaussian noise variance $1/r^2$. Solid lines correspond to full domain knowledge, while dashed lines represent the scenario with partially known dynamics and a misspecified graph topology.

learning approaches. While Kalman-type filters offer proven optimality, recursive efficiency, and interpretability under known models and assumptions, they often fail under real-world conditions where those assumptions—such as linearity, Gaussian noise, or known dynamics—do not hold. Conversely, deep learning methods can learn from complex, high-dimensional data, but often at the expense of interpretability, robustness, and computational efficiency.

KalmanNet addresses these challenges by embedding learning into the structure of the filter itself, retaining its recursive nature while introducing the flexibility to adapt to model mismatches, unknown parameters, and nonlinearities. We begin this chapter by revisiting the classical theory of state estimation in linear and nonlinear SS models, laying a solid foundation for the discussion of hybrid filtering approaches. We then introduced the KalmanNet architecture in detail, showing how it leverages data to learn the KG while preserving the essential structure of the Kalman filter. Extensions such as RTSNet and Adaptive KalmanNet further demonstrated the versatility of this framework, enabling smoothing and online adaptation to time-varying noise environments, respectively.

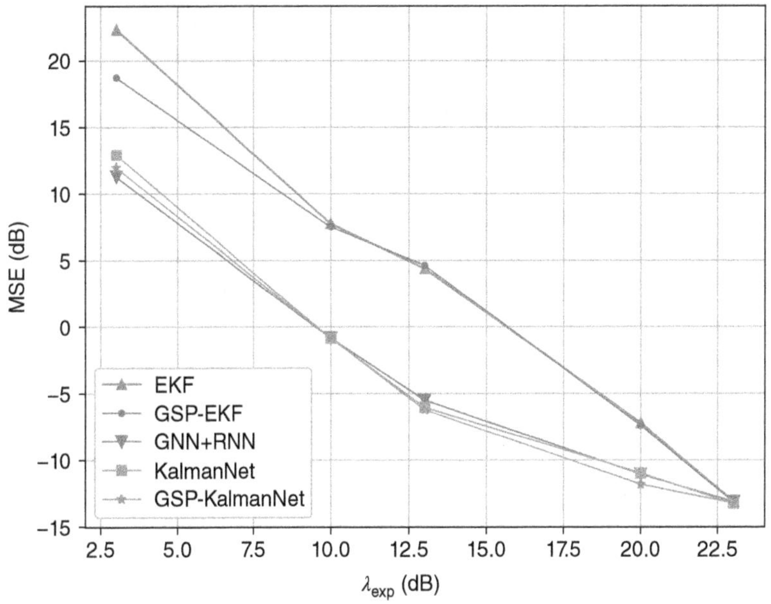

Figure 9.16 The measurement noise $\boldsymbol{v}_t$ follows an exponential distribution. We report the benchmark MSE as a function of the exponential noise parameter λ_{exp}, which is assumed Gaussian and selected via a grid search over the dataset.

In addition to algorithmic innovations, we discussed several conceptual and practical benefits of this hybrid design. These include interpretability and structure preservation, which are essential for deployment in regulated or safety-critical systems; uncertainty quantification, which enables risk-aware decision-making; unsupervised learning capabilities, which relax the need for labeled training data; and the ability to break the classical separation between preprocessing, estimation, and post-processing blocks, allowing for end-to-end optimization in complex pipelines. We also highlighted GSP-KalmanNet as a structure-aware variant that exploits prior knowledge about the domain—such as graph topology—to design more efficient and scalable models. Empirical studies further supported these claims, showing how KalmanNet outperforms both classical filters and fully data-driven models in challenging radar-based target tracking tasks under model mismatch and noise variability.

Looking forward, the KalmanNet paradigm opens numerous exciting directions for future research. One promising avenue involves the development of more advanced training strategies to improve generalization and adaptability in non-stationary environments. Techniques such as meta-learning—recently explored in [18]—can provide the model with the ability to quickly adapt to new regimes using only a few observations, making KalmanNet suitable for deployment in dynamic, real-time systems. Continual learning approaches may further enhance this adaptability, enabling models to refine themselves incrementally without catastrophic forgetting. In parallel, the integration of probabilistic machine learning tools—such as Bayesian neural networks or conformal prediction—can provide more reliable and calibrated uncertainty estimates, which are essential for safe decision-making in autonomous systems.

Another important research frontier is the extension of KalmanNet to more complex estimation tasks. These include multi-target tracking, multi-agent systems, and multi-sensor fusion, where state estimation must be performed jointly across multiple interacting entities or modalities. Here, the modularity and interpretability of KalmanNet offer an appealing starting point, but new innovations will be required to handle data association, inter-object dynamics, and scalability. Finally, there remains a strong need for rigorous theoretical understanding of hybrid filtering models. Formal guarantees on convergence, stability, and approximation error—as well as new tools to analyze the statistical efficiency and robustness of these algorithms—would enhance their reliability and help guide principled design choices. As hybrid AI methods continue to gain traction across signal processing and control, establishing these theoretical foundations will be crucial for their adoption in safety-critical domains such as radar tracking, autonomous navigation, and cyber-physical systems.

References

1 S. Gannot and A. Yeredor. The Kalman filter. In J. Benesty, M. M. Sondhi, and Y. A. Huang, editors, *Springer Handbook of Speech Processing*, pages 135–160. Springer, Berlin, Heidelberg, 2008.
2 R. E. Kalman. A new approach to linear filtering and prediction problems. *Journal of Basic Engineering*, 82(1):35–45, 1960.
3 D. Simon. Kalman filtering with state constraints: A survey of linear and nonlinear algorithms. *IET Control Theory & Applications*, 4(8):1303–1318, 2010.
4 Y. Bar-Shalom, X. R. Li, and T. Kirubarajan. *Estimation with Applications to Tracking and Navigation: Theory Algorithms and Software*. John Wiley & Sons, 2004.
5 S. Särkkä and L. Svensson. *Bayesian Filtering and Smoothing*, volume 17. Cambridge University Press, Cambridge, UK, 2023.
6 S. F. Schmidt. The Kalman filter-its recognition and development for aerospace applications. *Journal of Guidance and Control*, 4(1):4–7, 1981.
7 J. Durbin and S. J. Koopman. *Time Series Analysis by State Space Methods*, volume 38. Oxford University Press, Cambridge, UK, 2012.

8 B. Lim and S. Zohren. Time-series forecasting with deep learning: A survey. *Philosophical Transactions of the Royal Society A*, 379(2194):1–13, 2021

9 I. Goodfellow, Y. Bengio, and A. Courville. *Deep Learning*. MIT press, Cambridge, UK, 2016.

10 A. Vaswani, N. Shazeer, N. Parmar, J. Uszkoreit, L. Jones, A. N. Gomez, Ł. Kaiser, and I. Polosukhin. Attention is all you need. In *Advances in Neural Information Processing Systems*, pages 5998–6008, 2017.

11 N. Shlezinger, J. Whang, Y. C. Eldar, and A. G. Dimakis. Model-based deep learning. *Proceedings of the IEEE*, 111(5):465–499, 2023.

12 N. Shlezinger and Y. C. Eldar. Model-based deep learning. *Foundations and Trends® in Signal Processing*, 17(4):291–416, 2023.

13 N. Shlezinger, G. Revach, A. Ghosh, S. Chatterjee, S. Tang, T. Imbiriba, J. Dunik, O. Straka, P. Closas, and Y. C. Eldar. Artificial intelligence-aided Kalman filters: AI-augmented designs for Kalman-type algorithms. *IEEE Signal Processing Magazine*, 42(3):52–76, 2025.

14 G. Revach, N. Shlezinger, X. Ni, A. L. Escoriza, R. J. G. van Sloun, and Y. C. Eldar. KalmanNet: Neural network aided Kalman filtering for partially known dynamics. *IEEE Transactions on Signal Processing*, 70:1532–1547, 2022.

15 G. Revach, X. Ni, N. Shlezinger, R. J. G. van Sloun, and Y. C. Eldar. RTSNet: Learning to Smooth in Partially Known State-Space Models. *IEEE Transactions on Signal Processing*, 71:4441–4456, 2023.

16 X. Ni, G. Revach, and N. Shlezinger. Adaptive Kalmannet: Data-driven Kalman filter with fast adaptation. In *ICASSP 2024-2024 IEEE International Conference on Acoustics, Speech and Signal Processing (ICASSP)*, pages 5970–5974. IEEE, 2024.

17 J. Wang, X. Geng, and J. Xu. Nonlinear Kalman filtering based on self-attention mechanism and lattice trajectory piecewise linear approximation. *arXiv preprint arXiv:2404.03915*, 2024.

18 S. Chen, Y. Zheng, D. Lin, P. Cai, Y. Xiao, and S. Wang. MAML-KalmanNet: A neural network-assisted Kalman filter based on model-agnostic meta-learning. *IEEE Transactions on Signal Processing*, 73:988–1003, 2025.

19 L. Cubillos, G. Revach, M. Mender, J. Costello, H. Temmar, A. Hite, D. A. K. Zutshi, D. Wallace, X. Ni, M. Kelberman, et al. Exploring the trade-off between deep-learning and explainable models for brain-machine interfaces. *Advances in Neural Information Processing Systems*, 37:133975–133998, 2025.

20 D. Yang, F. Jiang, W. Wu, X. Fang, and M. Cao. Low-complexity acoustic echo cancellation with neural Kalman filtering. In *IEEE International Conference on Acoustics, Speech and Signal Processing (ICASSP)*, 2023.

21 H. Zhang, Y. Zhang, M. Yu, and D. Yu. Enhanced acoustic howling suppression via hybrid Kalman filter and deep learning models. *IEEE/ACM Transactions on Audio, Speech, and Language Processing*, 32:2828–2840, 2024.

22 A. Milstein, G. Revach, H. Deng, H. Morgenstern, and N. Shlezinger. Neural augmented Kalman filtering with Bollinger bands for pairs trading. *IEEE Transactions on Signal Processing*, 72:1974–1988, 2024.

23 L. Ketashvili, S. Ashkenasy, I. Nisky, and N. Shlezinger. Learning-aided Kalman tracking in biased dynamic systems: The case of cable-driven robots for surgery. In *IEEE International Conference on Acoustics, Speech and Signal Processing (ICASSP)*, 2025.

24 A. L. Escoriza, G. Revach, N. Shlezinger, and R. J. G. van Sloun. Data-driven Kalman-based velocity estimation for autonomous racing. In *IEEE International Conference on Autonomous Systems (ICAS)*, 2021.

25 J. Zhang, Y. Huang, C. Masouros, X. You, and B. Ottersten. Hybrid data-induced Kalman filtering approach and application in beam prediction and tracking. *IEEE Transactions on Signal Processing*, 72:1412–1426, 2024.

26 S. N. Aspragkathos, G. C. Karras, and K. J. Kyriakopoulos. Event-triggered image moments predictive control for tracking evolving features using UAVs. *IEEE Robotics and Automation Letters*, 9(2):1019–1026, 2024.

27 A. Bensoussan. *Stochastic Control of Partially Observable Systems*. Cambridge University Press, Cambridge, UK, 1992.

28 H. E. Rauch, F. Tung, and C. T. Striebel. Maximum likelihood estimates of linear dynamic systems. *AIAA Journal*, 3(8):1445–1450, 1965.

29 S. J. Julier and J. K. Uhlmann. Unscented filtering and nonlinear estimation. *Proceedings of the IEEE*, 92(3):401–422, 2004.

30 I. Arasaratnam and S. Haykin. Cubature Kalman filters. *IEEE Transactions on Automatic Control*, 54(6):1254–1269, 2009.

31 P. M. Djuric, J. H. Kotecha, J. Zhang, Y. Huang, T. Ghirmai, M. F. Bugallo, and J. Miguez. Particle filtering. *IEEE Signal Processing Magazine*, 20(5):19–38, 2003.

32 S. SÄrkkÄ. Unscented Rauch–Tung–Striebel Smoother. *IEEE Transactions on Automatic Control*, 53(3):845–849, 2008.

33 N. J. Gordon, D. J. Salmond, and A. F. M. Smith. Novel approach to nonlinear/non-Gaussian Bayesian state estimation. In *IEE Proceedings F (Radar and Signal Processing)*, volume 140, pages 107–113. IET, 1993.

34 I. Buchnik, G. Sagi, N. Leinwand, Y. Loya, N. Shlezinger, and T. Routtenberg. GSP-KalmanNet: Tracking graph signals via neural-aided Kalman filtering. *IEEE Transactions on Signal Processing*, 72:3700–3716, 2024.

35 N. Shlezinger, Y. C. Eldar, and S. P. Boyd. Model-based deep learning: On the intersection of deep learning and optimization. *IEEE Access*, 10:115384–115398, 2022.

36 N. Shlezinger and T. Routtenberg. Discriminative and generative learning for linear estimation of random signals [lecture notes]. *IEEE Signal Processing Magazine*, 40(6):75–82, 2023.

37 G. Choi, J. Park, N. Shlezinger, Y. C. Eldar, and N. Lee. Split-KalmanNet: A robust model-based deep learning approach for state estimation. *IEEE Transactions on Vehicular Technology*, 72(9):12326–12331, 2023.

38 L. Xu and R. Niu. EKFNet: Learning system noise statistics from measurement data. In *IEEE International Conference on Acoustics, Speech and Signal Processing (ICASSP)*, pages 4560–4564, 2021.

39 N. Shlezinger, M. Ma, O. Lavi, N. T. Nguyen, Y. C. Eldar, and M. Juntti. Artificial intelligence-empowered hybrid multiple-input/multiple-output beamforming: Learning to optimize for high-throughput scalable MIMO. *IEEE Vehicular Technology Magazine*, 19(3):58–67, 2024.

40 G. Sagi, N. Shlezinger, and T. Routtenberg. Extended Kalman filter for graph signals in nonlinear dynamic systems. In *IEEE International Conference on Acoustics, Speech and Signal Processing (ICASSP)*, 2023.

41 Y. Gal and Z. Ghahramani. Dropout as a bayesian approximation: Representing model uncertainty in deep learning. In *International Conference on Machine Learning*, pages 1050–1059. PMLR, 2016.

42 J. Gawlikowski, C. R. N. Tassi, M. Ali, J. Lee, M. Humt, J. Feng, A. Kruspe, R. Triebel, P. Jung, R. Roscher, et al. A survey of uncertainty in deep neural networks. *Artificial Intelligence Review*, 56(Suppl 1):1513–1589, 2023.

43 Y. Dahan, G. Revach, J. Dunik, and N. Shlezinger. Bayesian KalmanNet: Quantifying uncertainty in deep learning augmented Kalman filter. *IEEE Transactions on Signal Processing*, 73:2558–2573, 2025.

44 L. Lindemann, Y. Zhao, X. Yu, G. J. Pappas, and J. V. Deshmukh. Formal verification and control with conformal prediction. *arXiv preprint arXiv:2409.00536*, 2024.

45 Y. Dahan, G. Revach, J. Dunik, and N. Shlezinger. Uncertainty quantification in deep learning based Kalman filters. In *2024 IEEE International Conference on Acoustics, Speech and Signal Processing (ICASSP)*, pages 13121–13125, 2024.

46 G. Revach, N. Shlezinger, T. Locher, X. Ni, R. J. G. van Sloun, and Y. C. Eldar. Unsupervised learned Kalman filtering. In *European Signal Processing Conference (EUSIPCO)*, pages 1571–1575, 2022.

47 I. Buchnik, G. Revach, D. Steger, R. J. G. van Sloun, T. Routtenberg, and N. Shlezinger. Latent-KalmanNet: Learned Kalman filtering for tracking from high-dimensional signals. *IEEE Transactions on Signal Processing*, 72:352–367, 2023.

48 S. G. Casspi, O. Hüsser, G. Revach, and N. Shlezinger. LQGNet: Hybrid model-based and data-driven linear quadratic stochastic control. In *IEEE International Conference on Acoustics, Speech and Signal Processing (ICASSP)*, 2023.

49 A. N. Putri, C. Machbub, D. Mahayana, and E. Hidayat. Data driven linear quadratic Gaussian control design. *IEEE Access*, 11:24227–24237, 2023.

50 W. Gilpin. Chaos as an interpretable benchmark for forecasting and data-driven modelling. In *Thirty-fifth Conference on Neural Information Processing Systems Datasets and Benchmarks Track (Round 2)*, 2021.

51 V. G. Satorras, Z. Akata, and M. Welling. Combining generative and discriminative models for hybrid inference. In *Advances in Neural Information Processing Systems*, pages 13802–13812, 2019.

52 S. Kanna, D. H. Dini, Y. Xia, S. Y. Hui, and D. P. Mandic. Distributed widely linear Kalman filtering for frequency estimation in power networks. *IEEE Transactions on Signal and Information Processing over Networks*, 1(1):45–57, 2015.

53 L. Dabush, A. Kroizer, and T. Routtenberg. State estimation in partially observable power systems via graph signal processing tools. *Sensors*, 23(3):1387, 2023.

54 M. Halihal, T. Routtenberg, and H. V. Poor. Estimation of complex-valued laplacian matrices for topology identification in power systems. *arXiv preprint arXiv:2308.03392*, 2023.

55 A. Abur and A. Gomez-Exposito. *Power System State Estimation: Theory and Implementation*. Marcel Dekker, 2004.

10

Physics-aware Deep Learning Techniques for Radar-based Human Activity Recognition

Sevgi Zubeyde Gurbuz

Department of Electrical and Computer Engineering, North Carolina State University, Raleigh, NC, USA

*Corresponding author: szgurbuz@ncsu.edu

10.1 Introduction

Although recent advancements in artificial intelligence and machine learning (AI/ML) have enabled significant advances across many applications, the design of accurate and robust deep neural networks (DNNs) that are effective across a wide range of dynamically changing operating conditions and potential targets remains an elusive goal. Deep learning in radar applications faces several fundamental and critical challenges [1]. First, we must take into account that radar data is not inherently an image, but instead a complex stream of in-phase (I)/quadrature (Q) data, from which various dimensionalities (D)—2D, 3D, and even 4D data representations of range, Doppler, and angle versus time—are computed using radar signal processing techniques. However, many DNNs—such as convolutional neural networks (CNNs)—capture spatial correlations that may not exist between pixels in computed radar data representations. As an example, consider a radar micro-Doppler (μD) signature, where each pixel in the image represents a velocity at a given time. While the interconnectedness of the human body may result in some correlation between adjacent pixels comprising the target's motion, there is no correlation between potentially adjacent target and nontarget pixels in the image.

Second, how do we decide what the best way to provide the radar data at the input of a DNN is? Do we achieve the best recognition performance by using range-Doppler maps, range-angle maps, μD signatures, point clouds, or the raw radar data stream itself? While DNNs that take multiple radar data representations at their input have been shown to offer greater accuracy than one using just a single representation [2], computing these representations incurs additional computational costs, e.g., latency and memory consumption. Conversely, directly providing as input the radar I/Q data into a complex neural network may offer better real-time performance, but typically results in lower accuracy, especially under more challenging use cases where there is low signal-to-noise ratio (SNR). Thus, there are tradeoffs between computation time, network complexity, and the resulting accuracy and performance of the DNN model.

10.1.1 Adapting AI/ML Techniques to the Underlying Physics of the Signal Improves Performance

Let us consider an example based on the analogy of spectrograms to represent the radar μD signature and spectrograms for voice recognition. This similarity prompted the proposal of utilizing mel-frequency cepstral coefficients (MFCCs) as features for μD signature classification [3, 4]. MFCCs are computed by first filtering the discrete Fourier transform (DFT) of the windowed signal using a filter bank defined according to the mel-scale, after which the discrete cosine transform (DCT) of the logarithm is taken to yield the MFCC features. Appropriately for speech processing, the mel-scale is a frequency scale based upon the perception of equal distance between pitches and mimics the logarithmic perception of the human auditory system. But the mel-scale is actually quite unsuitable and mismatched to the Doppler frequencies encountered in radar data. While radar micro-Doppler often contains negative frequencies due to the times when there is motion away from the radar, the mel-scale filter banks only pass positive frequencies, discarding significant portions of the Doppler spectrum. Furthermore, based on the radar transmit frequency and pulse repetition frequency, there may also be a mismatch between the frequency range of Doppler shifts as compared to the intervals of the mel-scale. This mismatch results in no signal energy passing through many of the mel-scale filter banks, resulting in the generation of irrelevant and redundant cepstral features. A simple solution can be devised for this problem; however, by simply optimizing the filter banks for radar data. This can be done by utilizing a hyperbolic warping function to define the filter bank [5] or genetic algorithms to separately optimize the location and bandwidth of each filter bank [6]. The resulting generalized frequency warped cepstral coefficients (FWCC) have been shown to yield a *25% performance improvement* over MFCC for a 4-class human activity recognition (HAR) scenario and comparable performance to a convolutional autoencoder (CAE) on a 11-class HAR scenario [7].

We can expand upon the role of physics in AI/ML further if we examine this concept from a knowledge-aided statistical signal processing perspective. In classical estimation, we assume that we know a parametric model that represents the underlying relationship of the data, and we strive to estimate parameters such that the error is minimized. Alternatively, in Bayesian estimation, we assume that we know a statistical model for the distribution of the data and seek to estimate statistical model parameters such that the error is minimized.

What do we know when we use AI/ML? A pure data science perspective would boast of an ability to learn from the data only without knowing anything *a priori*. But in sensing problems, we *do know* quite a lot about the relationship underlying the data: we know the physics behind how the transmitted electromagnetic waves interact with matter and backscatter to the receiver. In human sensing problems, we also know the kinematics of how the human body moves. The body is comprised of interconnected parts, which in turn introduces kinematic correlations. We could try to apply AI/ML by ignoring our prior knowledge of physics and kinematics; but this would result in a more complex network that essentially is reinventing the wheel to figure out from the data what we already know. Furthermore, we would require massive amounts of training data to enable such a model to have a chance to adequately learn intricate relationships. Model performance would be limited by (1) the amount of training data we have available and (2) whether or not the data spanned all possible scenarios. In practice, radar datasets are significantly smaller than those available in computer vision (thousands of samples versus millions), and it is not feasible to acquire samples of all the different possible environments, targets, kinematics, distances, angles, and people involved in real-world human sensing scenarios. These factors would degrade model performance both in terms of accuracy and robustness.

10.1.2 Physics-aware AI/ML Represents a Tradeoff Between Knowledge and Data

While physics-based models represent a high level of domain knowledge gained over the years through science, they are less adept at capturing the nuances of environment-specific, sensor-specific, or subject-specific properties. Here, deep learning can provide tremendous insight through data-driven learning. Essentially, physics-aware AI/ML [8, 9] (see Figure 10.1) aims to integrate physics-based models with data-driven deep learning to reap the benefits of both approaches. While the integration of physics reduces the demands on the model and provides constraints, data-driven learning enables the model to bridge the gap to the real world, which may be dynamic and unpredictable.

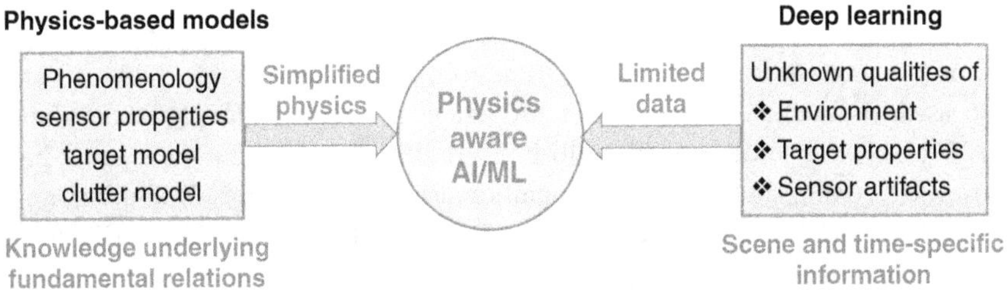

Figure 10.1 Physics-aware AI/ML.

This chapter provides insights into how the concept of physics-aware AI/ML has been applied in the context of target recognition, drawing extensively from examples of human motion recognition applications—recognition of activities of daily living, gestures, and sign language, as well as skeleton estimation. In particular, each phase of the model development and testing process is considered:

- *Pre-deployment Source Data:* What data is available prior to deployment of the radar in an operational setting? We may not have access or opportunity to collect data from the test environment. For example, in military scenarios, the operation zone is not accessible prior to model training, while in civilian applications we cannot know in advance who will purchase our product or where it will be utilized. Thus, it is almost certain that any source data acquired prior to deployment for model training will differ from that of the test data. The differences in the distribution of training and testing data are a cause of performance degradation that must be addressed in the model development process.
- *Data Augmentation and Diversification:* As data acquisition can be a costly proposition, even for low-cost radio frequency (RF) sensors where human participation incentives are involved, effective model training typically will require the use of data augmentation and diversification techniques to not only generate a larger amount of training data but also generate data that spans all probably target profiles.
- *DNN Design:* The DNNs designed for target classification should provide effective tradeoffs between performance, computational costs, and latency, while bearing in mind the relationship between radar transceiver architecture, transmit waveform parameters, and resulting RF data representations obtained.

Thus, in Section 10.2, we first introduce various methods for initializing DNNs under low training sample support, such as transfer learning and unsupervised pretraining. Physics-aware methods for generating synthetic RF data are presented, with a focus on the impact of kinematics and language on the efficacy of acquired and synthesized datasets for training. Next, in Section 10.3, we examine various DNN architectures for classification developed so as to incorporate an understanding of the signal structure as determined by physics, biomechanics, or language to improve classification accuracy. Three notable networks are discussed in detail: (1) PLFNet, which operates directly on complex radar data, (2) JD-MIMTL, a joint-domain multi-input multitask learning network that has proven effective in both activity recognition and sign language recognition, and (3) kinematic cycle consistency (KCC) in DNNs for human skeleton estimation. Section 10.4 concludes the chapter.

10.2 Training DNNs for Radar

The amount of data that is considered sufficient to train a model depends on the model capacity, often quantified using the Vapnik–Chervonenkis [10] (VC) dimension. The VC dimension provides an upper bound on the test error of a model:

$$\Pr\left(e_{\text{test}} \leq e_{\text{train}} + \sqrt{\frac{1}{N}\left[\text{VC}\left(\log\left(\frac{2N}{\text{VC}}\right)+1\right) - \log\left(\frac{\eta}{4}\right)\right]}\right) = 1 - \eta, \quad (10.1)$$

where $\Pr(x)$ denotes the probability of argument x, e_{train} and e_{test} are the training and testing errors, respectively, η is a number between $0 \leq \eta \leq 1$ and N is the size of the training dataset. To prevent overfitting, the previous inequality is valid only when $\text{VC} \ll N$; this leads to a criterion on training sample size in terms of the VC dimension as:

$$N = \Theta\left(\frac{\text{VC} + \ln(1/P_{\text{err}})}{\epsilon}\right), \quad (10.2)$$

where the Big-theta notation $f(n) = \Theta(g(n))$ indicates that $g(n)$ is an asymptotically tight bound on $f(n)$. As a practical rule of thumb, however, it has been recommended that the amount of training data should be at least 10 times the number of dimensions of the model [11]. Thus, the deeper and wider a network, the greater the number of parameters and, accordingly, the greater the amount of samples needed for training. Typical radar datasets, however, are on the order of several thousands of samples—a sharp contrast to that typical of computer vision applications. The ImageNet [12] database, for example, is comprised of over 1.2 million samples spanning 1,000 classes.

How can we overcome this data gap? CNNs are typically randomly initialized. However, because the objective function of CNNs is highly nonconvex, i.e., the parameter space of the model contains many local minima, gradient-based optimization algorithms may converge to a local minima that is not optimal in a global sense [13]. The fewer the amount of training samples, the more likely the network will converge to a point distant from the global optimum. There are three main ways we can mitigate this issue when training data is limited: transfer-learning, unsupervised pretraining, and training with massive amounts of synthetic samples.

Transfer learning is the phenomenon of transferring the weights of a trained model to initiate a better optimization space for another "related" problem. Because the trained model from which the weights are transferred is related to the given task, once the network is initialized using many samples of data set A, fine-tuning of the network with a few data samples from data set B can yield improved classification results. For example, Park et al. [14] used the ImageNet data set to pre-train deep CNN models to classify 625 data samples of five different aquatic activities; Li et al. [15] proposed an instance-based transfer learning technique that yielded high performance with just 100 samples per class. Although transfer learning itself does not require any radar data, the differences in the phenomenology of radar and optical data limit its efficacy. Consequently, when used on radar datasets, transfer learning is usually applied as a pretraining step before fine-tuning network weights with a small amount of real radar data.

Alternatively, the small amount of available real radar training data can be used for *unsupervised pretraining*, a process that initializes the network at a better local minimum that is closer to the global minimum and serves as a regularizer. One way of implementing unsupervised pretraining is to utilize an autoencoder, which is trained to learn an identity mapping. Afterward, the decoder is stripped away, and the remaining encoder is an initialized network that can be later fine-tuned in a supervised fashion if desired. A 2017 study [16] found that for an 11-class human activity recognition dataset, when the samples per class exceeded 60–70 samples, unsupervised pretraining with a CAE [17] outperformed transfer learning.

Simulating or generating synthetic samples of radar data is another way to boost performance and enable the training of deeper, more accurate networks. While such methods have the benefit of significantly increasing the amount of data available for training, discrepancies between synthetic and real samples can limit performance. In the next subsection, methods for generating synthetic radar data are examined in detail, focusing on physics-aware AI/ML methods to synthesize more accurate and realistic data.

10.2.1 Physics-based Synthesis of RF Data

Although computational electromagnetic and finite difference time domain (FDTD) techniques [18] can be utilized on a highly detailed model of the human body to generate highly accurate models of the RF backscatter from the human body, a simpler model representing the total human return as the superposition or sum of the returns from point targets modeling each body part has been shown [19] to be quite effective and practical [20]. Suppose we define a finite number of K point targets on the human body. Then the total radar backscatter from the body may be written as:

$$s_h(t) = \sum_{k=1}^{K} a_{t,k} e^{j\left[(2\pi f_0)t - \frac{4\pi}{\lambda} R_{t,k}\right]}, \tag{10.3}$$

where f_0 is the transmit frequency, λ is the wavelength, t is time, $R_{t,k}$ is the time-varying range of each point target and $a_{t,k}$ is the amplitude as computed from the radar range equation:

$$a_{t,k} = \frac{G\lambda\sqrt{P_k\sigma_k}}{(4\pi)^{1.5}R_{t,k}^2\sqrt{L_s}\sqrt{L_a}}. \quad (10.4)$$

Here, G is the antenna gain, P_k is the transmitter power, σ_k is the radar cross-section (RCS) for each point target, and L_s and L_a represent the system and atmospheric losses, respectively.

The key to the simulation is providing accurate representation of the time-varying range for each point target. But given the required ranges, these equations can be used to compute the estimated RF backscatter for any human activity, transmit radar parameters (e.g., center frequency, bandwidth, and pulse repetition frequency), and antenna-target geometry. That last point is essential, as one of the key factors limiting generalization is that typically we cannot collect real radar data across all possible scenarios, geometries, body types, and activities. With simulations, however, we can span a much wider range of the probable target space.

The requisite time-varying ranges can be obtained by utilizing optical motion capture (MOCAP), which can be acquired for any progression of motions. The U.S. Air Force Research Laboratory [21, 22] used the PoserTM system to animate human figures, while Ram et al. [23–25] used data from a MOCAP database to simulate not only human μD signatures, but also that of animals, such as dogs and horses. While nowadays many MOCAP databases are available, one of the popular sources of MOCAP data has been the Carnegie Mellon University Graphics Laboratory MOCAP Database [26], which has numerous samples spanning a variety of activities, including walking, running, crawling, jumping, boxing, dancing, a variety of sports, and more.

Barring the use of publicly available MOCAP databases, acquiring high-quality MOCAP data requires utilization of a markered, professional optical tracking system comprised of multiple cameras positioned around a room. Markerless systems, such as the KinectTM sensor [27], have also been proposed for RF data simulation. While substantially less expensive than optical tracking systems, markerless systems have greater estimation error, which in turn are reflected as kinematic errors in the simulation data and degrade performance.

To improve model generalization to people with various heights, walking speeds, and walking styles, a diversification methodology [28] involving transformations of the underlying skeleton and trajectories of the keypoints can be applied. On human activity datasets, whether it be a μD signature, range-Doppler, or range-angle map, it is critical that classical computer vision-based data augmentation methods be avoided. Processes like rotating or flipping the signature to increase the number of training samples change the underlying physics represented in the RF data image. For example, if a μD signature were rotated, the average velocity would increase linearly with time, whereas we know that for most ambulation average

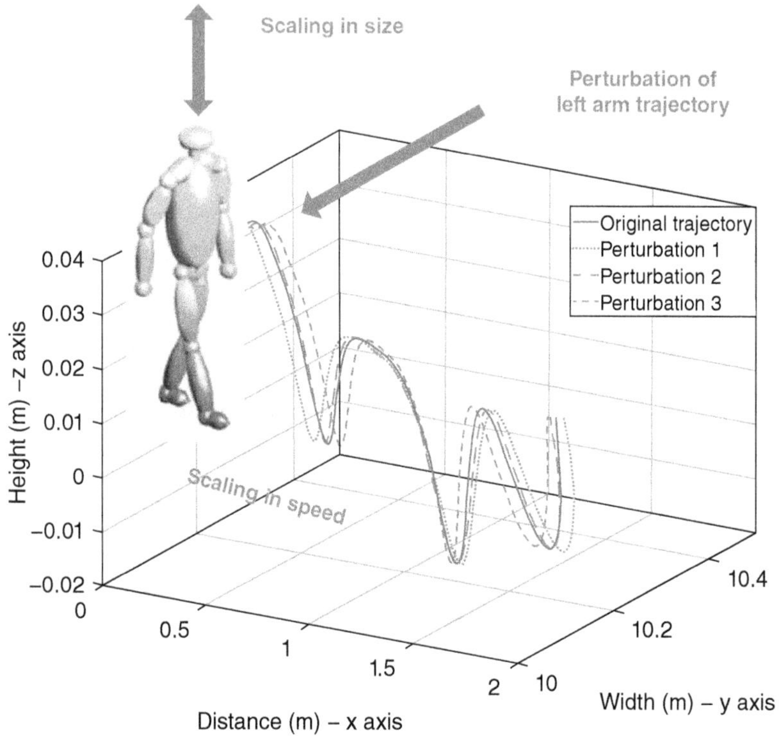

Figure 10.2 Skeletal diversification for kinematically consistent radar data augmentation.

speed remains constant over short intervals. Any process that changes the relationships between physical variables measured by the radar will result in training data that actually hurts rather than helps model training. The proposed skeleton-level diversification technique, illustrated in Figure 10.2, can take just a single MOCAP recording and generate a nearly unlimited amount of data by stretching the timescale to slow down or speed up the motion, scaling the distances between keypoints in the skeleton to account for people of different heights, and randomly perturbing the parameters of a Fourier parametrization of the trajectories of each keypoint. In this way, kinematically consistent variants are generated from just a small number of MOCAP measurements.

In [28], just 55 MOCAP measurements—a single recording for 5 participants enacting 11 different motion classes—were used to synthesize a training database of 32,000 samples. With this significantly larger training dataset, a 15-layer residual network, dubbed DivNET-15, was trained and shown to yield a classification accuracy that surpassed that of a CAE [29].

An important benefit of physics-based simulations using MOCAP data is that the expected raw RF backscatter itself is being emulated, allowing the generation of the entire set of computed RF data representations: micro-Doppler, range-Doppler, range-angle, and point clouds. One disadvantage, however, is that these simulations only include the expected target backscatter, whereas the backscatter from stationary surfaces or objects is not captured. Although there are physics-based methods for simulating clutter [30, 31], such models are somewhat generic estimates that are not site-specific: environment backscatter will vary dependent on the specific location, orientation, and positioning of the radar. Moreover, clutter backscatter is often dynamically variable, so that even with terrain matching, the resulting clutter simulation may not match that of the real RF backscatter. Consequently, there has been increasing interest in developing data-driven methods for representing not only the target signature but also sensor-specific and environment-specific backscatter, as discussed next.

10.2.2 Domain Adaptation and Generative Adversarial Networks

Generative adversarial networks (GANs) have been proposed for synthesizing realistic images in a variety of applications [32]. The first work to utilize GANs on radar μD data [33] evaluated the efficacy of adversarial learning for addressing the open-set problem—the case where the training dataset does not include all the classes as the test dataset. Subsequent studies [34, 35] investigated GANs to reduce the requirements for large amounts of real data by synthesizing training data. To bridge the gap between physics-based simulations and real-world data, adversarial domain adaptation methods, originally proposed for style transfer in computer vision, were leveraged to essentially add in the real-world factors not captured by target simulations [36] and to enable the synthesis of μD signatures acquired at different aspect angles [37]. However, because the pixels in computed images, such as μD signatures or range-Doppler maps, do not relate to embody the same spatial correlations as those present in adjacent pixels of optical images, oftentimes, GANs can generate synthetic samples that contain gross errors incompatible with target kinematics. Such errors degrade classification performance when GAN-synthesized samples are utilized for model training.

Consider, for example, the μD signatures synthesized by an auxiliary conditional GAN (ACGAN) using real data measurements of a 25 GHz continuous wave (CW) radar recording a person "falling," as shown in Figure 10.3. Falls are an unintentional, rapid motion typically characterized by a strong impulsive peak. During data collection, after the fall, participants remained in the prone position on a mattress (to avoid injury). Thus, the real data samples contained only the burst of positive micro-Doppler incurred during the fall toward the radar. If we look at the GAN-synthesized signatures, we can observe that, indeed, the GAN was able

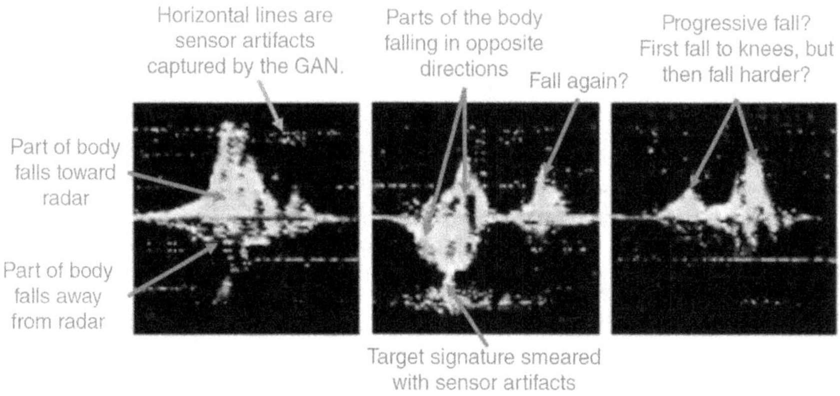

Figure 10.3 Examples of kinematic errors observed in ACGAN-synthesized μD signatures of a person falling toward the radar.

to capture the horizontal lines introduced into the real data by a quirk of the specific RF transceiver used during this data acquisition. But, if we examine the target component of these signatures, numerous kinematically inaccurate characteristics may be observed. The most egregious flaw is that the synthetic samples include strong peaks of velocity in both the positive (upper half of image) and negative (lower half of image) directions. This essentially means that the entire body is not falling in one direction but, is in fact, splitting apart and falling in two opposing directions simultaneously. This, of course, is completely impossible in real life. Furthermore, some samples involve the generation of an additional burst, which would physically be interpreted as the person falling again or progressively (e.g., first falling to the knees, then falling flat to the ground). Essentially, the GAN is introduction *additional movements* not present in the original, real data samples. Finally, in some cases, we can see smearing between the target the sensor-artifact components of the synthetic samples, which represents yet again an unrealistic, impossible phenomenon.

The significance of these kinematically erroneous samples is underscored by a study [38] involving classification of 8 different activities, where it was found that *kinematically sifting* the data by discarding 9,000 of the 40,000 synthesized samples—identified as outliers relative to the convex hull prescribed by the real data feature distribution—actually resulted in a boost of classification accuracy by 10%.

10.2.3 Physics-aware GANs

The question of whether GAN-generated samples conform to physical constraints has been raised in a number of contexts, not just radar. For example, in turbulent flow simulations, recent studies [39, 40] have shown that the generation of samples that violate physics can be improved by incorporating deterministic constraints (based on conservation laws) and statistical constraints (based on the energy spectrum of turbulent flows) into the loss function of GANs. With this approach, not only were the synthesized samples able to more accurately emulate certain physical properties of the system, but training time was also significantly reduced (by up to 80%).

What are the essential physics-based constraints on radar micro-Doppler? One of the most salient characteristics of any dynamic system is its maximum velocity. In the case of human motion, it is the envelopes of the μD signature, which bound the maximum and minimum possible velocity incurred during movement. Arguably, any discontinuity or trajectory surpassing the bounds of the envelopes represents a kinematic inconsistency.

Accurately representing target kinematics in synthetic μD signatures is especially critical because critical gait information is embodied by the motion of the legs, as reflected by the trajectory of the feet, which also exhibits the fastest speed among all body parts. Important μD features are computed from the envelope of the μD signature, for example: average velocity, maximum foot swing velocity, period, the phase between leg and torso, foot swing time, the range profile Doppler frequency, total bandwidth of the Doppler signal, offset of the total Doppler, normalized standard deviation of the Doppler signal strength, and period of the limb motion [41]. Thus, ensuring consistency in the envelope of synthesized signatures with that of measured data is essential for enabling accurate recognition of human gait.

One way to inform GANs about the physics-based constraints imposed by μD envelopes is to augment the discriminator with additional branches, which take as input the upper and lower envelope [42]. In this way, the discriminator will base its decisions on whether a sample is real or fake by taking into consideration enveloped features. In the single-envelope multi-branch GAN (MBGAN), shown in Figure 10.4, only one additional auxiliary branch taking the upper envelope as input is added. In the dual-envelope MBGAN, both upper and lower envelopes are provided as inputs to two additional auxiliary branches in the discriminator. It should be noted that the selection of the single versus dual-branch variants of MBGAN should be based upon the salient characteristics of the μD signatures of the activity classes involved. In many ambulatory classes, for example, the lower envelope does not embody that much information—the fluctuations in the envelope are relatively less than that exhibited in the upper envelope. In this case, just using a single-envelope MBGAN may outperform the dual-envelope MBGAN. In contrast, in sign language, the frequency three-dimensional hand and arm movements generate rich envelopes in both the positive and negative directions, resulting in better performance when the dual-envelope MBGAN is utilized [43].

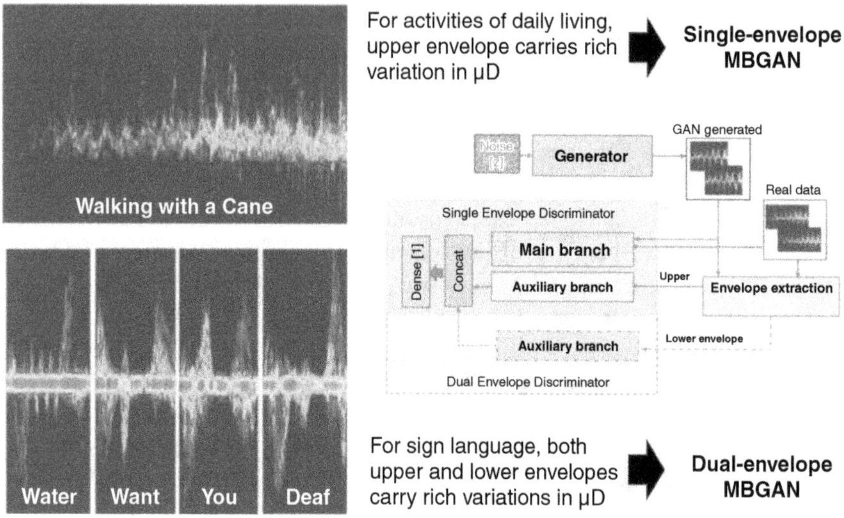

Figure 10.4 MBGAN architecture with single-envelope and dual-envelope auxiliary branches, selected based on μD characteristics of activity classes.

The generator of the MBGAN takes at its input a 100-dimension noise vector and provides at its output synthetic samples. The discriminator then aims to determine whether the samples are real or fake. The micro-Doppler is provided as a 2D input to the main discriminator branch, while the envelope is given as an input to the auxiliary discriminator branch. Both branches extract features, which are then flattened and concatenated before passing through a final dense layer. The discriminator then scores the samples on a scale of 0 to 1 based on a comparison between the distributions of real and fake samples. Within this framework, various neural networks can be used in the generator and discriminator. In [42], the generator consists of 10 2D-convolutional layers, where each of the layers is followed by a batch normalization with 0.9 momentum and a rectified linear unit (ReLU) activation function with the exception of the last convolutional layer, which is only followed by a *tanh* activation function. The main branch of the discriminator is a 7-layer CNN, where each layer is followed by a Leaky-ReLU activation function, while the auxiliary branch is comprised of three 1D-convolutional layers.

Another way to guide GANs with physics-based constraints is to define a new physics-based loss term in addition to currently used statistical measures, such as the distance between real and fake sample distributions. Statistical similarity does not guarantee consistency with physics because the feature spaces computed from images embody only spatial properties. Thus, distance metrics are indicative only of spatial similarity. Loss functions designed to include physics-based metrics increase the cost when kinematic errors are incurred in synthetic samples. An example of a physics-aware loss function would thus be:

$$L_D = \{D(x) - D(G(z))\} + \text{GP} + \alpha_P L_{\text{physics}}, \tag{10.5}$$

where L_{physics} is the physics-based loss term regulated with the hyperparameter α_P, and GP is the gradient penalty given as:

$$\text{GP} = \alpha_{\text{GP}}\left(||\nabla_{\hat{x}} D(\hat{x})||\right)^2. \tag{10.6}$$

In the above expression, α_{GP} is a hyperparameter for regularization; $D(x)$ is the discriminator's estimate of the probability that real data instance x is real; $G(z)$ is the generator's output when given noise z; $D(G(z))$ is the discriminator's estimate of the probability that a fake instance is real.

Conceptually, the physics-based loss is defined as any computable metric that is indicative of how well the synthetic samples resemble the real samples. Because activity recognition requires accurate representation of physical bounds on velocity, as determined by the μD signature envelopes, the physics-based loss term is designed as a metric that captures the degree of similarity between the envelopes of synthetic and real μD signatures. Thus, if we consider the envelope as a time series or a curve, metrics such as the dynamic time warping (DTW) distance, which measures the similarity between two temporal sequences that do not sync up with each other perfectly due to the variation in time or speed, can be used to quantify envelope similarity.

The MBGAN architecture has been shown to yield improved results over conventional GANs and domain adaptation methods in both activity recognition [42] and sign language datasets [43, 44]. Significantly, an inverse correlation was found between the magnitude of the kinematic errors seen in the synthetic samples and classification accuracy. In other words, the lesser the error, the better the performance.

10.2.4 Encoding Physics in DNN Layers

Another effective way to integrate physics into DNNs is to use physics-based models as part of a pretraining process that encodes the physics into some of the neural network layers. Rather than focus on envelope consistency, for example, suppose we instead utilized the idea of semantic segmentation to teach a neural network which pixels in an image comprise the target signature and which contain clutter or noise. If we were to use real radar data to train a DNN for semantic segmentation, we would be faced with a huge labeling and annotation problem. Instead, let us use physics-based simulations to synthesize a range of probable signatures for which we definitively know which pixels belong to the target response and which contain clutter or noise. To enable robustness in the resulting network, Gaussian

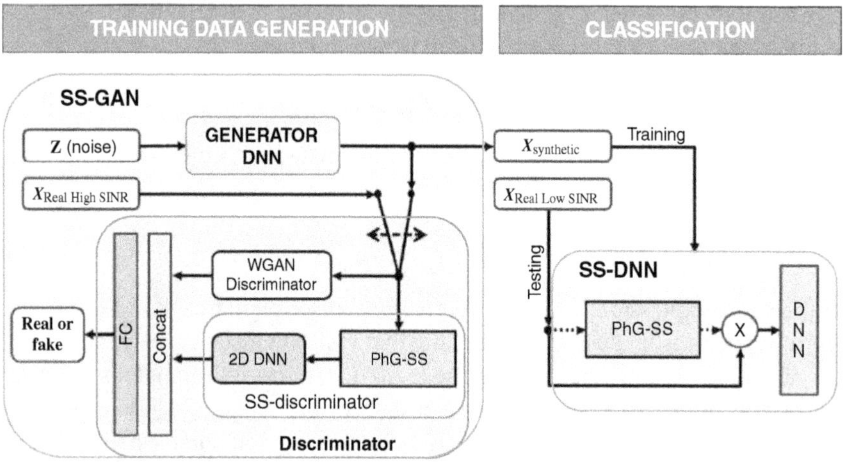

Figure 10.5 Integration of physics-based semantic segmentation (PhSS) encoded block into GAN architecture and DNN classifier.

noise at varying SNR levels can be injected into the data. This process would thus generate challenging use cases for the network to determine a target/nontarget designation.

The resulting PhSS network could then be integrated into the auxiliary branch of the discriminator of a GAN as a physics-encoded block interpreting the kinematic fidelity of the synthesized samples, as shown in Figure 10.5. In low SNR scenarios, the PhSS network also has the ability to be used effectively as a denoiser, where the physics-based training serves as a guide to mask out nontarget pixels and enhance classification accuracy by as much as 16% [45]!

10.2.5 Cross-frequency Training

In recent years, another topic that has garnered increased is the idea of utilizing "datasets of opportunity" for training—in other words, data acquired by a different RF sensor (waveform, bandwidth, pulse repetition frequency, etc.) at a different location from different participants. This is a situation that can also arise if, for example, you have trained a model for a particular application but now wish to change the RF sensor you deploy. Perhaps a newer, higher resolution RF sensor has hit the market, or you need to replace a burnt-out sensor, but the current version is out-of-date and a different model must be deployed. Or, maybe you have discovered a publicly available database of RF data, which you would like to exploit for training your own model. As low-cost, embedded RF sensors have become more ubiquitous, an increasing number of RF databases are becoming publicly available—we should use them!

But *can a model trained on one RF sensor translate to another RF sensor?*

This is known as the cross-frequency training problem—and the short answer, is no, not directly.

Cross-frequency compatibility is a key aspect of sensing that has seen limited discussion in the literature [46, 47], as most works focus on single-frequency systems. Vishwakarma et al. [47] proposed a dictionary learning approach for classification across multiple frequencies, which was motivated by cognitive radar applications. However, only four easily discernable classes were considered with transmit frequencies varying by just 2.5 GHz. When the difference between transmit frequency is great, important differences in μD signatures measured can be observed. While the general shape of motion may be the same, the signatures will differ in resolution, Doppler spread, electronic noise, clutter, and sensor artifacts due to the change in scattering mechanism, e.g., how moving targets, objects, and clutter reflect waves differently at different wavelengths.

So how significant to classification are the differences in data acquired across different radar transmit waveforms? A study [48] using a three-radar multifrequency network comprised of a 10 GHz ultra-wideband impulse radar (UWB-IR), a 24 GHz frequency-modulated continuous wave (FMCW) radar with 1.5 GHz bandwidth, and 77 GHz FMCW radar with 4 GHz bandwidth showed that the performance degradation incurred when training and testing across different RF transmit waveforms can be *as much as 70%!*

To date, only a few works have been published in the literature to address this challenge. In [49], a technology-agnostic approach for RF-based HAR (TARF) was proposed, which was shown to maintain performance across 4 different RF sensing technologies (2.4 GHz Wi-Fi, 5 GHz Wi-Fi, 77 GHz FMCW radar, and RFID). The proposed method involved first applying a data generalization technique to mitigate the disparity in measured data across the different RF devices, followed by a domain adversarial neural network. Similarly, in [50], an adversarial training approach of a gradient reversal layer (GRL) and ACGAN is proposed as a radar system-agnostic (RSA) learning architecture. Nevertheless, this topic remains an open area of research, which can benefit from physics-aware ML concepts, given the common aspects of the underlying human motion and ability to model phenomenological differences across sensors.

10.3 Physics-aware Classification

In this section, we examine various DNN architectures that have been designed so as to incorporate an understanding of the signal structure as determined by physics, biomechanics, or language to improve classification accuracy. The conventional processing chain involves two-steps: first, signal processing techniques are applied to convert the data into the 2D or 3D representations needed for input to a DNN. Second, the DNN takes the inputs and provides a prediction at its output. Thus, the overall latency is comprised of the time needed for processing as well as prediction. This multistep processing with parameter optimization consumes a lot of computational resources while also having high temporal latency, which imposes limitations in real-time sensing applications. This has motivated interest in the development of classifiers that directly take as input the raw, complex RF data stream. However, traditional complex-valued neural networks do not offer sufficient accuracy or robustness for most radar applications. Thus, the tradeoffs involved with signal processing—what the best RF representation for input to a DNN is complexity, latency, and performance—have driven research into novel DNN architectures for radar applications.

10.3.1 DNNs for Raw Radar Data

A variety of studies have proposed classification techniques that directly take as input the raw, complex RF data stream. In [51], the I and Q channels are provided as two separate inputs to a real recurrent neural network (RNN), resulting in an average classification accuracy of 93%. However, these methods proved to be computationally quite costly: the network required over 240 hours for training using four Nvidia GeForce GTX 1080 GPUs. In [52], a similar approach was used with a long short-term memory (LSTM) network but with the slight modification of the range profile also being provided as an input in addition to the raw I/Q data. In [53], real 2D sinc filters and 2D wavelet filters were utilized as the first layer of a DNN operating directly on 2D raw radar data to classify six different hand gestures—the real and imaginary components of the data are separated to enable utilization of a real-valued DNN.

Alternatively, complex-valued CNN (CV-CNN) architectures have been proposed for several radar-based HAR problems. In [54, 55], the authors proposed a CV-CNN architecture that trains on the complex-valued spectrograms instead of real-valued spectrogram images calculated in the logarithmic scale. The authors in [55] also introduced FourierNet, where they used a windowed Fourier transform that operates on the 1D-time series data before the short-time Fourier transform (STFT) operation. In [56], the authors used CV-CNN techniques on complex-valued 2D representations such as range-time, range-Doppler maps, or spectrograms.

A persistent question that relates to how we might improve the performance of complex-valued neural networks that operate directly on raw radar data is the question of whether or not we can indeed replace certain fundamental signal processing steps with neural networks. Because frequency is so fundamental to signal representation and radar measurements of range, Doppler, and direction-of-arrival, there has been specific attention to the implementation of Fourier processing with neural networks. For example, Velik [57] proposed a single-layer neural network with a linear transfer function to implement a DFT, while Akhtar [58] proposed a training methodology for neural networks with nonlinear activation functions to replicate and approximate the DFT. Meanwhile, Akasha et al. [59] compared the time required to execute a DFT when implemented using a fast Fourier transform (FFT) versus a neural network. Once trained, neural network prediction times tend to be quicker than the time needed to apply signal processing for input representation computation. But there has not been a comprehensive study exploring the tradeoffs between training data (size and type), DNN complexity, accuracy under varying SNRs, and computation time.

Nevertheless, in the context of radar-based μD classification, there have been several works demonstrating the computational benefits of replacing the computation of the STFT with physics-aware DNNs. Inspired by learnable filters for audio signal classification [60, 61], Biswas et al. [62, 63] developed a generalized parameterized learnable filter (PLF) approach for learning parameterized bandpass filters, such as Sinc or Gaussian filters, which can then be integrated with a CNN to directly and effectively operate on time-domain RF signals. In this way, the model can learn filters that have direct physical frequency interpretations, leading to more interpretable architectures. In addition, instead of learning all the weights of a convolutional filter, the model only learns the parameters of the structured filters, which reduces the number of learnable parameters, and, hence, the resulting models are more suitable when only a small amount of training data is available.

As shown in Figure 10.6, the PLF-based CV-CNN architecture, PLFNet, is developed to directly process the 1D complex raw RF data. The first layer is comprised of J complex-valued PLFs, each with a length of N. This layer is then followed by a complex max-pooling layer of size 1×4, batch normalization, ReLU activation, and dropout of 0.2. This sequence of layers comprises the PLF block (shown in Figure 10.6), which is followed by a series of convolutional blocks. Each convolutional block has a convolutional layer with 64 filters and a kernel size of 1×5 for each filter. This is followed by a max-pooling layer of size 1×3, batch normalization, ReLU activation, and dropout of 0.2. The outputs of the convolutional blocks are flattened before being input to a dense layer. After the application of dropout of 0.2, the network uses a softmax layer for classification. All of the aforementioned layers were developed to work on complex-valued data, including the complex-valued operations for convolution, fully connected layers, activation functions, pooling, and normalization [63]. These operations were implemented by representing complex numbers as their real and imaginary components, allowing for efficient computations within a standard deep learning framework.

The results of utilizing learnable parameterized filters as an expression of physics-aware ML can be seen in Figure 10.7, which compares the accuracy of PLFNet architectures with conventional 2D CNNs applied to μD signatures and

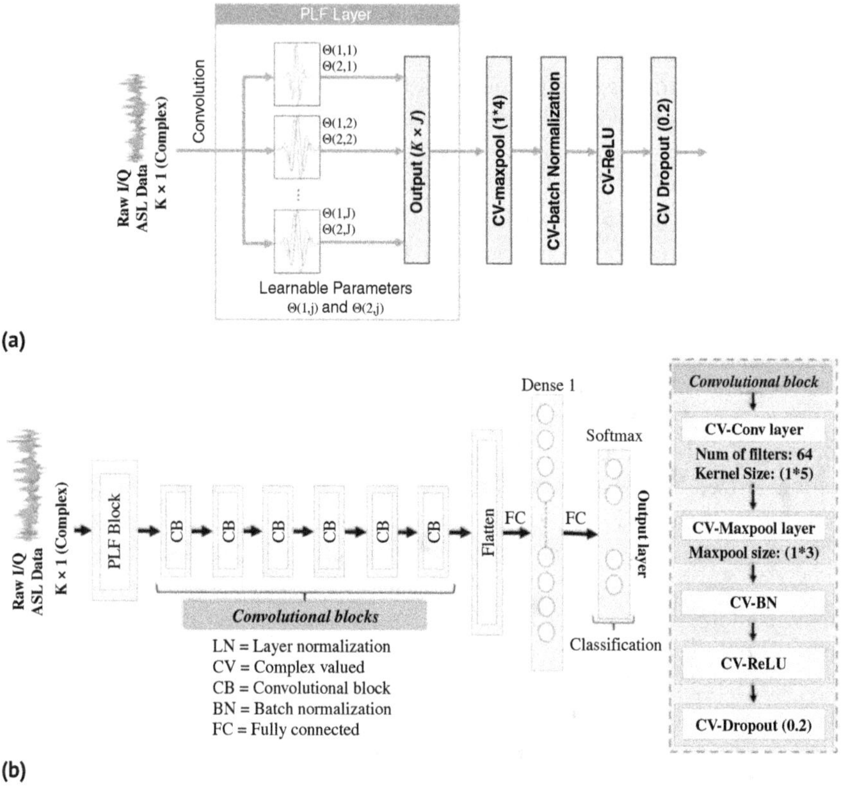

Figure 10.6 PLFNet architecture [62]. (a) PLF block architecture, (b) PLFNet architecture.

1D real-valued and complex-valued CNNs on the raw RF data stream. Note that the PLFNets (CV-SincNet and CV-GaussNet) significantly outperform the 1D CV-CNN baselines. Moreover, the PLFNets even outperform the results obtained with 2D CNNs on μD signatures, yielding a 2% boost in top-1 and top-3 accuracy and comparable top-5 accuracy. But there is a drastic savings in overall latency: the difference between the time the data is acquired and the time the prediction is obtained. While it takes 2.6 seconds to compute the μD signature and just 0.032 seconds for the 2D-CNN to provide a prediction, the processing time for the CV-GaussNet is just 0.377 seconds with a prediction time of 0.275 seconds. In other words, classification of μD signatures has a total latency of 2.633 seconds, while the CV-GaussNet needs just 0.652 seconds—only a quarter of the time!

However, when using image-based input representations, physics-aware GANs (PhGANs), such as MBGAN, can be utilized to enlarge the amount of kinematically accurate training data. Notice that training with PhGAN-synthesized data

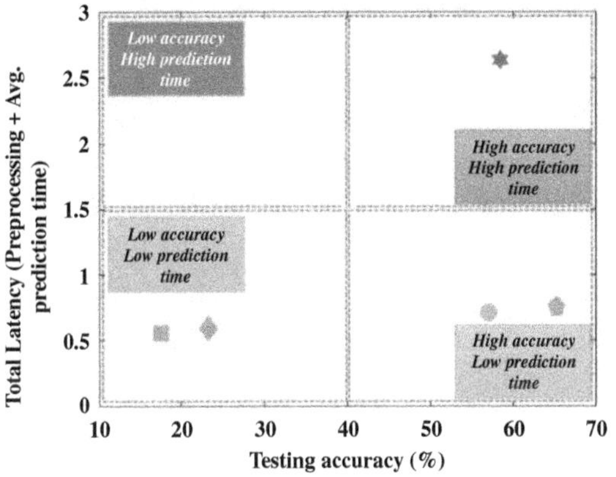

Performance on 100-sign ASL dataset

Network	Testing Accuracy		
	Top1	Top3	Top5
CNN-2D	63.95	81.46	89.92
CNN-1D	10.82	21.79	29.16
CVCNN-1D	17.63	27.94	38.89
Sinc	56.26	75.03	81.92
CV-Sinc	64.8	80.32	86.23
Gaussian	55.7	75.73	81.64
CV-Gauss	65.97	83.33	89.76
CNN-2D (w/PhGAN)	77.51	89.26	93.00

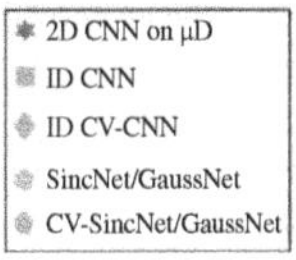

Figure 10.7 Latency/accuracy tradeoff in PLFNet for a 100-class dataset [43].

results in a 13% jump in accuracy. Presumably, we could apply physics-based models to synthesize the complex, raw RF data stream, thereby boost the performance of the PLFNets in similar fashion. However, accurate synthesis requires accurate time-varying ranges for a skeletal model of the motion. For sign language, high resolution skeleton tracking of the fingers, hands, and upper arms is required. This remains an area of ongoing research as commercially available video-to-skeleton software does not reliably extract fine-grain skeletons in a stable fashion. Alternatively, RGB-D cameras, which include infrared sensors to estimate depth (D) and acquire red, green, blue (RGB) color images, can be used to obtain more accurate fine-grain skeletal estimates from which simulated complex RF data streams can be obtained for model training.

10.3.2 Learning from Joint-domain Inputs

When we utilize 2D computed radar data representations, such as micro-Doppler, on the input of a DNN, we are actually only using part of the measurements acquired by the radar, e.g., supplying velocity versus time, whereas radar also measures time-varying range and angle as well. Joint-domain DNNs take as inputs all the physical variables measured by the radar. One way this can be accomplished is by using 3D data representations at the input, such as the radar data cube (RDC), which provides range and Doppler as a function of time. It has become common to view the RDC as a sequence of range-Doppler maps, which are akin to a "video"—but for radar data. Both range-Doppler videos [64–66] as well as μD videos [67] have been applied to radar μD classification of human activities and gestures. Other types of 3D joint-domain representations have also been proposed, such as *GaitCube* [68], which is comprised of micro-Doppler and micro-range signatures over time.

Additionally, the availability of different data representations has led to the proposal of fusing classification results achieved with different 2D representations. Li et al. [69] first extract physical features from both radar μD signatures and range-Doppler maps for the purpose of active shooter recognition. Here, the physical features extracted are determined specific to the classification problem at hand:

- *Frequency Gap:* When a person moves toward the radar with a concealed weapon, arm movement is limited, and the power density due to human movement is narrowed. In contrast, the muzzle of a gun produces a high-energy signature with large Doppler shifts. This results in a frequency gap between the human response and that of the tip of the gun. This frequency gap is thus a physical feature specifically relevant to the armed/unarmed recognition problem.
- *Total Doppler Bandwidth:* the difference between the largest positive frequency shift and the lowest negative frequency shift in the μD signature.
- *Torso Speed:* Magnitude of the velocity of the peak power trajectory within the μD signature, which is typically attributed to the torso, due to its large surface area.
- *Limb Speed Difference on Different Directions:* This is computed as the difference between the largest positive or negative Doppler frequency minus the torso speed.
- *Echo Aspect Ratio:* This feature is computed from the range-Doppler maps, where the area having the highest power density is considered as the echo corresponding to the subject's torso. The echo aspect ratio is calculated as the extent in the velocity direction divided by the extent in the range direction.

These physical features are the provided to the input of a feedforward artificial neural network to predict eight possible activities, one of which is walking with a concealed rifle under trench coat, while the remaining seven classes are different types of aided/unaided ambulatory motions that might be confused with carrying a concealed weapon.

More recently, it has become common place to instead use neural networks for automatic feature extraction, rather than compute physical features [70, 71]. For example, Jokanovic et al. [72], utilized three types of 2D RF data representations, namely, time-frequency, time-range, and range-Doppler, at the input of stacked autoencoders for human activity recognition. Some authors have also proposed novel ways of fusing the RF data representations prior to classification with a DNN. For example, in [73], the time-Doppler map (TDM), time-range map (TRM), and cadence velocity diagram (CVD) are merged into a fusion map, which is then given to the input of a MS-Net, a feature fusion CNN that is composed of two lightweight networks, MobileNetV3-large and ShuffleNetV2. Similarly, the micro-Doppler and radar point clouds gait recognition (mDS-PCGR) framework [74] fuses gait features derived from both μD signatures and 4-D radar point cloud sequences.

A unique concept applicable to data acquired from a multichannel radar is the concept of multiview fusion. In [75, 76], angular subspace projections of the RDC are computed for targets at different angles to boost the SNR of the target signature across different views of the RDC. Each view is then provided to the input of a DNN for feature extraction. Feature-level fusion is then applied via concatenation prior to input to the final softmax layer of the network. The multiview DNN was shown to have advantageous properties when trying to identify two-handed gestures, as the process of angular subspace projection independently enhances the SNR from each hand, resulting in extraction of unique multiview features that boost classification. A multiview processing perspective is also pursued in [77], where a multiview CNN-LSTM architecture is proposed for human activity recognition. First, CNNs are trained to extract optimal frame-based features from the time-range, time-Doppler and range-Doppler projections of the RDC. Next, the pretrained parameters of the encoder are fine-tuned to extract intermediate frame-based representations, which are subsequently provided to LSTM networks for sequential classification.

One way to incorporate physics-based constraints in DNNs is through the use of multitask learning (MTL). Although MTL traditionally involves a single data representation (image, text, speech signal) at the input, MTL networks jointly optimize multiple objectives by exploiting domain-specific information contained in commonalities and differences across tasks. By sharing representations among related (auxiliary) tasks, the generalization capability of the model can be improved on the main task. For example, in [78], a time-frequency transform of the radar signal was provided at the input, while the MTL network optimized across the two tasks of feature enhancement and signal recognition. In [79], a MTL radar transformer is proposed, which exploits the correlation between the personal identification task and the fall detection task to enhance the performance of discrimination for both tasks. In [80], the tasks are designed based on the physical properties of the drone that are critical to recognition. In particular, a MTL network is trained to estimate two or more of the following four indicators: a drone's wing type, its number of rotors, the presence of a payload, and its mean rotor rotation rate. Their results showed improved performance as compared to separately trained single-task neural networks.

Joint-domain multi-input MTL (JD-MIMTL) [2] expands the multitask learning framework to incorporate learning from multiple RF domain representations at the input: e.g., range-Doppler videos, range-Angle videos, and μD signature, as shown in Figure 10.8. The joint feature space derived from multiple input

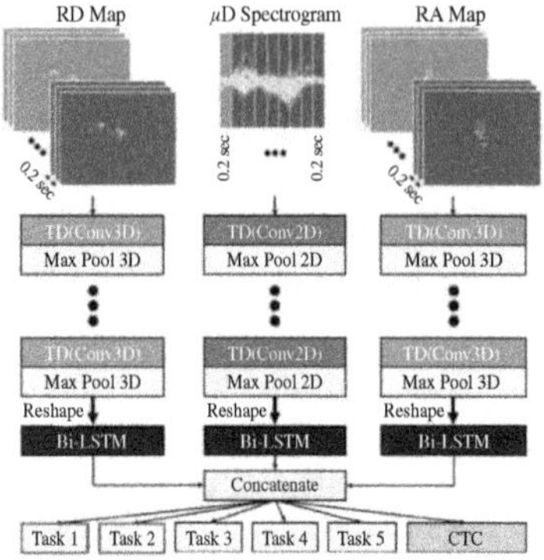

(a)

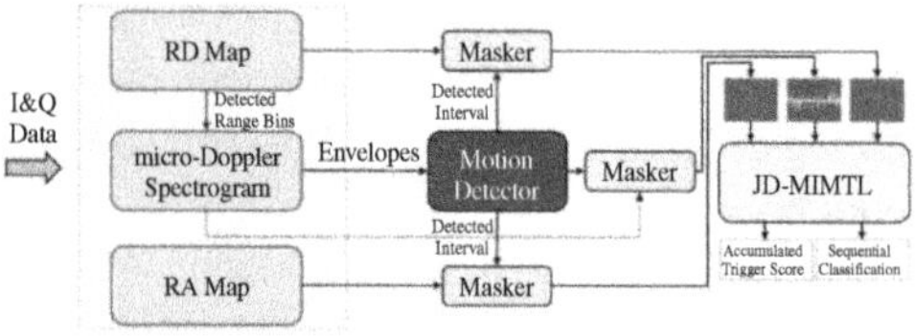

(b)

Architecture	μD	RD Map	RA Map	Feature-Level Fusion
CNN + BiLSTM	78.8%	72.8%	67.5%	84.3%
CNN + BiLSTM + CTC	80.6%	78.4%	71.3%	87.5%
CNN + BiLSTM + CTC + MTL	**83.6%**	**78.6%**	**71.4%**	JD-MIMTL 92%

(c)

Figure 10.8 (a) JD-MIMTL architecture and (b) its usage in device trigger detection and command gesture recognition system. (c) JD-MIMTL performance comparison relative to input of just single domain representation [2].

representations is enriched by fusing in a concatenation layer. The tasks themselves are selected based on attributes relevant and impactful on classification. Thus, JD-MIMTL enables the enforcement of physics-based relationships through the tasks specified. As an example, consider the problem of word-level American Sign Language (ASL) recognition. Five auxiliary tasks can be defined, based on our scientific knowledge of how kinematics relates to sign production:

- Task 1: One versus two handedness
- Task 2: Major location of hands
- Task 3: Movement type
- Task 4: Daily activity versus ASL sign
- Task 5: Number of strokes

The overall loss function, L_{total}, utilized in the JD-MIMTL framework is the weighted sum of the connectionist temporal classification (CTC) loss, α_{ctc}, and the loss L_i specific to each task i:

$$L_{\text{total}} = \alpha_{\text{ctc}} L_{\text{ctc}} + \sum_{i=1}^{I} \alpha_i L_i, \tag{10.7}$$

where α_i are the weights assigned to the various loss terms. Since each task has its own loss function and, hence, varying convergence times, the weights α_i need to be jointly optimized. After training the model, all of the auxiliary task and CTC output layers are removed, and the model is augmented with a softmax layer for classification.

The original work proposing JD-MIMTL [2] utilized this architecture as part of a device trigger/command gesture recognition system for radar-based human–computer interaction, as shown in Figure 10.8b. First, the raw radar I/Q data is processed to compute the range-Doppler and range-angle maps and μD signatures. This is then put at the input of a motion detection step so that periods of inactivity are discarded. JD-MIMTL is then utilized both for trigger detection and for recognition of sequential ASL commands. The performance of JD-MIMTL is compared for various networks (with/without CTC and with/without MTL) for single-input cases and joint-domain inputs (feature-level fusion). JD-MIMTL outperforms all variants in the ablation study.

A variant of JD-MIMTL was proposed in [81], where only the μD signature and range profile were provided to a JD-MIMTL network that jointly optimized just two tasks: activity and agility. Although current literature focuses on the recognition of activities for health monitoring—e.g., falls, gait abnormality, fall risk assessment—the onset of acute disease, chronic disease, or changes to a person's mental and physical health are often indicated by gait changes, especially ones ability to transition between static postural states—not just activity itself. Thus, characterization of not just activity but agility—the speed with which one moves or transitions between postures—is critical. Using JD-MIMTL, Kurtoglu et al. [81] showed that both activity and agility levels (low, medium, high) could be predicted from sequential articulation of activities involving both persistent motions, like walking, and transitional activities, such as sitting down and standing up.

10.3.3 AI/ML in Radar Networks

Multistatic radar networks are related areas, where the issue of fusion and how to best present data to the input of neural networks arises. Various types of fusion can be utilized to increase the performance afforded by each sensor individually. In decision fusion [82], the received return from each sensor is first separately classified and then an overall decision made through majority voting. In feature-level fusion [83, 84], separate networks are used to extract features from each sensor, concatenated, and then supplied a classifier. In [85], a two-step training process called *modality tuning* is utilized in which the shared layers are initially frozen while the input modality-specific layers are trained. In this way, the specific layers can be more optimally trained to decode the data from specific sensors. After training the sensor-specific layers, the shared layers are unfrozen and the entire network is trained jointly. The shared layers are thus optimized to boost feature-level fusion. Modality tuning was shown to significantly surpass both decision fusion and feature fusion (without modality tuning) by 20 and 12%, respectively, when three radars at different transmit frequencies (10, 24, and 77 GHz) are used as inputs.

Alternatively, cross-modal fusion networks [86, 87] can be used to capture both shared features in the data, while also separately extracting features specific to each modality. This approach is particularly well-suited for fusion in RF sensor networks because the common observations will result in shared target-specific features, while the phenomenological differences across frequency create sensor-specific differences in the data. One example of a uniquely designed network for fusion of radar data is MAEF-Net [88], a multiattention enhanced fusion network designed to improve robustness in omnidirectional human activity recognition. In [89], a Stockwell transform is used to extract the μD features of data obtained from two different radars, after which the ridgelines from each radar's μD are fused in the time-frequency domain. A CNN is then used for fall detection. A recent study compares the efficacy of three different DNN-based fusion strategies for continuous HAR [90].

10.3.4 RF Skeleton Estimation

Up until now, our discussion has focused exclusively on classification for target recognition; however, another important application of DNNs is for skeletal estimation: the process of estimating the position in x, y, and z for points on the body as a function of time. The idea of radar-based skeleton estimation was initially explored with RF-Capture [91], a system that utilized 5.4 to 7.2 GHz FMCW signals via an antenna array to detect coarse body part positions, subsequently reconstructing a human figure by piecing together these detected parts. In 2018, RF-Pose3D [92] advanced this framework by employing a T-shaped 12-element antenna array for FMCW signal transmission and reception at a 6.3 GHz center frequency and 1.8 GHz bandwidth. This system feeds range-azimuth and range-elevation heatmaps into a ResNet-based encoder neural network, coupled with 12 camera nodes capturing RGB video to collect labelled key points from OpenPose for training a region proposal network (RPN) and a ResNet-architecture CNN. This network focuses on RF data from individual persons to extract 3D skeletons from regions of interest, reporting average localization errors of 4.2, 4.0, and 4.9 cm in the x, y, z axes, respectively, with OpenPose estimated key points. Despite its groundbreaking demonstration of RF skeleton estimation's feasibility, the method's reliance on over 17 million data samples and 16 hours of recordings underscored its significant data and computational demands, limiting practical application.

In 2020, mmPose [93] was proposed, which predicted more than 15 joints using two specially oriented IWR1443 radars. This method, feeding point clouds into a bifurcated CNN, did not utilize the radar's Doppler and signal intensity data, leading to jitter in skeleton animation. To reduce these jitters, the use of additional filters [94] was proposed to obtain a more stable skeletal representation. The network mmPose-NLP [95] offered further improvements (<3 cm error in

25 skeletal keypoints based on Kinect estimates as ground truth) using a natural language processing (NLP) inspired sequence-to-sequence (Seq2Seq) skeletal keypoint estimator. In 2021, a millimeter-wave assistive rehabilitation system (MARS) [96] was proposed, which employed IWR1443 radar and standard software for point cloud data, including Doppler and intensity information, reporting an average mean absolute error (MAE) of 5.8 cm accuracies in 19-point predictions compared to Microsoft Kinect v2 camera estimates. They further investigated joint angle estimation from the predicted skeleton and reported the average MAE of MARS in estimating left elbow angle, right elbow angle, left knee angle, and right knee angles are 12°, 13°, 7°, and 6°, respectively.

In recent years, there has been a significant surge in publications focused on mm-wave-based human pose estimation, e.g., [41, 97–101]. Although a popular approach is to employ radar-generated range-azimuth and range-elevation heatmaps, with some works also incorporating radar point clouds, as data inputs for training DNN estimators, a few works estimate skeletons from the raw radar data [102] or using only μD signatures [103, 104].

For validating their findings, the bulk of these studies have predominantly leaned on either the Kinect system or the multicamera-based OpenPose model for ground truth, yet neither of these systems is accepted as gold standard optical tracking systems and incur significant estimation errors themselves [105, 106]. A pilot study examining micro-Doppler-based gait parameter estimates and skeleton tracking accuracy against a Vicon motion capture system is given in [41].

Ultimately, the skeleton estimation accuracies reported in the literature would not be considered as sufficient from a biomechanics and gait analysis perspective. To advance radar technologies such that they can have a greater impact on gait analysis-based health monitoring, we must consider how to improve DNN-based estimators. Here, physics-aware AI/ML perspectives play a critical role as purely data-driven perspectives have not been able to achieve desired accuracy. However, the human body itself provides an important factor in constraining the locations and kinematics of skeletal keypoints—a form of knowledge that should be taken into account in the design of the DNN itself. For example, in [107], a kinematically constrained learning architecture was proposed, where new loss functions—the distance loss and the angle loss with respect to the parent-child nodes of skeleton joints—were defined to facilitate improved skeleton reconstruction.

Kinematic cycle consistency-based skeleton estimation (KCC-Sk) [108, 109] is a physics-aware DNN that integrates physics-based models into both the DNN architecture and loss function to achieve improved estimation accuracy. The key concept behind KCC is that we wish to form a closed loop in the estimation process that requires the learning of an identity mapping. In other words, the cycle consists of a DNN for *estimation* (of the skeleton from RF data) and a physics-based model for *reconstruction* of the RF data from the estimated skeleton, as depicted in Figure 10.9.

The KCC framework produces the benefit of not only reducing skeleton estimation error, but also of eliminating the need for any measured data during the model training phase. In practice, before a radar is deployed in a home, we will have no prior knowledge of who— specifically—will be monitored. Consequently, any radar will need to be trained in a generic fashion to estimate the skeleton of *anyone* as best as possible. Given that there will always be a difference in distribution between training and testing data, even when using real measurements, due to difference in environment and the person being monitored, we focus instead on training the network (pre-deployment) to learn the fundamental physics-based relationship between the input radar data and the skeletal keypoints. In other words, pre-deployment, we *only use physics-based simulations* to train the DNN-based estimator. This can be accomplished using any source of MOCAP data. In initial work [109], the Carnegie Mellon University Motion Capture Library [26] was utilized as the source for simulated RF data. Note that we *know* the correct ground truth trajectories of the skeleton keypoints, and we can therefore use that knowledge in the training process when computing the loss function in addition to the KCC loss. Furthermore, because we are using simulations for training, the error obtained is solely due to DNN estimation error, not due to any radar system parameters (e.g., transmit frequency, bandwidth, etc.). Once the radar has been deployed in an environment, and we begin collecting data from the person being monitored. We can continue to fine-tune our DNN-based estimator using the physics-based KCC loss term—no MOCAP data is needed to accomplish this fine-tuning. This approach for utilizing KCC for online learning with no prior real data for training is summarized in Figure 10.10.

In the pre-deployment phase, when were are training on simulated data only, the KCC loss term combines positional, velocity, and envelope errors to ensure both spatial accuracy and temporal coherence. Positional error, calculated as the mean squared error (MSE), minimizes differences between predicted and actual joint positions, ensuring spatial accuracy (Equation 10.8).

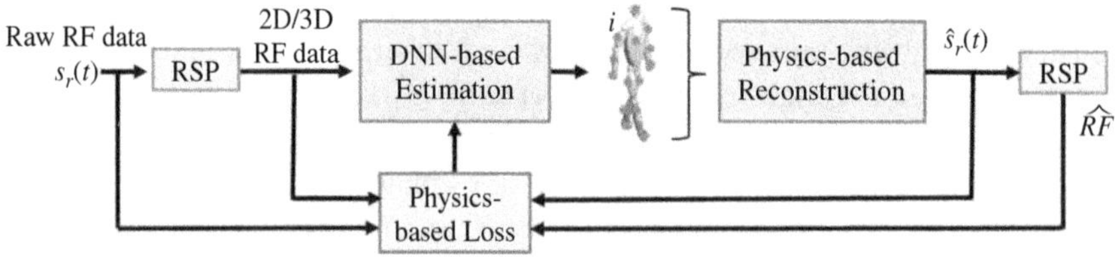

Figure 10.9 KCC framework.

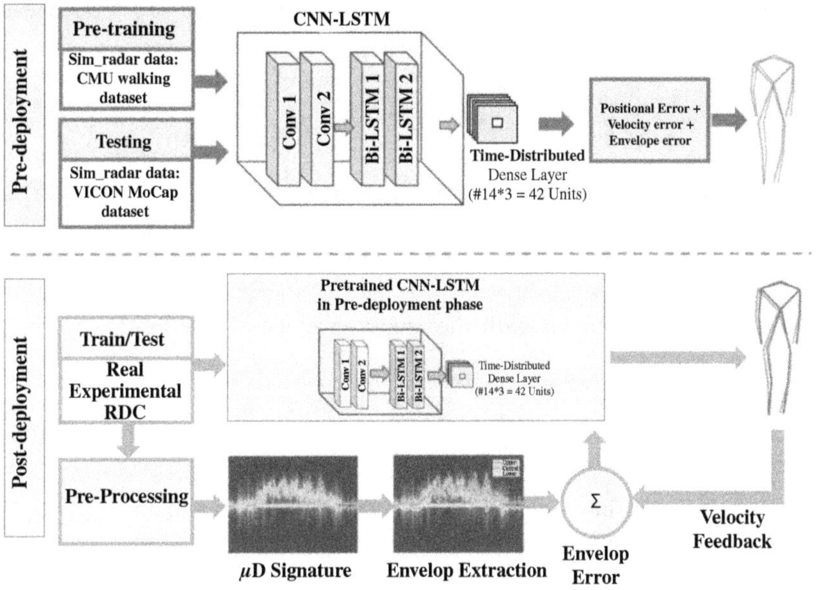

	Method/ Framework	Train	Test	Ground Truth	Mean Abs. Error		
					X (cm)	Y (cm)	Z (cm)
Pre-deployment	CNN + Bi-LSTM	Simulated Radar Data	Simulated Radar Data	Vicon MOCAP Data	3.66	5.81	1.89
	CNN + Bi-LSTM with KCC Loss				3.07	4.41	1.34
Post-deployment	Pre-Trained CNN + Bi-LSTM with Envelope Loss	Real Radar Data	Real Radar Data		4.97	5.82	5.22
Prior Work	CNN + Bi-LSTM	Real Radar & Vicon MOCAP Data	Real Radar Data		5.63	7.3	7.81

Figure 10.10 Online learning of skeletal keypoints using KCC and no prior real data [109].

$$K_{\text{points\_loss}} = \frac{1}{3K}\sum_{k=1}^{K}\left((x_k-\hat{x}_k)^2+(y_k-\hat{y}_k)^2+(z_k-\hat{z}_k)^2\right). \tag{10.8}$$

Here, K represents the total number of skeleton joints. The variables x_k, y_k, and z_k are the ground truth positions of the kth joint, while $\hat{x}_k$, $\hat{y}_k$, and $\hat{z}_k$ are the predicted positions of the kth joint.

The velocity error ensures the predicted skeleton adheres to realistic velocity patterns. Defined as the difference in joint positions over consecutive time steps, it compares the true velocity $v_{\text{true},k}(t)$ with the predicted velocity $v_{\text{pred},k}(t)$ for each joint k and time frame t (Equation 10.9).

$$v_{\text{loss}} = \frac{1}{3K} \sum_{k=1}^{K} \sum_{t=2}^{T} \left(v_{\text{true},k}(t) - v_{\text{pred},k}(t)\right)^2, \tag{10.9}$$

where T represents the total number of time frames, and $v_{\text{true},k}(t)$ and $v_{\text{pred},k}(t)$ are the true and predicted velocities of the kth joint at time t.

The envelope error ensures consistency in joint movement amplitudes by comparing the maximum velocities of true and predicted motions. It aligns the predicted joint movements with the upper envelope of true motion velocities (Equation 10.10).

$$v_{\text{env\_loss}} = \frac{1}{K} \sum_{k=1}^{K} \left(v_{\text{true\_ max},k} - v_{\text{pred\_ max},k}\right)^2, \tag{10.10}$$

where $v_{\text{true\_ max},k}$ is the maximum true velocity of the kth joint, and $v_{\text{pred\_ max},k}$ is the maximum predicted velocity of the kth joint.

While v_{loss} minimizes frame-to-frame velocity errors, the additional $v_{\text{env\_loss}}$ term ensures that the maximum movement speed aligns with realistic biomechanical constraints. This prevents unnatural motion artifacts and enhances model generalization. The final $\text{KCC}_{\text{train}}$ function integrates these three terms, weighted by factors γ and β, to form a comprehensive loss function that accounts for both positional accuracy and temporal coherence, as shown in Equation 10.11:

$$\text{KCC}_{\text{train}} = K_{\text{points\_loss}} + \gamma \cdot v_{\text{loss}} + \beta \cdot v_{\text{env\_loss}}, \tag{10.11}$$

where γ and β are the weights for the velocity and envelope loss terms, respectively.

Once the radar is deployed, computation of the points loss term $K_{\text{points\_loss}}$ will not be possible because MOCAP data simultaneously acquired will not be available. However, online learning from newly acquired radar data can be utilized for online learning to drive down estimation errors through the use of a KCC loss term that employs only envelope errors. However, now we consider not just the upper envelope, but also the peak of the μD signature (corresponding to the torso trajectory) and the lower envelope. Thus, the post-deployment KCC loss is defined as:

$$\text{KCC}_{\text{tune}} = \beta_{\text{upp}} \cdot v_{\text{upp\_env\_loss}} + \beta_{pk} \cdot v_{pk\_\text{loss}} + \beta_{\text{low}} \cdot v_{\text{low\_env\_loss}}, \tag{10.12}$$

based on envelopes computed from the radar μD signature as the "true" kinematics.

The results of an ablation study comparing pre-deployment and post-deployment skeleton estimation accuracy, with and without KCC loss terms, is given in Figure 10.10. Notice the error in the x-direction is reduced by 0.66 cm, in the y-direction by 1.48 cm, and in the z-direction by 2.59 cm.

10.4 Conclusion

In conclusion, physics-aware AI/ML techniques are essential to radar applications, especially human activity recognition, as it represents an extension of the knowledge-aided signal processing paradigm and can be used to achieve greater accuracy and robustness in real-world environments, while minimizing training data requirements.

Acknowledgments

The author and her work was supported in part by National Science Foundation (NSF) awards #1932547, #2233503, #2238653, the Air Force Office of Scientific Research (AFOSR) award #FA9550-22-1-0384, and an American Association of University Women (AAUW) Research Publication Grant in Engineering, Medicine and Biology. Human studies research was conducted under U. Alabama IRB Protocols #18-06-1271 and #23-04-6553.

References

1 S. Z. Gurbuz. *Deep Neural Network Design for Radar Applications.* IET/SciTech Publishing, London, 2020.
2 E. Kurtoğlu, A. C. Gurbuz, E. A. Malaia, D. Griffin, C. Crawford, and S. Z. Gurbuz. ASL trigger recognition in mixed activity/signing sequences for RF sensor-based user interfaces. *IEEE Transactions on Human-Machine Systems,* 52(4):699–712, 2022.
3 D. Yessad, A. Amrouche, M. Debyeche, and M. Djeddou. Micro-doppler classification for ground surveillance radar using speech recognition tools. In *Progress in Pattern Recognition, Image Analysis, Computer Vision, and Applications,* volume 7042, pages 280–287, 2011.
4 H. Tian, C. Dong, L. Yuan, and H. Yin. Motion state classification for micro-drones via modified mel frequency cepstral coefficient and hidden markov mode. *Electronics Letters,* 58(4):164–166, 2022.
5 B. Erol and S. Z. Gürbüz. Hyperbolically-warped cepstral coefficients for improved micro-doppler classification. In *2016 IEEE Radar Conference (RadarConf),* pages 1–6, 2016.
6 B. Erol, M. G. Amin, and S. Z. Gurbuz. Automatic data-driven frequency-warped cepstral feature design for micro-doppler classification. *IEEE Transactions on Aerospace and Electronic Systems,* 54(4):1724–1738, 2018.
7 B. Erol, M. S. Seyfioglu, S. Z. Gurbuz, and M. Amin. Data-driven cepstral and neural learning of features for robust micro-Doppler classification.

In K. I. Ranney and A. Doerry, editors, *Radar Sensor Technology XXII*, volume 10633, page 106330J. International Society for Optics and Photonics, SPIE, 2018.

8 J. D. Willard, X. Jia, S. Xu, M. S. Steinbach, and V. Kumar. Integrating physics-based modeling with machine learning: A survey. *ArXiv, abs/2003.04919*, 2020.

9 C. Meng, S. Griesemer, D. Cao, S. Seo, and Y. Liu. When physics meets machine learning: A survey of physics-informed machine learning. *Machine Learning for Computational Science and Engineering*, 1:20, 2025.

10 V. Vapnik. *The Nature of Statistical Learning Theory*. Springer, New York, NY, 2000.

11 Y. S. Abu-Mostafa, M. Magdon-Ismail, and H. -T. Lin. *Learning From Data*. AMLBook, 2012.

12 J. Deng, W. Dong, R. Socher, L. J. Li, K. Li, and L. Fei-Fei. Imagenet: A large-scale hierarchical image database. In *2009 IEEE Conference on Computer Vision and Pattern Recognition*, pages 248–255, 2009.

13 Y. Bengio and Y. Lecun. Scaling learning algorithms towards AI. *Large-Scale Kernel Machines*. MIT Press, Cambridge, UK, pp. 321–360, 2007.

14 J. Park, R. J. Javier, T. Moon, and Y. Kim. Micro-doppler based classification of human aquatic activities via transfer learning of convolutional neural networks. *Sensors*, 16(12):1990, 2016.

15 X. Li, Y. He, F. Fioranelli, X. Jing, A. Yarovoy, and Y. Yang. Human motion recognition with limited radar micro-doppler signatures. *IEEE Transactions on Geoscience and Remote Sensing*, 59(8):6586–6599, 2021.

16 M. S. Seyfioğlu and S. Z. Gürbüz. Deep neural network initialization methods for micro-doppler classification with low training sample support. *IEEE Geoscience and Remote Sensing Letters*, 14(12):2462–2466, 2017.

17 M. S. Seyfioğlu, A. M. Özbayoğlu, and S. Z. Gürbüz. Deep convolutional autoencoder for radar-based classification of similar aided and unaided human activities. *IEEE Transactions on Aerospace and Electronic Systems*, 54(4):1709–1723, 2018.

18 T. Dogaru and L. Nguyen. Fdtd models of electromagnetic scattering by the human body. In *2006 IEEE Antennas and Propagation Society International Symposium*, pages 1995–1998, 2006.

19 P. van Dorp and F. C. A. Groen. Human walking estimation with radar. *IEE Proceedings - Radar, Sonar and Navigation*, 150:356–365, 2003.

20 S. S. Ram, S. Z. Gurbuz, and V. C. Chen. Modeling and simulation of human motions for micro-doppler signatures. In *Radar for Indoor Monitoring: Detection, Classification, and Assessment*. CRC Press, Cambridge, UK, 2017.

21 E. Blasch, U. Majumder, and M. Minardi. Radar Signals Dismount Tracking for Urban Operations. In I. Kadar, editor, *Signal Processing, Sensor Fusion, and Target Recognition XV*, volume 6235, page 623504. International Society for Optics and Photonics, SPIE, 2006.

22 U. Majumder, M. Minardi, E. Blasch, L. Gorham, K. Naidu, T. Lewis, and R. Williams. Radar signals dismount data production. In E. G. Zelnio and F. D. Garber, editors, *Algorithms for Synthetic Aperture Radar Imagery XIII*, volume 6237, page 623704. International Society for Optics and Photonics, SPIE, 2006.

23 S. S. Ram and H. Ling. Simulation of human microdopplers using computer animation data. In *2008 IEEE Radar Conference*, pages 1–6, 2008.

24 S. S. Ram and H. Ling. Microdoppler signature simulation of computer animated human and animal motions. In *2008 IEEE Antennas and Propagation Society International Symposium*, pages 1–4, 2008.

25 S. S. Ram, C. Christianson, Y. Kim, and H. Ling. Simulation and analysis of human micro-dopplers in through-wall environments. *IEEE Transactions on Geoscience and Remote Sensing*, 48(4):2015–2023, 2010.

26 Carnegie Mellon University Graphics Lab. CMU graphics lab motion capture database, https://mocap.cs.cmu.edu/, 2024.

27 B. Erol and S. Z. Gurbuz. A kinect-based human micro-doppler simulator. *IEEE Aerospace and Electronic Systems Magazine*, 30(5):6–17, 2015.

28 M. S. Seyfioglu, B. Erol, S. Z. Gurbuz, and M. G. Amin. Dnn transfer learning from diversified micro-doppler for motion classification. *IEEE Transactions on Aerospace and Electronic Systems*, 55(5):2164–2180, 2019.

29 S. Z. Gurbuz and M. G. Amin. Radar-based human-motion recognition with deep learning: Promising applications for indoor monitoring. *IEEE Signal Processing Magazine*, 36(4):16–28, 2019.

30 D. K. Barton. Land clutter models for radar design and analysis. *Proceedings of the IEEE*, 73(2):198–204, 1985.

31 C. C. Lin and J. P. Reilly. A site-specific model of radar terrain backscatter and shadowing. *Johns Hopkins APL Technical Digest*, 18(3):433, 1997.

32 A. Creswell, T. White, V. Dumoulin, K. Arulkumaran, B. Sengupta, and A. A. Bharath. Generative adversarial networks: An overview. *IEEE Signal Processing Magazine*, 35(1):53–65, 2018.

33 Y. Yang, C. Hou, Y. Lang, D. Guan, D. Huang, and J. Xu. Open-set human activity recognition based on micro-doppler signatures. *Pattern Recognition*, 85:60–69, 2019.

34 B. Erol, S. Z. Gurbuz, and M. G. Amin. Gan-based synthetic radar micro-doppler augmentations for improved human activity recognition. In *2019 IEEE Radar Conference (RadarConf)*, pages 1–5, 2019.

35 I. Alnujaim, D. Oh, and Y. Kim. Generative adversarial networks to augment micro-doppler signatures for the classification of human activity. In *IGARSS 2019 - 2019 IEEE International Geoscience and Remote Sensing Symposium*, pages 9459–9461, 2019.

36 Y. Yang, Y. Zhang, H. Ji, B. Li, and C. Song. Radar-based human activity recognition under the limited measurement data support using domain translation. *IEEE Signal Processing Letters*, 29:1993–1997, 2022.

37 I. Alnujaim, S. S. Ram, D. Oh, and Y. Kim. Synthesis of micro-doppler signatures of human activities from different aspect angles using generative adversarial networks. *IEEE Access*, 9:46422–46429, 2021.

38 B. Erol, S. Z. Gurbuz, and M. G. Amin. Motion classification using kinematically sifted ACGAN-synthesized radar micro-doppler signatures. *IEEE Transactions on Aerospace and Electronic Systems*, 56(4):3197–3213, 2020.

39 Z. Yang, J. L. Wu, and H. Xiao. Enforcing imprecise constraints on generative adversarial networks for emulating physical systems. *Communications in Computational Physics*, 30(3):635–665, 2021.

40 J. L. Wu, K. Kashinath, A. Albert, D. Chirila, Prabhat, and H. Xiao. Enforcing statistical constraints in generative adversarial networks for modeling chaotic dynamical systems. *Journal of Computational Physics*, 406:109209, 2020.

41 S. Z. Gurbuz, M. M. Rahman, Z. Bassiri, and D. Martelli. Overview of radar-based gait parameter estimation techniques for fall risk assessment. *IEEE Open Journal of Engineering in Medicine and Biology*, 5:735–749, 2024.

42 M. M. Rahman, S. Z. Gurbuz, and M. G. Amin. Physics-aware generative adversarial networks for radar-based human activity recognition. *IEEE Transactions on Aerospace and Electronic Systems*, 59(3):2994–3008, 2023.

43 M. M. Rahman, E. A. Malaia, A. C. Gurbuz, D. J. Griffin, C. Crawford, and S. Z. Gurbuz. Effect of kinematics and fluency in adversarial synthetic data generation for asl recognition with rf sensors. *IEEE Transactions on Aerospace and Electronic Systems*, 58(4):2732–2745, 2022.

44 E. Kurtoğlu, K. DeHaan, C. K. Pezzarossi, D. J. Griffin, C. Crawford, and S. Z. Gurbuz. Interactive learning of natural sign language with radar. *IET Radar, Sonar & Navigation*, 18(8):1203–1216, 2024.

45 S. Kearney and S. Z. Gurbuz. Semantic segmentation guided RF micro-doppler synthesis and UAV classification in low SINR. In *International Radar Conference*, 2025.

46 A. Shrestha, C. Murphy, I. Johnson, A. Anbulselvam, F. Fioranelli, J. L. Kernec, and S. Z. Gurbuz. Cross-frequency classification of indoor activities with DNN transfer learning. In *2019 IEEE Radar Conference (RadarConf)*, pages 1–6, 2019.

47 S. Vishwakarma and S. S. Ram. Dictionary learning with low computational complexity for classification of human micro-dopplers across multiple carrier frequencies. *IEEE Access*, 6:29793–29805, 2018.

48 S. Z. Gurbuz, M. M. Rahman, E. Kurtoglu, T. Macks, and F. Fioranelli. Cross-frequency training with adversarial learning for radar micro-Doppler signature classification (Rising Researcher). In K. I. Ranney and A. M. Raynal, editors, *Radar Sensor Technology XXIV*, volume 11408, page 114080A. International Society for Optics and Photonics, SPIE, 2020.

49 C. Yang, X. Wang, and S. Mao. Tarf: Technology-agnostic rf sensing for human activity recognition. *IEEE Journal of Biomedical and Health Informatics*, 27(2):636–647, 2023.

50 Y. Ding, P. Lv, R. Liu, Y. Peng, and M. Ding. A radar system-agnostic (RSA) learning architecture for human activity recognition. *IEEE Sensors Journal*, 25(10):18492–18502, 2025.

51 S. Yang, J. L. Kernec, F. Fioranelli, and O. Romain. Human activities classification in a complex space using raw radar data. In *2019 International Radar Conference (RADAR)*, pages 1–4, 2019.

52 C. Loukas, F. Fioranelli, J. L. Kernec, and S. Yang. Activity classification using raw range and i & q radar data with long short term memory layers. In *2018 IEEE 16th Intl Conf on Dependable, Autonomic and Secure Computing, 16th Intl Conf on Pervasive Intelligence and Computing, 4th Intl Conf on Big Data Intelligence and Computing and Cyber Science and Technology Congress(DASC/PiCom/DataCom/CyberSciTech)*, pages 441–445, 2018.

53 T. Stadelmayer, A. Santra, R. Weigel, and F. Lurz. Data-driven radar processing using a parametric convolutional neural network for human activity classification. *IEEE Sensors Journal*, 21(17):19529–19540, 2021.

54 X. Yao, X. Shi, and F. Zhou. Human activities classification based on complex-value convolutional neural network. *IEEE Sensors Journal*, 20(13):7169–7180, 2020.

55 D. A. Brooks, O. Schwander, F. Barbaresco, J. Y. Schneider, and M. Cord. Complex-valued neural networks for fully-temporal micro-doppler classification. In *2019 20th International Radar Symposium (IRS)*, pages 1–10, 2019.

56 X. Yang, R. G. Guendel, A. Yarovoy, and F. Fioranelli. Radar-based human activities classification with complex-valued neural networks. In *2022 IEEE Radar Conference (RadarConf22)*, pages 1–6, 2022.

57 R. Velik. Discrete Fourier transform computation using neural networks. In *2008 International Conference on Computational Intelligence and Security*, volume 1, pages 120–123, 2008.

58 J. Akhtar. Discrete Fourier transform with neural networks. In *2021 IEEE 94th Vehicular Technology Conference (VTC2021-Fall)*, pages 1–5, 2021.

59 A. Akasha and K. Kulkarni. Comparing signal processing and neural networks for DFT applications. In *2022 Second International Conference on Advanced Technologies in Intelligent Control, Environment, Computing & Communication Engineering (ICATIECE)*, pages 1–5, 2022.

60 M. Ravanelli and Y. Bengio. Speaker recognition from raw waveform with SincNet. In *2018 IEEE Spoken Language Technology Workshop (SLT)*, pages 1021–1028, 2018.

61 H. Khan and B. Yener. Learning filter widths of spectral decompositions with wavelets. In S. Bengio, H. Wallach, H. Larochelle, K. Grauman, N. Cesa-Bianchi, and R. Garnett, editors, *Advances in Neural Information Processing Systems*, volume 31. Curran Associates, Inc., 2018.

62 S. Biswas, C. O. Ayna, and A. C. Gurbuz. Plfnets: Interpretable complex-valued parameterized learnable filters for computationally efficient rf classification. *IEEE Transactions on Radar Systems*, 2:1102–1111, 2024.

63 S. Biswas, C. O. Ayna, S. Z. Gurbuz, and A. C. Gurbuz. CV-SincNet: Learning complex sinc filters from raw radar data for computationally efficient human motion recognition. *IEEE Transactions on Radar Systems*, 1:493–504, 2023.

64 B. Erol and M. G. Amin. Radar data cube processing for human activity recognition using multisubspace learning. *IEEE Transactions on Aerospace and Electronic Systems*, 55(6):3617–3628, 2019.

65 S. Wang, J. Song, J. Lien, I. Poupyrev, and O. Hilliges. Interacting with soli: Exploring fine-grained dynamic gesture recognition in the radio-frequency spectrum. In *Proceedings of the 29th Annual Symposium on User Interface Software and Technology*, UIST '16, pages 851–860. Association for Computing Machinery, New York, NY, USA, 2016.

66 W. Y. Kim and D. H. Seo. Radar-based human activity recognition combining range–time–doppler maps and range-distributed-convolutional neural networks. *IEEE Transactions on Geoscience and Remote Sensing*, 60:1–11, 2022.

67 Z. Zhang, Z. Tian, and M. Zhou. Latern: Dynamic continuous hand gesture recognition using FMCW radar sensor. *IEEE Sensors Journal*, 18(8):3278–3289, 2018.

68 M. Z. Ozturk, C. Wu, B. Wang, and K. J. R. Liu. Gaitcube: Deep data cube learning for human recognition with millimeter-wave radio. *IEEE Internet of Things Journal*, 9(1):546–557, 2022.

69 Y. Li, Z. Peng, R. Pal, and C. Li. Potential active shooter detection based on radar micro-doppler and range-doppler analysis using artificial neural network. *IEEE Sensors Journal*, 19(3):1052–1063, 2019.

70 Y. Kim, S. Ha, and J. Kwon. Human detection using doppler radar based on physical characteristics of targets. *IEEE Geoscience and Remote Sensing Letters*, 12:289–293, 2015.

71 S. Z. Gürbüz, B. Erol, B. Çağlıyan, and B. Tekeli. Operational assessment and adaptive selection of micro-doppler features. *IET Radar, Sonar & Navigation*, 9(9):1196–1204, 2015.

72 B. Jokanovic, M. Amin, and B. Erol. Multiple joint-variable domains recognition of human motion. In *2017 IEEE Radar Conference (RadarConf)*, pages 0948–0952, 2017.

73 W. Zhang and L. Hong. Radar-based human activity recognition using multi-domain maps and feature fusion convolutional neural network. In *Proceedings of the 2024 3rd Asia Conference on Algorithms, Computing and Machine Learning*, CACML '24, pages 277–281. Association for Computing Machinery, New York, NY, USA, 2024.

74 C. Ma and Z. Liu. mDS-PCGR: A bimodal gait recognition framework with the fusion of 4-D radar point cloud sequences and micro-doppler signatures. *IEEE Sensors Journal*, 24(6):8227–8240, 2024.

75 E. Kurtoğlu, A. C. Gurbuz, E. Malaia, D. Griffin, C. Crawford, and S. Z. Gurbuz. RF micro-doppler classification with multiple spectrograms from angular subspace projections. In *2022 IEEE Radar Conference (RadarConf22)*, pages 1–6, 2022.

76 E. Kurtoğlu, S. Biswas, A. C. Gurbuz, and S. Z. Gurbuz. Boosting multi-target recognition performance with multi-input multi-output radar-based angular subspace projection and multi-view deep neural network. *IET Radar, Sonar & Navigation*, 17(7):1115–1128, 2023.

77 H. U. R. Khalid, A. Gorji, A. Bourdoux, S. Pollin, and H. Sahli. Multi-view CNN-LSTM architecture for radar-based human activity recognition. *IEEE Access*, 10:24509–24519, 2022.

78 H. Li, D. Quan, F. Zhou, F. Ren, K. Yu, N. Jin, and J. Wu. Radar signal recognition based on multi-task learning. *IEEE Access*, 12:153209–153220, 2024.

79 X. Jiang, L. Zhang, and L. Li. Multi-task learning radar transformer (MLRT): A personal identification and fall detection network based on IR-UWB radar. *Sensors*, 23(12):5632, 2023.

80 A. Pappas, J. J. M. de Wit, F. Fioranelli, and B. Jacobs. Multitask learning approaches towards drone characterisation with radar. *IET Radar, Sonar & Navigation*, 19(1):e70012, 2025.

81 E. Kurtoğlu, S. Salehin, M. G. Amin, I. P. Kan, M. A. McKay, and S. Z. Gurbuz. Ethogram-based personalization of human activity and agility from radar μD signatures. In *2024 IEEE EMBS International Conference on Biomedical and Health Informatics (BHI)*, pages 1–8, 2024.

82 H. Li, A. Mehul, J. L. Kernec, S. Z. Gurbuz, and F. Fioranelli. Sequential human gait classification with distributed radar sensor fusion. *IEEE Sensors Journal*, 21(6):7590–7603, 2021.

83 S. Z. Gurbuz, A. C. Gurbuz, E. A. Malaia, D. J. Griffin, C. S. Crawford, M. M. Rahman, E. Kurtoglu, R. Aksu, T. Macks, and R. Mdrafi. American sign language recognition using rf sensing. *IEEE Sensors Journal*, 21(3):3763–3775, 2021.

84 F. J. Abdu, Y. Zhang, and Z. Deng. Activity classification based on feature fusion of FMCW radar human motion micro-doppler signatures. *IEEE Sensors Journal*, 22(9):8648–8662, 2022.

85 S. Z. Gurbuz, M. M. Rahman, E. Kurtoglu, E. Malaia, A. C. Gurbuz, D. J. Griffin, and C. Crawford. Multi-frequency rf sensor fusion for word-level fluent asl recognition. *IEEE Sensors Journal*, 22(12):11373–11381, 2022.

86 Y. Aytar, L. Castrejon, C. Vondrick, H. Pirsiavash, and A. Torralba. Cross-modal scene networks. *IEEE Transactions on Pattern Analysis and Machine Intelligence*, 40(10):2303–2314, 2018.

87 J. Gao, P. Li, Z. Chen, and J. Zhang. A survey on deep learning for multimodal data fusion. *Neural Computation*, 32(5):829–864, 2020.
88 X. Li, Y. Qiu, Z. Deng, X. Liu, and X. Huang. Lightweight multiattention enhanced fusion network for omnidirectional human activity recognition with fmcw radar. *IEEE Internet of Things Journal*, 12(5):5755–5768, 2025.
89 T. Yang, F. Meng, Q. Xu, and Y. X. Guo. Fall feature enhancement and fusion using the stockwell transform with dual mmWave radars. *IEEE Sensors Journal*, 23(2):1368–1376, 2023.
90 S. Zhu, R. G. Guendel, A. Yarovoy, and F. Fioranelli. Continuous human activity recognition with distributed radar sensor networks and CNN–RNN architectures. *IEEE Transactions on Geoscience and Remote Sensing*, 60:1–15, 2022.
91 F. Adib, C. Y. Hsu, H. Mao, D. Katabi, and F. Durand. Capturing the human figure through a wall. *ACM Transactions on Graphics*, 34(6):1–13, 2015.
92 M. Zhao, Y. Tian, H. Zhao, M. A. Alsheikh, T. Li, R. Hristov, Z. Kabelac, D. Katabi, and A. Torralba. RF-based 3D skeletons. In *Proceedings of the 2018 Conference of the ACM Special Interest Group on Data Communication*, SIGCOMM '18, pages 267–281. Association for Computing Machinery, New York, NY, USA, 2018.
93 A. Sengupta, F. Jin, R. Zhang, and S. Cao. mm-pose: Real-time human skeletal posture estimation using mmWave radars and CNNs. *IEEE Sensors Journal*, 20(17):10032–10044, 2020.
94 S. Hu, A. Sengupta, and S. Cao. Stabilizing skeletal pose estimation using mmWave radar via dynamic model and filtering. In *2022 IEEE-EMBS International Conference on Biomedical and Health Informatics (BHI)*, pages 1–6, 2022.
95 A. Sengupta and S. Cao. mmpose-NLP: A natural language processing approach to precise skeletal pose estimation using mmWave radars. *IEEE Transactions on Neural Networks and Learning Systems*, 34(11):8418–8429, 2023.
96 S. An and U. Y. Ogras. Mars: mmWave-based assistive rehabilitation system for smart healthcare. *ACM Transactions on Embedded Computing Systems*, 20(5s):1–22, 2021.
97 Y. Wu, Z. Jiang, H. Ni, C. Mao, Z. Zhou, W. Wang, and J. Han. mmHPE: Robust multiscale 3-D human pose estimation using a single mmWave radar. *IEEE Internet of Things Journal*, 12(1):1032–1046, 2025.
98 C. Xie, D. Zhang, Z. Wu, C. Yu, Y. Hu, and Y. Chen. RPM 2.0: RF-based pose machines for multi-person 3D pose estimation. *IEEE Transactions on Circuits and Systems for Video Technology*, 34(1):490–503, 2024.
99 Z. Cao, W. Ding, R. Chen, J. Zhang, X. Guo, and G. Wang. A joint global–local network for human pose estimation with millimeter wave radar. *IEEE Internet of Things Journal*, 10(1):434–446, 2023.

100 G. W. Kim, S. W. Lee, H. Y. Son, and K. W. Choi. A study on 3D human pose estimation using through-wall IR-UWB radar and transformer. *IEEE Access*, 11:15082–15095, 2023.

101 M. M. Rahman, D. Martelli, and S. Z. Gurbuz. Radar-based human skeleton estimation with CNN-LSTM network trained with limited data. In *2023 IEEE EMBS International Conference on Biomedical and Health Informatics (BHI)*, pages 1–4, 2023.

102 L. Chen and G. Wang. CPFormer: End-to-end multi-person human pose estimation from raw radar cubes with transformers. *IEEE Sensors Journal*, 25(7):12466–12478, 2025.

103 C. Tang, W. Li, S. Vishwakarma, F. Shi, S. Julier, and K. Chetty. MDPose: Human skeletal motion reconstruction using WiFi micro-doppler signatures. *IEEE Transactions on Aerospace and Electronic Systems*, 60(1):157–167, 2024.

104 X. Zhou, T. Jin, Y. Dai, Y. Song, and Z. Qiu. MD-Pose: Human pose estimation for single-channel UWB radar. *IEEE Transactions on Biometrics, Behavior, and Identity Science*, 5(4):449–463, 2023.

105 Q. Wang, G. Kurillo, F. Ofli, and R. Bajcsy. Evaluation of pose tracking accuracy in the first and second generations of microsoft kinect. In *2015 International Conference on Healthcare Informatics*, pages 380–389, 2015.

106 M. Ota, H. Tateuchi, T. Hashiguchi, T. Kato, Y. Ogino, M. Yamagata, and N. Ichihashi. Verification of reliability and validity of motion analysis systems during bilateral squat using human pose tracking algorithm. *Gait & Posture*, 80:62–67, 2020.

107 W. Ding, Z. Cao, J. Zhang, R. Chen, X. Guo, and G. Wang. Radar-based 3D human skeleton estimation by kinematic constrained learning. *IEEE Sensors Journal*, 21(20):23174–23184, 2021.

108 S. Salehin, S. Kearney, and Z. Gurbuz. Kinematic cycle consistency using wearables and RF data for improved human skeleton estimation. In *2024 58th Asilomar Conference on Signals, Systems, and Computers*, pages 1215–1219, 2024.

109 S. Salehin, S. Kearney, and Z. Gurbuz. Radar-based online learning of human skeleton keypoints without any prior real data. In *IEEE International Radar Conference*, 2025.

11

Data-driven Resource Allocation in Distributed Radar Network

Batu K. Chalise,[1] Jahi Douglas,[1] Benjamin H. Kirk,[2] and David McNamara[2]

[1] *Department of Electrical and Computer Engineering, New York Institute of Technology, Old Westbury, NY, USA*
[2] *DEVCOM Army Research Laboratory, U.S. Army, Adelphi, MD, USA*

11.1 Introduction

Radar systems transmit electromagnetic signals and process signals reflected from targets/objects for detection, tracking, and parameter estimation. The objective of detection is to determine whether a target is present or not in a given search space and observation period. The parameter estimation includes the estimation of distances (ranges), velocities, locations, angles, and the shapes of targets. Tracking includes estimating the change in target parameters over the course of a time. Initial development of radar technology was mainly confined to military applications, which include, but not limited to, detection and tracking of aircraft, guided missiles, spacecraft, ships, and ground vehicles [1]. This initial development later spurred the development of radar technology for numerous civilian applications, such as air traffic control and navigation, weather forecasting, border patrolling, crop monitoring, fire detection, landmine detection, and radio glaciology. In many of these applications, radar systems are often bulky and need huge antennas with large power transmission capabilities. However, recent advancements in hardware technology and signal processing have made the deployment of power-efficient low-cost radar systems possible in consumer applications, such as wireless health monitoring, vital sign monitoring, wearable sensors, assisted living, autonomous driving, automated crack detection on bridges, and roads, etc.

The aforementioned radar applications often employ a monostatic configuration, in which the radar transmitter and receiver are colocated on the same platform. The detection and tracking performance of monostatic radar systems can be significantly enhanced with the deployment of multi-static (single transmitter with multiple receivers or vice versa) and multiple monostatic/bistatic radar configurations [2, 3]. Although these configurations can be viewed as the radar systems with diversely located and spatially distributed nodes, such systems cannot be considered as truly distributed if detection, estimation, tracking, and resource allocation algorithms are implemented in a centralized manner. In a centralized network, either one of the nodes or a designated coordinator collects complete in-phase/quadrature (I/Q) signals from all other nodes for executing the algorithms. This type of sharing complete I/Q data with the central coordinator (CC) demands a significant amount of communications bandwidth and power. Moreover, such sharing is subject to significant performance loss, especially when nodes are far away from the coordinator and observe fading wireless links. However, in order to leverage the spatial diversity of target scattering for extracting important target features, the nodes are required to be widely separated. As such, high-quality robust wireless links are needed for centralized implementation, which significantly increases cost in practice.

One way for overcoming this bottleneck was proposed in [4, 5] for target detection. It requires sharing of only binary information between each node and the CC. Each radar node computes a likelihood ratio test (LRT) using its received signals, compares the computed LRT with a certain threshold, and makes a binary decision depending on whether the LRT is above or below the threshold. When the LRT is above the threshold, the radar node transmits bit "1" to the coordinator, confirming the presence of a target in a given observation. Otherwise, the node transmits bit '0' to the coordinator, confirming the target's absence. The CC then makes a global decision on detection by fusing or combining binary decisions from all nodes [4]. These methods, however, assume that binary information from all radar nodes can be correctly decoded at the central controller, which is possible only in error-free communications links. But such communication is not possible due to wireless channel impairments inevitable in practice, which include path loss, fading, interference, and noise. The effects of Rician fading in the design of a fusion rule at a CC have been investigated for multiple-input multiple-output (MIMO) channels [6]. However, this work is limited to a wireless sensor network and does not involve active radar operation.

Target detection and parameter estimation techniques for a distributed radar network without a central controller (DNwoCC) were proposed in [7–9]. In the case of target detection for the DNwoCC, the nodes aim to achieve a network-wide consensus on target-related information, which can be quantified in terms of generalized LRT (GLRT). This consensus can be achieved by minimizing average disagreement among nodes on their acquired GLRT values. This minimization leads to an iterative local update of each node's GLRT value with the knowledge of the GLRT values from only the neighboring nodes. A key advantage of this method, referred to as *average consensus-based* distributed detection, is that it avoids the need for having the CC, because, upon convergence of the iterative approach, each node can achieve the global estimate despite that only the neighboring nodes are sharing information with one another. The convergence of the iterative process to the global estimate, for example, global GLRT value in the case of distributed detection, is possible under certain channel conditions. This type of average consensus-based distributed estimation [10, 11] was proposed for multi-agent systems [12, 13], wireless sensor networks [14, 15], and radar networks [16]. An extension of the consensus-based estimation method was proposed in [7] for distributed detection in the presence of radar clutter, which was modeled as a spherically invariant random process [17, 18]. The case of noisy communications links was considered in [7], whereas the impacts of quantization errors on the probability of detection and probability of false alarm were analyzed in [19]. These papers show that the consensus-based distributed estimation/detection cannot be guaranteed to converge to the global estimate in the case of noisy communications links with quantization errors. The software-defined radio-based approach for demonstrating the proof of concept of the distributed detection mechanism in the presence of noise and quantization errors was proposed in [20].

The consensus-based distributed direction-of-arrival (DoA) and delay-Doppler estimation was proposed in [8, 9]. Assuming that all nodes monitor the same target scene and signals observed by each node can be compensated with respect to the location and orientation of a reference node, the DoA estimation was formulated as a distributed sparse signal recovery problem. More specifically, consensus-based approach was employed to achieve consensus among nodes, on their estimates of the sparse vectors, where a sparse vector associated with each node determines its individual estimates of the DoAs of targets. This approach enables each node to solve the local sparse recovery optimization problem for updating its sparse vector with the knowledge of the sparse vectors from the neighboring nodes. This update process continues iteratively until the distributed algorithm converges to the global DoA estimates of the targets. The similar approach was applied for the distributed delay and Doppler estimation in [9], where the estimated sparse vectors determine delay and Doppler associated with the targets. The distributed DoA estimation in the presence of noisy communications links was proposed in [21]. In this paper, two distributed nodes, each having a uniform linear array, form a co-prime array when viewed from a common reference and transmit single-bit quantized signals to the CC, which then solves the sparse recovery problem to estimate DoAs of the sources.

In all the aforementioned papers [8, 9], the distributed detection and estimation of target angles, ranges (delays), and velocities (Doppler) assumed equal but non-overlapping bandwidth allocation to the radar nodes for their monostatic operations such that inter-node interference can be avoided. Such equal bandwidth allocation (EBWA) scheme may become suboptimal, especially when nodes observe different signal-to-interference-and-noise ratios (SINRs) associated with targets. To overcome the limitation of EBWA scheme, bandwidth and carrier frequency optimization and allocation (BWCFOA) algorithms were proposed in [22, 23] for a distributed radar network. With the motivation of prioritizing the node that observes the lowest SINR or is in the least advantageous situation, the harmonic mean of the SINRs (HM-SINRs) was maximized in [22] under constraints on minimum range resolution and available bandwidth. The model-based and data-driven neural network-based approaches were proposed with the objective of maximizing the worst-case SINR under the constraints on range resolution and total bandwidth. But a key limitation of these papers is not considering BWCFOA for communications, which becomes an integral part of the distributed radar systems because nodes need to share the acquired target related information (TRI) with each other or with the central controller via communications channels. Therefore, [24] extended [22, 23] to BWCFOA for jointly supporting radar and communications in a distributed network. The objective in [24] was to maximize the minimum signal-to-clutter-and-noise ratio (SCNR), while ensuring constraints on each node's minimum range resolution and channel capacity, and the total available bandwidth. However, [24] considered the fixed allocation scheme. In general, the BWCFOA is an intertwined problem of finding optimum bandwidths and carrier frequencies, and then determining optimal assignment between carrier frequencies and nodes. The problem can be solved by solving the bandwidth and carrier frequency optimization (BWCFO) for all possible combinations of node-bandwidth assignment and then finding the combination that gives the best performance. This means that, for a network with Q nodes, we need to solve BWCFO problem for $Q!$ combinations. The computational complexity of this approach increases exponentially when Q increases and can be computationally prohibitive even for a moderate value of Q. To overcome this issue, joint BWCFOA and radar waveform optimization for radar and communications was proposed in [25], which aimed to maximize the harmonic-mean of the channel capacities under constraints on signal-to-clutter-and-interference-plus-noise ratio (SCINR) and the minimum range resolution requirement for each node. For a given bandwidth allocation, the joint radar and communications BWCFO problem was solved as a geometric programming (GP) problem [26], and for the given radar and communications bandwidths and carrier frequencies, the allocation problem was efficiently solved as a mixed-integer convex programming (MICP) problem, leveraging McCormick relaxation technique [27]. The latter approach avoided the exponential increase in computational complexity when the number of nodes increased.

In this chapter, we consider the deep learning-based data-driven approaches for joint radar and communications BWCFOA in a distributed radar network.

Considering that DNwoCC has stricter synchronization requirements than a distributed network with the CC, this chapter focuses on the network with a central controller. The main motivation is to overcome the limitations of conventional model-based optimization methods, which are subject to suboptimal performance when assumptions on target dynamics and clutter statistics change. Moreover, the model-based approaches are computationally costly even for finding the suboptimal solutions when network size increases. However, model-based suboptimum solutions can be employed as labels while training deep neural network in an offline mode. The trained network can then be efficiently and effectively utilized for determining optimum radar and communications bandwidths and carrier frequencies, and their allocations to the nodes. In this context, we first train a long-short-term memory (LSTM) network [28–30] for a distributed radar system consisting of distributed nodes, a CC, a target, and a moving interferer. The LSTM network is proposed because it has a capability to learn from the time dependency in the sequence of locations, especially under conditions that the interferer locations follow certain trajectories and are not independent in practice. The input to the LSTM is an assumed trajectory of the moving interferer, i.e., interferer locations in a given time frame, and the labels are the solutions of the model-based BWCFOA method. During the test phase, the trained LSTM network outputs the bandwidths and their allocations to nodes for a given interferer location. The proposed supervised learning approach shows that the online runtime complexity of the optimization problem can be significantly reduced. We then propose an alternative approach for radar bandwidth allocation to nodes in a distributed network consisting of multiple nodes, a target, and multiple emitters or interferers. Given that the available bandwidth is divided into an equal number of subbands and the nodes (considering inter-node interference) and interferers can occupy any of these subbands, the objective is to determine the optimum allocation of bandwidths to nodes. In other words, the problem is to determine the subbands to which nodes should be assigned. In order to solve this combinatorial optimization problem, we consider the framework of reinforcement learning (RL) [31, 32], in which an agent (node) takes an action (i.e., chooses a subband), learns from the environment, and continues modifying its next action until a final goal is achieved. The overall objective is to find the optimum set of actions that leads to the maximum average reward. Considering that the computational complexity of Q-learning-based RL for a large-size network is prohibitive, we propose deep-Q neural network (DQNN), which uses deep neural network and can output Q-function values for the given state-action pairs, avoiding a need to tabulate such values for all possible states in a Q-learning framework. We then compare the performance and complexity tradeoff between the proposed DQNN and the exhaustive search method.

The remainder of this chapter is organized as follows. The model-based and data-driven approaches for joint radar and communications BWCFOA are presented in Section 11.2. The radar bandwidth allocation for a distributed radar system with DQNN approach is described in Section 11.3. The conclusions are drawn in Section 11.4.

Notations: Upper (lower) bold face letters will be used for matrices (vectors); $(\cdot)^{\mathrm{T}}$, $(\cdot)^{\mathrm{H}}$, and $\odot$, respectively, denote the vector transpose, hermitian transpose, and element-wise multiplication, respectively. $\mathbb{C}^{n\times n}/\mathbb{R}^{n\times n}$ denotes spaces of $n \times n$ complex/real entries and $\mathbb{E}\{\boldsymbol{x}\}$ denotes the expectation of $\boldsymbol{x}$.

11.2 Model-based and Data-driven Approaches for BWCFOA

This section deals with the joint radar and communications BWCFOA in a distributed radar network. Both model-based iterative alternating optimization and data-driven deep learning-based methodologies are discussed in detail and then compared in terms of performance and complexity. In the following subsection, we first review the system model for the distributed radar network and the corresponding performance metrics.

11.2.1 System Model for Distributed Radar with Communications

Consider a distributed radar network with Q nodes and a CC as shown in Figure 11.1. The total available bandwidth B is shared among nodes and is given by $B = B_U - B_L$, where B_U and B_L, are respectively, the upper and lower edge frequencies. Initially, all nodes are allocated with equal but non-overlapping and contiguous bandwidth of $\frac{B}{Q}$. This bandwidth for each node is then equally divided into radar and communications bands, where the radar band is used by each node for acquiring TRI and the adjacent communications band is used for sharing the acquired TRI, for example, target range, with the CC. During the monostatic radar operation phase, the qth node uses the allocated equal bandwidth and the most recently optimized radar waveform to acquire the target range, which is denoted by R_q. The nodes then send their range estimates, $\{\hat{R}_q\}_{q=1}^{Q}$, to the CC during communications phase. The CC decodes $\{\hat{R}_q\}_{q=1}^{Q}$ and utilizes the resulting estimates and the outcome of waveform optimization to solve the joint radar and communications BWCFOA problem. The optimization algorithm assumes that each node is allocated with the bandwidth B_q, which is then divided into radar bandwidth, $B_{r,q}$, and communications bandwidth, $B_{c,q}$, such that $B_q = B_{r,q} + B_{c,q}$. As shown

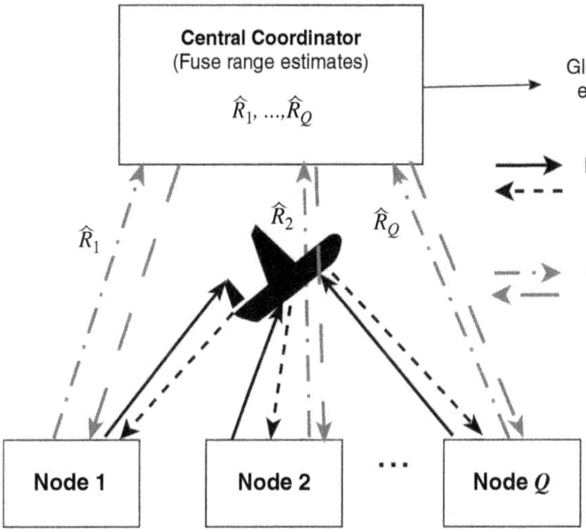

Figure 11.1 Distributed radar network with a CC.

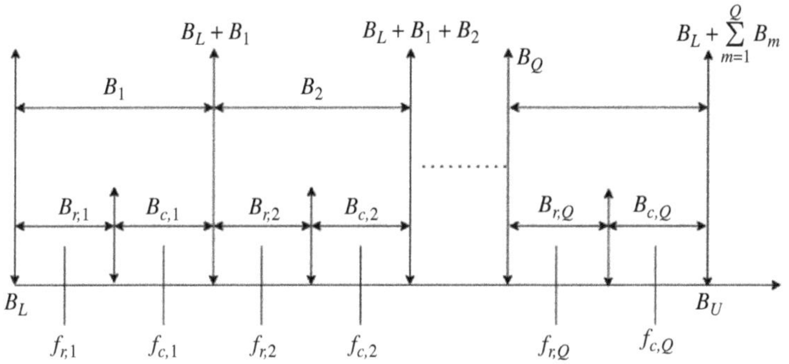

Figure 11.2 Bandwidth and carrier frequency relation.

in Figure 11.2, it is also assumed that the radar and communications carrier frequencies, $f_{r,q}$ and $f_{c,q}$, are at the center of the corresponding bands. After solving the BWCFOA problem, the CC communicates the nodes with the optimized bandwidths and carrier frequencies, which will then be used by each node to refine its target range estimate.

At each node, the target reflected radar signal is down-converted, sampled, and delay-Doppler compensated. The resulting L samples of the received radar signal at the qth node can be expressed as:

$$\boldsymbol{y}_q = \beta_{t,q}\boldsymbol{x}_q + \boldsymbol{c}_q + \boldsymbol{z}_q + \boldsymbol{n}_q, \tag{11.1}$$

where $\boldsymbol{x}_q = [x_{q,1}, x_{q,2}, \dots, x_{q,L}]^{\mathrm{T}}$ is the discrete form of the radar waveform used by the qth node. The waveform satisfies the unit-modulus constraint, i.e., $|x_{q,l}| = 1, \forall l = 1, \dots L$. $\boldsymbol{c}_q \in \mathbb{C}^L$ denotes clutter as seen by the qth node, $\boldsymbol{z}_q \in \mathbb{C}^L$ is the signal from an interferer, and $\boldsymbol{n}_k \in \mathbb{C}^L$ is complex-valued additive white Gaussian noise with zero-mean and covariance matrix $\sigma_{r,q}^2 \boldsymbol{I}_L$. The effects of both signal attenuation and target radar cross section (RCS) in a monostatic radar configuration are included into $\beta_{t,q}$, which according to radar range equation [1], can be expressed as:

$$\beta_{t,q} = \sqrt{\frac{P_{T,q} G_{T,q} G_{R,q} \sigma_{t,q} c^2}{f_{r,q}^2 (4\pi)^3 R_q^4}} = \sqrt{\frac{\Gamma_{r,q}}{f_{r,q}^2}}, \tag{11.2}$$

where $\Gamma_{r,q} \triangleq \frac{P_{T,q} G_{T,q} G_{R,q} \sigma_{t,q} c^2}{(4\pi)^3 R_q^4}$, $P_{T,q}$ is the radar's transmit power, $G_{T,q}$ is the transmit antenna gain, $G_{R,q}$ is the receive antenna gain, $\sigma_{t,q}$ is the target RCS as observed by the qth node, R_q is the monostatic target range, and $c = 3 \times 10^8$ m/s is the speed of light. After match filtering $\boldsymbol{y}_q$ with $\boldsymbol{x}_q$ and noting that $\boldsymbol{x}_q^{\mathrm{H}} \boldsymbol{x}_q = L$, the SCINR at the qth node can be expressed as:

$$\mathrm{SCINR}_q = \frac{\beta_{t,q}^2 L^2}{\beta_{c,q}^2 \boldsymbol{x}_q^{\mathrm{H}} \boldsymbol{R}_{c,q} \boldsymbol{x}_q + L(\sigma_{z,q}^2 + \sigma_{r,q}^2)}, \tag{11.3}$$

where $\boldsymbol{R}_{c,q} = \mathbb{E}\{\boldsymbol{c}_q \boldsymbol{c}_q^{\mathrm{H}}\}$ is the clutter covariance matrix, $\beta_{c,q}^2$ is the clutter power observed by the qth node, and $\sigma_{z,q}^2$ is the interference power. The exact modeling of $\beta_{c,q}^2$ depends on the type of clutter and radar application scenario. In this chapter, we consider the case of a surface clutter [33], and therefore $\beta_{c,q}$ can be expressed as:

$$\beta_{c,q} = \sqrt{\frac{P_{T,q} G_{T,q} G_{R,q} \bar{\sigma}_{c,q} c^2}{f_{r,q}^2 (4\pi)^3 R_q^4}} = \sqrt{\frac{\Gamma_{c,q} \bar{\sigma}_{c,q}}{f_{r,q}^2}}, \tag{11.4}$$

where $\Gamma_{c,q} \triangleq \frac{P_{T,q} G_{T,q} G_{R,q} c^2}{(4\pi)^3 R_q^4}$ and $\bar{\sigma}_{c,q}$ is the clutter RCS. It can be expressed as [33]:

$$\bar{\sigma}_{c,q} = \frac{\sigma_{c,q}^0 R_q \theta_{3,q} \sec(\Psi_{g,q})}{2B_{r,q}}, \tag{11.5}$$

where $\Psi_{g,q}$ is the grazing angle, $\theta_{3,q}$ is the 3dB beamwidth (azimuth angle), and $\sigma_{c,q}^0$ is the clutter scattering coefficient. Substituting $\bar{\sigma}_{c,q}$ from (11.5) into (11.4), $\beta_{c,q}$ can be rewritten as:

$$\beta_{c,q} = \sqrt{\frac{\Gamma_{ac,q}}{B_{r,q} f_{r,q}^2}}, \text{ with } \Gamma_{ac,q} \triangleq \frac{\Gamma_{c,q} \sigma_{c,q}^0 R_q \theta_{3,q} \sec(\Psi_{g,q})}{2}.$$

The interference power observed by the qth node within the radar bandwidth of $B_{r,q}$ can be expressed as:

$$\sigma_{z,q}^2 = \frac{\bar{P}_I B_{r,q} G_I G_{R,q} c^2}{\left(4\pi d_{I,q}\right)^2 f_{r,q}^2} \triangleq \bar{\Gamma}_{I,q} \frac{B_{r,q}}{f_{r,q}^2}, \tag{11.6}$$

where $\bar{\Gamma}_{I,q} = \dfrac{\bar{P}_I G_I G_{R,q} c^2}{\left(4\pi d_{I,q}\right)^2}$, and $\bar{P}_I$ and G_I, respectively, denote the power spectral density (PSD) and transmit antenna gain of the interferer. $d_{I,q}$ denotes the distance between the interferer and the qth node. We consider a worst-case scenario in which the interferer's carrier frequency in a given bandwidth overlaps with the node's bandwidth. Therefore, $f_{r,q}$ is also the carrier frequency of the interferer occupying the bandwidth $B_{r,q}$. The noise variance for the radar bandwidth can be expressed as $\sigma_{r,q}^2 = k_B T_0 F_q B_{r,q}$, where $k_B = 1.38 \times 10^{-23}$ J/Kelvins is the Boltzmann constant, T_0 is the receiver noise temperature in degree Kelvins, and F_q is the receiver noise figure. Hence, SCINR_q can be re-expressed as:

$$\text{SCINR}_q = \frac{\bar{\Gamma}_{r,q} B_{r,q}}{\Gamma_{ac,q} \boldsymbol{x}_q^{\text{H}} \boldsymbol{R}_{c,q} \boldsymbol{x}_q + \tilde{\Gamma}_{I,q} B_{r,q}^2 + \bar{F}_q B_{r,q}^2 f_{r,q}^2}, \tag{11.7}$$

where $\bar{\Gamma}_{r,q} = L^2 \Gamma_{r,q}$, $\bar{F}_q = k_B T_0 F_q L$, and $\tilde{\Gamma}_{I,q} = L \bar{\Gamma}_{I,q}$.

The quality of the communications link between each node and the CC is measured in terms of achievable capacity in bits/sec, which can be expressed as:

$$C_q = B_{c,q} \log_2 \left(1 + \frac{P_C G_C G_{R,q}}{\left(\frac{4\pi d_q}{\lambda_{c,q}}\right)^2 \left(\bar{\sigma}_{z,q}^2 + \sigma_{c,q}^2\right)} \right), \tag{11.8}$$

where P_C and G_C denote the transmit power and antenna gains of the CC, respectively. The distance between the CC and the qth node is denoted by d_q and $\lambda_{c,q}$ is wavelength of the carrier over which communications between the CC and the qth node takes place. The interference power observed by the node within the communications bandwidth of $B_{c,q}$ is given by $\bar{\sigma}_{z,q}^2$, and the corresponding noise variance is given by $\sigma_{c,q}^2 = k_B T_0 F_q B_{c,q}$. Under similar assumptions as in the case of $\sigma_{z,q}^2$, $\bar{\sigma}_{z,q}^2$ can be expressed as $\bar{\sigma}_{z,q}^2 = \bar{\Gamma}_{I,q} \dfrac{B_{c,q}}{f_{c,q}^2}$. The qth node's channel capacity can be re-expressed in a concise form as:

$$C_q = B_{c,q} \log_2 \left(1 + \frac{\bar{\Gamma}_{cc,q}}{\bar{\Gamma}_{I,q} B_{c,q} + \frac{\bar{F}_q}{L} B_{c,q} f_{c,q}^2} \right), \tag{11.9}$$

where $\bar{\Gamma}_{cc,q} = \dfrac{P_C G_C G_{R,q} c^2}{(4\pi d_q)^2}$. It is worthwhile to emphasize that SCINR, as seen from (11.7), is a monotonically decreasing function of bandwidth and carrier frequency. But a monotonic increment of channel capacity with respect to communications bandwidth cannot be ascertained because the argument of the logarithm function in (11.9) decreases with both bandwidth and carrier frequency.

11.2.2 Model-based Optimization Approach and Results

The objective of the proposed optimization is to maximize the network-wide channel capacity while ensuring that the requirements on SCINR and range resolution are fulfilled for each node. With the motivation of giving priority to the nodes that observe large attenuation, i.e., the nodes that are far away from the CC, we propose maximizing the harmonic mean of the node channel capacities. The harmonic mean is dominated by the lowest channel capacity value. As such, the following optimization problem is proposed:

$$
\begin{aligned}
\max_{\{B_{r,q}, B_{c,q}, f_{r,q}, f_{c,q}, \mathbf{x}_q\}} \quad & \frac{Q}{\sum_{q=1}^{Q} \frac{1}{C_q}} \\
\text{s.t.} \quad & \text{SCINR}_q \geq \gamma_{th}, \forall q, \\
& B_{r,q} \geq \frac{c}{2\bar{R}_q}, \forall q, \\
& \sum_{q=1}^{Q} (B_{r,q} + B_{c,q}) \leq B, \\
& |x_{q,l}| = 1, \forall q, l = 1, \dots, L,
\end{aligned}
\tag{11.10}
$$

where the first constraint ensures that each node achieves SCINR above the threshold value γ_{th}. The second constraint ensures that each node achieves range resolution smaller than $\bar{R}_q$ in meters. The third constraint ensures that the sum of the radar and communications bandwidths of all nodes is less than the total available bandwidth. The final constraint enforces that the radar waveform satisfies unit-modulus constraint. However, this problem optimizes and determines bandwidths, carrier frequencies, and waveforms for a given allocation scheme. To jointly optimize the resources and their allocation, we introduce allocation vectors, $\{\eta_q\}_{q=1}^{Q}$, to the optimization problem (11.10). Each allocation vector consists of only ones and zeros. Thus, the SCINR and channel capacity can be, respectively, expressed as:

$$\text{SCINR}_q = \frac{\bar{\Gamma}_{r,q}\boldsymbol{\eta}_q^{\mathrm{T}}\boldsymbol{b}_r}{\Gamma_{ac,q}\boldsymbol{x}_q^{\mathrm{H}}\boldsymbol{R}_{c,q}\boldsymbol{x}_q + \boldsymbol{\eta}_q^{\mathrm{T}}(\bar{\Gamma}_{I,q}\bar{\boldsymbol{b}}_r + \bar{F}_q\tilde{\boldsymbol{b}}_r)},$$

$$C_q = \boldsymbol{\eta}_q^{\mathrm{T}}\boldsymbol{b}_c \log_2\left(1 + \frac{\bar{\Gamma}_{cc,q}}{\boldsymbol{\eta}_q^{\mathrm{T}}\left(\bar{\Gamma}_{I,q}\boldsymbol{b}_c + \frac{\bar{F}_q}{L}\tilde{\boldsymbol{b}}_c\right)}\right), \tag{11.11}$$

where

$$\boldsymbol{b}_r \triangleq [B_{r,1}, \dots, B_{r,Q}]^{\mathrm{T}},\ \tilde{\boldsymbol{b}}_r \triangleq [B_{r,1}^2 f_{r,1}^2, \dots, B_{r,Q}^2 f_{r,Q}^2]^{\mathrm{T}},$$
$$\boldsymbol{b}_c \triangleq [B_{c,1}, \dots, B_{c,Q}]^{\mathrm{T}},\ \tilde{\boldsymbol{b}}_c \triangleq [B_{c,1} f_{c,1}^2, \dots, B_{c,Q} f_{c,Q}^2]^{\mathrm{T}},$$

and $\bar{\boldsymbol{b}}_r = \boldsymbol{b}_r \odot \boldsymbol{b}_r$. Note that only a single element of $\boldsymbol{\eta}_q$ can take the value of one. Introducing new auxiliary variables, $\{t_q\}_{q=1}^{Q}$, the joint resource optimization and allocation can be expressed as:

$$\begin{aligned}
\min_{\{\boldsymbol{\eta}_q \in [0,1], B_{r,q}, B_{c,q}, f_{r,q}, f_{c,q}, t_q, \boldsymbol{x}_q\}} \quad & \sum_{q=1}^{Q} t_q^{-1} \\
\text{s.t.} \quad & C_q \geq t_q, \forall q,\ \text{SCINR}_q \geq \gamma_{th}, \forall q, \\
& \boldsymbol{\eta}_q^{\mathrm{T}}\boldsymbol{b}_r \geq \frac{c}{2\bar{R}_q}, \forall q,\ \sum_{q=1}^{Q}(\boldsymbol{\eta}_q^{\mathrm{T}}\boldsymbol{b}_r + \boldsymbol{\eta}_q^{\mathrm{T}}\boldsymbol{b}_c) \leq B, \\
& |x_{q,l}| = 1, \forall q, l = 1, \dots, L, \\
& \sum_{q=1}^{Q} \boldsymbol{A}(:, q) = 1,\ \sum_{q=1}^{Q} \boldsymbol{A}(q, :) = 1,
\end{aligned} \tag{11.12}$$

where $\boldsymbol{A} = [\boldsymbol{\eta}_1, \dots \boldsymbol{\eta}_Q]$ is a $Q \times Q$ matrix of allocation vectors, and the last two constraints ensure that only one element of $\boldsymbol{\eta}_q$ is one and a given bandwidth is allocated to only one node. This is a combinatorial non-convex optimization problem with both continuous and binary variables. Therefore, its global optimum solution cannot be guaranteed with polynomial computational complexity. It can be observed from (11.12) and the expression for SCINR that $u_q \triangleq \boldsymbol{x}_q^{\mathrm{H}}\boldsymbol{R}_{c,q}\boldsymbol{x}_q$ is the only variable that depends on $\boldsymbol{x}_q$, although the overall BWCFOA problem depends on u_q. It can be readily observed that the SCINR in (11.12) can be maximized with the minimum value of u_q. Therefore, we can first optimize $\boldsymbol{x}_q$ under unit-modulus constraints and obtain the minimum value of u_q (denoted by $\bar{u}_q$), which can then be used for solving the remaining BWCFOA in (11.12). Consequently, this means that (11.12) can be decoupled into two problems: waveform optimization and BWCFOA. Although the waveform optimization problem itself is non-convex due to constant modulus constraints, as shown in [24], it can be efficiently solved with Riemannian manifold optimization approach [34, 35]. Moreover, assuming that $\{f_{r,q}, f_{c,q}\}_{q=1}^{Q}$ are in the ascending order (as shown in Figure 11.2), the carrier frequencies can be expressed in terms of bandwidths as:

$$\begin{aligned} f_{r,q} &= B_L + \frac{B_{r,q}}{2} + \sum_{m=1}^{q-1} (B_{r,m} + B_{c,m}), \forall q, \\ f_{c,q} &= B_L + \frac{B_{c,q}}{2} + \sum_{m=1}^{q} B_{r,m} + \sum_{m=1}^{q-1} B_{c,m}, \forall q. \end{aligned} \tag{11.13}$$

The remaining BWCFOA problem in (11.12) can be solved in terms of $B_{r,q}$ and $B_{c,q}$ using (11.13). The resulting values of $B_{r,q}$ and $B_{c,q}$ can then be used for calculating the carrier frequencies. Therefore, the BWCFOA problem in (11.12) can be divided into two sub-problems: a) optimizing bandwidths for a given allocation, i.e., BWCFO and b) optimizing allocation for the given bandwidths, i.e., BWCFA. For the given $\{\eta\}_{q=1}^{Q}$ and $\{\bar{u}_q\}$, the objective function and the constraints of the optimization problem (11.12) satisfy the requirements of a standard GP problem [26], except the constraint, $C_q \geq t_q$, which can be approximated as a convex constraint as shown in [24]. The GP problem can be efficiently solved with the optimization toolbox CVX [36]. For the given bandwidths, the BWCFA problem can be reformulated as a MICP problem after multiple reformulation steps and the application of McCormick relaxation technique that enables efficient approximation of the bilinear non-convex constraint, such as $x_1 x_2 \leq c$, where x_1 and x_2 are optimization variables and c is some constant [25]. Such MICP problem can also be solved by calling MOSEK package from the CVX toolbox [36]. Finally, the BWCFO and BWCFA problems can be iteratively solved until some convergence accuracy is achieved for obtaining the solutions of BWCFOA problem (11.12).

As shown in Figure 11.3, we consider a simulation scenario with a single target, four nodes, and a moving interferer. Nodes 1–4 are located at $[0, 0, 1]^{\mathrm{T}}$ m, $[80, 0, 1]^{\mathrm{T}}$ m, $[80, 80, 1]^{\mathrm{T}}$ m, and $[0, 80, 1]^{\mathrm{T}}$ m, respectively. The stationary target and the CC are, respectively, located at $[40, 40, 1]^{\mathrm{T}}$ m and $[20, 40, 1]^{\mathrm{T}}$ m. The interferer is initially located at $[0, 40, 1]^{\mathrm{T}}$ m, which follows linear trajectories to reach locations $[40, 80, 1]^{\mathrm{T}}$ m, $[80, 40, 1]^{\mathrm{T}}$ m, $[40, 0, 1]^{\mathrm{T}}$ m, and finally returns to the original position $[0, 40, 1]^{\mathrm{T}}$ m. In each step of the interferer's movement, the x- and y-coordinates change by ± 1 m. We take $B_L = 2.5$ GHz and $B_U = 4.5$ GHz (i.e., $B = 2$ GHz). For all $q = 1, \ldots, Q$, we set $\bar{R}_q = 5$ m, $P_{T,q} = 10$ dBW, $G_{T,q} = G_{R,q} = 10$ dB, $F_q = 5$ dB, and $\sigma_{t,q} = 10\ m^2$. The clutter parameters are chosen as follows: $\Psi_{g,q} = 20°$, $\theta_{3,q} = 0.02$ radians, $\sigma^0_{c,q} = 0.0135$, and $\sigma_{c,q} = 0.1\ m^2$ for all $q \in [1, \ldots, Q]$. The CC transmit power and transmit antenna gains are set to $P_C = 10$ dBW and $G_C = 10$ dB, respectively. The interferer PSD is set to -97 dBW/Hz, and it is set to be active on bands 2 (communications band), 3 (radar band), 4 (communications band), and 5 (radar band). Throughout the simulations, we compare

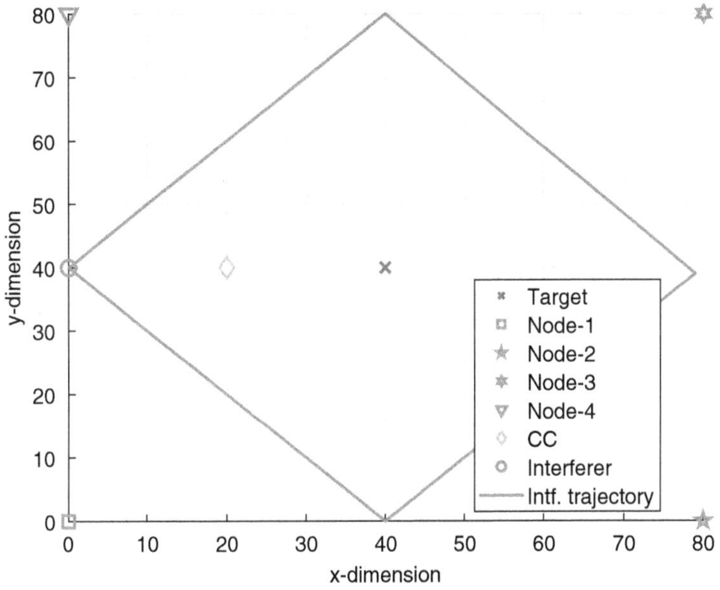

Figure 11.3 Network scenario with moving interferer.

the following three methods: a) the proposed waveform optimization plus BWCFOA (WO+BWCFOA), b) waveform optimization with equal bandwidth allocation (WO+EBWA), and c) random waveform plus BWCFOA (RW+BWCFOA). In the case of RW+BWCFOA, the waveform is not optimized, but a waveform that satisfies unit-modulus constraint is randomly chosen for each node. The SCINR for each node is then calculated by averaging over 10^4 values of u_q. In the simulations, the network-wide radar utility function is defined as $U_R = \sum_{q=1}^{Q} \log(1 + \text{SCINR}_q)$ and the sum channel capacity is defined as $U_C = \sum_{q=1}^{Q} C_q$.

The individual SCINRs obtained with the proposed WO+BWCFOA algorithm are compared with those obtained from the WO+EBWA algorithm in Figure 11.4. It can be observed from this figure that the SCINRs of nodes 1 and 4, for both methods, do not change when the interferer location changes, whereas those of nodes 2 and 3 fluctuate. This is due to our simulation setup in which interference is active on the radar bands of only nodes 2 and 3. With the WO+EBWA algorithm, nodes 2 and 3 observe much smaller SCINRs than with the proposed algorithm. But nodes 1 and 4, which already observe high values of SCINRs, benefit more from the WO+EBWA approach than the proposed WO+BWCFOA algorithm. This shows that the proposed method protects nodes that are in a disadvantageous situation in a given network scenario.

In Figure 11.5, we compare the proposed WO+BWCFOA algorithm with RW+BWCFOA in terms of the SCINRs achieved by all nodes. The SCINRs of the

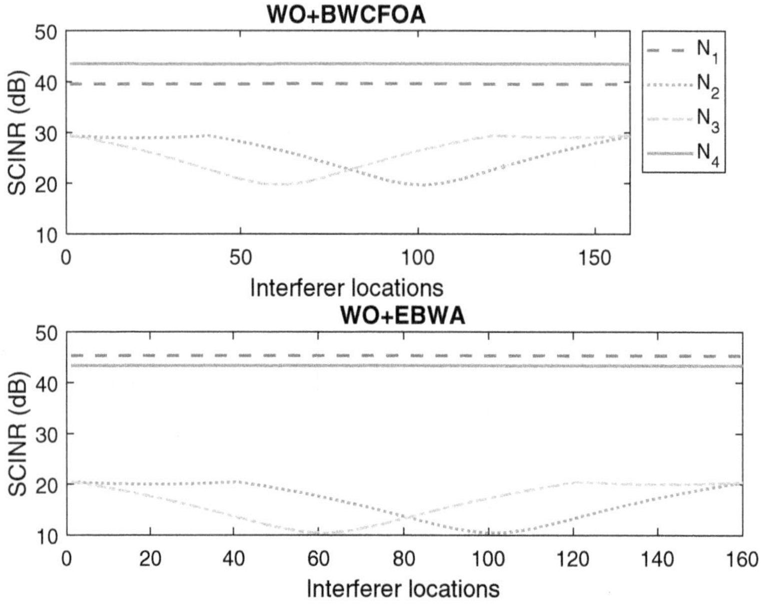

Figure 11.4 Comparison of the proposed WO+BWCFOA with WO+EBWA.

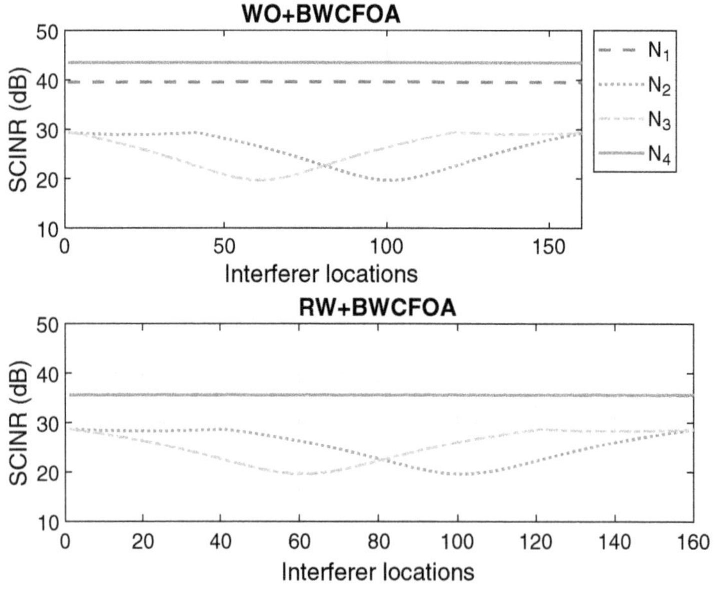

Figure 11.5 Comparison of the proposed WO+BWCFOA with RW+BWCFOA.

nodes 1 and 4 do not depend on the interferer location, whereas those of nodes 2 and 3 fluctuate with the change in interferer location. It can also be observed from this figure that the proposed algorithm ensures higher SCINR values for all nodes than the RW+BWCFOA.

The change in network-wide radar utility function value, U_R, when interferer location changes is shown in Figure 11.6. It can be observed from this figure that the proposed method gives significantly larger radar utility function than the other two methods, showing the superiority of the proposed method in the considered network scenario.

In Figure 11.7, the proposed WO+BWCFOA method is compared with WO+EBWA method in terms of the achieved network-wide sum channel capacity, U_C. It can be observed from this figure that the proposed method significantly outperforms WO+EBWA method in terms of communications performance.

11.2.3 Data-driven Optimization and Allocation with LSTM and Results

Although the iterative BWCFOA problem can be solved efficiently in polynomial time for a given network configuration, for example, for the given target and/or interferer location, the overall computational complexity of such an iterative method can be prohibitive if the target or interferer moves and the solutions of the BWCFOA need to be updated for an entire duration of the target's or interferer's trajectory. The computational complexity of the iterative optimization can be confined to an offline mode, and consequently, the run-time complexity of the bandwidth and carrier frequency allocation can be further reduced with

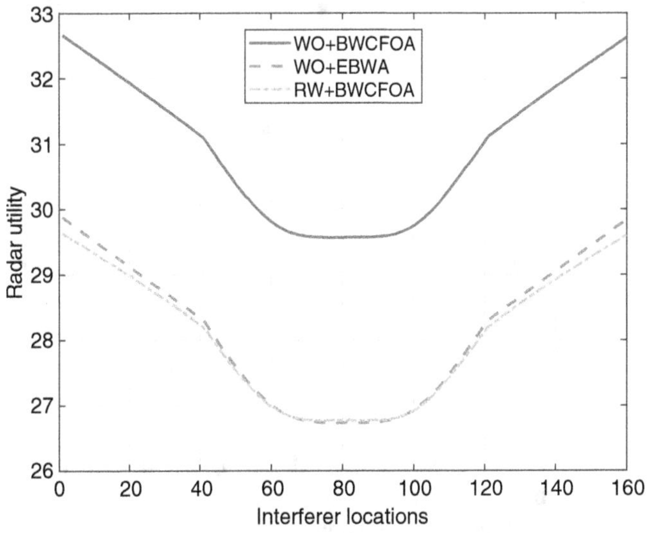

Figure 11.6 Radar network utility versus interferer locations.

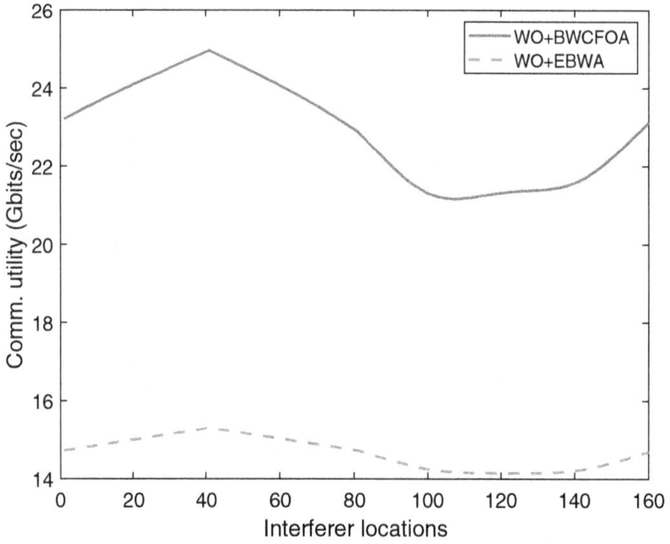

Figure 11.7 Network-wide channel capacity versus interferer locations.

the help of supervised deep learning approach. The key idea can be clearly and concisely discussed with a specific scenario in which interferer moves around the network, causing different levels of interference to the nodes. In such a scenario, we can train a deep neural network with interferer locations corresponding to a certain trajectory as inputs, and bandwidths and corresponding allocations to nodes obtained after solving (11.12) as labels. Typically, the interferer location at a given time instant may not be completely independent of its locations at previous time instances. As such, we propose employing recursive neural network (RNN) that has a capability to learn from temporal dependency in a data sequence. Considering that the RNNs often suffer from the vanishing gradient problem, we propose LSTM network [28], which, although being a type of RNN, avoids such a problem. One of the key differences between the RNN and LSTM is that the hidden layer processing of RNN is replaced by a memory cell unit in LSTM. The level of information flow to and out of this memory cell can be controlled with *forget*, *input*, and *output* gates. Each gate consists of a sigmoid (σ) neural network layer followed by an element-wise multiplication operator ($\odot$). The forget gate determines information that will be discarded from the cell state, and the input gate determines new information that is going to be stored in the cell. The output gate determines the new state.

Let the input signal vector at a time instance t be denoted by $\boldsymbol{x}_t \in \mathbb{R}^d$, where d is the number of features. Let the hidden state vector at the previous time instance $t-1$, which is also the output of the LSTM, be denoted by $\boldsymbol{h}_{t-1} \in \mathbb{R}^h$, where h is the number of hidden states. Let us denote the current and previous cell states by $\boldsymbol{c}_t$ and $\boldsymbol{c}_{t-1}$, respectively. Therefore, the internal operations of the LSTM can be described with the following equations [23]:

$$\begin{aligned}
\boldsymbol{f}_t &= \sigma\left(\boldsymbol{V}_\mathrm{f}\left[\boldsymbol{h}_{t-1}^\mathrm{T}, \boldsymbol{x}_t^\mathrm{T}\right]^\mathrm{T} + \boldsymbol{b}_\mathrm{f}\right),\\
\boldsymbol{i}_t &= \sigma\left(\boldsymbol{V}_\mathrm{i}\left[\boldsymbol{h}_{t-1}^\mathrm{T}, \boldsymbol{x}_t^\mathrm{T}\right]^\mathrm{T} + \boldsymbol{b}_\mathrm{i}\right),\\
\boldsymbol{o}_t &= \sigma\left(\boldsymbol{V}_\mathrm{o}\left[\boldsymbol{h}_{t-1}^\mathrm{T}, \boldsymbol{x}_t^\mathrm{T}\right]^\mathrm{T} + \boldsymbol{b}_\mathrm{o}\right),\\
\boldsymbol{c}_t &= \boldsymbol{f}_t \odot \boldsymbol{c}_{t-1} + \boldsymbol{i}_t \odot \tilde{\boldsymbol{c}}_t,\\
\tilde{\boldsymbol{c}}_t &= \tanh\left(\boldsymbol{V}_\mathrm{c}\left[\boldsymbol{h}_{t-1}^\mathrm{T}, \boldsymbol{x}_t^\mathrm{T}\right]^\mathrm{T} + \boldsymbol{b}_\mathrm{c}\right),\\
\boldsymbol{h}_t &= \boldsymbol{o}_t \odot \tanh(\boldsymbol{c}_t),
\end{aligned} \tag{11.14}$$

where $\boldsymbol{V}_\mathrm{f}, \boldsymbol{V}_\mathrm{i}, \boldsymbol{V}_\mathrm{o}$, and $\boldsymbol{V}_\mathrm{c}$, respectively, denote $h \times (d + h)$ weight matrices associated with forget, input and output gates, and the cell state. The corresponding bias vectors are denoted by $\boldsymbol{b}_\mathrm{f}, \boldsymbol{b}_\mathrm{i}, \boldsymbol{b}_\mathrm{o}$, and $\boldsymbol{b}_\mathrm{c}$, each having a length h. $\{\boldsymbol{f}_t, \boldsymbol{i}_t, \boldsymbol{o}_t\} \in (0, 1)^h$ represent activation vectors of the forget, input, and output gates, respectively. $\boldsymbol{c}_t \in \mathbb{R}^h$ and $\tilde{\boldsymbol{c}}_t \in (-1, 1)^h$ denote cell state vector and cell input activation vector, respectively.

The training and test data are generated employing the scene as shown in Figure 11.3. We made an approximation that the harmonic mean of the channel capacities of nodes in (11.12) is a monotonically increasing function of the communications bandwidth. Because the target associated SCINR at each node is a monotonically decreasing function of radar bandwidths, it can be readily shown that the optimum values of radar bandwidths are $B_{r,q} = \frac{c}{2\bar{R}_q}, \forall q$. The remaining available bandwidth for communications is given by $B - \sum_{q=1}^{Q} B_{r,q}$. Because each node is assigned adjacent radar and communications bands, the remaining BWCFOA problem in (11.12) reduces to the optimization of communications bandwidths, $\{B_{c,q}\}$, and the allocation vectors, $\{\boldsymbol{\eta}_q\}$. Let P be the number of interferer locations, each denoted by $\boldsymbol{v}_p$, where $p = 1, \dots, P$, for the interferer trajectory considered in Figure 11.3. Out of P interferer locations, 80% are randomly chosen for generating training dataset, whereas the remaining are used for generating test dataset. Let $\bar{P}$ and $\tilde{P}$ be the numbers of interferer locations in the training and test datasets, respectively, such that $P = \bar{P} + \tilde{P}$. Similarly, let $\bar{\mathcal{P}}$ and $\tilde{\mathcal{P}}$ be the sets of interferer location indexes chosen for the training and test datasets, respectively. For the interferer locations in the training set, i.e., for $\{\boldsymbol{v}_{\bar{p}}\}_{\bar{p}\in\bar{\mathcal{P}}}$, we solve the optimization problem (11.12) to find the communications bandwidths, $\{B_{c,q,\bar{p}}\}_{q=1,\bar{p}\in\bar{\mathcal{P}}}^{Q}$ and allocation vectors, $\{\boldsymbol{\eta}_{q,\bar{p}}\}_{q=1,\bar{p}\in\bar{\mathcal{P}}}^{Q}$. These bandwidths and allocation vectors are the labels for our supervised learning method and $\{\boldsymbol{v}_{\bar{p}}\}_{\bar{p}\in\bar{\mathcal{P}}}$ are the input. The input and the labels are pre-processed for training the LSTM network. The pre-processing is defined as follows: $\boldsymbol{X}_{\text{norm}} = \dfrac{\boldsymbol{X} - \min(\boldsymbol{X})}{\max(\boldsymbol{X}) - \min(\boldsymbol{X})}$, where $\boldsymbol{X}$ denotes the data matrix. To form the test set, the same normalization is carried out for $\left\{\boldsymbol{q}_{\tilde{p}}\right\}_{\tilde{p}\in\tilde{\mathcal{P}}}$ and $\{B_{c,q,\tilde{p}}\}_{q=1,\tilde{p}\in\tilde{\mathcal{P}}}^{Q}$, where $\boldsymbol{q}_{\tilde{p}}$ is the interferer's location from the test set. In the training setup, the number of features in the input data is 3, the number of LSTM responses is $2Q$ (Q responses for communications bandwidths and Q responses for allocations), and the data sequence length is $\bar{P}$. The LSTM network and the corresponding hyperparameters are as follows.

For the simulations, we select the network with 12 LSTM hidden layers and a fully connected regression output layer. The input training data is a $3 \times \bar{P}$ normalized matrix (with $\bar{P}$=128) formed from $\{\boldsymbol{v}_{\bar{p}}\}$, whereas the label is a $2Q \times \bar{P}$ normalized matrix formed from the solutions obtained from (11.12). The network is trained with the Adam algorithm, which uses the following parameters: gradient threshold value of 1, initial learning rate of 0.005, learning rate drop period of 125, learning drop factor of 0.25, and the maximum number of epochs of 500. The responses (communications bandwidths and allocations to four nodes) are predicted for a $2Q \times \tilde{P}$ normalized test data matrix (with $\tilde{P}$=32) formed from $\{\boldsymbol{q}_{\tilde{p}}\}$. The same radar network parameters, as mentioned at the beginning of this section, are used.

The normalized mean-square estimate (NMSE) of the predicted communications bandwidths versus $\tilde{P}$ interferer locations of the test dataset is shown in Figure 11.8 (top). The NMSE in this figure is defined as

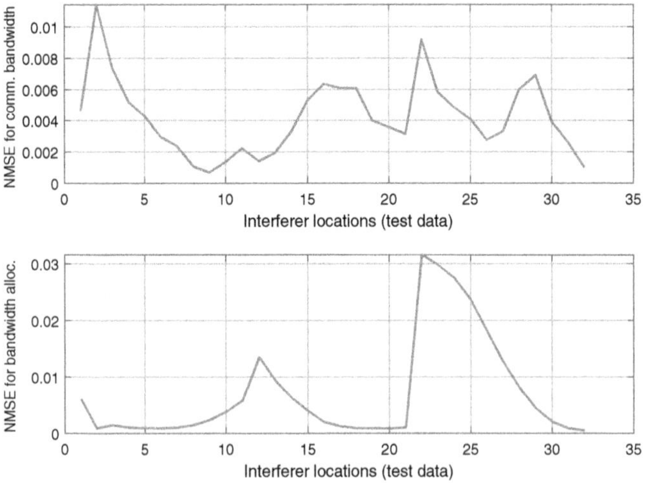

Figure 11.8 NMSE for comm. bandwidths and NMSE for allocation.

$e_1 = \frac{1}{Q}\sum_{q=1}^{Q}\frac{|B_{c,q,\tilde{p}} - \hat{B}_{c,q,\tilde{p}}|^2}{B^2_{c,q,\tilde{p}}}$, where $B_{c,q,\tilde{p}}$ are the actual bandwidths obtained from (11.12) and $\hat{B}_{c,q,\tilde{p}}$ are the corresponding predicted values with the LSTM network. It can be observed from this figure that the NMSE fluctuates depending on the interferer location, but, except for few outliers, NMSE values for most of the data points are below 6×10^{-3}. This shows that LSTM network predicts the bandwidths with reasonably good accuracy despite the fact that only a limited set of data, i.e., $\bar{P} = 128$, is used for training. Figure 11.8 (bottom) shows the NMSE of the predicted allocation versus $\tilde{P}$ interferer locations from the test dataset. In this case, the NMSE is defined as $e_2 = \frac{1}{Q}\sum_{q=1}^{Q}\frac{|\bar{a}_{q,\tilde{p}} - \hat{\bar{a}}_{q,\tilde{p}}|^2}{\bar{a}^2_{q,\tilde{p}}}$, where $\bar{a}_{q,\tilde{p}}$ are the actual allocations and $\hat{\bar{a}}_{q,\tilde{p}}$ are the corresponding estimated ones. Note that $\bar{a}_{q,\tilde{p}}$ is the position of the nonzero value (binary 1) on the qth row of $\boldsymbol{A}$ defined in (11.12), and therefore takes values between 1 and Q. As in the case of Figure 11.8 (top), the NMSE of the predicted allocation is less than 0.01 for most of the data. This result also indicates the good accuracy of the prediction, which is worthwhile to note when considering only limited data is used for training.

We also compare the optimization problem (11.12) and the proposed LSTM in terms of online runtime complexity. A laptop with 32 GB RAM and 1.30 GHz Intel (R) Core (TM) I7-1065G7 processor is used for solving (11.12) and estimation with LSTM for $\tilde{P}$ interferer locations in the test dataset. It requires more than 2.16 hours to solve (11.12), whereas the LSTM-based approach can solve the estimation problem within 0.5 seconds. This is a significant reduction in online runtime complexity, which shows the importance of deep learning-based approach in bandwidth and carrier frequency allocation.

11.3 Radar Bandwidth Allocation with DQNN

In this section, we propose bandwidth allocation for distributed radar nodes in the presence of both inter-node and external interference with DQNN-based reinforcement learning approach. Note the radar performance metrics, such as probability of detection for a given false alarm probability and range-velocity estimation error can be, respectively, maximized and minimized by maximizing the SINR at each node. However, this approach leads to a non-tractable multi-objective optimization problem, and therefore a network-wide performance metric is needed. Its typical examples are (a) radar utility function: $f_1 = \sum_{q=1}^{Q}\log(1 + \mathrm{SINR}_q)$, (b) sum SINR: $f_2 = \sum_{q=1}^{Q}\mathrm{SINR}_q$, and the (c) worst-case SINR: $f_3 = \max_q\{\mathrm{SINR}_q\}_{q=1}^{Q}$, where SINR_q is the SINR observed by the qth radar node. The first two performance metrics are dominated by the largest SINR value, which means that the maximization of f_1 and f_2 favors the node that is near from the target and/or far away from the interferer. On the other hand, maximization of the worst-case SINR unnecessarily emphasizes the nodes that are far away from the target and/or near the interferer, resulting in a reduction of the individual SINRs observed by the nodes that are near the target and/or far away from the interferer. Therefore, HM-SINRs is used as a network-wide performance metric. Despite being dominated by the lowest SINR value, harmonic mean's maximization can be considered as a compromise between f_1/f_2 and f_3. In this section, we first derive a harmonic mean SINR for the considered network configuration and discuss how its change from one action to another action can be transformed to a reward, which is key for the implementation of the DQNN. Here, action is defined as the set of bands assigned to the nodes.

11.3.1 System Model for Modeling Reward for DQNN

We consider a network with Q radars nodes, a single target, and N_I interferers or emitters. The total bandwidth, B, is divided into $Q + N_I$ subbands, each having the same bandwidth. The inter-node interference is taken into account and the external interferers can occupy any N_I subbands out of the $Q + N_I$ subbands. The main objective of the DQNN approach is to find the optimum bandwidth allocation for the radar nodes in this scenario with the interferers. Let B_L and B_U denote, respectively, the lower and upper band edge frequencies such that $B = B_U - B_L$. The bandwidth of each subband is given by:

$$\Delta B = \frac{B}{Q + N_I}. \tag{11.15}$$

The carrier frequency and the wavelength corresponding to the mth sub-band can be expressed as:

$$f_m = B_L + \frac{\Delta B}{2} + (m-1)\Delta B,\ \lambda_m = \frac{c}{f_m},\ m = 1, \dots, Q + N_I. \tag{11.16}$$

Let the 3-dimensional positions of the qth radar node and the target be, respectively, denoted by $[x_q, y_q, z_q]^{\mathrm{T}}$ and $[x_{\mathrm{t}}, y_{\mathrm{t}}, z_{\mathrm{t}}]^{\mathrm{T}}$, where x_q/x_{t}, y_q/y_{t}, and z_q/z_{t} represent the $x-$, $y-$, and $z-$ coordinates, respectively. The lth interferer location is denoted by $[x_{\mathrm{I},l}, y_{\mathrm{I},l}, z_{\mathrm{I},l}]^{\mathrm{T}}$. The distance between the target and the qth radar node, i.e., target range as seen by the qth node is given by:

$$R_q = \sqrt{(x_q - x_{\mathrm{t}})^2 + (y_q - y_{\mathrm{t}})^2 + (z_q - z_{\mathrm{t}})^2},\ q = 1, \dots, Q. \tag{11.17}$$

The distance between the qth and jth nodes is given by:

$$d_{q,j} = \sqrt{(x_q - x_j)^2 + (y_q - y_j)^2 + (z_q - z_j)^2}, \forall(q, j),\ q \neq j. \tag{11.18}$$

Similarly, the distance between the qth node the lth interferer is given by:

$$\bar{d}_{q,l} = \sqrt{(x_q - x_{\mathrm{I},l})^2 + (y_q - y_{\mathrm{I},l})^2 + (z_q - z_{\mathrm{I},l})^2}, \ \forall q, l. \tag{11.19}$$

Each radar operates in a mono-static mode. We assume that each node transmits the same power across all subbands, and the transmit and receive antenna gains of each node are same for all subbands. The target RCS is also considered to be invariant over the considered total bandwidth. The useful signal power received by the qth node on the mth subband can be expressed according to radar range equation as:

$$P_{m(q)} = \frac{P_{T,q} G_{T,q} G_{R,q} \sigma_{c,q} \lambda_m^2}{(4\pi)^3 R_q^4}, \tag{11.20}$$

where $m(q)$ indicates assignment of the qth node to the mth subband. $P_{T,q}$ is the transmit power, $G_{T,q}$ is the transmit antenna gain, and $G_{R,q}$ is the receive antenna gain, all corresponding to the qth node. $\sigma_{c,q}$ is the target RCS as seen by the qth node and λ_m is the wavelength corresponding to the mth subband. Let $\mathcal{Q}_m$ denote the set of other radar nodes that transmit on the same mth subband. The corresponding inter-node interference power observed at the qth node can be expressed as:

$$\bar{P}_{m(q)} = \sum_{j \neq q, j \in \mathcal{Q}_m} \frac{P_{T,j} G_{T,j} G_{R,q} \lambda_m^2}{(4\pi d_{q,j})^2}. \tag{11.21}$$

Similarly, let $\mathcal{N}_{I,m}$ denote the set of emitters that transmit on the same mth subband. The corresponding interference power at the qth node, received from the emitters, can be expressed as:

$$P^I_{m(q)} = \sum_{l \in \mathcal{N}_{I,m}} \frac{P_{I,l} G_{I,l} G_{R,q} \lambda_m^2}{(4\pi \bar{d}_{q,l})^2}, \tag{11.22}$$

where $P_{I,l}$ and $G_{I,l}$ denote, respectively, the transmit power and transmit antenna gain of the lth emitter. Consider an action in which the qth radar node uses the mth subband, which is then also occupied by the nodes from set $\mathcal{Q}_m$ and the interferers from set $\mathcal{N}_{I,m}$. Clearly, the SINR observed by each node will be action dependent. Therefore, the SINR at the qth node due to the kth action is given by:

$$\mathrm{SINR}_{q,k} = \frac{P_{m(q),k}}{\bar{P}_{m(q),k} + P^I_{m(q),k} + \sigma^2_{m(q),k}}, \forall k \in \mathcal{A}, \tag{11.23}$$

where $\mathcal{A}$ is the set of actions, $[m(q), k]$ denotes the assignment of the qth node to the mth subband corresponding to action k, and $\sigma^2_{m(q),k} = k_B T_0 F_q \Delta B$ is the thermal noise power. The network-wide SINR is measured in terms of the harmonic mean SINR, which for the kth action, can be expressed as:

$$\mathrm{SINR}_k = \frac{Q}{\sum_{q=1}^{Q} \frac{1}{\mathrm{SINR}_{q,k}}}. \tag{11.24}$$

This network-wide harmonic mean SINR plays a key role in the design of the DQNN for bandwidth allocation. When the agent moves from the current action to the next action, the accumulated reward will be calculated depending on whether the harmonic mean SINR increases or decreases. More specifically, if the harmonic mean SINR increases, the agent accumulates a high or positive reward, and if the harmonic mean SINR decreases, the agent will be penalized for the new action.

11.3.2 DQNN for Radar Bandwidth Allocation

The allocation of subbands to nodes can be conventionally found by iterating through all possible combinations and finding the optimal combination that yields the maximum network-wide radar performance, such as the harmonic mean SINR. However, this process is computationally expensive. We leverage a DQNN reinforcement learning for finding the optimal combination. For a small number of nodes, the exhaustive approach is comparable to the DQNN approach, but for a large number of nodes, the exhaustive approach will have a significantly high computation time compared to the DQNN. The foundation of DQNN is Q-learning, which is a model-free method and enables an agent to learn optimal policy by estimating the Q-value function. The Q-value function represents the reward for taking an action at a state and adhering to the optimal policy. In other words, Q-learning is represented by a [state, action] table of possible states as the rows and actions as the columns within an environment as an agent traverses the environment.

An example of an environment is shown in Table 11.1 and an example of the Q-table is shown in Table 11.2. The agent in Table 11.1 is located at the start position (State 0), and it will traverse the environment with the objective of reaching the goal (State 15) while avoiding the traps marked with "X". The goal and the

Table 11.1 Example of an environment.

State 0 (Agent starts here)	State 1	State 2	State 3
State 4	State 5 (X)	State 6	State 7 (X)
State 8	State 9	State 10	State 11 (X)
State 12	State 13	State 14	State 15 (Goal)

Table 11.2 Q-table for environment represented by Table 11.1.

	Action 1 (left)	Action 2 (down)	Action 3 (right)	Action 4 (up)
State 0	0	0	0	0
State 1	0	0	0	0
⋮				
State 15	0	0	0	0

"X" markings are referred to as terminal states, because when the agent lands on them it ends the iteration of traversing the environment. If the agent reaches the goal, there will be a positive reward and a negative reward for reaching the traps ("X" markings). Suppose the agent here has four actions: moving left, down, right, and up.

Table 11.2 initially has the Q-table values set to 0. These values are updated as follows

$$Q(s,a) = Q(s,a) + \alpha\left(R + \gamma \max_{a'} Q(s',a') - Q(s,a)\right), \tag{11.25}$$

where $Q(s, a)$ is the Q-value for the action a taken in state s and α is the learning rate that controls how much the new information overrides the old one. R is the reward for choosing action a in state s and γ is the discount factor, which determines how much an agent values future rewards compared to immediate rewards, influencing its learning process and decision-making. $\max_{a'} Q(s', a')$ is the maximum Q-value for the new state s', which is the state that is reached after taking the action at the current state. The maximum Q-value here represents the best possible reward that can be achieved from state s'.

As an example consider calculating the Q-value in Table 11.2 for the State 14 where the agent takes action 3 (moves right). The update process can be represented by

$$\begin{aligned} Q(14,3) &= Q(14,3) + \alpha\left(R + \gamma \max_{a'} Q(s',a') - Q(s,a)\right) \\ Q(14,3) &= 0 + 0.001 \times (1 + 0.9 \times 0–0) = 0.001, \end{aligned} \tag{11.26}$$

where the learning rate and discount factor are, respectively, chosen as 0.001 and 0.9, the reward is 1 since the objective is to reach the goal, and the new state is 15 since moving right at state 14 moves the agent to state 15. $Q(14, 3)$ is initially set to 0 and the $\max\{Q(15, 3)\}$ is 0 since for terminal states all the action Q-values are 0. Then, if we calculate the Q-value at state 13 taking action 3 using the same update process, the Q-value, $Q(13, 3) = 9 \times 10^{-7}$. As such, the updated Table 11.2 can be shown as Table 11.3.

Table 11.3 Updated Q-table for Table 11.2.

	Action 1 (left)	Action 2 (down)	Action 3 (right)	Action 4 (up)
State 0	0	0	0	0
State 1				
⋮				
State 13	0	0	9×10^{-7}	0
State 14	0	0	0.001	0
State 15	0	0	0	0

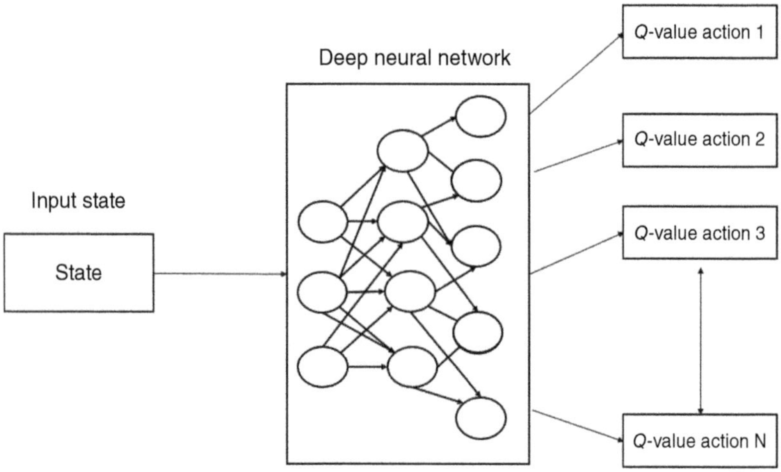

Figure 11.9 DQNN architecture.

The Q-table will continue to be updated, and after enough training there will be an optimal path for the agent to take to reach the goal while avoiding the traps. This optimal path will be found by taking the best action at a given state row.

Q-learning works well for small discrete state-action environment spaces, but like the exhaustive search approach, struggles with high-dimensional state-action spaces such as continuous states. Q-learning requires storing Q-values in a Q-table, which is complex and computationally expensive for large environments. The DQNN applies deep neural networks with Q-learning to estimate the Q-function value instead of updating a Q-table. DQN networks also allow them to generalize and solve non-convex problems that are difficult for Q-learning.

A typical DQNN architecture is shown in Figure 11.9. A state is passed as an input into the DQNN, which then outputs the Q-values for each action. The deep Q-learning update used in DQNN is given by

$$Q(s,a) = Q(s,a) + \alpha\left(R + \gamma \max_{a'} Q(s',a';\theta') - Q(s,a;\theta)\right), \tag{11.27}$$

where θ are the weights of the main Q-network (policy) and θ' are the weights of the target Q-network.

11.3.2.1 Training Steps in DQNN

Initialization: The DQNN training involves first creating two identical deep Q-networks. One is for the policy network with weights (θ) and the other is for the target network with weights (θ'). The target network uses (11.27) to compute the target Q-values during updates. Backpropagation is performed on the policy network during training. The weights and bias of the policy network are initially copied to the created target network to make them identical ($\theta = \theta'$).

The agent starts training by iterating through episodes. In each episode, the agent starts at a position and traverses through the environment. The agent deals with exploration, where the agent traverses the environment through random actions, versus exploitation, where the agent selects the best action from its trained policy. In order to handle exploration and exploitation, the algorithm uses an epsilon-greedy policy to balance exploration and exploitation. The epsilon greedy policy has epsilon initially set to a value between 0 and 1, but typically 1. Then, for a training episode, with probability ϵ, the agent will select a random action to explore the environment; otherwise, the agent will choose the action with the highest Q-value based on the network's output. After each episode or N number of episodes, the ϵ value will decrease. The results are that the agent initially takes random actions in the environment in early episodes but then acts based on its learned policy.

Experience Memory Buffer: After each agent's environment traversal or step, the experience is stored in a memory buffer. The agent's state, action, reward, new state, and termination flag are stored inside the replay memory buffer. The termination flag is binary, typically 0, and is 1 only if the new state reached is a terminated state. An element inside the buffer is the 5-tuple of (state, action, reward, new state, termination flag). The memory buffer is used for experience replay where the elements or experiences saved inside the buffer are sampled and used in training the policy network. The size of the memory buffer is typically fixed, while the data object can include deques, lists, binary heaps, etc. An example of memory buffer is shown in Figure 11.10.

One-hot Encoding: One-hot encoding (OHE) is performed during exploitation and experience replay while training the policy network. During exploitation, before the action is decided by the agent and the agent reaches the new state,

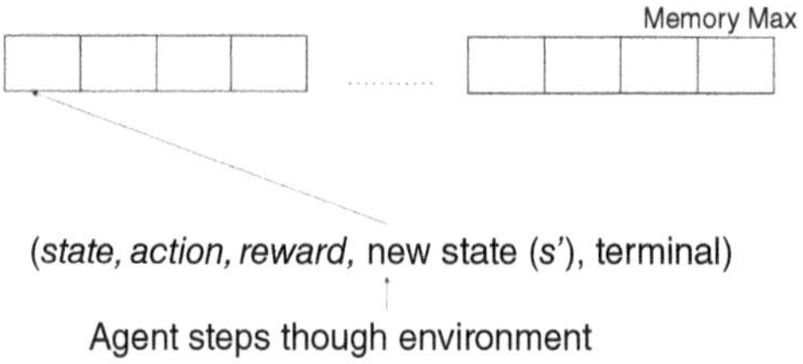

Figure 11.10 Experience memory deque example.

Figure 11.11 Example of OHE.

the current state is one-hot encoded before passing through the neural network. Then the best Q-value index output from the neural network is selected, which corresponds to the best action at the current state from the trained model. OHE transforms categorical data into a numerical format that the neural network can understand, preventing it from assuming an order between categories, which can improve model performance and accuracy. For example, consider a case with 16 possible states. Therefore, the length of the encoding is 1×16 of ones and zeros. The state 1 then can be represented by [0 0 0 0 0 0 0 0 0 0 0 0 0 0 0 1] as shown in Figure 11.11.

Experience Replay: The experiences saved in the memory buffer will be sampled based on the hyperparameter batch size. The sampled experiences will be used to train the policy network by first going through each experience (5-tuple) and taking the state. The state will be OHE and forward passed through the policy and target networks.

Initially, since both policy and target networks have the same weights and biases, they will output the same Q-values from the networks. After the forward pass of the state, the Q-values of policy network and target network are stored in temporary variables (e.g., policy output and target output). Then the target value is calculated from (11.27) where the $\max_{a'} Q(s', a'; \theta')$ is the maximum value of the Q-value outputs of the target network from a forward pass at the given new state. The target value calculated from (11.27) will override the target network output vector (e.g., target output) at the index of the action number. The updated target network Q-value output will be used to train the policy network by calculating the loss between the policy network and the updated target network outputs. The loss is used to backpropagate and update the parameters of the policy network using gradient descent. The process of the agent traversing with epsilon-greedy approach, the forward pass of the state to the networks, calculating the target value using (11.27), replacing the action index of the target network output with the target value, and updating the policy network from the loss between the network outputs will be repeated for the number of training episodes set. After several steps through the environment or episodes, the target network parameters will sync with the policy network. The number of steps or episodes to sync the networks is referred to as the sync rate.

11.3.2.2 Simulation Settings for DQNN

In our simulations, each state for the DQNN is represented by $1 \times Q$ vector, where Q is the number of nodes. Each element in the array is one-hot encoded to a binary vector of the size equal to the number of available subbands. The encoded states are concatenated to produce a binary vector of size $1 \times (Q(Q + N_I))$, where $Q + N_I$ is the number of subbands. An example of the OHE process is shown in Figure 11.12. The architecture of the DQNN has 5 hidden linear layers with the output layer equal to the number of nodes. The linear layers have rectified linear unit (ReLU) activation functions applied to their outputs. The output layer also represents the Q-values for the possible actions. Starting from an initial action in which all nodes occupy the same topmost subband (i.e., subband $Q + N_I$), the possible actions are formed by allowing a random node to decrease its subband by a step, i.e., by choosing the subband $Q + N_I - 1$. Figure 11.13 shows the structure of the DQNN for the policy and target networks used in the simulations.

All nodes start at the highest subband for an episode, and the agent will take the actions of moving a node down a band to reach the optimal allocation of bandwidths to nodes. The simulation involves testing the exhaustive approach and the DQNN approach for finding the optimal state of the node-bandwidth allocation. The simulated cases include 4-node case with 6 available subbands, 6-node case

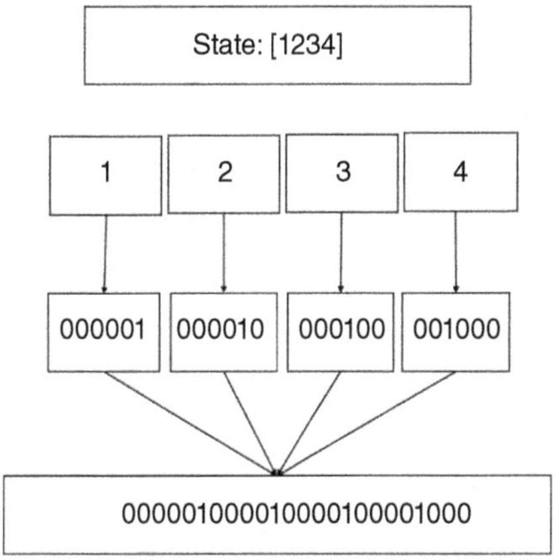

Figure 11.12 Example of OHE of a state in the environment of the simulation.

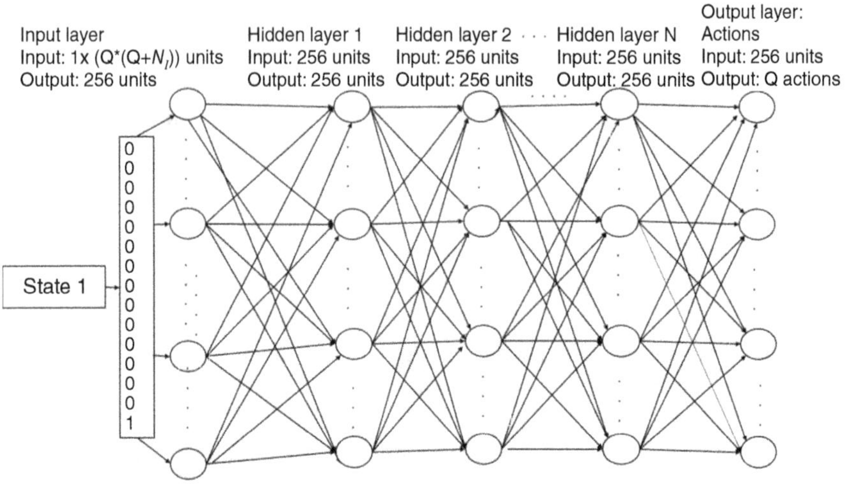

Figure 11.13 The structure of the DQNN used for the policy and target network simulation.

with 8 available subbands, 8-node case with 10 available subbands, and the 10-node case with 12 available subbands. In all simulations, 2 interferers occupy the lowest two subbands, i.e., subbands 1 and 2.

Hyperparameters: The following hyperparameters are used for training the DQNN. The learning rate is set to 0.001. The discount factor is set to 0.9, and the sync rate is chosen as 70 steps before copying the policy weights to target network weights. The memory size is 2,000 and the batch size is 32. The truncation step count is used as a hyperparameter in simulation to end the episode early if the agent commits an action that doesn't increase the harmonic mean SINR. The truncation step is set to 0, to ensure the agent's first action that decreases the harmonic mean SINR, to end the episode. The reward value for the agent reaching an improved harmonic mean SINR is set to 20, and the penalty for reaching a decreased harmonic mean SINR is set to −1. The parameters mentioned so far are the hyperparameters kept constant across simulations with different nodes. The number of training episodes scales directly depending on the number of nodes simulated. The larger the number of nodes, the larger the number of possible node-to-bandwidth allocation. The 4-node case requires 2,000 episodes, the 6-node case requires 5,000 episodes, the 8-node case requires 10,000 episodes, and the 10-node case requires 50,000 episodes.

Hardware: All simulations are conducted using a computer equipped with a 3.4 GHz AMD Ryzen 7 5700X 8-core processor having 16 logical processors. The graphics card used is an NVIDIA GeForce RTX 3070. The CUDA libraries are installed and used during simulation for its improved speed of utilizing the GPU compared to using just the CPU. The benefits of using CUDA libraries include significant performance gains for computationally intensive iterations through the training of different combinations of radar nodes to bandwidth allocations.

Radar Network Parameters:

The network-wide performance, i.e., harmonic mean SINR depends on radar nodes locations. For the 4-node case, nodes 1 to 4, are respectively, located at $[-40, -40, 1]^T$, $[-20, -20, 1]^T$, $[0, 0, 1]^T$, and $[120, 120, 1]^T$ (in meters). The locations of nodes 1 to 6, in the 6-node case, are respectively, taken as $[-40, -40, 1]^T$, $[-30, -30, 1]^T$, $[-20, -20, 1]^T$, $[-10, -10, 1]^T$, $[0, 0, 1]^T$, and $[120, 120, 1]^T$. For the 8-node case, nodes 1 to 8 are, respectively, located at $[-40, -40, 1]^T$, $[-30, -30, 1]^T$, $[-20, -20, 1]^T$, $[-10, -10, 1]^T$, $[0, 0, 1]^T$, $[10, 10, 1]^T$, $[20, 20, 1]^T$, and $[120, 120, 1]$. The locations of nodes 1 to 10, in the 10-node case, are respectively, set to $[-40, -40, 1]^T$, $[-30, -30, 1]^T$, $[-20, -20, 1]^T$, $[-10, -10, 1]^T$, $[0, 0, 1]^T$, $[10, 10, 1]^T$, $[20, 20, 1]^T$, $[50, 50, 1]^T$, $[70, 70, 1]^T$, and $[120, 120, 1]^T$. The example network scenario with 8 nodes, 2 interferers, and a target is shown in Figure 11.14.

The available numbers of subbands during simulations range from 6 to 12, where the interferers occupy the subbands 1 and 2. There are two interferers at $[40, 40, 1]^T$ and $[80, 80, 1]^T$. A single target is located at $[150, 150, 1]^T$. It is assumed that each node observes this target with the same RCS of 10 m^2. Each node's transmit power is 10 W and each interferer's transmit power is 100 W. The transmit

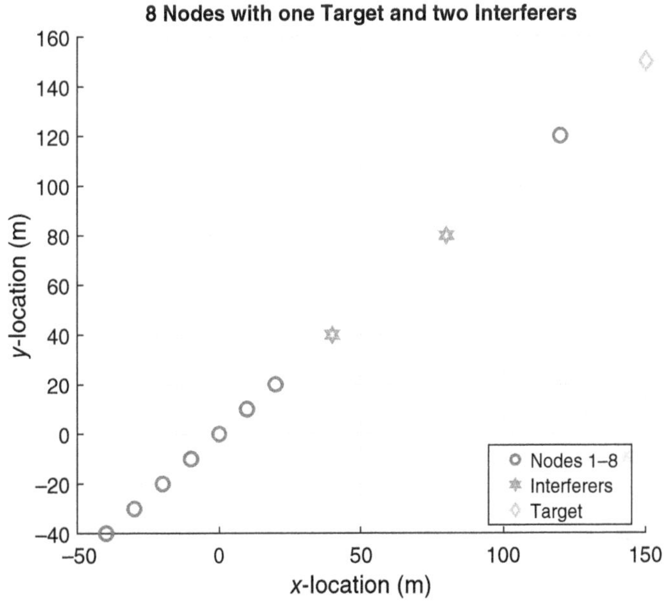

Figure 11.14 Network with 8 nodes, 2 interferers, and a target.

antenna gains of all nodes and interferers as well as the receive antenna gains of all nodes are set to 5 dB. The total available bandwidth is 0.5 GHz between 2 and 2.5 GHz. The total bandwidth is divided among the number of nodes plus the two interferers. The SINR calculations for a given node include interference power from nodes occupying the same band (inter-node interference) and the interference from emitters occupying the same subband.

11.3.2.3 Results with DQNN

The simulation results for the DQNN approach with different numbers of nodes are compared with the exhaustive search approach for finding the optimal bandwidth allocation. Figures 11.15 and 11.16 show the results after training the DQNN for the 2,000 episodes. Figure 11.16 displays the harmonic mean SINR for the different number of nodes, where the displayed harmonic mean SINR corresponds to the optimal state reached after training. Initially, for a small number of nodes, the exhaustive search offers lower runtime with higher harmonic mean SINR than the proposed DQNN. However, with an increasing number of nodes, the exhaustive approach is computationally expensive and struggles to find the optimal solution. In Figures 11.15 and 11.16, for 10-node case with 12 available subbands, the exhaustive search couldn't be executed, because it requires iterations over 61 billion possible combinations. Therefore, these plots do not show

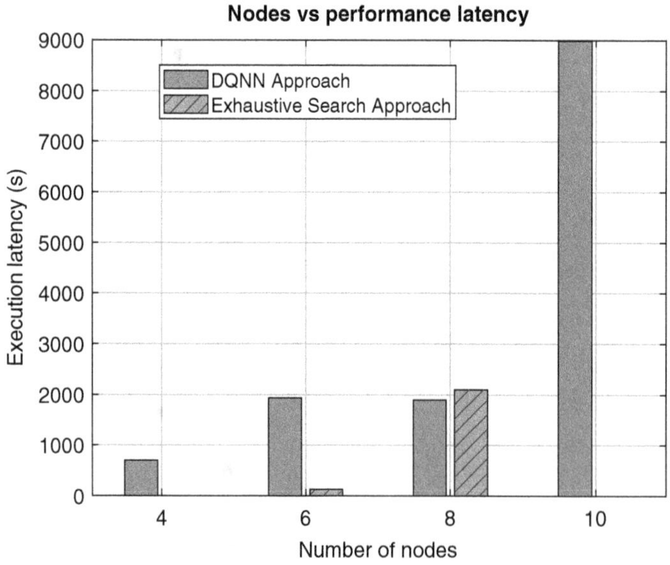

Figure 11.15 Execution time for the exhaustive search and DQNN for different numbers of nodes. The exhaustive search could not be executed for the 10-node case.

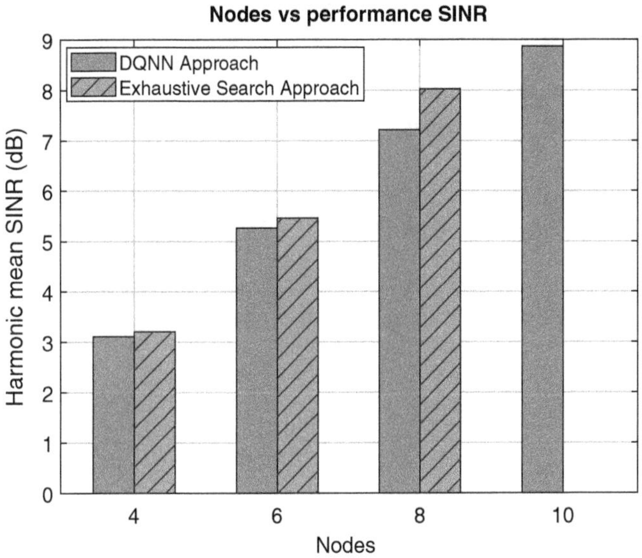

Figure 11.16 The harmonic mean SINR vs the number of nodes. The exhaustive search could not run for the 10-node case due to memory issues due to $6.19173642 \times 10^{10}$ combinations.

the results for the exhaustive search approach. The DQNN performs better than the exhaustive approach at a higher number of nodes in terms of runtime complexity, as seen from Figure 11.15. The state the DQNN approach reaches isn't the absolute optimal state, as seen from the performances in 4-node and 6-node cases, compared to the optimal state obtained with the exhaustive approach, but the difference in harmonic mean SINR is small, approximately from 0.5 to 1 dB. This small loss in dB but improved execution times in 8- and 10-node cases highlights the benefits of the DQNN approach compared to the exhaustive approach.

The training results for the 6-node case with 8 available subbands are shown in Figure 11.17. Similarly, the training results for the 10-node case with 12 available subbands are shown in Figure 11.18. In each figure, the left plot shows the total rewards accumulated by the agent when iterating in an episode. The right plot shows the epsilon decay, which starts at 1 and decreases toward 0. The epsilon decay is used to track if the agent is learning and traversing properly during simulation. Similar results have been obtained for the 4-node case with 6 available subbands and 8-node case with 10 subbands, but they have been skipped for conciseness.

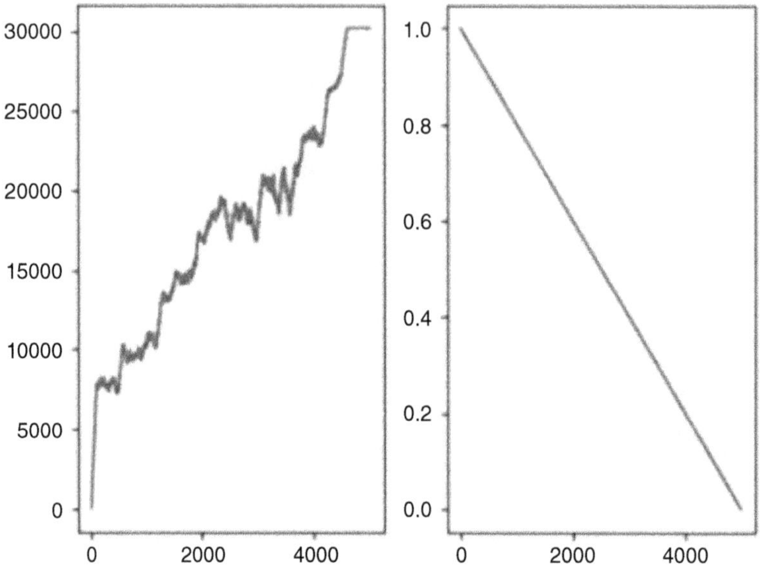

Figure 11.17 Rewards per episode (left) and the epsilon decay (right) for the 6-node scenario with 2 interferers.

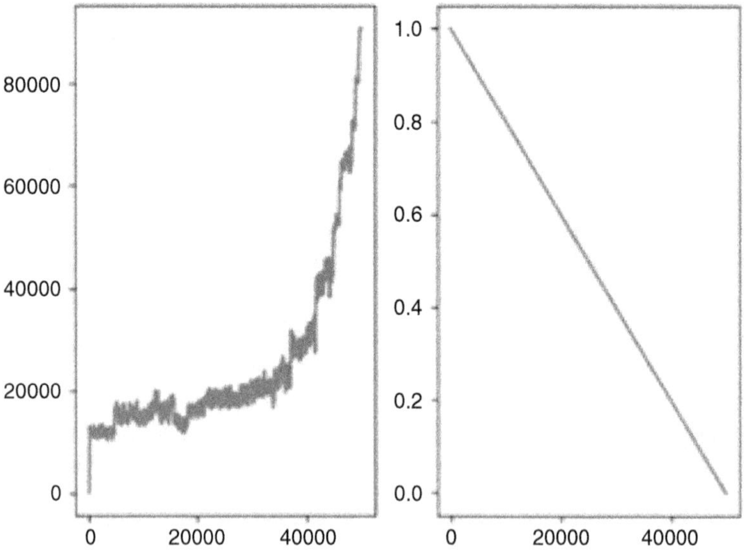

Figure 11.18 Rewards per episode (left) and the epsilon decay (right) for the 10-node scenario with 2 interferers.

11.4 Conclusions

In the first section of this chapter, we proposed model-based joint radar and communications BWCFOA, along with waveform optimization. Given available bandwidth, the key objective was to maximize the network-wide channel capacity, measured in terms of the harmonic mean of the individual node channel capacities, while ensuring that each node achieves target associated minimum SCINR and the minimum range resolution. The waveform optimization, which involved non-convex unit modulus constraint on each element of the radar waveform vector, was solved locally at each node with Riemannian manifold-based approach. The BWCFOA problem was solved by iteratively solving the BWCFO for a given allocation and the BWCFA for the given bandwidths and carrier frequencies. The BWCFO was solved as a GP problem, and the BWCFA was reformulated as the MICP problem, utilizing McCormick relaxation techniques for convexifying bilinear constraints. Simulations showed the superior performance of the proposed method over two other benchmark approaches: (a) WO+EBWA and (b) RW+BWCFOA. Because this type of iterative approach suffers from high runtime complexity, the LSTM-based supervised learning was proposed for estimating bandwidths and allocation. More specifically, the LSTM was trained with the locations of a moving interferer as input and the solutions of the model-based iterative approach as labels. This allowed us to significantly decrease the runtime complexity of bandwidth and carrier frequency allocation because the complexity of iterative approach was confined to offline training stage. Simulation results demonstrated the advantage of the proposed LSTM-based approach over the model-based approach. When the available bandwidth was divided equally into a given number of subbands, where some subbands were occupied by the interferers and inter-node interference was also possible, the problem of optimal bandwidth allocation to radar nodes was solved under a framework of reinforcement learning using DQNN. The HM-SINR at radar nodes was employed to determine the reward. More specifically, if the harmonic mean SINR decreased when taking a next action from the current one, the agent accumulated a low or negative reward. On the other hand, if the harmonic mean SINR increased when moving from one action to another, the agent accumulated high reward. The proposed DQNN-based method was compared with an exhaustive search approach with simulations. The results showed that the DQNN provides a better performance and complexity tradeoff than the exhaustive search.

References

1 M. A. Richards, J. A. Scheer, and W. A. Holm. *Principles of Modern Radar*, volume 1. Basic Principles. Scitech Publishing Inc., 2010.

2 D. E. Hack, L. K. Patton, B. Himed, and M. A. Saville. Centralized passive MIMO radar detection without direct-path references. *IEEE Transactions on Signal Processing*, 62(11):3013–3023, 2014.
3 S. Pak, B. K. Chalise, and B. Himed. Target localization in multistatic passive radar systems with artificial neural networks. In *Proceedings IEEE International Radar Conference*, pages 1–5. Toulon, France, 2019.
4 R. Srinivasan. Distributed radar detection theory. *IEE Proceedings*, 133(1):55–60, 1986.
5 S. C. A. Thomopoulous, R. Vishwanathan, and D. C. Bougoulias. Optimal decision fusion in multiple sensor systems. *IEEE Transactions Aerospace and Electronic Systems*, 23(5):644–653, 1987.
6 D. Ciuonzo, A. Aubry, and V. Carotenuto. Rician MIMO channel-and jamming-aware decision fusion. *IEEE Transactions Signal Processing*, 65(15):3866–3880, 2017.
7 B. K. Chalise and K. T. Wagner. Distributed GLRT-based detection of target in SIRP clutter and noise. In *Proceedings IEEE Radar Conference*, pages 1–6. Atlanta, GA, USA, 2021.
8 B. K. Chalise, D. M. Wong, M. G. Amin, R. M. Buehrer, A. F. Martone, B. H. Kirk, and K. Sherbondy. Sparsity-based distributed DoA estimation in radar networks. In *Proceedings IEEE Radar Conference*, pages 1–6. NYC, NY, USA, 2022.
9 B. K. Chalise, D. M. Wong, M. G. Amin, A. F. Martone, and B. H. Kirk. Detection, mode selection, and parameter estimation in distributed radar networks: Algorithms and implementation challenges. *IEEE Aerospace and Electronic Systems Magazine*, 37(11):4–22, 2022.
10 L. Xiao, S. Boyd, and S. Kim. Average consensus algorithms robust against channel noise. *Journal of Parallel and Distributed Computing*, 67(1):33–46, 2007.
11 L. Pescosolido, S. Barbarossa, and G. Scutari. Average consensus algorithms robust against channel noise. In *Proceedings IEEE 9th Workshop on Signal Processing Advances in Wireless Communications*, pages 261–265. Recife, Brazil, 2008.
12 R. Olfati-Saber and R. M. Murray. Consensus problems in networks of agents with switching topology and time-delays. *IEEE Transactions on Automatic Control*, 49(9):1520–1533, 2004.
13 W. Ren and R. W. Beard. Consensus seeking in multiagent systems under dynamically changing interaction topologies. *IEEE Transactions on Automatic Control*, 50(5):655–661, 2005.
14 L. Xiao, S. Boyd, and S. Lall. A scheme for distributed sensor fusion based on average consensus. In *Proceedings 4th IEEE International Conference on Information Processing in Sensor Networks (IPSN)*, pages 63–70. Los Angeles, CA, USA, 2005.
15 I. D. Schizas, A. Ribeiro, and G. B. Giannakis. Consensus in ad hoc WSNS with noisy links - part i: Distributed estimation of deterministic signals. *IEEE Transactions Signal Processing*, 56(1):350–364, 2008.

16 L. Pescosolido, S. Barbarossa, and G. Scutari. Radar sensor networks with distributed detection capabilities. In *Proceedings IEEE Radar Conference*, pages 1–6. Rome, Italy, 2008.

17 K. Yao. A representation theorem and its applications to spherically invariant random processes. *IEEE Transactions on Information Theory*, 19(5):600–608, 1973.

18 B. Liu, B. Chen, and J. H. Michels. A GLRT for radar detection in the presence of compound-gaussian clutter and additive white noise. In *Proceedings IEEE Sensor Array and Multichannel Signal Processing Workshop*, pages 87–91. Rosslyn, VA, USA, 2002.

19 B. K. Chalise, D. M. Wong, K. T. Wagner, and M. G. Amin. Enhancing distributed radar detection performance with reliable communications. *IEEE Transactions on Aerospace and Electronic Systems*, 59(4):4118–4133, 2023.

20 D. M. Wong, B. K. Chalise, A. F. Martone, and B. H. Kirk. Distributed radar network implementation using software defined radio. In *Proceedings IEEE LISAT*, pages 1–6. Old Westbury, NY, USA, 2022.

21 B. K. Chalise, M. G. Amin, A. F. Martone, and B. H. Kirk. Effect of communications link impairments on the sparsity-based distributed DoA estimation. In *56th Asilomar Conference on Signals, Systems, and Computers*, pages 481–485. Pacific Grove, CA, USA, 2022.

22 B. K. Chalise, A. F. Martone, and B. H. Kirk. Harmonic mean SINR maximization-based bandwidth and carrier frequency allocation for distributed radar networks. In *Proceedings IEEE Radar Conference*, pages 1–6. San Antonio, TX, USA, 2023.

23 B. K. Chalise, A. F. Martone, and B. H. Kirk. Model- and deep learning-based bandwidth and carrier frequency allocation in distributed radar networks. *IEEE Transactions on Aerospace and Electronic Systems*, 59(6):8022–8036, 2023.

24 B. K. Chalise, B. H. Kirk, and K. Ranney. Bandwidth, carrier frequency, and waveform optimization for sensing and communications in a distributed radar network. In *Proceedings IEEE Radar Conference*, pages 1–6. Denver, CO, USA, 2024.

25 B. K. Chalise, B. H. Kirk, and A. F. Martone. Resource allocation for sensing and communications in a distributed radar network. In *Proceedings IEEE Radar Conference*, pages 1–6. Atlanta, Georgia, USA, 2025.

26 S. Boyd, S.-J. Kim, L. Vandenberghe, and A. Hassibi. A tutorial on geometric programming. *Optimization and Engineering*, 8(1):67–127, 2007.

27 J. K. Scott, M. D. Stuber, and P. I. Barton. Generalized mccormick relaxations. *Journal of Global Optimization*, 51(x):569–606, 2011.

28 F. A. Gers, J. Schmidhuber, and F. Cummins. Learning to forget: Continual prediction with LSTM. In *Proceedings Ninth International Conference on Artificial Neural Networks (ICANN)*, pages 850–855. Edinburgh, UK, 1999.

29 Z. Yu, V. Ramanarayanan, D. Suendermann-Oeft, X. Wang, K. Zechner, L. Chen, J. Tao, A. Ivanou, and Y. Qian. Using bidirectional LSTM recurrent neural networks to learn high-level abstractions of sequential features for automated scoring of non-native spontaneous speech. In *Proceedings IEEE Workshop on Automatic Speech Recognition and Understanding (ASRU)*, pages 338–345. Scottsdale, AZ, USA, 2015.

30 M. M. Rahman, Y. Watanobe, and K. Nakamura. A bidirectional LSTM language model for code evaluation and repair. *Symmetry*, 13(2):247, 2021.

31 A. Plaat. *Deep Reinforcement Learning*. Springer Nature, 2022.

32 V. Francois-Lavet, P. Henderson, R. Islam, M. G. Bellemare, and J. Pineau. An introduction to deep reinforcement learning. *Foundations and Trends in Machine Learning*, 11(3-4):219–354, 2018.

33 B. R. Mahafza. *Radar Systems Analysis and Design with MATLAB*. Chapman and Hall/CRC, 2000.

34 P.-A. Absil, R. Mahony, and R. Sepulchre. *Optimization Algorithms on Matrix Manifolds*. Princeton University Press, Cambridge, UK, 2008.

35 N. Boumal. *An Introduction to Optimization on Smooth Manifolds*. Cambridge University Press, Cambridge, UK, 2023.

36 M. Grant and S. Boyd. CVX: Matlab software for disciplined convex programming, version 2.0 beta. https://cvxr.com/cvx, 2013.

Index

n

o

S

www.ingramcontent.com/pod-product-compliance
Lightning Source LLC
LaVergne TN
LVHW011304051126
851579LV00058B/22